Quality management in food chains

Quality management in food chains

edited by:

Ludwig Theuvsen

Achim Spiller

Martina Peupert

Gabriele Jahn

Wageningen Academic Publishers

ISBN: 978-90-76998-90-9

First published, 2007

Wageningen Academic Publishers
The Netherlands, 2007

Table of contents

Invited speakers

Firm's costs and benefits of quality management and quality assurance systems

Strategic, organisational and human resource management aspects of quality management

Benchmarking and harmonisation methods and their effects on the effectiveness of quality assurance schemes

Farmers' perceptions and motivation on food safety and quality

Quality management: a supply chain perspective

Perspectives of quality management in modern agribusiness

Ludwig Theuvsen and Achim Spiller
University of Goettingen, Department of Agricultural Economics and Rural Development, Platz der Goettinger Sieben 5, 37073 Goettingen, Germany; theuvsen@uni-goettingen.de

The following book includes 38 papers contributed by participants in the EAAE Seminar on Quality Management and Quality Assurance in Food Chains held at the Georg-August-University of Goettingen in 2005. For the past decade, quality management has been one of the most controversial topics in agribusiness research, and it is always a risk to add another conference to such a hot topic. However, the interesting discussions in Goettingen and the wide spectrum of papers from different research streams have demonstrated that the risk was manageable and that it is still possible to add new insights after more than ten years of research in the field.

The unique approach of the conference was to introduce a broader management perspective that confronted widespread technical views on quality assurance and quality management in food chains with diverse images from strategic, organisational, marketing, motivational and social theories. Like the conference, the book is divided into seven main chapters:

- Invited speakers from business and research institutions.
- Firm's costs and benefits of quality management and quality assurance systems.
- Strategic, organisational and human resource management aspects of quality management.
- Consumers' expectations of quality assurance and management systems.
- Benchmarking and harmonisation methods and their effects on the effectiveness of quality assurance schemes.
- Farmers' perceptions and motivation regarding food safety and quality.
- Quality management: a supply chain perspective.

Quality control and quality management have a long tradition. For many decades statistical quality approaches have been used to guarantee a certain quality level in almost every industrial production process (Deming 1986). In the food business, quality control can be traced back to early regulations concerning the production of, for instance, wine and beer in the 15th and 16th century. More systematic approaches for safeguarding consumers against fraud and food hazards were introduced in the late 19th century (Theuvsen, 2007). New developments like Total Quality Management or Six Sigma have enriched quality management through more motivation and participation-oriented approaches (Pfeifer, 2002).

In recent years third-party control (certification) has come to the fore (Böcker *et al.*, 2003; Hatanaka *et al.*, 2005; Jahn *et al.*, 2005; Newslow, 2001). One of the main features of certification systems is that inspections are carried out by (accredited) independent bodies grounded on standards laid down by different external organisations (standard owners) (Meuwissen *et al.*, 2003). By means of regular control and - whenever necessary - additional sampling, neutral inspection institutions monitor the entire supply chain. Once they have been awarded the certificate, companies are entitled to make use of the quality label for marketing purposes. The supplier provides a certificate issued by a neutral certifier based on the quality and certification standard laid down by the standard owner; this certificate serves as a quality signal (Luning *et al.*, 2002).

Due to the dominance of small and medium-sized companies, in many cases the agribusiness sector has adopted quality management approaches a little bit (too) late. Therefore, although once a pioneer in the field of quality assurance and quality management, in the late 20th century, agriculture and the food industry have dropped behind and suffered severe food quality crises. Other reasons for the prevalence of quality crises in the food sector are unclear responsibilities for food safety, which are somewhat arbitrarily and inefficiently divided between government, public authorities and private business, and the high level of political regulation, which lowers the intrinsic motivation of farmers and processors for improving food safety through advanced quality assurance and quality management approaches. On the other hand consumer awareness and uncertainty rise through new technologies, such as genetically modified organisms (Napier *et al.*, 2004). A major problem affecting communication with the broader public is that lay perceptions of risks often differ significantly from expert views (Brennan *et al.*, 2004). Furthermore, in many countries governmental institutions have lost reputation. The same is true for the industry, whereas nongovernmental institutions, like Greenpeace or consumer protection agencies, are gaining relevance for many consumers (Dierks, 2006). Thus, quality communication and trust management are necessary parts of modern quality management approaches (Fritz *et al.*, 2006).

A closer look at the papers contributed to the conference and this book makes it clear that three perspectives are explicitly or implicitly included in nearly all presentations: certification schemes, supply chain approaches and consumer perceptions and behaviour towards food-related lifestyles and risks.

Certification schemes

The complex evolving system of certification schemes, a new and fast-growing business, is the topic of Walgenbach, Möller, Neuendorff and Fischer, Theuvsen and Peupert, and Lazo *et al.*[1]. During the last few years, it has become more and more obvious that the term certification scheme covers a wide spectrum of alternative governance mechanisms. The systems developed so far can be categorised as follows (Theuvsen *et al.*, 2007):

- *Target:* Consumer-oriented schemes like organic farming, fair trade and PDO, PGI and TSG, can be distinguished from business-to-business tools such as EurepGAP, the International Food Standard, the BRC Global Standard and the ISO family (9000, 9001, 22000).
- *Focus:* Many schemes focus on process characteristics with regard to basic production and documentation standards (like Qualität und Sicherheit in Germany or the British Assured Farm Standards), improved animal welfare standards (for instance the Neuland scheme in Germany) or more environmentally-friendly and sustainable production (organic farming schemes). Other schemes focus on product characteristics, such as region of origin (PDO, PGI) or traditional character (for example, the Dutch Boerenkaas label).
- *Goal:* Basic level schemes try to guarantee legal minimum requirements in a mass market (such as IKB in the Netherlands) whereas marketing-oriented approaches deal with niche markets and product differentiation (like PDO, PGI, TSG and organic farming schemes). The latter typically rely on labelling to inform consumers willing to pay for special process or product characteristics, whereas in the case of the mass marketing approach, labels are a rare exception (See, for example, the German system Qualität und Sicherheit).
- *Content:* Quality assurance schemes can focus on safety (for example, IKB), quality (for instance, PDO schemes), or production standards (like organic farming standards).
- *Number of stages of the food supply chain involved:* Standards for one stage of the supply chain involve only direct suppliers and customers, for instance, food manufacturers and retailers in the case of the International Food Standard or farmers and retailers in the case of the British

[1] Quotations without date always refer to articles in this book.

Assured Farm Standards. On the other hand, chain-wide concepts include all stages of the food supply chain. The German Qualität und Sicherheit GmbH, for instance, was founded by associations in the meat industry representing all stages from feed production to the retailer. In the same way the EU organic approach covers the whole food chain.

- *Standard owner:* A wide range of schemes can be differentiated according to the respective standard owner. The spectrum embraces completely state-run systems (like organic farming in Denmark), governmental schemes with private inspections (such as organic farming schemes in most European countries), systems developed by international standardisation organisations (for instance, ISO 9000, 9001 and 22000), stakeholder approaches (for example, Word Wildlife Fund/WWF in the case of the Marine Stewardship Council; Fairtrade), schemes founded by producer associations (such as farmers associations in the case of the British Assured Farm Standards), standards by private inspection bodies (like Vitacert by the German Technical Monitoring Institution/TÜV) and retailer-driven schemes (such as BRC Global Standard and International Food Standard).
- *Degree of harmonisation:* Original schemes (like EurepGAP) are sometimes used as blueprints for national adoptions of a broader approach (such as ChileGAP). This process reflects the conflict between international acknowledgement by customers and special national requirements.
- *Geographic focus:* With regard to geographic focus regional approaches (for instance quality assurance systems developed and run by the German federal states such as Geprüfte Qualität Bayern in Bavaria), national systems (like Certus in Belgium), international approaches (like the French-German International Food Standard or the European regulations on organic farming) and global standards (such as EurepGAP or ISO 22000) can be distinguished.
- *Number of participating firms:* Small regional or highly differentiated schemes often have only several dozen or a few hundred participants. Medium-sized schemes have several hundred to several thousand members (3,200 farms and firms in the case of the Demeter organic farming scheme, for instance) whereas large standards can have tens of thousands of members (the German Qualität und Sicherheit system, for example, encompasses about 80,000 farms and firms).

All in all, the fast growing certification business causes numerous difficult research problems. These include companies' cost-benefit ratios (Gellynck *et al.*; Peris Moll and Igual; Mora and Menozzi), competitive advantage (Arellanes and O'Reilly), effects on transaction costs (Villalobos *et al.*), farmer acceptance due, for instance, to certification costs (Enneking *et al.*; Beletti *et al.*), impact on food safety (Theuvsen and Peupert) and the prevention of fraud (Hirschauer and Mußhoff; Neuendorff and Fischer). The most obvious topic is multiple audits at the same site, which provoke attempts to benchmark different standards (Möller; Mazé *et al.*; Krieger and Schiefer; Rother) or to support multiple audit procedures through IT solutions (Binner and Jansen). Whereas food processors are used to coping with different standards, for small and medium-sized companies and especially for European family farms, certification is fraught with fears of bureaucratic burden and doubts about the real quality impact (Enneking *et al.*; Jahn and Spiller; Lazo *et al.*). The conference papers reflect the ongoing discussion about quality assurance schemes in European agriculture and the food industry, applying a variety of research perspectives, including new institutional economics, management theory and case study and survey approaches.

Supply chain perspective

A supply chain consists of activities, organisations and resources that are necessary for bringing a product or service to the final customer. In the food business, the entities of a supply chain are, for instance, farm suppliers, farmers, traders, processors, wholesalers, restaurants and retailers. In recent years the supply chain perspective has gained much relevance in agribusiness research since it is considered paramount for the understanding of various current issues in the management of the agri-food sector such as traceability (Theuvsen and Hollmann-Hespos, 2005), transparency (Frentrup and Theuvsen, 2006), logistics (Fritz and Hausen, 2006) and governance (Schulze *et al.*, 2006).

Food supply chains are characterised by more or less intensive division of labour. The division of labour results in efficiency gains through specialisation and economies of scale, but also in a need for improved coordination and for a solution to agency problems (Theuvsen, 2004). The coordination problems can be traced back to separate decision-making by different companies in the food supply chain. Since each decision by a farm or firm has effects on all other companies in the supply chain, there has to be a certain amount of communication in order to coordinate activities. Agency problems are a second consequence of the division of labour in food supply chains. Supply chains are characterised by reciprocal multi-stage agency relationships in which companies delegate tasks to each other. Farmers, for instance, delegate processing to food manufacturers, who in turn delegate production of agricultural raw materials to farmers. Principals and agents in food chains behave opportunistically, that is, they act in self-interest with guile (Williamson, 1985). For this reason, the correctness and completeness of information transferred throughout the food supply chain cannot be taken for granted. Besides opportunistic behaviour, agency relationships are characterised by information asymmetries. Opportunistic behaviour and information asymmetries result in agency problems known as hidden characteristics, hidden action and hidden intention (Akerlof, 1970; Arrow, 1985); these problems influence the amount of information shared in food chains (Theuvsen, 2003).

Coordination and agency problems in food supply chains are highly relevant for quality assurance and quality management due to their influence on the amount and reliability of quality-related information shared between farms and firms. Therefore, the growing legislation by EU and national authorities and the above-mentioned certification schemes focus primarily on the quality-related information stored and transmitted in food supply chains, and supply chain approaches have become an integral part of nearly all quality management literature. Papers presented at the Goettingen conference focus on a variety of supply chain-related quality aspects. One major focus is the use of state-of-the-art quality techniques such as Quality Function Deployment (Peupert and Theuvsen) or Failure Mode and Effect Analysis (Gödderz et al) for improved chain-wide quality communication and information systems. Techno-organisational solutions to quality-related coordination and agency problems in food supply chains are also discussed (Dries and Swinnen; Hannus *et al.*; Fritz *et al.*). Other research topics are the impact of network structures on information exchange and network management on quality at large (Poignée and Schiefer; D'Haese *et al.*; Engler *et al.*). More specifically, the roles of contracts (Hinrichs), incentives (Hirschauer and Mußhoff), quality assurance schemes (Canavari *et al.*) and relationship management (Gerlach *et al.*) on quality and customer-supplier relationships are analysed.

Consumer perceptions and behaviour

The certification as well as the supply chain approaches sketched so far focus on how quality improvements in food supply chains can be achieved. This supply-side perspective has to be complemented by an analysis of how consumers perceive food quality and how these perceptions influence consumer decisions and behaviour. Grunert (2005; Grunert *et al.*, 1996) suggests the Total Quality Food Model as a guiding framework for organising knowledge about quality-related consumer perceptions and behaviour (Figure 1). The model proposes a distinction between two major lines of analysis: a vertical and a horizontal dimension.

In the vertical dimension, the model hypothesises how consumers infer quality from quality signals and cues and how they form judgments on whether and to what extent the perceived properties of a food product help to attain basic life values such as being responsible or having fun, which serve as basic motivators of human behaviour. From this means-end perspective, quality is an intermediate concept that relates concrete product attributes (like fat content or packaging) to abstract life values. Quality influences consumers' intention to buy only when it is expected to contribute to purchase motive fulfilment and when perceived quality is high enough to justify the (perceived) costs, that is, the price demanded by the retailer.

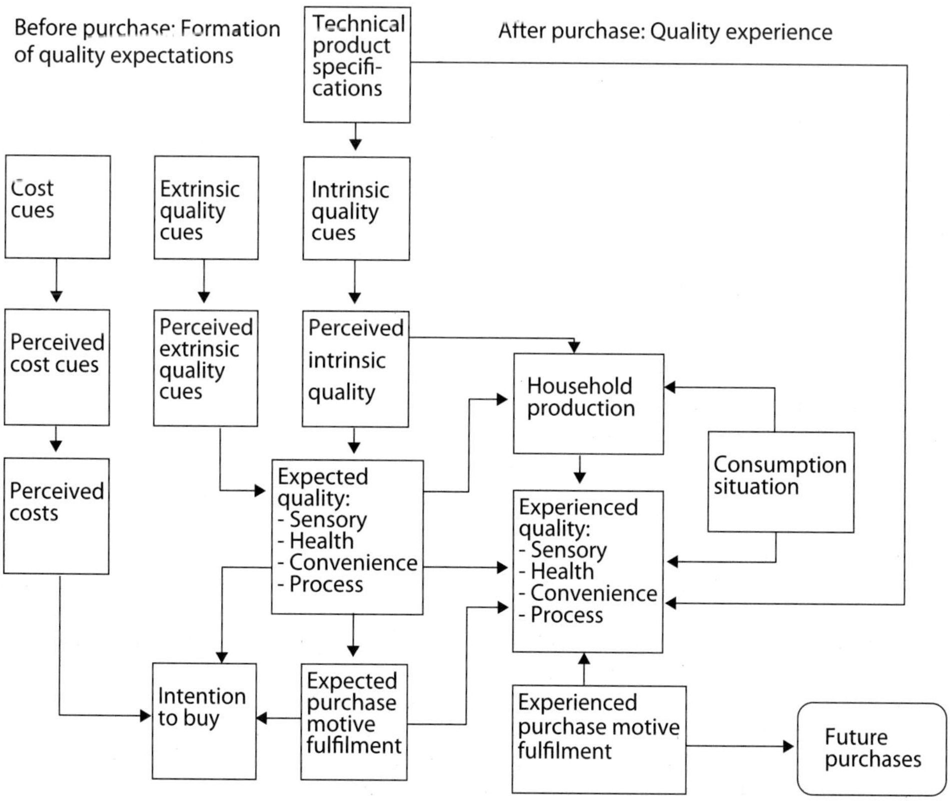

Figure 1. The Total Quality Food Model (Grunert, 2005: 373).

In the horizontal dimension Grunert considers time aspects, namely the formation of quality expectations in the pre-purchase phase and quality experiences after purchase. How consumers perceive quality (including safety and risks) before and after purchase depends strongly on whether they judge search, experience or credence attributes of food products. It is generally accepted that perceived risks are related to the trend towards more hidden characteristics of food, or credence attributes, which are connected to the production process and cannot be controlled at the end product level or after consumption. The comparison of quality expectations and actual quality experiences is a major determinant of consumer (dis-)satisfaction, which strongly influences future buying decisions.

The papers presented at the EAAE Seminar in Goettingen present new insights into several of the aspects summarised in the Total Quality Food Model. Lüth and Spiller and Kühl and Schulz highlight the importance of quality signals in food marketing. Franz and his co-authors: Wirthgen and Faße; Bruhn and Grebitus; and Zenner, Wirthgen and Altmann shed some light on various aspects of consumer perceptions with regard to food quality and safety. Saggau analyses on an aggregate level the role of information releases concerning food scares and their impact on consumers' buying decision. Consumers' perceptions and buying decisions strongly determine whether or not food quality has the potential to create a competitive advantage (Hanf and Hanf). As is typical of for means-end analyses in the field of consumer perceptions and behaviour (Grunert, 2005), several authors apply qualitative research methods, such as association techniques (Bruhn and Grebitus; Wirthgen and Faße). A different methodological development is the use of experimental (Lüth and Spiller) or formal (Saggau) approaches.

All in all, the broad range of methods and theories demonstrate the success of the 92nd EAAE Seminar, which contributed not only to presenting new research results but also to successfully bridging the gap between varying theoretical and methodological research streams in agricultural economics.

References

Akerlof, G.A., 1970. The market for 'lemons': Quality uncertainty and the market mechanism. Quarterly Journal of Economics 84, 488-500.

Arrow, K.J., 1985. The economics of agency. In: Pratt, J.W. and R.J. Zeckhauser (eds.), Principals and agents: The structure of business. Harvard University Press, pp. 37-51.

Böcker, A., M.E. Bredahl and J. Northen, 2003. ISO 9000 certification in British agribusiness: Motivations and performance impacts. In: Schiefer, G. and U. Rickert (eds.), Quality assurance, risk management and environmental control in agriculture and food supply networks, Volume A. ILB-Press, pp. 51-61.

Brennan, M., C. Ritson, M. Ness, S. Kuznesof, L. Frewer and S. Miles, 2004. Communicating food risk uncertainty with the public. In: Schiefer, G. and U. Rickert (eds.), Quality assurance, risk management and environmental control in agriculture and food supply networks, Volume B. ILB-Press, pp. 433-442.

Deming, W.E., 1986. Out of the crisis: Quality, productivity and competitive position. Cambridge etc., Cambridge University Press.

Dierks, L.H., 2006. Trust as a determinant of consumer behaviour in Germany. In: Fritz, M., U. Rickert and G. Schiefer (eds.), Trust and risk in business networks. ILB-Press, pp. 263-274.

Frentrup, M. and L. Theuvsen, 2006. Transparency in supply chains: Is trust a limiting factor? In: Fritz, M., U. Rickert and G. Schiefer (eds.), Trust and risk in business networks. ILB-Press, pp. 65-74.

Fritz, M. and T. Hausen, 2006. Tailored improvement of supply chain processes in agri-food networks. In: Bijman, J., O. Omta, J. Trienekens, J. Wijnands and E. Wubben (eds.), International agri-food chain and networks: Management and organisation, Wageningen Academic Publishers, pp. 179-195.

Fritz, M., U. Rickert and G. Schiefer (eds.), 2006. Trust and risk in business networks, ILB-Press.

Grunert, K., 2005. Food quality and safety: Consumer perception and demand. European Review of Agricultural Economics 32, 369-391.

Grunert K., H.H. Larsen, T.K. Madsen and A. Baadsgaard, 1996. Market orientation in food and agriculture. Kluwer Academic Publishers.

Hatanaka, M., C. Bain and L. Busch, 2005. Third-party certification in the global agrifood system. Food Policy 30, 354-369.

Jahn, G., M. Schramm and A. Spiller, 2005. The reliability of certification: Quality labels as a consumer policy tool. Journal of Consumer Policy 28, 53-73.

Luning, P.A., W.J. Marcelis and W.M.F. Jongen, 2002. Food quality management: A techno-managerial approach. Wageningen Academic Publishers.

Meuwissen, M.P.M., A.G.J. Velthuis, H. Hogeveen and R.B.M. Huirne, 2003. Technical and economic considerations about traceability and certification in livestock production chains. In: Velthuis, A.G.J., L.J. Unnevehr, H. Porkeveen and R.B.M. Huirne (eds.), New approaches to food safety economics. Kluwer Academic Publishers, pp. 41-54.

Napier, T.L., M. Tucker, C. Henry and S.R. Whaley, 2004. Consumer attitudes toward GMOs: The Ohio experience. Journal of Food Science 69, 69-76.

Newslow, D.L., 2001. The ISO 9000 quality system: Applications in food and technology. Wiley.

Pfeifer, T., 2002. Quality management: Strategies, methods, techniques, 3rd ed., Hanser.

Schulze, B., A. Spiller and L. Theuvsen, 2006. Is more vertical integration the future of food supply chains? Empirical evidence and theoretical considerations from German pork production. In: Bijman, J., O. Omta, J. Trienekens, J. Wijnands and E. Wubben (eds.), International agri-food chain and networks: Management and organisation, Wageningen Academic Publishers, pp. 49-63.

Theuvsen, L., 2003. Motivational limits to tracking and tracing: Principal-agent problems in meat production and processing. In: Schiefer, G. and U. Rickert (eds.), Quality assurance, risk management and environmental control in agriculture and food supply networks, Vol. A. ILB-Press, pp. 223-230.

Theuvsen, L., 2004. Transparency in netchains as an organisational phenomenon: Exploring the role of interdependencies. Journal on Chain and Network Science 4, 125-138.

Theuvsen, L., 2007. Von der Gewährleistungs- zur Initiativfunktion: Neue Handlungsstrategien des Staates am Beispiel des gesundheitlichen Verbraucherschutzes. In: Wissenschaftlicher Beirat der GÖW (ed.), Die Zukunft der öffentlichen Dienstleistungen. Berlin, pp. 40-68

Theuvsen, L. and T. Hollmann-Hespos, 2005: The economics of traceability: A model of investments in tracking and tracing systems in agriculture and the food industry. In: Boaventura Cunha, J. and R. Morais (eds.), Proceedings of the EFITA/WCCA 2005 Joint Conference, Vila Real (Portugal), pp. 914-921.

Theuvsen, L., J.-C. Gawron and C.-H. Plumeyer, 2007. Qualitätsanforderungen in Zertifizierungssystemen: Ansatzpunkte für die Messung von Qualität. In: Linß, G. (ed.), Messbare Qualität. Shaker, pp. 180-201.

Williamson, O.E., 1985. The economic institutions of capitalism. Free Press.

Economics of standard owners: competition as barrier to global harmonisation of food assurance systems

Kristian Möller
EurepGAP Secretariat c/o FoodPLUS GmbH, Spichernstr. 55, 50672 Köln, Germany; moeller@foodplus.de

Abstract

Since the early 1990`s, there has been a growing number of emerging Food Assurance Systems (FAS). Their scope cover agricultural production as well as food manufacturing, and range from local, single firm schemes to national and global standards. Whereas the inward motivation always contains the harmonisation objective, external objectives often reveal market differentiation as leading motive. The paper observes objectives of FAS standard owners from the perspective of a member of the group of standard owners and proposes solution strategies to reduce the competition factor among FAS as barrier to global harmonisation.

Keywords: EurepGAP, certification, harmonisation, food Assurance Systems

1. Introduction

Since the early 1990's one can observe the creation of a globally growing number of Food Assurance Systems (FAS). Today these schemes cover all sectors and processes of the agriculture-food-supply chain: feed, fertiliser, agrochemical industries, agricultural production, transport, trade, processing and distribution. The geographic dimension ranges from single local one-firm schemes to national systems and globally operating standards. And all FAS have an identified standard owner, mostly as legal entity, residing with the private sector or the public sector.

The list of incentives for the creation and implementation of FAS is long, but one motive can be seen as common: internal harmonisation among the users of the FAS. On the other hand, there is evidence of competition across different FAS, using differentiation as a marketing tool.

This competition element leads to higher costs on the user level due to duplication of control and membership fees and bears the threat of negative image effects for the entire sector, if marketing communication focuses upon revealing safety and integrity differences to competing FAS. So far there has not yet been a fully completed harmonisation process in the market.

2. Objective and procedure

The objective of this paper is to analyse the incentives of standard owners of a number of relevant FAS in order to propose strategies to reduce market inefficiencies as the result of FAS competition as barrier to global harmonisation to raise harmonisation benefits to the farm and food sector.

After selecting a few FAS examples for this purpose, and describing their main policies, we highlight how these policies are linked to their objectives. The implementation of those policies creates sunk costs for FAS standard owners which can be seen as barriers to exit.

A comparison of the incentives of FAS owners identifies conflicting objectives which reflect their behaviour in the harmonisation efforts. A final section attempts to assess the probability for a realisation of a globally harmonised solution for FAS.

3. Characteristics of Food Assurance Systems

The FAS for this paper have been selected on the basis of their relevance and their existing overlap in the European farm and food sector, thus already creating a degree of duplication of compliance costs for farm and food industry operators (Table 1).

The main cluster differences among the selected FAS can be interpreted as policies, whether the FAS …

a. attempt to cover all stages of the food chain or just a part of it ('All stage vs. part stage');
b. manage the supply chain information with central database systems to exercise control and support the interface to upstream and downstream systems as well as traceability ('central database');
c. intend to build up a final consumer marketing or remain in the business-to-business communication in order not to create more labels in the market;
d. are owned publicly or privately ('public vs. private ownership');
e. are driven by a supply chain partnership or by a single sector.

The above policies can be linked to underlying incentives of FAS (Table 2).

Some vertically integrated systems with a national base like AFS and Q&S follow the objective to level imports and integrate retail customers into their control system. This objective contradicts with the supplier control and market access motives of global standards that cover only parts of the food supply chain, like IFS or BRC and EurepGAP.

In other words the following conflict scenarios can be identified:

1. The retailer policy to communicate with a differentiated private retailer label for a product that is globally sourced conflicts with the objective of national schemes to use the (national) standard logo aiming at increasing consumer demand for a nationally sourced product.
2. Controlling the customer, i.e. retailer, to maintain a level of product quality and integrity via good distribution practices is a legitimate objective of suppliers in a food chain, but the enforcement of this measure against given market forces of multiple retailers and global buying groups is limited, unless political and consumer support can be assured. In return, supplier control objectives and outsourcing of the control activities to FAS are implemented more easily, since they follow demand oriented market forces.
3. A little different - and not visible in Table 2 - are conflicts between horizontal standards of the same stage like SQF 1000 (Agricultural standard of the Food Marketing Institute in the US) versus EurepGAP, and International Food Standard (IFS) versus British Retail Consortium International Technical Standard (BRC). The nature of these conflicts of interest is similar with those among all-stage feed to retail schemes. The different policies lead to competition and duplication on the FAS user level, especially when the peer FAS have overlapping geographic scopes.

Table 1. Selected FAS and their main policies.

	SQF 1000/2000(USA)	IFS and BRC	EurepGAP	KAT (Kontrollierte alternative Tierhaltung)	Q&S, Qualität und Sicherheit (Germany)	AFS, Assured Farm Standard (UK)	ISO 22000
Vertical Scope	pre-farm gate (1000)/ post-farm gate (2000)	post-farm gate	pre-farm gate	feed to retail	feed to retail	feed to retail	feed to retail
Geographic Scope	global	global	global	European (global)	one country (European)	one country	global
Product Scope	several	all foods	all farmed products	eggs	beef, pork, fruit and vegetable	several	all foods
Ownership	private	private	private	private	private	private	public (ISO)
Consumer Marketing	consumer label	business to business	business to business	consumer label	consumer label	consumer label	consumer possible
System Management	database	database	database	database	database	database	de-centralised
Driving industry	supply chain partnership	retailer driven	supply chain partnership	supply chain partnership	supply chain partnership	supply chain partnership	supply chain partnership

Table 2. Objectives of standard owners and their policies.

	Feed to retail	Central database	Consumer label	Public Ownership	Supply Chain partnership
Improve sector image	+	+	+	±	+
Customer (retailer) control	+	+	+	±	+
Supplier control	-	+	-	-	-
Import levelling	+	+	+	-	-
Export promotion	-	±	±	±	-
Market access	-	+	-	±	+
Brand support	-	+	±	-	-
Legal compliance	-	+	-	-	-
Due diligence	-	+	-	-	-
Supply chain efficiency	-	+	-	-	+
Outsourcing	-	+	-	-	-
Food chain communication	+	+	-	-	+
Political lobbying	-	-	-	-	-

+: yes; ±: partly yes; -: no

4. Sunk costs as barriers to harmonisation

Any harmonisation would mean the reduction of duplication of control activities on FAS user level, thus ultimately the reduction of the number of FAS operating directly with the same user, either by a market segregation and cooperation among FAS or by FAS leaving the market. Both strategic options lead to reduced revenue and not recovering investments of FAS owners.

Many FAS owners have created sunk costs during their development and implementation phase. Those costs are barriers to exit from the FAS market and tend to be higher, where the FAS owners

- had high development and/or purchase/license costs;
- cover the entire supplier chain, thus had higher coordination efforts;
- invested into a central database;
- involved consumer and non-government organisations in its development (gained reputation out of a successful multi-stakeholder process);
- are directly or indirectly linked to governments;
- communicate with a consumer logo (sunk marketing costs);
- have a long lifetime of the standard;
- created a close link to national legal requirements;
- have created a strong positive interdependence to other activities of the standard owner, like association membership or contribution to association revenue.

5. Strategies to reduce barriers to harmonisation

The scenarios (1) to (3) of Section 3 can be solved following certain strategies:

To (1): National vertically integrated feed to retail schemes use the global horizontal standards and build an add-on module for their extra requirements. Examples are:
- Q&S Meat uses the IFS for their processing level;
- AFS uses BRC for their processing level;
- KAT uses IFS for packing houses.

To (2): Consumer oriented schemes can open up and allow the labelled products also to be sent to non-FAS members, i.e. a retailer that is not a member of the consumer logo scheme, can use the audit system of the consumer labelling scheme for his/her supplier control without being forced to be signatory of the national scheme and to communicate the logo.

This strategy can also be used among national feed-to-retail FAS, where *bilateral* recognition may lead to the acceptance of the other country's label products entering the national scheme. An example is the recognition of the Dutch IKB system for pigs by Q&S Meat. A prerequisite for a high integrity of that strategy is the harmonised content and control system as well as an operational database interface between the two schemes. The disadvantage of such bilateral agreements is high transaction costs, since there is a need to signed those agreements where ever product flows are.

Alternatively, there is a system that can recognise the globally agreed core content and control system requirements via a centralised assessment and comparison tool. Once those elements of FAS are recognised and a globally accepted database interface and interoperability code is accepted, goods can flow across FAS without audits duplicated on farm and processing level, only because there are different country and FAS of destination requirements.

One of the most transparent examples is the *benchmarking* system of EurepGAP. National schemes can continue to exist, build own add-on requirements into their content, e.g. to reflect national legislation and/or achieve a marketing differentiation effect, and receive their global *multilateral* recognition of the core content and control system by the global partnership of all participating FAS owners and buying groups. The database interface is also addressed and its development almost completed.

To (3): Competing FAS on the same level will need to adopt strategies that reflect their high sunk costs. Whereas there is already a movement towards a reduced number of globally operating post-farm gate FAS - today only IFS and BRC and probably SQF2000 (Safe Quality Food from the Food Marketing Institute in the US) have continued retailer recognition - those remaining FAS are in ongoing competition. Mutual recognition agreements among these standard owners have not yet been signed.

On the other hand, the development and introduction of the ISO 22000 will create a new FAS with largely public support, but so far no visible support of the leading retailers. Sunk cost of ISO 22000 are high, since it is a globally developed system in the ISO family, driven by organisations and governments that like to regain control of food safety issues. It can be observed that the high sunk costs for setting up a new FAS are high, and restricts new entry for those schemes that have the intention to compete and not to cooperate with other reference standards.

In this market situation of competing FAS, market participants can only adopt a sector sub-optimal strategy to reduce costs of duplication: contracting control organisations (certification bodies) that can conduct combi-audits in parallel and issue two or more different FAS compliance certificates after a one-stop-shop audit. Examples are:

- combi-audit of IFS and BRC;
- combi-audit of EurepGAP and Q&S.

6. Conclusion

Duplication of FAS audits on market participant level is an evidence of exiting market competition among FAS and can be seen as a sub-optimal market solution for firms in the agri-food sector due to the lack of global harmonisation of those standards. The comparison of the adopted policies and underlying objectives of FAS owners reveal the existence of conflicts of interest, thus a low incentive to cooperation and harmonisation. Sunk costs in the market are a further barrier to exit of FAS, but also a barrier to entry of new FAS that wish to compete. This is the main reason that the situation of competition is likely to continue among a limited number of globally operating horizontal FAS. On the other hand, there is a successful tool of cooperation that has achieved global harmonisation on the agricultural sector: the benchmarking procedure of EurepGAP, the Global Partnership of Safe and Sustainable Agriculture, where national FAS can exploit there differentiation strategies and achieve global recognition at the same time, thus contribute to global harmonisation of FAS.

Façade and means of control: the use of ISO 9000 standards

Peter Walgenbach
University of Erfurt, Chair of Organisational Theory and Management, Nordhäuser Str. 63, 99089 Erfurt, Germany; peter.walgenbach@uni-erfurt.de

Abstract

ISO 9000 has become an institutionalised structural element of organisations. Firms are said to implement ISO 9000 because it is expected by customers or public authorities. However, the general technical efficiency of the standards is often questioned. It is this tension which was the starting point for an empirical study on the adoption of the standards. An important question addressed in this study is whether and to what extent organisations decouple the documented quality system based on ISO 9000 from their work-activities. The study thus addresses a core argument of institutional theory that has only recently received attention in empirical investigations.

Keywords: ISO 9000, institutional theory, decoupling

1. Introduction

In recent years, especially since the late 1980s, 'quality' has been hailed as an important, if not the 'ultimate goal' of organisations and a number of quality management tools and techniques have been promoted as rational means to increase organisational performance and to secure organisational survival. The ISO 9000 standards have been particularly prominent and have been pushed by a number of actors (Walgenbach, 2000; Walgenbach and Beck, 2002). In fact, the ISO 9000 standards have become an institutionalised practice in the last 15 years all over the world (Guler *et al.*, 2002). Today, it is taken for granted that firms are able to boast an ISO certificate. Worldwide more than 550,000 ISO 9000 certificates have been issued (ISO, 2002).

The ISO 9000 standards published in 1994 'represent three distinct forms of quality system requirements suitable for the purpose of a supplier demonstrating its capability, and for the assessment of the capability of a supplier by external parties'. The ISO standards contain demands concerning the documentation of the quality system. The ISO 9001 standard is the most extensive, consisting of 20 elements. The elements of the ISO 9001:1994 standard relate to:

1. management responsibility;
2. quality system;
3. contract review;
4. design control;
5. document and data control;
6. purchasing;
7. control of customer-supplied product;
8. product identification and traceability;
9. process control;
10. inspection and testing;
11. control of inspection, measuring and test equipment;
12. inspection and test status;

13. control of nonconforming product;
14. corrective and preventive action;
15. handling, storage, packaging, preservation and delivery;
16. control of quality records;
17. internal quality audits;
18. training;
19. servicing; and
20. statistical techniques.

What is important is that it is not the quality system itself which is to be standardised by implementing the standards. Instead, the standards lay down requirements relating to the documentation of the quality system. Thus, it is the documentation of an organisation's quality system, the congruence of the quality system's documentation with the practices within the organisation, which are audited and certified.

However, the continuous growth in the number of organisations acquiring an ISO 9000 certificate is somewhat surprising, since the ISO 9000 standards have their source in standards which were made for military products, nuclear power plants, electrical industry and the aerospace industry. Increasingly the standards are spreading in industries they were not originally made for, for example in the service sector, and where their implementation may lead to massive 'translation' problems (Czarniawska and Joerges, 1996). They are also applied in organisations which do not deliver any primary products or services to other organisations. Originally the standards were made to facilitate the relationship between company customers and suppliers. Today, the ISO 9000 standards are also used in firms which sell their products or services to end-users.

This is not yet sufficiently explained. After being 'translated', the standards may also lead to an increase in the efficiency of work-activities and transactions in industries they were not originally made for. However, we may also be dealing with what Selznick (1957) calls institutionalisation. To institutionalise, according to Selznick, means infusing with value beyond the technical requirements of the task at hand. Organisations adopt practices, techniques or procedures not because they are required in terms of a technical efficiency, but because they have acquired a level of importance which exceeds their technical value with respect to efficient production or exchange. The suspicion that the ISO 9000 standards are being institutionalised and the scepticism expressed appear to be reasonable because of doubts as to the technical efficiency of the ISO 9000 standards:

- A certificate which confirms that an organisation fulfils the requirements of the standards is no guarantee for products or services being of high quality (Kamiske *et al.*, 1994; Zuckerman, 1997). As long as customers are aware of this, the acquisition of an ISO 9000 certificate will not necessarily be followed by an increase in turnover that is sufficient to compensate for the costs of implementing the system and acquiring a certificate.
- It costs a great deal to implement a quality system according to the requirements of the ISO 9000 standards, but it does not guarantee a reduction in the cost of defects. In a study conducted by Kamiske *et al.* (1994), 192 of 325 quality managers mentioned that in their firm no or only minor reductions in the costs of defects were achieved by implementing a quality system based on the standards. 104 quality managers said that the costs of defects decreased, but 50 of them were not able to quantify the reductions.
- As the standards require the formalisation of procedures, tasks and responsibilities, the ISO standards are inconsistent with other popular approaches to 'rational' organisation and management, for example lean management.

Invited speakers

Despite all the criticism, some advantages were also mentioned in the literature:

- According to many authors the most important argument for achieving an ISO 9000 certificate is that it may become a prerequisite for the survival of an organisation (see, for example, Jackson and Ashton, 1995: 32). According to them 'ISO 9000 is a convenient way for large company buyers to keep their lists to manageable proportions. [...] Whether or not it is a thorough mechanism, an ISO 9000 policy automatically shortens the list of potential suppliers.'
- Further, if an organisation acquires a certificate confirming that the quality system fulfils the requirements of the ISO 9000 standards, customers may be expected to cancel their audits. But, this positive effect has not materialised in actual fact. In a study conducted by the Federation of German Industries (BDI, 1992) 32 percent of a sample of 908 firms answered that despite the certificate their customers were continuing to audit the complete quality system, and 52 percent said that parts of the quality system are still audited by customers.

Interestingly, most of the pro-arguments in the literature do not address the technical requirements about effectively assuring quality, but make reference to the relationship between an organisation and its environment. Of course, some arguments which relate to the quality system and the quality of the products and/or services are also listed in the literature. For example, it is argued that fulfilling the requirements of the ISO standards, a third party audit, and the attempt to acquire a certificate should be considered as opportunities to reflect on and reconsider the quality system of an organisation.

To summarise the literature, it seems as if there are only two arguments which appear to be plausible motives for the implementation of the standards and the acquisition of a certificate:

- Management is interested in rethinking the quality system in order to enhance the efficiency of the system. But to do this there is no imperative to have recourse to the ISO 9000 standards; nor does this necessarily require a third party audit or a certification of the quality system.
- The organisation is exposed to external pressures. For example, customers or public authorities make demands for a certificate which confirms that the quality system conforms to the requirements of the standards.

Indeed, as the literature indicates, an increasing number of firms at home and abroad are now demanding a certificate (ISO, 2002).

2. The effects of the ISO 9000 standards on the formal structure and the work-activities of organisations: theoretical considerations and research questions

In attempting to explain the phenomenon of the adoption of the ISO standards, it soon becomes obvious that most of the common organisation theories, for example contingency theory (Donaldson, 2001) or transaction cost theory (Williamson, 1985), are not suited for this endeavour. In these theories it is generally argued that the success and the survival of an organisation depend on efficiently controlling the work-activities and on efficient business-transactions. Elements of the formal structure of organisations, for example rules and procedures, are regarded as technical solutions to technical problems of production and exchange.

Another perspective is presented by institutional theory (Powell and DiMaggio, 1991; Scott, 2001). Institutionalists argue that the formal structure of organisations is less a means of coordinating and controlling the relationship between organisation and environment and of coordinating and controlling work-activities, but rather that the elements of the formal structure are manifestations of rules and expectations within the environment of organisations, which become obligatory (Meyer and Rowan, 1977). Institutionalists argue that general assumptions and belief-systems in the environment of organisations define what firms, schools or hospitals should look like (Scott and Meyer, 1994). Thus, many of the positions, departments or routines and procedures, which are part of an organisation are adopted because of 'public opinion', the expectations of influential customers or the state. These elements of the formal structure are adopted by organisations irrespective of their effect on the performance of the organisation. They are adopted to achieve legitimacy. Thus, in institutional terms, firms acquire a certificate not because the adoption of the standards improves technical efficiency, but because customers expect it; increasingly a certificate has become a matter of course, or, in other words, a prerequisite for the survival of an organisation.

Meyer and Rowan (1977) point to two problems an organisation will, in general, be confronted with, if its success depends on the adoption of institutionalised elements, rules or structures: (1) Technical activities and demands for efficiency create conflicts arising from the organisation's efforts to conform to the institutionalised rules. The adoption of institutionalised rules may, from the point of efficiency, give rise to sheer cost. (2) Because institutionalised elements are based on and transmitted by myths of 'rational' organisation that arise from different parts of the environment, the rules may conflict with each other, for example lean management and ISO 9000. However, organisations seeking external support and stability in institutional environments have to incorporate all sorts of incompatible structural elements. The resulting inconsistencies make a concern for efficiency and tight control of the core-activities of an organisation problematic.

According to Meyer and Rowan (1977) organisations can resolve conflicts between institutionalised rules and efficiency by decoupling institutionalised rules from each other and from the work-activities of the organisation. Coordination, interdependencies, and mutual adjustments among structural units are handled informally. The advantages of decoupling are that organisations are enabled to maintain standardised and legitimate formal structures, while the activities vary in response to practical considerations.

The proposition that organisations often decouple institutionalised structural elements from work-activities appears to be highly plausible. However, this proposition has with a few exceptions not been empirically tested in for-profit-organisations (Boiral, 2003; Fiss and Zajac, 2004; Walgenbach, 2001; Westphal and Zajac, 1998). Hence, the questions whether and to what extent certified quality systems are decoupled from work-activities became core issues of this study.

The question as to what extent the quality system is decoupled indicates that further theoretical considerations have slipped into the study. So far, it has been assumed that firms adopt the quality system and acquire a certificate because of pressure within their environment. Furthermore, it has been assumed that the quality system is decoupled from the work-activities. However, a closer look at the standards revealed that the quality system may enable management to enhance its control of the work-activities. Because the ISO standards not only require the definition of job-related responsibilities, but also make demands in terms of the formalisation of routines and procedures, management has the opportunity not only to lay down the individual responsibilities (as in job-descriptions), but also the chance to increase the transparency of organisational work-processes (Tuckman, 1994). Thus, if a firm implements a quality system based on the ISO 9000 standards, processes which are not yet very transparent for management can be made explicit and transformed into impersonal or technocratic instruments of coordination.

Against this background it seemed to be important to investigate whether the opportunities to enhance management control are used in organisations. It also seemed to be important to find out whether these opportunities were taken into consideration from the outset, or whether they became an important goal later on, that is, as the standards were being implemented.

3. Sample and methods

Due to the questions raised in this study, it did not seem appropriate to collect data based on a standardised questionnaire. It seemed unlikely that a standardised questionnaire sent to a large number of firms would produce reliable answers to questions concerning the decoupling of the quality system from the work-activities of an organisation. Collecting data in a less standardised manner generally implies smaller sample sizes, because the time needed to collect and analyse the data increases dramatically. Thus, the composition of the sample became a crucial step in order to make the study as representative as possible. The following considerations influenced the selection of the firms: The ISO 9000 standards distinguish between hardware, software, processed materials and services. It seemed meaningful to use these categories as well, rather than, for example, industries, for the selection of the sample. The study was conducted in 37 organisational units. The products of eleven organisational units were mainly 'hardware'; in nine units the main product was 'software'; the products of seven organisational units were predominantly 'processed materials'; and in ten organisational units it was 'services'. Besides the criterion of the main product/service offered, other categories were applied. As it seemed to be more likely that before the implementation of the ISO standards the degree of formalisation was higher in larger organisations (Kieser and Walgenbach, 2003), the sample selection was also based on the number of employees in the organisational units. Furthermore, an attempt was made to include organisations which were involved in the development of the ISO standards. Interestingly, in the analysis of the data it became clear that the factors mentioned had, with very few exceptions, no or only a minor influence on the implementation and the use of the quality system, rather there are some general tendencies.

Interviews were conducted in 1996 with quality managers of the 37 organisational units. In the following sections we refer to transcriptions of these interviews (for instance, interview 24, p. 3). The quality managers interviewed were responsible for implementing the standard and updating the documentation of the quality system. With the exception of two interviews all were tape-recorded, and then entirely transcribed. Interviews lasted between one and a half to four hours. Whenever possible, additional interviews with top or senior managers of the particular organisational unit were conducted. The analysis of the interview was based on the framework for analysing qualitative data by Miles and Huberman (1994).

4. The effects of the ISO 9000 standards on the formal structure and the work-activities of organisations - some empirical observations

4.1. Triggers for implementing the ISO 9000 standards and acquiring a certificate

The first area of interest was to analyse why the implementation of a quality system based on the ISO standards was considered. It is worth noting that in most cases a certificate had not been required. In 13 interviews it is explicitly stated that customers did not make any demands. In three other cases it is unlikely that it was the expectations of customers which triggered the adoption of the quality system and the acquisition of a certificate. Two of these organisational units only had internal customers. It is reasonable to assume that internal customers have other, better opportunities to check their suppliers' quality system. The customers of the third organisational unit are end-users only. It is very unlikely that consumers make any demands on the organisational unit to have a documented quality system or a certificate.

On the other hand, 17 managers reported that customers demanded a certificate or asked whether the organisational unit had acquired a certificate. However, the assertion that a certificate was made into a condition by customers can be found in only eight interviews. In one of these eight cases the manager interviewed said that a certificate was demanded by customers only once - after they had acquired a certificate. In another interview the quality manager talked about customers' expectations and the requirements of the market initially, but it became clear that these far-reaching demands were only based on assumptions regarding future developments. Further, three of these organisational units were operating in industries, i.e. automobile, electrical-engineering, and nuclear-power generation, in which quality management handbooks and external audits of the quality system were common long before the ISO 9000 standards came into existence. Ten quality managers mentioned, however, that general inquiries as to whether the firm had acquired a certificate were customary. But these questions did not entail a direct demand for a certificate.

The argument that customers' demands were not the main trigger is supported by an inverse conclusion too: the fact that a certificate plays no or only a minor role in the relationship to the suppliers of the organisational units. 32 out of 35 quality managers who expressed an opinion on this issue said that a certificate is not a prerequisite for suppliers of their firm.

Other triggers were more important. These triggers are best described as: an assumed diffusion of an idea or a concept, a general, but vague atmosphere, speculation, hearsay or the impression of an emerging trend. In 15 interviews there are statements which can be summarised under these categories. Two characteristics of all these descriptive expressions, namely uncertainty and ambiguity, are likely to have forced the pace of institutionalisation. The extent to which uncertainty and ambiguity were experienced was very different in the various organisational units. However, 13 quality managers said that in retrospect the decision to implement the standards as well as the decision to acquire a certificate were embedded in a messy and confusing flow of information full of uncertainties and ambiguities.

> 'I believe, well, I don't want to call it herd instinct, but at the time when the ISO descended on Germany like a cloud, when everybody talked about ISO, and nobody knew what it was, and you need two years to implement it - or three or five or even more: 'It's about time to begin' [...] "O.K., let's do it!" And suddenly you were doing it, you contacted the certifying organisations. [...] I don't want to say that ISO had acquired a life of its own, but somehow that is the way it was.' (Interview 26, p. 5)

On the other hand, there is a much smaller group of two organisational units in which the development within the organisation's environment was experienced differently. The quality managers argued that the trend was obvious.

What we have seen so far may have created the impression that customers played only a minor role in the process of institutionalisation of the ISO 9000 standards. This impression would be wrong. Customers were important triggers, but in a way other than is assumed in the literature (see, for example, Jackson and Ashton, 1995). In 15 interviews the following reasons for the adoption of the ISO 9000 standards were mentioned: They concerned themselves with the standards, because the customers of the organisation were dealing with them (mentioned three times), or because a marketing effect or a competitive advantage was expected (mentioned ten times). Sometimes, however, the answers of the quality managers appear a little cynical.

> 'Exclusively - no, about 80 percent marketing. That this junk may be useful - probably. And then I said to myself: "Compared to other trainers in management you are far ahead. You know all about it. You will be the first to pick up that junk. I don't believe in it, but if you want me to have it [...]!"' (Interview 23, p. 1)

Competitors were another important trigger. In six organisational units a decisive factor in the adoption of the ISO 9000 standards was that competitors had already acquired a certificate. One gets the impression that sometimes the orientation of organisational units to other organisational units on the same horizontal level, for example competitors, other divisions or even departments within the company, was more important than the consideration of customers' needs or demands. Obviously, there was competition to be the first organisational unit within an industry, a holding or a company to be able to boast the certificate. In 14 interviews mention was made of the fact that the value of a certificate is determined by the moment in time at which it is acquired.

> 'If we were going to have a certificate, then it should be based on 9001. Thus, we were the first to acquire 9001. [...] Shortly before (X) (a competitor, P.W.) had acquired a certificate based on 9002 [...] of course, this is something you don't like, coming second [...] in principle, it means playing around intellectually to include "design control" [...]' (Interview 24, p. 3)

Another important trigger for implementing the ISO 9000 standards and acquiring a certificate were instructions from the board of the company or holding. In seven interviews orders from top management were named as the main reason for the adoption of the standards. The orders from top management were not welcome in all these organisational units; actually, they were perceived as coercion.

A further trigger for the adoption of the standards was EU directives. They were mentioned in four interviews, but three times only after further probing.

To summarise, the implementation of the standards and the acquisition of a certificate were triggered externally (see Table 1). Coercion was as important as imitation. Often both mechanisms were operating at the same time, and thus reinforced each other. In addition, and in line with institutional theory (DiMaggio and Powell, 1983), the pace at which certified quality systems spread increased as a result of the high degree of uncertainty and ambiguity concerning their future importance.

4.2. Objectives of the implementation of the ISO 9000 standards and the acquisition of a certificate

The main objective of the organisational units investigated was 'to get the certificate', but due to different motives. In seven units it was an order from the board. In 15 organisational units it was the expected marketing effect, the competitive advantage, or because of the opportunity to signal competence to the organisation's environment. In four organisational units it was to meet customers' demands. In one organisation it was because of EU directives, and in three organisational units the acquisition of a certificate was an end in itself. In the answers contained in the last category in particular it became clear that in these organisations the process of institutionalisation had already reached its final stage - an ISO 9000 certificate is taken for granted.

Table 1. Triggers for implementing ISO 9000 and the acquisition of a certificate (multiple answers were possible). The table contains only those triggers which were mentioned more than three times.

Trigger	Number
Assumed trend (uncertainty, ambiguity)	15
Marketing/competitive advantage	14
Competition for the first certificate	14
Behaviour of competitors, customers, other organisational units within the company	13
Questions from customers	10
Demands from customers	8
Orders	7
EU directives	4

'Well, frankly we just did it.' *'Really?'* 'Yes. The motto was: "What? That's what we need?" "Come on, let's do it!" "Well, why not!" And: "When will we get it?" And so on. But there was no goal. The only goal we had was that we didn't want to be one of those who don't have it; we wanted to be among those who have it - the certificate. But nobody had any other goals in mind. No, that was really not the case.' (Interview 27, pp. 5-6)

However, the managers mentioned internal objectives more often than when asked what triggered the adoption of the quality system. At least three quality managers named an orderly and workable quality system as the main objective; one manager said it was a question of optimising business processes; and another manager mentioned reductions of friction at the intersection of departments (see Table 2).

Interestingly, after having named the main objective, the quality managers often without any further probing mentioned additional objectives which were of importance. Furthermore, a sequence of the objectives regarded as important became evident. Whilst in the beginning, it was to get a certificate 'no matter what the cost', it was internal objectives, that is objectives relating to the work-activities of the organisational unit which came to the fore later. In 19 out of

Table 2. Main objective of implementing a quality system according to the ISO 9000 standards and of acquiring a certificate.

Main objective	Number
To use it for marketing	15
To acquire a certificate in order to carry out orders	7
To meet customers' demands	4
To signal conformity with institutionalised expectations	3
Orderly/workable quality system	3
To reduce frictions	1
To optimise processes	1
EU directives	1

34 interviews in which the topic of a succession of objectives was addressed, the managers stated that internal objectives came to the fore for the first time when the standards were implemented, or when a certificate was already acquired. Only ten quality managers said that all the objectives they mentioned were of importance from the very outset. Five quality managers said that there is still no goal other than having a certificate for its own sake.

There are two reasons why internal objectives became important later. One is the enormous costs of creating and maintaining the documentation of the quality system. According to the quality managers these costs would not be justified if a documented quality system did not entail benefits. The second reason emphasised is the experience of the organisation during and since the implementation of the standards, which revealed to them, the full potential of the quality system. In the majority of the organisational units investigated something became evident which Ortmann (1995) calls the slow fabrication of objectives while acting.

One important objective, which came to the fore, was to harmonise, standardise and optimise processes within the organisational unit. Other important goals were to make the work-processes of the organisation more transparent, to define and optimise interactions between departments, to define rules, to lower costs, to enhance productivity, to secure organisational know-how, to enhance the potential for monitoring employees' work, and to motivate employees. Thus, in the process of implementing the ISO 9000 standards, those responsible referred to familiar (traditional) goals and patterns of organising (see Table 3). Thus the majority of the firms in the sample did not adopt quality systems for reasons of internal efficiency, but it was while the institutionalised element was being implemented that technical efficiency became an important consideration.

Table 3. Further objectives of realising a documented quality system according to the ISO 9000 standards and of acquiring a certificate (multiple answers were possible).

Further objectives	Number[a]
Standardisation/harmonisation of work-processes	12 (3)
To enhance transparency	11 (6)
Optimisation of intersection between departments	10 (4)
To lower costs, to increase productivity	9 (3)
To improve the quality of products/services	6 (1)
Systematise, organise	5 (2)
To define rules, to formalise	5
No further objectives	5
Customer satisfaction	4
To secure organisational know-how	4
To use the certificate in order to fulfil demands of EU-directives	3
Control (checking employees)	3 (2)

[a]Data in brackets refer to the number of interviews in which is explicitly stated that the objectives came to the fore once the standards had been implemented.

4.3. The implementation of the ISO 9000 standards

A substantial argument within institutional theory is that the formal structure of an organisation is designed so that external constituents can apply their criteria of inspection and thus evaluate the organisation (Meyer and Rowan, 1977). However, most of the quality managers reported that the standards entailed no or no fundamentally new demands relating to the design of the formal structure of the organisation. Whenever a new demand arose, it was mostly an increase in documentation. Furthermore they said that no or no fundamental organisational changes were required. Whatever had to be changed 'only' meant: collecting rules, which had already been written; trimming the formal structure of the organisation to fit in another type of systematisation; and formalising employees' knowledge of interdependencies within the organisational units. It became clear only after further probing that in most units much has changed.

To begin with, in all organisational units the position and the job-title 'quality management representative' was created, which did not exist before the adoption of the ISO 9000 standards. A set of tasks and responsibilities is now coupled with the position of a 'quality management representative', which also did not exist before the standards were applied. These tasks consumed a considerable proportion of the quality managers' resources. Most quality managers were fully or partly exempt from other tasks and duties while the standards were being implemented. In addition to the quality manager interviewed, other people within the organisational unit were often extensively involved. Expenditure fell once the certificate had been issued. However, especially in large organisational units, the quality managers still spent their working-days maintaining the quality system full-time or to a large extent. Furthermore, tasks and functions, for example responsibilities for a specific process, managerial responsibilities for an area within the organisational unit, or managerial responsibilities for an element within the ISO standards were frequently introduced.

Thus, implementing the standards not only entailed documenting responsibilities, activities and procedures within the organisational unit, but also establishing new functions and responsibilities. First, the 'infrastructure' to implement and maintain the quality system had to be set up. Second, due to the requirements of the standards in most of the units, completely new activities had to be introduced, for example internal quality audits, which in general had not been previously conducted. It also became clear that in most organisational units some of the elements required by the standards had to be partly or entirely introduced from scratch. 29 quality managers reported that they had to implement or at least fundamentally adapt one element of the quality system. In 21 organisational units it was not only one element, but two or more. An extreme case was a manager who reported eight elements which had to be introduced or adapted to the requirements of the standards (see Table 4).

Due to the fact that in most organisational units some elements had to be implemented more or less in their entirety, the question arises why these elements did not exist previously, or at least not as required by the standards before the system was introduced. One answer to this question is that they were unknown or not regarded as a means of increasing the efficiency of the organisation. Having to tackle the requirements of the standards, one could argue, provided management with an opportunity to implement useful solutions to solve organisational problems and to enhance efficiency. The empirical data presented so far provides some evidence for this. However, another answer is that achieving the requirements of the standards is problematic or does not produce tangible benefits.

Table 4. Number of elements introduced or adapted to fulfill the requirements of the ISO 9000 standards.

Number of elements introduced or adapted according to the requirements of the ISO 9000 standards	Number
0	5
1-2	15
3-4	10
5 and more	4
No statement	3

29 of the quality managers interviewed said that at least one element of the standards was difficult to fulfil. Only four quality managers reported that they had no problems fulfilling the requirements of the standards. Moreover, 25 quality managers said that at least one element of the ISO 9000 standards was not relevant for their organisation, or that at least one element appeared not to be applicable. Only seven quality managers said that all elements of the standards were relevant for their organisational unit.

But even if it was difficult to fulfil the requirements, or if it was considered not to be useful or sometimes meaningless, most of the organisational units tried to fulfil the requirements of the standards. Sometimes this meant extensively explaining why a certain element could not be applied. But often it led to an attempt to fulfil the demands 'creatively' or by 'stretching the imagination'. The consequences of this approach were some remarkable capers. The objectives of the standards were distorted or even inverted. For example, one organisation fulfilled the requirements of the element 'purchasing' by describing the acquisition of writing desks. This solution had to be 'invented', because there were no primary products which had to be purchased for the services offered to the customers of the organisation. Thus, often the adoption of the quality system immediately involved decoupling the formal structure from work-activities. Formal structures were introduced for work-activities which did not exist or which were of minor importance.

> 'You're tempted to say, this is artificial, this is a complete fabrication. There were situations when I had no other explanation than: 'Well, it is required by the standards!' The goal of getting a certificate - is this really necessary?' (Interview 30, p. 12)

4.4. The documentation of the quality system and work-activities

Even if parts of the documented quality system based on the ISO 9000 standard were decoupled form the work-activities of the organisations, it is important to mention that, according to the interviewees, there were also aspects of the quality system, in which a tight coupling of formal structure and work-activities was intended or seemed desirable. Thus the question arises how the documentation of the quality system was used in the daily work-processes after the system had been implemented. For management the value of the documented quality system here depends on the congruence between documentation and actual work-activities.

In 25 interviews it was said that even in areas where a congruence of the documentation of the quality system and the actual work-activities was intended, the actual work-activities tended to deviate from the documented ones. According to the quality managers an indication that the system was not 'lived and breathed' was that the documentation was not read by the employees. Employees must be familiar with the documentation to recognise differences between actual and documented work-activities. Likewise, not accepting the system wholeheartedly is expressed in non-conformist behaviour as well as not reporting changes in the actual work-activities.

However, quality systems based on ISO 9000 standards include an instrument, the internal audit, which should be used to inspect the congruence between the documentation and the actual work-activities. Without being asked about the internal audit, 16 quality managers emphasised this function. 'Inspection' has two connotations. Firstly, the term involves documentation, which is checked to see to what extent it reflects the actual work-activities. Thus, internal audits are used to update and to optimise the documentation (mentioned twelve times). A sub-function of checking the documentation is to prevent the system from petering out (mentioned six times). The internal audit is used to signal to employees that the quality system is a serious matter. Moreover, it is a means of monitoring whether employees stick to the rules as laid down in the documentation of the quality system (mentioned twelve times).

> 'Internal auditing - that it is what is essentially new in a quality system based on the standards compared with other management techniques. That there is something like, in Lenin's terms, "supervision is better than trust", something which enables to give feedback to the members of the board about what is going on on the shop floor. Have the rules you made been achieved? Do they stick to the regulations or does their behaviour deviate? For us, this will be the main function of quality management in the future.' (Interview 8, p. 5)

The advantage of the internal audit is that this control instrument 'has to be applied because of external demands'. The ISO standards incorporate the element 'internal quality audits'. If top management were able to convince employees that a quality system based on the ISO 9000 standards was required because of customers' demands, EU directives, or competitive advantages which in the long run will secure jobs, then the application of internal audits would be externally justified. Thus, the application of internal audits appears legitimised and easier than in a situation where internal audits seem to be only motivated by a desire to improve supervision. As a result, techniques of behavioural control, standardisation and the depersonalisation of tasks and activities are no longer restricted to the lower levels of organisational hierarchies as was envisaged by scientific management (Taylor, 1911), but can now be applied and intensified in the layers of middle and senior management as well.

5. Conclusion

Quality systems can acquire a wide range of forms and functions. They range from a façade to attain legitimacy to an instrument for monitoring work-behaviour. A quality manager expressed this finding more metaphorically:

> 'From my point of view, the philosophical core beyond the phenomenology, that is to say, beyond these 20 elements, these regulations, quality management, procedures, process-descriptions, documentation and organisation handbook and all these surface regulations - all this will remain wasted paper if the employees can not be motivated to apply it. Because, as long as they are not motivated, you only

> have two possibilities: The whole thing becomes either show biz hype or rowing on the slave galley!' (Interview 30, p. 4)

The findings are of importance both for the development of institutional theory and for the practice of organisational design. For the development of institutional theory the importance of the findings is that they show on the one hand that, as predicted by the theory, decision-makers in organisations make an effort to design the formal structure of an organisation in accordance with what they suppose is expected in the institutional environment. On the other hand they reveal that this should not be equated with a blind knee-jerk adoption of a particular institutionalised element. The potential and the benefits of adopting the instrument were discovered during the very process of implementation. In the process of reinterpreting or 'translating' (Czarniawska and Joerges, 1996) the functions of the standards well-known objectives of applying technocratic instruments of control became clear and dominant. Furthermore, the findings are of importance because they show that elements of the formal structure of an organisation can be both a façade for acquiring legitimacy and an instrument for enhancing technical efficiency.

For practical organisation design, the findings are of importance because they highlight a problem that organisations may be confronted with. A structural element, which is adopted because the institutional environment expects it, may resist any attempt to be reinterpreted. It may not be possible to use it in a way which enhances the technical efficiency of the organisation. The adoption of a particular institutional element may even impair the technical efficiency of an organisation. Even if it is possible to decouple formal structures from work-activities, the implementation of structural elements ties up organisational resources and consequently entails costs. A general implication of this argument is that firms should deal more carefully with ideas of organisation and management, which are tending to become, but are not yet fully, institutionalised. Thus, as long as internal benefits are not discernible, the strategy of 'wait and see' seems to be more advisable than adopting the idea or approach.

References

BDI, 1992. BDI Firmenumfrage über die Nachfrage nach zertifizierten Qualitätssicherungssystemen von ausländischen Kunden. Bundesverband der Deutschen Industrie e.V., Köln.

Boiral, O., 2003. ISO 9000: Outside the iron cage. Organisation Science 14, 720-737.

Czarniawska, B. and B. Joerges, 1996. Travel of ideas. In: B. Czarniawska and G. Sevón (eds.), Translating Organisational Change. De Gruyter, pp. 13-48.

DiMaggio, P.J. and W.W. Powell, 1983. The iron cage revisited: Institutional isomorphism and collective rationality in organisational fields. American Sociological Review 48, 147-160.

Donaldson, L., 2001. The contingency theory of organisations. Sage.

Fiss, P.C. and E.J. Zajac, 2004. The diffusion of ideas over contested terrain: The (non)adoption of a shareholder orientation among German firms. Administrative Science Quarterly 49, 501-534.

Guler, I., M.F. Guillén and J.M. Macpherson, 2002. Global competition, institutions, and the diffusion of organisational practices: The international spread of ISO 9000 certificates. Administrative Science Quarterly 47, 207-232.

ISO, 2002. The ISO Survey of ISO 9000 and ISO 14000 certificates. Eleventh Cycle. International Organisation for Standardization, Genevea.

Jackson, P. and D. Ashton, 1995. Achieving BS EN ISO 9000. Kogan Page.

Kamiske, G., K. Kassebohm, C. Malorny, H. Michael and H.D. Schulze, 1994. Qualifizierung, Auditierung und Zertifizierung im Meinungsspektrum zertifizierter Unternehmen - Resultate und Konsequenzen einer empirischen Studie. In: H.C. Riekhof (ed.), Praxis der Strategieentwicklung. 2nd ed., Schäffer-Poeschel, pp. 357-385.

Kieser, A. and P. Walgenbach, 2003. Organisation. 4th ed., Schäffer-Poeschel.
Meyer, J.W. and B. Rowan, 1977. Institutionalized organisations: Formal structure as myth and ceremony. American Journal of Sociology 83, 340-363.
Miles, M.B. and A.M. Huberman, 1994. Qualitative data analysis. 2nd ed., Sage.
Ortmann, G., 1995. Formen der Produktion - Organisation und Rekursivität. Westdeutscher Verlag.
Powell, W.W. and P.J. DiMaggio (eds.), 1991. The new institutionalism in organisational analysis. University of Chicago Press.
Scott, W.R., 2001. Institutions and organisations. 2nd ed., Sage.
Scott, W.R. and J.W. Meyer, 1994. Developments in institutional theory. In: W.R. Scott and J.W. Meyer (eds.), Institutional environments and organisations - structural complexity and individualism. Sage, pp. 1-8.
Selznick, P., 1957. Leadership in administration. University of California Press.
Taylor, F.W., 1911. The principles of scientific management. Norton.
Tuckman, A., 1994. The yellow-brick road: Total Quality Management and the restructuring of organisational culture. Organisation Studies 15, 727-751.
Walgenbach, P., 2000. Die normgerechte Organisation. Schäffer-Poeschel.
Walgenbach, P., 2001. The production of distrust by means of producing trust. Organisation Studies 22, 693-714.
Walgenbach, P. and N. Beck, 2002. The institutionalization of the quality management approach in Germany. In: K. Sahlin-Andersson and L. Engwall (eds.), The expansion of management knowledge: Carriers, flows, and sources. Stanford University Press, pp. 145-174.
Westphal, J.D. and E.J. Zajac, 1998. The symbolic management of stockholders: Corporate and governance reforms and shareholder reactions. Administrative Science Quarterly 43, 127-153.
Williamson, O.E., 1985. The economic institutions of capitalism. Free Press.
Zuckerman, A., 1997. International standards desk reference. American Management Association.

Firm's costs and benefits of quality management and quality assurance systems

Firm's costs of traceability confronted with consumer requirements

Xavier Gellynck, Renata Januszewska, Wim Verbeke and Jacques Viaene
Gent University, Department of Agricultural Economics, Coupure Links 653, 9000 Gent, Belgium; Xavier.Gellynck@Ugent.be

Abstract

This paper addresses food safety and traceability costs on the one hand and repeated consumer research about needs and perception of traceability on the other. Compared to 2001, overall meat quality perception scores better in 2004. Respondents are segmented based on their subjective perception of meat quality, and differences in the perceived need for traceability systems between the various segments are investigated. Functional traceability attributes, such as organisational efficiency, chain monitoring and individual responsibility, are important to all consumers. Extensions with respect to process attributes, such as production methods, are less relevant to the general population, being only of interest to specific market segments, i.e. consumers with a more negative perception of meat quality and lower consumption levels. In conclusion, it is recommended that public policy focuses on the level of functional traceability attributes, whereas extensions with respect to process attributes are left to private initiatives, which focus on specific market segments.

Keywords: costs, traceability, investments, requirements

1. Introduction

Increased knowledge about food-borne illnesses and consequently growing consumer concerns about food safety in recent years forced both the food industry and the public authorities to develop quality and safety assurance systems (Bredahl *et al.*, 2001). Shogren (2004) mentions 300,000 hospitalisations and 5,000 deaths in the USA annually, with associated costs estimated to be between 3 and 7 billion US$. Experts claim that the risks posed by food-borne disease will increase because of changes in elements such as climate, microbiological systems, water supplies, urbanisation, population greying and food trade intensity (Kaferstein and Abdussalam, 1999). Policymakers have responded to these concerns by creating new policies for safer food with the ultimate goal of increasing consumer health. Food safety crises of recent years such as BSE, dioxin and MPA also led consumers to rethink their attitudes to and behaviour towards food consumption in general and meat consumption in particular (Burton and Young, 1996; Latouche *et al.*, 1998; Verbeke and Viaene, 1999; Henson and Northen, 2000; Verbeke *et al.*, 2000; Buzby, 2001). The emerging issues and related problems pertain to different production stages within the food supply chain.

Several authors (Downey, 1996; Fearne, 1998) believe that governmental regulations and more specifically in the case of the EU, regulations such as EU Directive 89/397 on food safety and hygiene standards, followed up by EU Directive 93/43 (Food Hygiene Regulation) and more recently the EU Regulation 178/2002 known as the General Food Law, are the single most important factor contributing to the change in food quality systems in general and partnerships arrangements in particular. However, the introduction of new regulations does not always happen as smoothly as expected and often affects the competitiveness of individual companies, sectors, even member states or the EU as a whole (Gellynck *et al.*, 2004). It is linked with the differences in capacity of food firms to meet the new requirements and standards on the one hand and with the way food processors are controlled and penalised, both at the national and international level, on the other. Anticipating the introduction of new food regulations does not happen quite often at EU level.

A major challenge for today's food industries pertains to controlling costs associated with new food safety regulations. The recent developments in food regulations have a direct impact on food quality management requirements and consequently on costs at the level of food companies. The increased complexity both at consumer level (risk aversion and sensitivity) and processor level (risk management and communication) requires adaptation of strategies and of institutional organisation in the food supply chains. It is within such a climate of change that value-laden concepts like authentic, healthy and safe possibly offer opportunities to add extra legal food product attributes. In this way, food companies or even entire food chains may be able to differentiate from one another and may compensate increasing quality management costs. Food safety perception differs between countries (Henson and Traill, 2000; Buzby, 2001). Consequently, consumer concerns and acceptance of measures to monitor food safety can be expected to differ. Frenzen *et al.* (2000) have even identified differences within a single country in terms of acceptance and willingness to pay for measures that reduce food safety risks. Gilg and Battershill (1998) find some consumer segments being interested in knowing inside-out how the food is produced, whereas others do not care as long as e.g. hedonic benefits (good taste) are supplied. Such a changing environment offers opportunities for producers, who manage to add value to their products through adapted competitive strategies.

Given the above background, the objective of this paper is twofold:
- Explore differences in general food safety and traceability costs.
- Assess consumer segments related to meat quality perception and their possible interest in functional and process attributes associated with traceability systems as well as the evolution compared to 2001.

The structure of the paper is as follows. Section 2 presents an overview of the relevant recent literature on traceability in food chains. While there appear to be strong beliefs in the potential benefits and costs of traceability systems, few empirical studies have been carried out. Section 3 sets out the research methodology and framework for the present analysis. Qualitative and exploratory research was used to identify costs. Cross-sectional consumer data are used to investigate consumer perception of traceability in beef, pork, poultry and meat mixture chains. Section 4 presents and discusses the empirical findings. First, food safety costs are discussed. Second, a segmentation analysis is carried out, based on the perception of meat quality. Associations between consumer segments, personal consumer characteristics, attitudes to and behaviour towards meat on the one hand, and the perception of traceability on the other, are analysed and compared to the findings of 2001. The final section presents the conclusions and recommendations regarding the development of future traceability systems and traceability-based meat marketing.

2. Traceability systems

Wilson and Clarke (1998) and Jack *et al.* (1998) define food traceability as the information necessary to describe the production history of a food crop and any subsequent transformation or process the crop might undergo on its journey from the grower to the consumer's plate. Traceability means that companies must be able to identify the suppliers of its raw materials and the customer of its end products on a transaction basis. It includes both tracking and tracing. Tracking refers to the determination of the ongoing location of items during their way through the supply chain. Tracing relates to defining the role the composition and the treatments a food product has received during the various stages in the production life cycle. Most of the traceability systems that are operational today used the existing systems of identification and registration as a starting point. The development of traceability systems gained momentum largely as a result of changes at the consumer level (Downey, 1996), and received further impetus through the rapid development of hardware, software and information technology since the 1980s (Oude Luttinghuis, 2000; Wortmann, 2000).

Originally, traceability systems were basically concerned with animal health, disease and food safety control. However, they are gradually extending into proactive management and marketing tools, either through the feedback of information upstream or the introduction of labelling schemes with the traceability system as backbone. All traceability initiatives employ similar principles, in that they lay down standards and procedures which must be observed by members and which are monitored to ensure compliance. In the case of livestock and meat schemes, these standards, procedures and controls embrace all stages from the farm, including feeding, livestock handling and transportation, slaughtering and meat processing, to distribution (Leat *et al.*, 1998). Key elements include the identification and registration of animals, herds, meat processors, exporters, data capture, communication, and data management and verification. Although the key issues of traceability are reasonably straightforward, their implementation is complicated, primarily by the number of levels within the chain and the numbers of producers supplying the chain (Timon and O'Reilly, 1998). A further obstacle to installing traceability systems is the low degree of vertical integration in certain livestock and meat chains in specific countries or regions (Porin and Mainsant, 1998).

The introduction of the General Food Law and consequently the obligation to install traceability systems from January 1st, 2005 on in each food company in the EU, makes traceability system as such no longer a unique selling proposition in the market. This process is enforced by the development of sector-oriented quality assurance schemes rather than enterprise-level quality management approaches (Schiefer, 2004). To overcome this inconvenience, companies can either form supply chain networks to gain operating efficiencies or add extra legal aspects to the traceability system. Operating efficiencies refer to the installation of compatible IT infrastructures throughout the chain (Hanf and Kühl, 2002), enhancing competition between chain networks rather than individual companies. Some studies examined possible consumer interests in traceability. Dickinson and Von Bailey (2002) illustrate differences between consumers in both the UK and the USA related to their willingness to pay for traceability. It is concluded that traceability attributes are important marketing characteristics in both countries but different strategies would be needed in the different countries since concerns about food-born illness are more general in the UK than they are in the USA. General food safety is a main concern in the USA while traceability appears to be a specific topic for British consumers. Meuwissen *et al.* (2004a) studied multiple pork chain attributes, which possibly can be used as traceability extensions. They found that different consumer segments set different priorities to attributes such as animal housing and production related aspects of food safety.

Traceability schemes may vary from an individual company organising its own traceability as the one extreme to interrelated and complete traceability in the whole supply chain as the other. The selection of a traceability scheme depends on the trade-off between costs and benefits. Van der

Vorst (2004) defines costs related to the redesign of infrastructures, processes, ICT-systems and co-ordination costs. Possible benefits are associated with limited recall, risk reduction, process improvement, competitive advantage and fulfil demand. While the fact that traceability involves costs is generally acknowledged (Calder and Marr, 1998; Hobbs and Young, 2000), there is no mention of the level, quantification or distribution of these costs among chain participants. According to Meuwissen *et al.* (2004b), economic design of traceability and quantification of the consequences of vertical co-ordination and traceability is rare and largely incomplete.

3. Framework and methodology

To evaluate at food company level quality costs in general and traceability costs in particular, we relied on two sources of information. First, the data and analysis present in the food quality management literature was analysed. Second, we conducted a qualitative and exploratory research at food company level. Based on a topic list, 17 food companies in Belgium were interviewed about their investments and costs related to food quality management. Food quality managers where asked for the reasons for realising the investments and costs, focusing on competitive, consumer, retailer or regulatory pressures. The data related to costs and investments were collected from the internal cost price calculations. It means that during most interviews both the quality and financial manager participated in the discussion.

To assess consumer perception of traceability in the meat supply chain, the research framework presented in Figure 1 is used (see also Gellynck and Verbeke, 2001). The meat chain from producer to consumer constitutes the core of the framework. Tracking meat products within this chain focuses on two types of attribute, namely functional attributes such as organisational efficiency and meat chain monitoring on the one hand, and process attributes such as origin and production method on the other. Functional attributes are linked with the intrinsic opportunities of a traceability system, while the process attributes deal with characteristics of the production process along which the tracking is organised. The tracking serves as a kind of peg for potential consumer benefits.

The consumer study is based on cross-sectional consumer data collected from a sample of 170 meat consumers in Belgium in June 2001 and from a sample of 155 in November 2004. The socio-demographic and behavioural characteristics of the samples are presented in Table 1. Respondents were selected based on convenience sampling, with the restriction that they were the main person responsible for buying meat in the household. This resulted in a gender distribution of 60% female and 40% male respondents in the sample. Compared to 2001, consumption frequency of all meat types increased. The highest meat consumption frequencies are reported for beef, with 40% of the respondents saying that they eat beef several times a week. Contrary to 2001, the lowest consumption frequencies are reported for pork, with one third of the respondents saying that they eat this less than once a week. The average age of the sample is 36 years. The elderly (>50 years) and less educated are slightly underrepresented in both samples.

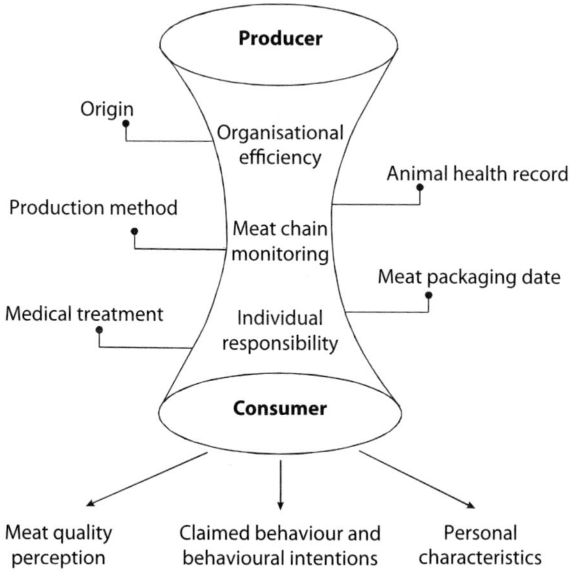

Figure 1. Framework for research into consumer perception of traceability.

Table 1. Socio-demographic and behavioural characteristics of the samples, % of respondents (n=169 in 2001; n=155 in 2004).

Consumption frequency	Beef		Pork		Poultry		Meat mixtures	
	2001	2004	2001	2004	2001	2004	2001	2004
> 1 / week	39.6	40.6	28.7	31.4	22.5	28.4	22.4	32.3
Once a week	33.7	41,3	38.9	35.3	42.0	51.0	29.8	38.1
< 1 / week	26.6	18.1	32.3	33.3	35.5	20.6	47.8	29.7
Socio-Demographic characteristics	2001	2004					2001	2004
Gender								
Female	60.0	55.0						
Male	40.0	45.0						
Education			Place of residence					
Under 18	29.2	49.0			urban		45.6	43.1
Over 18	70.8	51.0			rural		54.4	56.9
Age (mean = 37)			Children under 10					
< 30	38.8	49.7			yes		30.0	15.0
30 - 50	40.6	31.0			no		70.0	85.0
> 50	20.6	19.4						

4. Empirical results

4.1. Costs

Investments and costs related to food safety are presented in Table 2. Hereby, a distinction is made between small (less than 20 employees), medium (between 20 and 100 employees) and large (over 100 employees). Consequently the capacity of meeting new legislative requirements differs. These differences find their origin in several aspects:

- When a 'quality philosophy', like a way of life, is present in the company, the focus on quality in general and food safety in particular is much more at the centre stage than in the opposite case. It is often linked with the type of customers food companies work for. When working with retail chains, such 'quality philosophy' is much more present than it is in the case of the catering sector or smaller, traditional shops.
- The efforts made for food safety are more important in sectors characterised by higher food safety risks (microbiological contamination) such as the dairy, meat or fish sector than in other food sectors such as the chocolate or sugar confectionery.
- The efforts made for food safety are relatively more important in small enterprises than in the larger ones. This is linked with the fact that larger enterprises benefit more from scale economies.

On domestic markets, some companies remain in business and continue to produce food products despite the fact that they do not comply with the regulations related to food safety (e.g. presence of HACCP-plan). Such companies did not make the necessary investments, can consequently work with other costs structures and compete on the same markets. On international markets, the interviewed food companies claim that differences in cost structures related to food safety exist between member states because of differences in the way food companies are controlled and penalised. The competitive position of food companies from member states where food authorities control and penalise more severely than in competing member states is weakened and not compensated by additional access to market as often claimed by advocates of rigid control.

In 2002, a survey with 50 companies illustrates (Table 3) differences between companies in traceability operating costs and investments (Deschoolmeester and Lootens, 2002). Hereby, a distinction is made between small (less than 100 employees), medium (between 100 and 500 employees) and large (over 500 employees). The most important drivers for installing traceability are food safety and meeting sector standards while the barriers are lack of financial means and problematic integration in the current business processes.

Table 2. Food safety investments and costs, 2002 in EUR per full time equivalent.

Type		GMP/GHP	HACCP	Audit	Investments	Total
Large	min	665	240	42	334	1,555
	max	4,694	1,980	1,109	3,100	8,755
Medium	min	2,029	260	37	423	2,748
	max	3,856	1,894	578	2,393	7,514
Small	min	3,189	611	159	0	4,997
	max	9,452	2,408	1,248	14,527	26,165
Total	min	665	240	37	0	1,555
	max	9,452	2,408	1,248	14,527	26,165

Table 3. Traceability investments and operating costs, in % of respondents.

Type	Investments in €1,000				Operating costs in €1,000			
	None	<100	100-250	>250	None	<25	25-250	>250
Large	-	-	11	89	-	11	67	22
Medium	8	31	39	23	8	31	58	4
Small	6	31	31	32	7	47	34	13
Total	6	25	31	37	6	32	52	10

4.2. Consumer segmentation

A hierarchical cluster analysis (Ward's method - squared Euclidean distance), followed by a k-means clustering on the perception of quality of fresh meat now compared with five years ago, results in a three cluster solution in 2001 and a four cluster solution in 2004 (Table 4). Quality perception is assessed using six fresh meat attributes, based on a 7-point semantic differential scale ranging from -3 to +3. Fresh meat includes beef, pork, poultry and mixtures such as hamburgers and brochettes. One-way ANOVA was carried out to illustrate the differences in quality perception between the clusters or consumer segments. The clusters are labelled based on an interpretation of their patterns of mean perception scores: enthusiasts, cautious and pessimists. In 2004, a new cluster can be identified and labelled as greens.

The total sample scores above the middle scale position of the semantic differential for the attributes safety, healthiness, animal welfare and environmental friendliness, whereas there are negative average scores for taste and price in 2001. The negative score for taste turns into a

Table 4. Present perception of meat quality compared with five years ago, average attribute ratings of semantic differential from -3 to +3 (n = 160 in 2001, n=151 in 2004).

Attribute	Enthusiasts		Cautious		Pessimists		Greens	Total	
	2001 n = 50	2004 n = 52	2001 n = 66	2004 n = 43	2001 n = 44	2004 n = 25	2004 n = 31	2001 n = 160	2004 n = 151
Unsafe - safe	1.94^a	1.96^a	0.97^b	1.00^b	-0.95^c	-1.08^c	0.81^b	0.74	0.98
Unhealthy - healthy	1.70^a	1.37^a	0.30^b	0.40^b	-1.09^c	-1.36^c	0.52^b	0.35	0.50
Not animal friendly - animal friendly	1.40^a	1.27^a	0.91^b	0.37^b	-0.70^c	-0.60^c	-0.68^c	0.28	0.32
Not environmentally friendly - environment friendly	1.52^a	1.50^a	0.18^b	0.67^b	-0.34^c	-0.56^{cd}	-0.87^d	0.46	0.46
Tasteless - tasty	0.40^a	1.04^a	-0.17^b	0.14^{bd}	-0.70^c	-0.92^c	-0.06^d	-0.14^*	0.25^*
Expensive - cheap	-1.30^a	-1.62^a	-0.15^b	-0.09^b	-1.50^a	-1.68^a	-1.55^a	-0.83^*	-1.19^*

$^{a-d}$The various superscripts indicate significant differences in the *post-hoc* Duncan test ($p<0.05$).
*= t-test with significant different means ($p<0.05$).

positive one in 2005. All attributes except price obtain a better score in 2004. Compared with 2001, there is a clear increase in overall meat quality perception which is translated into the enthusiasts becoming the most important segment with 34.4% of the respondents. The greens are a new segment representing 20.5% of the respondents and can be characterised as being positive related to safety and health, but negative towards the other attributes. Especially, the extremely low score for the environment strikes.

4.3. Consumer perception of traceability in the meat chain

The perception of traceability of fresh meat is determined through the evaluation of eight statements on a 7-point scale (Table 5). Consumers were asked to express the degree of importance they attach to possible attributes of traceability systems in the meat chain (see research framework). Thus, a distinction is made between process and functional attributes. The functional attributes obtain the highest scores in the overall sample. Most importance is attached to 'individual responsibility' and 'meat chain monitoring'. Organising the chain in a more efficient way is somewhat less important, but still more important than most of the process attributes. All the functional attributes score significantly higher than the process attributes in the t-test for paired comparison of means, except for 'meat packaging date'. This statement was assumed to be somewhat misleading, consumers confusing 'eat-by date' for 'meat packaging date'.

The importance consumers attach to the functional attributes does not significantly differ between segments. It can be concluded that organising traceability and related marketing efforts around the functional attributes cannot be considered an efficient instrument for changing perceptions of meat quality, since it does not address concerns about safety, healthiness, environment and animal friendliness. It also shows that organising traceability based on functional attributes will not meet consumer concerns, as the retail sector often pretends.

Contrary to the functional attributes, most of the process attribute perceptions differ significantly between the segments. The scores given by the pessimists and the greens are systematically higher than those of the other two segments in 2004, which shows that focusing on additional process attributes through traceability could at least meet the meat quality concerns of these consumers. However, it is important to remember that the pessimists and the greens constitute only a limited part of the sample (25% in 2001 and respectively 16% and 20% in 2004). It is therefore debatable whether it would be worthwhile organising such a traceability system (including opportunities for consumers to personally check process attributes) for the entire meat chain, though individual chain participants or private initiatives might find it useful to address the specific concerns of the market segments we call pessimists and greens.

4.4. Consumer perception of the urgency of introducing traceability systems

The questionnaire also focused on the perceived urgency of introducing a traceability system for the different meat types (Table 6). The system to be introduced was defined as one with the attributes considered important by the respondent. Consequently, respondents were asked to express the degree of urgency of introduction on a 7-point scale, ranging from 'not at all urgent' to 'extremely urgent'. The perceived need for a traceability system remains the highest in the case of meat mixtures compared with 2001. The differences in urgency between beef and poultry as well as between pork and poultry are not statistically significant. The top score for meat mixtures can be explained by the fact that evaluating mixture quality is perceived as more difficult than for the other meat types. Moreover, the risk of abuse is highest in the case of mixtures.

Table 5. Perception of traceability in the meat chain related to consumer segments, average scores on a 7-point scale (n = 158 respondents).

Statement	Enthusiasts		Cautious		Pessimists		Greens	Total	
	2001	2004	2001	2004	2001	2004	2004	2001	2004
Process Attributes									
I have access to information regarding the medical treatment of the animal.	3.57[a]	4.13[a]	3.94[a]	4.35[ab]	4.56[b]	4.68[b]	5.10[b]	3.99*	4.48*
I can check the animal production method.	4.22[a]	4.63[a]	4.35[a]	4.14[b]	5.00[b]	5.08[c]	5.03[c]	4.49	4.66
I can check the origin of the product (region, farmer and slaughterhouse).	4.31[a]	4.96[a]	4.44[ab]	4.33[a]	4.98[b]	4.84[a]	4.58[a]	4.55	4.73
I can check the meat packaging date.	5.53	5.49[a]	5.47	5.19[b]	5.84	5.96[a]	5.77[a]	5.59	5.57
I have access to information regarding the health record of the animal.	3.95[a]	4.71[a]	4.14[a]	4.65[a]	4.81[b]	5.16[b]	4.90[ab]	4.27*	4.81*
Functional Attributes									
Organisations responsible for monitoring public health can intervene in the event of a problem in the meat chain (e.g. dioxin scare: only contaminated products are removed from the shelves, not all products)	5.73	5.94	5.62	6.12	5.82	6.00	5.90	5.71	5.99
In the case of abuses, individuals responsible can be clearly identified and held accountable.	5.83	5.90	5.86	5.74	6.11	5.58	5.71	5.92	5.77
The meat chain (from animal feed to the consumer's plate) can be organised more efficiently to further reduce costs.	5.20	5.24	4.83	4.98	5.24	4.71	4.71	5.06	4.97

[a-c]The various superscripts indicate significant differences in the post-hoc Duncan test ($p<0.10$).

*= t-test with significant different means ($p<0.05$).

Table 6. Consumer segments and perception of the need for traceability according to meat types, average scores on a 7-point scale (n = 158 in 2001 and n = 150 in 2004).

Attribute	Enthusiasts		Cautious		Pessimists		Greens	Total	
	2001	2004	2001	2004	2001	2004	2004	2001	2004
Beef	5.02 a	4.65 a	5.14 a	4.53 a	5.89 b	5.45 b	4.79 a	5.31*	4.80*
Pork	4.75 a	4.75 a	5.17 ab	4.60 b	5.59 b	5.33 c	4.67 b	5.16	4.81
Poultry	4.88 a	4.96 a	5.35 ab	4.74 a	5.50 b	5.21 b	5.28 b	5.25	5.02
Meat mixtures	5.25 a	5.04 a	5.55 ab	5.05 a	5.86 b	5.46 b	4.93 a	5.55*	5.11*

[a,b]The various superscripts indicate significant differences in the post-hoc Duncan test ($p<0.05$).
*= t-test with significant different means ($p<0.05$).

Table 6 shows differences in the perceived urgency of introducing traceability between the identified consumer segments. Pessimists express the strongest need for traceability systems, which for all meat types is more urgent than for the other segments, both in 2001 and 2004. This means that consumers who are more concerned about meat quality and eat meat less frequently indicate a more urgent need for the introduction of traceability systems. The scores for all meat types are still higher than the middle scale position, but show a clear decline in 2004 compared to 2001.

5. Conclusion

Food safety costs in general and traceability costs in particular differ strongly between countries. It is mainly linked with differences in company size, activity and philosophy. With respect to traceability characteristics, there is a distinction between functional and process attributes, the former referring to the intrinsic opportunities of the systems, i.e. the ability to organise the chain more efficiently, monitor the chain, and assess individual responsibilities. These attributes can be regarded as the minimum requirements of a true 'traceability system'. Process attributes refer to characteristics of the production process at different levels of the chain, i.e. they can be regarded as resulting from extensions to the minimum requirements.

Compared to 2001, overall meat quality perception scores are higher than in 2004. It should be linked with the increased efforts related to quality assurance by both the private sector and public authorities. In 2004, the cluster analysis yielded a cluster, namely the greens, additional to the three-cluster solution (segments denoted enthusiasts, cautious and pessimists) found in 2001. None of the clusters can be typified through socio-demographic characteristics. Our empirical findings contribute to the debate about who is responsible, and to what extent, for providing meat quality and safety. Since functional attributes are broadly supported by all consumer groups, public policy plays an important role in guiding and monitoring this aspect of traceability. Extensions with respect to process attributes, such as production methods, are less relevant to the broad public and only interest specific market segments. Government intervention or regulation on the process attribute side of traceability is thus less evident. These attributes are more appropriate for private initiatives of chain participants.

The introduction of traceability is regarded as the most urgent in the case of meat mixtures. However, organising traceability for mixtures in terms of functional attributes is the most difficult, because different meat types can be an ingredient in one and the same product. Despite the fact that the pessimists consume meat the least frequently, they regard the introduction of traceability in the meat chain as the most urgent, which means that traceability could be an answer to their concerns.

Future research could focus on pessimists' willingness to pay for traceability systems, which are extended with process attributes on the one hand, and on a more precise characterisation of this market segment (e.g. purchase outlet, moment of meat consumption) on the other. At the level of the meat chain, it could be interesting to quantify costs and benefits for all participants in the meat chain. Another future research topic could be the distribution of these costs and benefits among chain participants, as well as the role of the retail industry as potential chain leader/gatekeeper.

Finally, this analysis shows that collecting market information (about consumer value attached to traceability in this case) reveals differences between consumers. These differences include opportunities for the meat chains to become more market oriented, i.e. to differentiate through the development of traceability based on process attributes. Such a market oriented approach should further enable companies to counterbalance increasing quality costs, create competitive advantage and consequently obtain better profitability in tomorrow's global markets.

References

Bredahl, M.E., J.R. Northen, A. Boecker and M.A. Normille, 2001. Consumer demand sparks the growth of quality assurance schemes in the European food sector. In Regmi, A. (ed.), Changing structure of the global food consumption and trade. Market and Trade Economics Division, Economic Research Service, US Department of Agriculture, Agriculture and Trade Report. WRS-01-1, May, 90-102.

Burton, M. and T. Young, 1996. The impact of BSE on the demand for beef and other meats in Great Britain. Applied Economics 28, 687-693.

Buzby, J.C., 2001. Effects of Food Safety Perceptions on Food Demand and Global Trade. In Regmi, A. (ed.), Changing structure of the global food consumption and trade. Market and Trade Economics Division, Economic Research Service, US Department of Agriculture, Agriculture and Trade Report. WRS-01-1, May, 55-66.

Calder, R. and P. Marr, 1998. A beef producer initiative in traceability: Scottish Borders TAG. Supply Chain Management 3, 123-126.

Deschoolmeester, D. and A. Lootens, 2002. Vlerick survey on material tracking and tracing. Seminar on material tracking and tracing: Necessary evil or business challenge. Vlerick Leuven Gent Management School, 27 November 2002, 8.

Dickinson, D.L. and D. von Bailey, 2002. A comparison between US and European consumer attitudes and willingness to pay for traceability, transparency, and assurance for pork products. In Trienekens, J.H. and S.W.F. Omta (eds.), Paradoxes in food chains and networks. Wageningen Academic Publishers, 229-237.

Downey, W.D., 1996. The challenge of food and agri products supply chains. In J.H. Trienekens and P.J.P. Zuurbier (eds.), Proceedings of the 2nd International Conference on Chain Management in Agri- and Food Business. Department of Management Studies, Wageningen Agricultural University, 3-13.

Fearne, A., 1998. Editorial. Supply Chain Management 3, 112-114.

Frenzen, P.D., A. Majchrowicz, J.C. Buzby and B. Imhoff, 2000. Consumer acceptance of irradiated meat and poultry products. Economic Research Service, US Department of Agriculture, AIB-757, August.

Gellynck, X. and W. Verbeke, 2001. Consumer perception of traceability in the meat chain. Agrarwirtschaft 50 (6), 368-374.

Gellynck, X., W. Verbeke and J. Viaene, 2004. Quality Management in the Food Supply Chain: How does the food industry interact with consumers, retailers and public authorities? In G. Schiefer and U. Rickert (eds.), Quality assurance, risk management and environmental control in agriculture and food supply networks (Volume B). ILB Press, 443-452.

Gilg, A.W. and M. Battershill, 1998. Quality farm food in Europe: A possible alternative to the industrialised food market and to current agri-environmental policies: Lessons from France. Food Policy 23 (1), 25-40.

Henson, S.J. and J. Northen, 2000. Consumer assessment of the safety of beef at the point of purchase: A pan-European study. Journal of Agricultural Economics 51, 90-105.

Henson, S.J. and W.B. Traill, 2000. Measuring perceived performance of the food-system and food-related welfare. Journal of Agricultural Economics 51 (3), 388-404.

Hanf, J. and R. Kühl, 2002. Consumer values versus economic efficiency in food chains and networks. In Trienekens, J.H. and S.W.F. Omta (eds.), Paradoxes in food chains and networks. Wageningen Academic Publishers, 35-43.

Hobbs, J.E. and L.M. Young, 2000. Closer vertical co-ordination in agri-food supply chains: A conceptual framework and some preliminary evidence. Supply Chain Management 5, 131-143.

Jack, D., T. Pardoe and C. Ritchie, 1998. Scottish quality cereals and coastal grains: Combinable crop assurance in action. Supply Chain Management 3, 134-138.

Kaferstein, F. and M. Abdussalam, 1999. Food Safety in the 21st century. Bulletin of the World Health Organisation, 77 (4), 318-338.

Latouche, K., P. Rainelli and D. Vermersch, 1998. Food safety issues and the BSE scare: Some lessons from the French case. Food Policy 23, 347-356.

Leat, P., P. Marr and C. Ritchie, 1998. Quality assurance and traceability - the Scottish agri-food industry's quest for competitive advantage. Supply Chain Management 3, 115-117.

Meuwissen, M.P.M., I.A. van der Lans and R.B.M. Huirne, 2004a. A synthesis of consumer behaviour and chain design. In Bremmers, H.J., S.W.F. Omta, J.H. Trienekens and E.F.M. Wubben (eds.), Dynamics in chains and networks. Wageningen Academic Publishers, 310-317.

Meuwissen, M.P.M., A.G.J. Velthuis, H. Hogeveen and R.B.M. Huirne, 2004b. Technical and economic considerations about traceability and certification in livestock production chains. In Velthuis, A.G.J., L.J. Unnevehr, H. Hogeveen and R.B.M. Huirne (eds.), New approaches to food-safety economics. Kluwer Academic Publishers, 49-62.

Oude Luttinghuis, P., 2000. ICT service infrastructure for chain management. In Trienekens, J. and R. Zuurbier (eds.), Proceedings of the Fourth International Conference on Chain Management in Agri- and Food Business. Agricultural University Wageningen, 275-282.

Porin, F. and P. Mainsant, 1998. Quelles stratégies pour les concurrents de la filière bovine dans le contexte de l'après ESB? In Viau, C. (ed.), Long-term prospects for the beef industry. INRA, 125-135.

Shogren, J.F., 2004. Food-safety economics: Consumer health and welfare. In Velthuis, A.G.J., L.J. Unnevehr, H. Hogeveen and R.B.M. Huirne (eds.). New approaches to food-safety economics. Kluwer Academic Publishers, 19-28.

Schiefer, G., 2004. Traceability and certification in food quality production: A critical view. In Velthuis, A.G.J., L.J. Unnevehr, H. Hogeveen and R.B.M. Huirne (eds.), New approaches to food-safety economics. Kluwer Academic Publishers, 63-68.

Timon, D. and S. O'Reilly, 1998. An evaluation of traceability systems along the Irish beef chain. In Viau, C. (ed.), Long-term prospects for the beef industry. INRA, 219-225.

Van der Vorst, J.G.A.J., 2004. Performance levels in food traceability and the impact on chain design: Results of an international benchmark study. In Bremmers, H.J., S.W.F. Omta, J.H. Trienekens and E.F.M. Wubben (eds.), Dynamics in chains and networks. Wageningen Academic Publishers, 175-183.

Verbeke, W. and J. Viaene, 1999. Beliefs, attitude and behaviour towards fresh meat consumption in Belgium: Empirical evidence from a consumer survey. Food Quality and Preference 10, 437-445.

Verbeke, W., R. Ward and J. Viaene, 2000. Probit analysis of fresh meat consumption in Belgium: Exploring BSE and television communication impact. Agribusiness 16, 215-234.

Wilson, N. and S. Clarke, 1998. Food safety and traceability in the agricultural supply chain: Using the Internet to deliver traceability. Supply Chain Management 3, 127-133.

Wortmann, J, 2000. Information systems for supply chain management. In Trienekens, J. and Zuurbier, R. (eds.), Proceedings of the Fourth International Conference on Chain Management in Agri- and Food Business. Agricultural University Wageningen, 3-9.

The effect of transaction costs associated with certification of exports on the profitability of farming systems for Chilean small farmers

Pablo Villalobos[1], *Michel Leporati*[2], *Luis Saez*[2] *and Carlos Padilla*[1]
[1]*Department of Agricultural Economics, University of Talca, Chile; pvillal@utalca.cl*
[2]*Institute of Agricultural Development, Ministry of Agriculture, Chile*

Abstract

To meet the standards of farm exports certification of the Chilean Agriculture and Livestock Service (Servicio Agrícola y Ganadero de Chile - SAG) represents, at the production level, incurring transaction costs related to meeting the requirements of biosecurity and traceability, among others. The purpose of this study is to evaluate the effect on the profitability for small farmers produced by the entrance of Farms with Official Certification (Planteles Bajos Certificación Oficial PABCO) into the farming system. A case study was done on a sample of 50 farming properties in the process of certification. This study of small farming operations is one of the first private economic studies done in Chile.

Keywords: transactions costs, farms with official certification system (PABCO), Chilean smaller farmers

1. Introduction

The Chilean meat industry is faced with a very complex scenario. On one hand, the domestic market can barely compete with the meat coming from MERCOSUR, the Southern Cone Market, a situation that is maintaining the profitability of this sector in a prolonged crisis. Simultaneously, the signing of free trade agreements with the principal economies of the planet, the advantages granted to Chile by its sanitary status as a country free from List A diseases (OIE, 2004) as well as the changes in consumer habits provoked by the episodes of Bovine Spongiforme Encephalitis (BSE) and others, have opened for Chilean meat interesting possibilities of participation in Foot and Moth Disease free market circuits, hopefully giving access to competitive prices for the national livestock sector.

Taking advantage of the external market opportunities makes necessary some structural adjustments in the chain that, up to this time, has been oriented to an essentially domestic market. The official Chilean Agricultural and Livestock Service (SAG) finds itself implementing a system of farm exports certification that gives guarantees of health and safety to the intended importing countries and, at the same time, is capable of being taken on, technically and economically, by the different agencies that make up the national meat industry.

As a result of the above, from 2002 to the present markets such as Israel, Cuba, Costa Rica, Libya, Germany, Japan, the United Kingdom, France and Mexico, among others have opened up. The opening of markets such as the North American one and others is currently in the negotiation stage (SAG, 2004). The possibility of advancing in a sustainable development in exports, which would turn around the current profitability crisis of the sector, would basically happen by increasing the quantity and the quality of the national product. In this context the participation of family operated small farms becomes strategically relevant due to the fact that 42% of the bovine population is concentrated therein (ODEPA, 2000), and for reasons of structure and of comparative advantages, these farms have naturally specialised in breeding, acting as suppliers of calves for the rest of the chain. However, this represents a small heterogeneous flock and their production systems are not adapted to international standards of animal health, safety and traceability, and biosecurity, all of which constitutes the main barrier to their incorporation in the chain.

This situation can be readily reversed in the measure that the small producers meet the official standards of export certification. To this end the Ministry of Agriculture, in collaboration with the various agencies that make up the meat chain, finds itself promoting the technical upgrading of the small farms producing livestock to meet the PABCO standards (Farms with Official Certification). Adapting the productive systems to PABCO involves the inclusion of costs that, although being associated with the improvement of the infrastructure, equipment, handling and technical consultation, do not essentially modify the production function but still answer to the requirements of the official service to guarantee the meeting of third party agreements. Thus, these costs, as seen by the authors, can be considered as 'transaction costs' in order to be able to access to the export market.

This paper forms part of an on going study between INDAP and the Department of Agricultural Economics of the Faculty of Agriculture Science at the University of Talca. It is intended to be contribution to the background information about the 'transaction costs' incurred by adapting farms to the PABCO standards and the repercussions on the profitability of the family operated small farm system. The hypothesis is put forward that the costs associated with adapting farms to export needs will be compensated for by the price differences generated in conjunction with the export chain, making the activity more profitable.

2. Objectives

The general objective is to evaluate the effects of 'transaction costs' associated with joining the system of livestock farms with official certification (PABCO) on the profitability of small farming operations. Specific objects are:

- To define the characteristics of the small producers attached to the Cattle Breeders Plan for family operated small farms in the 7th region.
- To identify and quantify the costs associated with meeting the requirements for entrance into the system of official certification (PABCO) of the national agricultural service (SAG).
- To economically evaluate the effects of the costs associated with official certification (PABCO) on profitability for the small producers within the Cattle Breeders Plan for family operated small farms in the 7th region.

3. Methodology

A case study was done of a sample of 50 small producers of livestock for meat consumption in the 7th region of Chile. Data from a questionnaire was tabulated and submitted to a descriptive statistical analysis with the purpose of defining these individuals from a socio-economic and production technology point of view.

In conjunction with this a model was developed for the costs related to implementing the requirements concerning infrastructure, equipment, handling, and technical assistance in order to reach the certification standard (PABCO) as specified by the Chilean Agricultural and

Livestock Service (SAG, 2004). This costs model was set out on a table of dynamic calculations that enabled simulation of investment, fixed costs and variable costs for different production situations (INDAP, 2004).

After fixing the input data and average costs, a 10- year flow was constructed determining the Net Present Value, the Internal Rate of Return, the Cost/Benefit ratio and the investment recovery time for a situation not adapting to the export standards and for a situation with the cost of adapting to the export standards. A discount rate of 12% was used maintaining constant the use of production factors and varying the market price according to the type of product (animals raised on certified lands or animals raised on uncertified lands).

4. Results

4.1. Description of analysed production systems

Characteristics of the family group and the type of work

The analysed production systems correspond to small farm livestock fattening or breeding units. The production units are exploited by small nuclear families with an average age of 34 years and with men predominating over women (see Table 1). As for the educational level of the family group it was observed that nearly 70% of the individuals were capable of using basic reading, writing and mathematic skills (elementary school complete, secondary school complete and incomplete). However, the scarcity of professional or technical studies along with the 28.8% instance of incomplete elementary education or no formal education by all means presents a barrier to technological innovation and entrepreneurial development (see Table 2).

Table 1. Socio-economic background of the family group.

Statistics	Number of members	Age
Max	6.0	75
Min	1.0	25
Mode	1.0	35
Mean	2.0	34
Stan. Dev.	1.4	20.7

Table 2. Level of formal education in the active population (%).

Category	%
Lacking formal education	11.9
Elementary incomplete	16.9
Elementary complete	33.9
Secondary incomplete	1.7
Secondary complete	28.8
Technical	5.1
Professional	1.7
	100.0

The economically active population represents 45% of the total population of the sample. The farm work force is largely represented by men with an average age of 42 years dedicated to agricultural work and the care of animals. Fifty-four percent of these are permanent full-time labourers, while 46% are seasonal workers. Of these workers, 74.3% receive a regular salary for their work, while 25.7% do not receive one regularly.

Farmland holdings

The average surface area of farmlands in the study was 23 hectares. The range is very wide varying from 8 hectares to 110 hectares. Of the total exploited surface, an average of 67% corresponds to land ownership and 33% to other forms of tenancy. The greater portion of the land surface is set aside as meadows even though there is always a smaller area set aside for crops and gardens, which explains that even though the tendency is to opt for a specific activity as the commercial centre of the productive unit, small farming continues to be a multifaceted activity.

System of production

Of the production systems surveyed 70% tend towards beef cattle, 15% to dairy cattle and 15% to both beef and dairy cattle. Within the beef cattle operations 60% are involved with breeding, 33.3% with fattening cattle, and 6.7% with both breeding and fattening. This study will take on the analysis of the first two, leaving the last aside because it is little representative of the sample.

Structure of the herd

Coinciding with the wide range of sizes of land holdings, the herd shows this same characteristic, with the most frequent composition in the breeding systems being 5 cows, 5 heifers, 8 calves, 2 bulls, and 1 ox while in the fattening operations the composition is made up of 5 cows, 8 heifers, 14 steers, and 2 bulls. As for the type of animal it was observed that dual purpose animals predominate over purely beef or dairy cattle (see Table 3).

Installations and farm equipment

Concerning the minimum infrastructure necessary to be able to adapt the farm to export standards, in general deficiencies are observed but these are not critical. As far as boundary enclosures are concerned, a third possess good fences, meeting the standard without any upgrading being necessary. Almost half have fencing of average condition and with a minor investment would meet the norm. Only a fifth of the cases showed serious faults or did not have fences, which

Table 3. Breed of cattle (%).

Breed	%
Hybrid	18.8
Holstein	6.3
Aberdeen Angus	12.5
Hereford	15.6
Overo Colorado	28.1
Frison	18.8

implies large investments to meet the required standard. The previous tendency is repeated for corrals and pens, there being a high proportion of the infrastructure in good or average condition and a low proportion in bad condition or non existent.

Regarding offices or special areas for record keeping and storage of medication, a general deficiency exists with 90% lacking this type of installation. A similar situation occurs with the existence of refrigerated units for handling biological products such as vaccines. A first aid kit is always found and in generally good condition.

Production management

In terms of production management it was observed that not only the breeding operations but also the fattening systems make similar use of veterinary and agricultural technical assistant services. In particular, the frequency of veterinary doctor visits is on average less than a third of the number that is established in the PABCO certification regulations. Likewise the visits of other professionals are rather scarce (see Table 4).

A high proportion of animals show some kind of tag for individual identification (52% tags, 48% brands). However, this does not necessarily mean that the tagging is accompanied by records and control systems for health, reproductive and other handling required for official certification (PABCO).

From the point of view of health a wide coverage of basic management was noticed, made up of periodic parasite control and vaccinations. Even so there is a large group (48.3%) that does only one of the two mentioned controls or none whatsoever.

4.2. Investment, expenses and income

For breeding operations an average investment of € 9,631.9 is estimated constituting breeders, fences and other installations[1]. In the case of the cattle fattening systems the average investment is estimated at € 62,469.9. The difference in investment for the two types of operation is basically explained by the level of installations since in relation to the extension of farmlands, there is no major difference being on average about 23 hectares.

For the cattle breeding operations a set of annual costs was identified that on average came to a total of € 2,639.4 for the handling of feed, reproduction and health. In general terms the costs from highest to lowest portions of the total were: paid labour (35.7%), feed (35.7%)[2], repair and maintenance of installations (11.9%), health management (8.8%), reproduction (3.1%), technical aid (2.6%) and freight (2.2%) (Table 5).

For the cattle fattening operations the average total cost reached € 40,743.7. On average, the costs from the highest to lowest portion of the total were: purchase of animals (65.8%), feed (23.3%)[3], paid labour (4.4%), health management (2.5%), repair and maintenance of installations (1.5%), freight (0.7%), and technical aid (0.5%) (Tables 6a and 6b).

---

[1] €1= $743 Chilean pesos.

[2] Includes machinery rental and use of agrochemicals and fertilizers.

[3] Includes machinery rental and use of agrochemicals and fertilizers.

Table 4. Use of technical consultants per month (No. of visits).

	Veterinarian	Agronomist	Agricultural technician
Max	4.00	2.00	4.00
Min	0.16	0.16	1.00
Mode	1.00	1.50	4.00
Mean	0.98	0.79	2.86
Stan. Dev.	1.10	0.72	1.35

The average annual income from the sale of animals for the cattle breeder was estimated at €5,413.1. This income corresponds to the following: sale of calves (41.2%), cows (27.3%), heifers (4.2%), and bulls (7.3%) (see Table 7).

The annual average income for the sale of animals for the cattle fattening operations was estimated at €25,495.3. This income was composed of the following: sale of steers (63.3%), heifers (27.4%) calves (3.8%), cows (3.0%), and bulls (2.5%) (see Table 8).

4.3. Determining 'transaction costs' to meet official certification requirements (PABCO)

Cattle breeding operations

The calculation of transaction costs associated with implementing the PABCO requirements in the cases studied, in accordance with the costs and technical standards for official certification designed in the context of the present study, where cattle breeders had an average of 17 head and an area of 23 hectares, concluded that the necessary repair to fences was around 1,200 linear meters. There was no need to build or repair corrals or gates, nor installations for storing medication. Indeed, the necessity of regular visits by an accredited veterinarian was established on a 3-month schedule, as well as the purchase of health supplies, and a refrigeration unit for handling biological products. All of this amounted to an investment of € 2,011.46; costs (fixed and variable) of € 487.2 (see Table 9)

Cattle fattening operations

The calculation of costs associated with meeting the standards for official certification for cattle fattening operations, in accordance with the costs and technical standards of PABCO, designed in the context of the present study, where these operations had, on average, 20 calves and an area of 23 hectares, found that the necessary repair to fences was around 1,200 linear meters. There was no need to build or repair corrals or gates, nor installations for storing medication. Indeed, the necessity of regular 3-monthly visits by an accredited veterinarian was established along with the purchase of health products and a refrigeration unit for the handling of biological products. All of this amounted to an investment of € 2,011.46; costs (fixed and variable) of € 446.9 (see Table 10).

Table 5. Cost structure for cattle breeding operations (in Euro).

	Feed	Agrochem.	Mach. rent	Health mgt.	Freight	Labour	Reproduction	Main. of install.	Tech. aid	Total
Max	431.3	404.4	673.9	404.4	97.0	943.5	148.3	673.9	80.9	3,857.7
Min	202.2	78.9	269.6	87.6	20.2	943.5	14.8	134.8	53.9	1,805.5
Mean	316.8	154.2	471.8	231.4	58.6	943.5	81.3	314.5	67.4	2,639.4

Table 6a. Cost structure for cattle fattening operations (in Euro).

	Feed	Agrochem.	Mach. rent	Health mgt.	Freight	Labour	Reproduction	Main. of install.	Tech. aid	Total
Max	17,885.2	3,437.1	145.6	2,911.4	539.2	4,852.4	552.6	2,021.8	539.2	32,884.5
Min	3,763.2	80.9	404.4	80.9	53.9	259.6	552.6	75.5	53.9	5,334.9
Mean	8,532.7	913.1	73.0	1,010.9	293.7	1,731.1	552.6	606.0	195.4	13,958.6

Table 6b. Cost structure for cattle fattening operations. Animal Purchase (in Euro).

	Steers	Calves	Total
Max	72,786.1	67,394.5	79,525.5
Min	1,771.1	53,915.6	71,626.9
Mean	20,780.0	6,065.5	26,845.6

Table 7. Income for sale of animals for cattle breeders (in Euro).

	Cows	Heifers	Calves	Bulls	Total
Max	3,154.1	2,830.6	6,556.1	3,942.6	12,935.0
Min	397.8	404.4	970.5	3,942.6	2,166.9
Mean	1,477.8	227.4	2,230.2	395.2	5,413.1

Table 8. Income for sale of animals in cattle fattening operations (in Euro).

	Cows	Heifers	Steers	Calves	Bulls	Total
Max	769.2	40,436.7	50,377.4	1,887.0	1,182.8	94,653.1
Min	175.2	218.4	505.5	271.7	103.8	1,274.6
Media	769.2	7,005.7	16,164.8	967.2	626.2	25,533.1

Table 9. Transaction costs PABCO requirements for cattle breeders (in Euro).

Item	
Investment	2,011.46
Variable costs	372.67
Fixed costs	114.57
Total	2,498.69

Table 10. Cost of implementing PABCO for cattle fattening operations (in Euro).

Item	
Investment:	2,011.46
Variable costs:	332.39
Fixed costs:	114.57
Total	2,458.41

4.4. Economic evaluation results: situation without project

Cattle breeding operations

The result of the economic evaluation for the cattle breeding operations without a project, came to an internal rate of return (IRR) of 21% with a net present value (NPV) for the operation of €10,573.8 calculated over 10 years and with a discount rate of 12% (see Table 11).

Table 11. Economic evaluation results for cattle breeders without project (in Euro).

Item	Year 0	Year 1	Years 2 to 9	Year 10
Income		5,421.2	5,421.2	5,421.2
Direct costs		2,167.7	2,167.7	2,167.7
Investment	9,646.2			
Flow		3,253.5	3,253.5	12,899.7
NPV	10,573.8			
IRR	21%			
Benefit/cost ratio	1.4			
Investment recovery	Year 4 (9,882.0)			

Cattle fattening operations

The result of the economic evaluation for the cattle fattening operations without a project, came to an internal rate of return of 49% with a net present value for the project of €124,113.7 calculated over 10 years and with a discount rate of 12% (see Table 12).

4.5. Economic evaluation results: situation with project

Cattle breeding operations

The result of the economic evaluation for the cattle breeder with a project, came to an internal rate of return of 31% with a net present value for the operation of 17,893.3, calculated over 10 years with a discount rate of 12% (see Table 13).

Cattle fattening operations

The result of the economic evaluation for the cattle fattening operations with a Project came to an internal rate of return of 58% with a net present value for the project of €160,779.9 calculated over 10 years with a discount rate of 12% (see Table 14).

Table 12. Economic evaluation results for cattle fattening operations without project (in Euro).

Item	Year 0	Year 1	Years 2 to 9	Year 10
Income		74,585.5	74,585.5	74,585.5
Direct costs		43,667.0	43,667.0	43,667.0
Investment	62,562.6			
Flow		30,918.5	30,918.5	9,481.1
NPV	124,113.7			
IRR	49%			
Benefit/cost ratio	1.36			
Investment recovery	Year 4 (74,261.0)			

Table 13. Economic evaluation results for cattle breeding operation with project (in Euro).

Item	Year 0	Year 1	Years 2 to 9	Year 10
Income		6,032.3	6,032.3	6,0324
Direct costs		2,654.9	2,654.9	2,654.9
Investment	11,657.7			
Flow		3,253.5	3,864.7	13,510.9
NPV	17,893.3			
IRR	31%			
Benefit/cost ratio	1.4			
Investment recovery	Year 5 (13,385.6)			

Table 14. Economic evaluation results for cattle fattening operations with project (in Euro).

Item	Year 0	Year 1	Years 2 to 9	Year 10
Income		85,602.9	85,602.9	85,570.6
Direct costs		44,105.6	44,105.6	44,105.6
Investment	10,658.5			
Flow		30,479.9	41,497.3	104,027.6
NPV	160,779.9			
IRR	58%			
Benefit/cost ratio	1.51			
Investment recovery	Year 3 (89,809.6)			

5. Comments and conclusions

The results obtained from the economic evaluation permit the preliminary confirmation of the proposed hypothesis in that:

- The economic and technical requirements associated with official certification (PABCO) do not present major difficulties to be incorporated by small farmers.
- The costs associated with the restructuring in order to achieve the required PABCO standard are recoverable through access to better market prices. This is noted not only in the calculation of the investment recovery but also in the benefit/cost ration for both breeding and fattening operations, seeing a slightly greater advantage for the latter.
- There is a substantial improvement in income for both operations that opt for the official certification standards. This is seen in the calculation of the NPV and of the IRR for both breeding and fattening operations.

Even though the above shows a trend, it is necessary to strengthen the quality and amount of data, especially those for the production costs, in the different types of operation in order to have a more precise result in relation to the impact on the income and competitiveness of the small cattle farmer in the meat export chains.

References

INDAP, 2004. Ficha técnica de estándares y costos para habilitación predial, según requisitos PABCO. Plan Ganadero INDAP.

ODEPA, 2000. Clasificación de las explotaciones agrícolas del VI censo nacional agropecuario según tipo de productor y localización geográfica. Documento de trabajo n°5. ODEPA.

OIE, 2004. Organisación Mundial de Sanidad Animal. Información sanitaria mundial. Lista de difusión OIE-Info, http://www.oie.int/esp/maladies/es_classification.htm#ListeA.

SAG, 2004. Servicio Agrícola y Ganadero, Ministerio de Agricultura, Chile. Manuales e instructivos PABCO, http://www.sag.cl/framearea.asp?cod=12.

Production costs of citrus growing in the Comunidad Valenciana (Spain): EurepGAP protocol versus standard production

Elena Maria Peris Moll and Juan Francisco Juliá Igual
Center for Research and Specialisation in Agrifood Business Management (CEGEA), Department of Economy and Social Sciences, Polytechnic University of Valencia,Camino de Vera, s/n. 46022, Valencia, Spain; elpemol@doctor.upv.es

Abstract

The first aim of this paper is to analyse the average production costs of citrus cultivated under EurepGAP protocol regulation in the Comunidad Valenciana (Spain). To this end, we shall study a sample of plots administered under the same management, specifically a citrus cooperative located in one of the most important production areas of the region. The results obtained shall then be compared with the average standard regional costs employing traditional production methods. The second aim of this study is to draw attention to the extraordinary effort made by the Spanish citrus cooperative sector to offer products to European consumers in compliance with the new standards of food quality and safety. The farmers who produce citrus following the EurepGAP protocol must face a series of extra costs (registration fees), or controls (analysis and certification costs), in addition to the obligation of implementing the code of Good Agricultural Practices (GAP).

Keywords: citrus, production costs, EurepGAP protocol, food quality and safety

1. Introduction

After recent crises in the European agrifood system, as the case of the mad cow disease (BSE), chicken dioxin or the overuse of antibiotics in animal production, food quality and safety has become one of the main concerns of European consumers.

Consequently, the concept of quality has experienced a change in recent years. In fact, food quality is not only limited to the absence of external failures as it was before. Consumers actually value other kinds of food attributes, including the use of good agricultural practices within the production process, a good market preparation, the absence of empty flavours, the best organoleptical qualities, the absence of genetic modifications in raw animal and plant materials or the implementation of traceability systems (Planells, 2003).

The aim of this paper is to highlight the effort made by the Spanish cooperative agrifood sector within the consecution of the above named values. The agrarian cooperative movement of the Comunidad Valenciana is widely known because of its initiatives regarding the introduction of the new concept of quality (Planells, 2003). Quality management systems based on integrated production should be named, as for example Naturane (i.e. the ANECOOP´s certification quality system), or the adoption of protocols as EurepGAP, BRC or the same Integrated Production normative of the Regional Department of Agriculture.

Our case study focuses on citrus cultivated under the EurepGAP quality protocol. First of all, some general information about citriculture in the Comunidad Valenciana is offered. Citrus fruits are the main crop of the region, with 22.4% of the Useful Agricultural Area (UAA, which is the area under cultivation, meadow and pastures), and 43.3% of the region's final agricultural output in 2000 (Peris and Juliá, 2005). One of the structural problems of agriculture at the regional level is the small size of holdings. In fact, after analysing the structure of property in the region, 57.5% of the holdings have a surface less than 2 Ha, while another 40% have a surface between 2 and 20 Ha. Both segments of holdings are equivalent to 62% of the UAA (Peris, 2002). The small size of the holdings leads to other negative consequences at first sight, such a considerable percentage of part time agriculture.

The adoption of integrated production techniques, even the compliance of quality protocols, involves a series of extra costs for producers. For instance, the required technical assistance, or the analysis and certification costs, both situations adding to the financial burden of small farmers. In this kind of context, cooperatives could play an important role, not only for the common resources shared and other advantages that social economy enterprises represent, but also to facilitate control of the whole production process under a common management (Fernández Zamudio *et al.*, 2004).

The field work for this study was done in the cooperative 'La Constancia', located in Pobla de Vallbona (Valencia). Established in 1978, it currently has 2,500 members, 450 of whom are farmers. All of them together bring to the enterprise an area of 340 hectares.

The products usually supplied by 'La Constancia' co-op are citrus, summer fruits (mainly watermelon) and vegetables (especially cauliflower). Citrus constitute their main product, as they are equivalent to 80% of the enterprise turnover and contribute a stabilised harvest of between 7 and 9,000 tones per season. Around 90% of this harvest is certified under ANECOOP´s quality system Naturane, although during the last two or three years there has been an increase in the offer of citrus certified under other quality protocols, as EurepGAP or the Integrated Production normative of the Regional Department of Agriculture.

The whole commodity production of 'La Constancia' co-op is traded through ANECOOP. ANECOOP is a second-tier cooperative, the first co-op in fresh fruit and vegetable trade in Spain. With 110 base members which are ordinary cooperatives and a turnover of 400 million €, it traded 600,000 tones of commodities during the 2002-03 season. ANECOOP´s core business is citrus fruit, followed by summer fruits, vegetables and wine, being the main destinations of their sales Germany and France (ANECOOP, 2005). 'La Constancia' co-op is also a member of the management board of AGRICONSA S.A., one of the main regional processors of fruits and vegetables, which is only supplied by coops. 'La Constancia' provides AGRICONSA with citrus for juice, as well as Satsuma selections to produce canned tangerines.

Of the entire cooperative land surface, the holdings taken as samples to analyse their production costs are completely managed by the co-op technicians (that is, not by the owners of the land, but by the cooperative staff). Consequently, not only inputs or technical skills, but also human resources are shared and optimised. Hence, producing under a common management system may solve the problem of the small sized holdings. In addition, the citrus produced are certified following the EurepGAP quality protocol, as described in the next section.

2. The EurepGAP protocol

EurepGAP is the abbreviation for 'Euro Retailer Group for Good Agricultural Practices'. This group of European retailers work to offer high quality agricultural products, grown and certified under their widely known protocol.

Firstly, Good Agricultural Practices (GAP) within the EurepGAP protocol express the need of including other kinds of crop management techniques that are not usually employed in conventionally growing systems, as for instance integrated pest management, or integrated crop management. In other words, the purpose is to cultivate the future marketable productions with integrated production methods in order to obtain healthy food, while respecting the environment and contribute to long-term sustainable agriculture (EurepGAP, 2001).

All farmers, retailers or trade operators who join the EurepGAP protocol are committed to five main principles (EurepGAP, 2001):
- To maintain consumer's confidence in the quality and safety of the Eurep certified food.
- To cultivate employing good agricultural practices.
- To minimise the use of pesticides and other chemical inputs as much as possible.
- To use non renewable resources (as soil, water, etc.) efficiently.
- To be responsible for the occupational health and safety of their workers.

Consequently, it can be easily deducted that the EurepGAP protocol is not only limited to the implementation of integrated production in the fields, but it is also concerned about the environment, even beyond the agricultural production process, or about socially related issues, as worker's health, safety and welfare.

Our case study considers the last EurepGAP update, which specifies the following normative documents to control the implementation of good agricultural practices:
- EurepGAP General Regulation Fruit and Vegetables (EurepGAP, 2004a), where the EurepGAP structure or the procedures needed to obtain and maintain certification of raw fruits and vegetables are explained.
- EurepGAP Control Points and Compliance Criteria Fruit and Vegetables (EurepGAP, 2004b), where the standard and requirements that farmers must comply are established.
- EurepGAP Checklist Fruit and Vegetables (EurepGAP, 2004c), the document that farmers must use to fulfil the annual internal audit requirements.

Focusing on our case study, 'La Constancia' co-op applied for EurepGAP certification as a Farmer Group, which is the second option considered within the General Regulation. Requirements are divided into the following points (EurepGAP, 2004a):
1. Internal management and control system:
    - A quality system with written controls and a procedure manual which guarantees that internal inspections are undertaken in a competent way must be implemented. Also a traceability system which allows to separate EurepGAP certified products from non certified, and also enables to trace back to the farm where the product was obtained.
    - Central administration and management: all registered members of the farmer group must be operating under the same management and subject to central management review.
    - Contract duration: farmers belonging to the farmer group must register for EurepGAP certification for at least one whole year.
    - Internal audit procedures: all farmer groups must have an internal audit procedure that establishes a minimum of an annual inspection for each registered farmer.

2. Farmer internal self-inspection:
   - A completed internal self-inspection based on the EurepGAP Checklist must be available on each registered farm to be reviewed by either the internal or the external inspector (EurepGAP, 2004c).
   - The internal self-inspection must be carried out at least once a year.
3. Farmer group internal inspection:
   - A minimum of one internal inspection per year of each registered farm must be carried out by qualified staff within the farmer group, or subcontracted to an external verification body different from the certification body responsible for the external verification.
   - The annual internal inspection must be based on the EurepGAP Checklist.
4. External verification by an EurepGAP approved Certification Body:
   - The external inspection is annual and made by taking a random sample that is, as a minimum, the square root of the total number of EurepGAP registered farmers within the farmer group.
   - The external inspection reports are prepared in accordance with the requirements of EN 45011 / ISO Guide 65.

Hence, after analysing the EurepGAP requirements, some of the extra costs that EurepGAP certified farmers have to assume, unlike a conventional farmer, include: qualified technical assessment to implement integrated pest or integrated crop management techniques, record keeping of all the agricultural practices done on each farm, EurepGAP registration fee (based on the number of farms registered), and last but not least, external audit and EurepGAP certification fees. The EurepGAP certification is valid for one year. After that period, the farmer group must be audited and certified again.

Apart from the record keeping of all the farming operations, the farmer group must have implemented a quality and traceability system and an EurepGAP procedure manual. Furthermore, three different inspections are needed to comply with all the protocol points: farmer internal self-inspection, farmer group internal audit and external verification. In addition, record keepings must be maintained for at least two years (EurepGAP, 2004a). So maybe the fact of being a member of a co-op could help farmers to face the requirements needed to produce this high quality fruit and vegetables, not only for the technical assistance or the coordination of the tasks needed to fulfil the certification process, but also for the externalisation of costs that an organisation could offer to their members.

3. Methodology

The production costs´ analysis has been carried out employing the full costing methodology, based in other studies published before (Caballero *et al.*, 1992; Juliá and Server, 2000).

The full costing methodology takes into account three types of costs: variable, fixed and opportunity costs. However, the production costs´ scheme managed in this case study was elaborated considering only fixed and variable costs, as opportunity costs did not add any new information to our purpose, that was to compare the economic performance of EurepGAP versus conventional citrus growing systems.

Consequently, the production costs´ scheme employed is divided into two parts: the first one for variable costs, which depend on the activity developed (in this case, on the way the citrus crop is managed), while the second part contains the fixed costs (those directly linked with the structure of the farm, as investments or facilities). Total variable costs will increase at the same time as the yields obtained. By contrast, fixed costs per unit will decrease proportionally to the positive growth of the yields.

Considering the citrus crop cycle, we should enhance the following items within the variable costs because of their economic impact inside the production costs´ scheme:
1. irrigation water;
2. fertilisers;
3. pesticides, fungicides, herbicides and foliar nutrients;
4. other inputs (bait-traps, props, etc.);
5. equipment operating costs (mainly fuels and lubricants); and
6. labour cost.

Regarding the fixed costs, we should distinguish between:
1. equipment ownership costs (depreciation, insurance, repairs and maintenance);
2. crop depreciation, the result of dividing the investment made until the plantation is economically viable into its operating life;
3. holding maintenance (tree substitution, structural repairs of retaining walls, etc.);
4. taxes and insurances (taxes on rural property, social security, crop insurance, etc.).

4. Results

Nine plots were selected from 'La Constancia' Co op in order to carry out the comparative cost study. All of them were cultivated with oranges, variety Navel, selection Navelina. Eight of the nine plots had a surface less than 1 hectare, while the ninth was classified between 1 and 2 hectares. Hence, the nine plots were small-sized holdings. They were also completely managed by the co-op technicians, and their harvests certified under the EurepGAP quality protocol. The irrigation system present in all of them was the traditional one, which is a flow irrigation system, and the costs´ relations those contained in Table 1.

The variable costs show a significant variability, as can be deducted from the data in Table 1; by contrast, the fixed costs are the same for the nine plots, as they are all conducted under the same management.

The first item to be analysed within the variable costs is the cost of the irrigation water, which depends on the location of the plot inside the village area. The minimum cost corresponds to those holdings which take their irrigation water from the public supply channel, with a cost of 144 €/ha (see Table 1). For those plots which take their irrigation water from privately-owned wells, the cost will depend on the quantities employed and the stipulated prices of water on each well, until the maximum value paid per hectare inside this sample, which is 495 €/ha (see Table 1).

Regarding the use of pesticides, foliar nutrients, and fertility and weed management practices (items 1.2 and 1.3), their cost will depend on the plan designed for each plot by the co-op technicians after studying their particular needs. There are no costs computed as other inputs (item 1.4), as any practices related to their use have been done (for example, placing of props or bait-traps).

The same that happens with fertility and weed management, equipment operational costs and labour costs (items 1.5 and 1.6), will vary depending on the particular needs of each plot. However, as said before, fixed costs are the same for all the sampled plots. Within the fixed costs, the equipment ownership costs (2.1) are the result of charging depreciations, taxes, insurances,

Table 1. 'La Constancia' Coop. Average production costs of Orange crop under EurepGAP protocol. Based on data provided for by 'La Constancia' Coop technicians.

1. Variable costs	Average (€/ha)	Maximum value	Minimum value
1.1. Irrigation water	259.91	495.00	144.00
1.2. Fertilisers	319.50	559.90	84.00
1.3. Pesticides, herbicides, fung.	220.95	833.58	0.00
1.4. Other inputs	0.00	0.00	0.00
1.5. Equipment operating costs	93.67	219.45	0.00
1.6. Labour cost	734.12	1,213.41	240.90
Total variable costs	1,628.15		
2. Fixed Costs	Average (€/ha)		
2.1. Equipment ownership costs	336.74		
2.2. Crop depreciation	360.79		
2.3. Holding maintenance	0		
2.4. Taxes & insurances	62.50		
Total fixed costs	760.03		
Total costs (without opportunity costs), €/ha	2,388.19		

repairing and maintenance of the cooperative pool proportionally to all the plots cultivated under the cooperative staff responsibility. Other fixed costs are crop depreciation (2.2), which depends on the citrus varieties grown and the irrigation system available (in this case, both variables are the same for all the sampled plots), or taxes on rural property, which are also the same as the amount of this tax depends on the village where the holdings are located. Holding maintenance (2.3) has no cost because of the lack of drip irrigation systems, containing walls, tree substitution, repair of facilities or any other operations related to this category in the sample.

Table 2 compares the average results obtained by the co-op technicians producing oranges certified under EurepGAP protocol, to the average regional costs of the same citrus selection and the same farm typology, but conventionally grown. The production cost scheme of the intensive citrus modality has been taken from Caballero and Fernández (2002).

The holdings that constitute the conventional sample present practically the same characteristics as those of the EurepGAP sample. Consequently, they are also classified as small sized, with flow irrigation systems and narrow planting distances. Their standard machinery pools are simple and small. Moreover, special treatments as pesticide sprays or chemical flowering controls are always contracted to cooperatives or agricultural services′ firms (Caballero and Fernández, 2002).

The most important difference found between both production systems is related to variable costs, which are 45% less in average in the integrated production scheme with regard to the conventional one. Within the conventional system, the costliest input is the irrigation water. The irrigation water is the most unpredictable cost, as it depends on the availability of the resource, and also on the village area where the samples have been taken. Regarding the cost of fertilisers, pesticides, herbicides or foliar nutrients, the integrated production figures are generally lower

Table 2. Average production costs of EurepGAP oranges versus conventionally grown oranges.

1. Variable costs	EurepGAP[1] 2003 (€/ha)	%	Conventional[2] 2002 (€/ha)	Conventional[3] 2003 (€/ha)	%
1.1. Irrigation water	259.91	10.88	668.48	685.19	17.48
1.2. Fertilisers	319.50	13.38	438.74	449.71	11.47
1.3. Pesticides, herbicides, etc.	220.95	9.25	453.27	464.60	11.85
1.4. Other inputs	0.00	0.00	60.85	62.37	1.59
1.5. Equipment operating costs	93.67	3.92	58.74	60.21	1.54
1.6. Labour costs	734.12	30.74	1,187.67	1,217.36	31.05
Total variable costs	1,628.15	68.18	2,867.75	2,939.44	74.97
2. Fixed costs					
2.1. Equipment ownership costs	336.74	14.10	262.94	269.51	6.87
2.2. Crop depreciation	360.79	15.11	346.42	355.08	9.06
2.3. Holding maintenance	0	0.00	56.86	58.28	1.49
2.4. Taxes & insurances	62.50	2.62	291.16	298.44	7.61
Total fixed costs	760.03	31.82	957.38	981.31	25.03
Total costs without opportunity costs (€/ha)	2,388.18		3,825.13	3,920.76	
EurepGAP certification and analysis costs (€/ha)	205.4				
Total costs (€/ha)	2,593.58			3,920.76	
Average production (kg/ha)	23,000			30,000	
Average costs (€/kg)	0.11			0.13	

[1]Data provided by 'La Constancia' Coop technicians.
[2]Data from Caballero and Fernandez (2002).
[3]Costs of column Conventional[2] capitalised with the Annual Variation rate of the Gross Domestic Product (INE, 2004).

than those of the conventional systems, mainly because the co-op technicians adjust the input doses as closer to minimum values as possible in order to comply with the EurepGAP standard specifications.

According to the co-op principles, the equipment ownership costs of the integrated production sample are those of the co-op pool, proportionally charged to each plot and depending on its surface. The equipment operating costs are higher in the co-op sample, as the co-op pool is wider than those of conventional farmers. In fact, conventional farmers hire most of the crop management tasks that require the use of special machinery. These hired tasks are reflected inside the labour cost item, which is also higher in the conventional sample (see Table 2).

Nevertheless, another reason for having lower labour costs in the integrated production scheme is that the co-op technicians coordinate the labour force under their responsibility, which also means an optimisation of timing and availability of human resources.

Within the fixed costs, the main difference found is related with the first item, equipment ownership costs. In the integrated production sample this item reflects the ownership costs of the co-op pool, constituted by a heavyweight tractor, a speed sprayer, an hydropneumatic tank, a mower and a groundwood crusher (to use after pruning). By contrast, conventional small farmers only own a lightweight tractor and a hydropneumatic tank adapted to their engine power. As a consequence, not only the equipment ownership costs, but also the equipment operating costs are higher in the EurepGAP sample than in the conventional, as the use of the owned machinery is more important in the former in economic terms.

Finally, once all the expenditures have been accounted, including the EurepGAP analysis and certification costs in the case of the co-op sample, it can be deducted that the total costs are in average 34% higher in conventionally grown systems than in the integrated production plots (see Table 2). It is important to consider that the average productivity rates of the area where the co-op is located are lower than the average productivity rates of the region (23,000 kg/ha versus 30,000 kg/ha). Despite this, the total production costs are lower in the certified co-op plots, with an average of 0.11 €/kg, while the average for conventionally grown systems is 0.13 €/kg.

5. Conclusions

After considering the results obtained, the important role that agrarian cooperatives play regarding the production of fruit and vegetables under quality protocols in the Comunidad Valenciana should be enhanced, mainly for two reasons:

- First of all, the advantage that being part of an organisation, in this case an agrarian cooperative, represents for farmers. The co-op management and technical assessment is essential to comply with the various requirements of quality protocols, for instance internal and external audits, implementation of quality and traceability systems, record keeping of field operations, or technical advice.
- Secondly, and concretely in our case study, the fact of having the whole citrus crop management done by the cooperative technicians is a good viable solution for problems as the small size of holdings, obtaining at the same time profitability of systems where not only inputs, but also human resources were not optimised.

Last but not least, it is known that the supply of high quality products is one of the keys to obtain profit in today's markets. Therefore, agrarian cooperatives should be an instrument to cultivate these kinds of products, in compliance with the quality and safety requirements demanded by the European markets, which are at the same time their main customers. Furthermore, the enterprise structure could be used to face other regional problems related with citriculture, as the small-sized holdings or the high production costs.

Acknowledgements

The authors wish to thank Rafael Cosme and Verónica Ferrandis, manager and quality technician of 'La Constancia' Co-op (Pobla de Vallbona - Valencia), for their help during the field work stage and the data set collection process. We also would like to express our appreciation to Debra Westall (UPV), for revising the final text.

References

ANECOOP (2005). ANECOOP Corporative Information. http://www.anecoop.com

Caballero, P., M.D. de Miguel and J.F. Juliá, 1992. Costes y precios en hortofruticultura. Mundi-Prensa.

Caballero, P. and M.A. Fernández Zamudio, 2002. Tiempos de obligado cambio en la citricultura. Levante Agrícola, No. 359, 1er. Trimestre 2002, 17-27.

EurepGAP, 2001. EUREPGAP Protocol for Fresh Fruit and Vegetables. Version September 2001, Rev. 02. Available in http://www.eurep.org

EurepGAP, 2004a. EUREPGAP General Regulation Fruit and Vegetables. Version 2.1-Jan04. Available in http://www.eurep.org

EurepGAP, 2004b. EUREPGAP Control Points & Compliance Criteria Fruit and Vegetables. Version 2.0-Jan04. Available in http://www.eurep.org

EurepGAP, 2004c. EUREPGAP Checklist Fruit and Vegetables. Version 2.0-Jan04. Available in http://www.eurep.org

Fernández Zamudio M.A., I. Pavía and P. Caballero, 2004. La producción integrada en el sector citrícola y su adopción por las pequeñas explotaciones. ITEA, Vol. 100V, No. 3, 147-166.

INE (INSTITUTO NACIONAL DE ESTADISTICA), 2004. Producto Interior Bruto a Precios de Mercado. Tasa de variación interanual. Available in http://www.ine.es/prensa/pib_tabla_cne.htm

Juliá J.F. and R. Server, 2000. Economic and financial comparison of organic and convencional citrus-growing systems. Study prepared for the Horticultural Products Group, Tropical and Horticultural Products Service, Commodities and Trade Division, FAO.

Peris, E., 2002. La citricultura y la aplicación de políticas agroambientales. Un análisis comparado de las regiones europeas de Calabria y la Comunidad Valenciana. Master Thesis. Polytechnic University of Valencia (Spain).

Peris, E. and J.F. Juliá, 2005. Effects of Regulation (EEC) 2078/92 on citrus growing in Calabria (Italy) and the region of Valencia (Spain). Spanish Journal of Agricultural Research 3, No.1, 34-42.

Planells, J.M., 2003. La calidad como objetivo de la agricultura y el cooperativismo valenciano ante el futuro. Congreso 60 aniversario UTECO-Valencia, Agricultura y cooperativismo. UTECO-Valencia, 71-91.

The role of ISO standards in micro and small-scale food ingredient companies

Miriam Quintero Arellanes and Seamus O'Reilly
Department of Food Business and Development, University College, College Road, Cork, Ireland; s.oreilly@ucc.ie

Abstract

The aim of this study was to assess the applicability of ISO standards in Irish micro and small-scale food ingredient companies and explore the reasons why ISO standards were or were not adopted and what benefits were gained. A case study methodology was employed to explore use of Quality Management Systems (QMS) in four micro/small-scale food ingredient firms. The two main reasons for implementing the ISO standard were customer satisfaction and pursuit of quality improvement. The evidence suggests that the ability to adopt systems and standards relevant to particular markets is more important in establishing competitive advantage than the achievement of any particular standard such as ISO.

Keywords: ISO, micro/small-scale food ingredient companies, competitiveness

1. Introduction

Many authors agree that there are several differences between large companies and SMEs (Small and Medium-sized Enterprises). The differences are in: company structure, responses to external factors, policy-making procedures, managerial style, management of resources and even how they compete in their environment. SMEs in general are more flexible, with less stability in the marketplace (Ghobadian and Gallear, 1996; Man *et al.*, 2002; Wiklund and Wiklund, 1999). ISO standards are often seen as rather bureaucratic systems more suited to the procedures and systems employed in large companies than the more informal procedures and communication systems employed in smaller companies. Therefore, the applicability of ISO standards in micro and small-scale food companies can be questioned.

Many small companies have argued that achieving ISO certification is too costly. However, ISO certification could be an instrument for differentiation for SMEs as well as essential for developing new business. For many SMEs the decision to implement a Quality Management System is due to customer pressures rather for benefits such as of improved efficiency and effectiveness. However, small manufacturing businesses have embraced ISO standards quicker than the services industry (Curkovic and Pagell, 1999; Escaciano *et al.*, 2002; Yusof and Aspinwall, 2000). Varzakas and Jukes (1997) stated that with the globalisation of food quality in Europe there has been a need to adopt quality systems such as ISO standards and Hazard Analysis and Critical Control Points (HACCP) to guarantee the quality and safety of food production. Thus many small companies have adopted these systems, however it is interesting to investigate why they have done so and how they use them.

The aim of this study was explore the reasons why ISO standards were or were not adopted, identify the approaches used to implement ISO standards and consider its role in Irish micro and small-scale food ingredient companies. Specific research propositions were tested and these are outlined below.

2. Research propositions and methodology

Research propositions formulated for this study were:

- ISO is essential for increased sales and market expansion;
- ISO certification does offer a significant competitive advantage;
- documentation generates bureaucracy rather than clear procedures and systems; and
- ISO does little to enhance internal communication.

A case study methodology was employed to explore use of QMS in four micro and small-scale food ingredient firms. The strength of the case study lies in its ability to explore on a micro level the daily life of organisations. Case studies help improve our understanding of events in their context. In the small business context a case study approach may help to reveal what decision-making strategies were taken, how they were carried out and with what effects (Cassell and Symon, 1994; Curran and Blackburn, 2001). Table 1 presents the main characteristics of the four firms selected.

From Table 1 it is evident that companies were selected across a range of products/services and were at different stages of implementing ISO standards, one had not implemented ISO at all, while another had first adopted it 20 years ago.

Qualitative data collection and analytical tools were used. The following data collection approach was used with each case: profile questionnaire, depth interviews with key informants, collection of documentary data and questionnaires (employee survey). By using different sources of evidence it is possible to obtain multiple measures of the same phenomenon and ensure the validity of the research (Yin, 1994). In order to analyse the role of ISO standards in the four companies the literature suggests using cross-case analysis to understand and explain the phenomenon in detail. In studying the similarities and differences across cases, it is possible to identify how certain elements and conditions may be related (Miles and Huberman, 1994). Analytical tools such as matrices and diagrams were used.

Table 1. Companies' main characteristics relevant for the study.

Company code	Founded in	No. of employees	ISO-9000 status	Food ingredient	Main products	Type of organisation
A	1991	20- 30	certified 2001 (all areas in the company)	manufacturer	animal feed supplements	private ltd. company
B	1993	20- 30	certified in 1994 (all areas except R&D department)	manufacturer	seasonings	private ltd. company
C	2002	< 10	non-ISO	service	consultancy	partnership
D	2000	< 10	implementing	manufacturer	colours	sole proprietorship

The interviews were conducted with key informants in each company. All interviews were conducted on company premises. All the interviews were recorded with previous consent of the interviewees. A survey questionnaire was applied according to the level of ISO implementation in the company. There were twenty-three questionnaires in total. Table 2 illustrates a summary of the data obtained from interviews and questionnaires. Table 3 summaries some of the organisational documentation collected from the companies.

Table 2. Summary of data obtained from interviews and survey.

Company code	Interviews	Survey questionnaire
A	1 - R&D Manager	1 - Top management 1 - Production department 1 - Quality department 1 - Sales department
	one interview in total.	4 questionnaires in all
B	1 - Quality Manager (QM) 1 - Sales Director (SD)	1 - Top management 6 - Production department 2 - Quality department 1 - Sales department
	in total 2 interviews.	10 questionnaires in all
C	1 - Business Director 2 - Research Director 1 - Technology Director	
	in total 4 interviews.	5 questionnaires in all
D	1 - Managing Director 1 - R&D / Quality Manager 1 - Customer service Manager	
	in total 3 interviews.	4 questionnaires in all

Table 3. Summary of some of the organisational documentation collected across companies.

Documentation	A	B	C	D
Quality policy	•	•		•
Organisational chart	•	•		•
Quality manual (table of contents)	•	•		•
Quality plan		•		
Business plan			•	
Customer complaint procedure	•	•		•
Employees responsibilities	•	•		•

Once data collection was completed the raw data was transcribed, coded and analysed. The data was analysed using the template approach. This was based on a number of themes relevant to the research (Cassell and Symon, 1994). QSR N6 version software was used to analyse the data. The development of categories for certain topics was used to organise and manage the data obtained. The software allowed the researcher to analyse the categories by 'search and compare nodes' by using the options of intersection, overlap and union of topics under study. Tables and figures were used to exemplify and link the evidence.

3. Findings

3.1. ISO implementation

Figure 1 illustrates the conditions and the process that influenced the implementation of ISO standards. Companies A, B and D had two main reasons for implementing the ISO standard: (1) customer satisfaction and (2) pursuit of quality improvement (both marked in grey in the diagram). For company C the main reason for not implementing ISO was that none of their customers were requesting for it. The other three companies had expectations (i.e. increased sales, recognition, credibility and traceability) once the company was engaged in an ISO project. During the implementation process the main factors that had an influence were company size, company's resources, employees' expertise, motivation, external consultant, training, documentation and internal audits. An external audit was conducted by a third party in both companies A and B once they had fully implemented the standard. After certification the companies (A and B) evaluated the expectations vs. benefits.

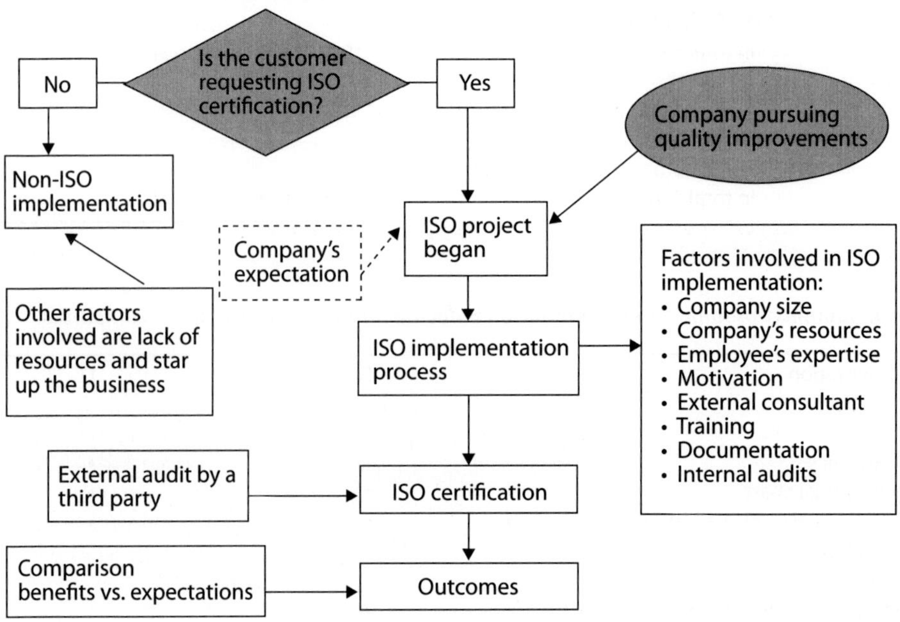

Figure 1. The role of ISO-9000.

The four companies were in different phases of ISO certification, company C was a non ISO company, while company B had evolved from ISO to the British Retail Consortium (BRC) standard. This could be explained by the differences that existed between companies, such as employees' expertise, years in business, business requirements, main products and markets, type of industry (manufacturing vs. service), size of the company (micro vs. small), technology applied in their processes, number of years with ISO, and competitors. Figure 2 illustrates the different stages of ISO in the companies under study. The remainder of this section considers each of the theoretical propositions.

3.2. Sales and markets

The literature suggests that implementation of quality systems could create a higher rate of profit by increasing market share (Hansen, 2001). Several authors recognise that ISO-9000 has benefits for small firms such as opportunities in new markets, increased profits and business retention among others (Escaciano *et al.*, 2002; Yusof and Aspinwall, 2000). However, researchers also recommend that ISO certification is part of a company's marketing strategy and is not seen as a short-term tactic to increase market share (Curkovic and Pagell, 1999).

The link between ISO certification and increase in sales and market share exists because a certified company is able to fulfil customer satisfaction in terms of proving company commitment to quality by having a quality system in place. Although ISO standard is now recognised worldwide as a quality standard, in 1994 certified companies, who achieved it, had a differentiation point. By 2004 the ISO certification has become an expectation rather than a differentiation point. The evidence found suggests that by achieving ISO-9000 certification, credibility and recognition were attained in 2001 in company A and in 1994 in company B. This led companies A and B to retain their business, increase their profits and explore new market opportunities. In the particular case of company B, by achieving the accreditation in 1994 the company differentiated itself from competitors. However, findings in this company suggested that in 1994 company B relied mainly on the ISO certification as a marketing strategy, rather than considering ISO as part of the strategy. In 2003 this company obtained credibility and recognition by updating ISO. However, the BRC standard became more important in delivering business retention and increased profits.

The evidence suggested that ISO had both direct and indirect influence on the factors related to the expansion in sales and markets. The direct influence is the one that is achieved at the moment of the certification such as business requirement, differentiation, credibility and recognition. The indirect influence not only depends on the accreditation but also on other factors such as

Non ISO company	ISO implementing company	ISO accredited company	ISO accredited company and moved on to BRC standard
C	D	A	B

Figure 2. Different stages of ISO across the companies under study.

company marketing strategy, including price and quality of the product. Table 4 illustrates the direct and indirect influence of the ISO certification on different factors that were linked to increased sales and markets expansion.

Note that only company B achieved differentiation by obtaining the ISO certification in 1994, even though the literature identified differentiation as a benefit of ISO. The evidence suggests that more recently ISO-9000 has become a prerequisite rather than a point of differentiation. For company D, ISO was part of a marketing strategy and they hoped to benefit from increased profits, business retention and new market opportunities. They also considered ISO as a business requirement or prerequisite and something that would enhance their credibility and recognition in the marketplace.

A comparison of customer profile and markets was necessary to understand the impact that ISO had on these companies. Although companies A, B and D were food ingredient manufacturers, each operated in a different market with different business requirements. Ireland was the main market for company B while companies A and D were mainly export driven. Furthermore, ISO-9000 established credibility and recognition in the market place for companies A and D, while BRC was more important for company B. As a consultancy firm company C did not identify ISO as a business requirement. Their business approach was conducted through agreements on a project basis. This is not to suggest that ISO does not have role in service companies, rather for this case their customers were not demanding it and the company had devised and implemented performance based systems to monitor and report on projects.

Thus the evidence supports the general proposition that 'ISO is essential for increased sales and market expansion', since not only is it important, but it appears to be a prerequisite. However, the type of business and market influenced company adoption of specific standards and benefits gained. The specific standard used may not be the most important factor; rather the ability to put the systems in place to achieve specific standards is of fundamental importance.

Table 4. Link between the ISO standard and increased sales and markets expansion. (+) Positive influence; (-) No influence; (☑) Expectation; (N/A) Not applicable.

Factor	Case evidence				
	A	B		C	D
		1994	2003		
ISO direct influence on					
Business requirement	+	+	-	N/A	☑
Differentiation	-	+	-	N/A	-
Credibility	+	+	+	N/A	☑
Recognition	+	+	+	N/A	☑
ISO indirect influence on					
Increase in profits	+	-	-	N/A	☑
Business retention	+	+	-	N/A	☑
New market opportunities	+	+	-	N/A	☑

3.3. Competitive advantage

The literature suggests that competitive advantage could be achieved when a company achieves the following factors: quality service, good organisational structure and continuous improvements being part of their regular practices. The literature also indicated that a company would become more cost-effective with ISO certification (Bolton, 1997; Escaciano *et al.*, 2002; Munro-Faure and Munro-Faure, 1992; Yusof and Aspinwall, 2000). According to Aghaide and Popplewell (1997) any company delivering the right product with quality would achieve a competitive advantage especially in industries where there are constant changes in the market.

The evidence suggested that ISO certification had a direct influence on the following factors: structure, continuous improvements and quality service. ISO certification removed informalities in companies A, B and D. Continuous improvements were initiated through the use of internal audits that were carried out on a regular basis. As ISO-9000:2000 had more emphasis on customer satisfaction; ISO certification had a direct influence on the quality of service in these companies. Table 5 illustrates the influence of ISO on these factors that were linked to competitiveness. Company A was the only company to report improved cost-effectiveness due to the quality improvements in the company. Company B did not achieve this benefit in 1994 and by 2003 the BRC standard improved company cost-effectiveness. As company B had already defined its structure before updating ISO in 2003, this standard had no substantial influence on company structure in 2003.

The findings suggest that the four companies considered delivering quality products and quality service as an important source of competitive advantage. Company A is an example of a company that placed great importance on product quality in order to achieve a better market position.

Companies B and D also considered the technology involved in their processes as a source of competitive advantage. These two companies were technology driven. After ISO certification, company A identified key company performance metrics, and these were considered as a source of competitive advantage. By documenting its systems, company A was able to compare its performance over time.

Table 5. Link between the ISO standard and competitive advantage. (+) Positive influence; (-) No influence; (☑) Expectation; (N/A) Not applicable.

Factor	Case evidence				
	A	B		C	D
		1994	2003		
Quality service	+	+	+	N/A	☑
Structure	+	+	-	N/A	☑
Cost-effective	+	-	-	N/A	☑
Continuous improvements	+	+	+	N/A	☑

Thus, there is some support for the proposition was that 'ISO certification does offer a significant competitive advantage', since the evidence suggests that ISO contributed to competitive advantage. This was important where companies took a proactive approach to the implementation of ISO and used it to inform continuous improvement.

3.4. Documentation

In order to achieve ISO certification companies must have certain documentation in place. The literature indicated that documentation has benefits as well of drawbacks. However, small business has an advantage in having less complex processes and fewer products. The benefits of documentation can be codification of knowledge, consistency of processes, traceability, clear company procedures and better training and commitment (Bolton, 1997; Curkovic and Pagell, 1999; Varzakas and Jukes, 1997). The same authors agreed that there is a link between documentation and continuous improvements. However, the most evident drawback of documentation is probably bureaucracy, followed by the time that it takes to comply with full documentation (Escaciano *et al.*, 2000;Yusof and Aspinwall, 2000).

The evidence found suggested that the ISO standard had a positive and direct influence on the codification of knowledge, in enhancing traceability and in developing company procedures in companies A, B and D. The indirect influence was on enhancing operational consistency and in creating efficiencies in the training of new employees. In these case studies all the informants associated documentation with bureaucracy, although all pointed out that in their own companies' bureaucracy was minimal due to the size of their business.

Table 6 illustrates the direct and indirect influence of the ISO standard on the different factors related to documentation. The literature stated that bureaucracy is a drawback of ISO documentation. However, bureaucracy had very little influence on companies A, B and D. Even though company D did not have ISO accreditation when the study was carried out, they did have all the documentation in place and were ready to be audited. Findings suggested that the documentation that was created for ISO implementation had a positive influence across companies. However, to comply with this requirement was not an easy task for the companies as there were time, training and new procedures involved. In companies A and D, an external consultant guided them through the documentation process. Company B did it internally for two reasons. Firstly, the quality manager had the expertise to update the ISO standard and secondly, company B had most of the documentation in place in May 2003, when they achieved the BRC standard.

In the particular case of company A, where the ISO certification did not include the R&D department, documentation had a positive influence in encouraging full internal traceability in this department even though it was not certified by third-party. In company C, where the filing system was mainly IT-oriented, employees recognised that there was a need to document and improve company procedures and the coding system. Therefore, the third proposition that 'documentation generates bureaucracy rather than clear procedures and systems' was not supported.

3.5. Internal communication

The literature indicates that communication through teamwork is a key for the success of a project such as ISO. Consequently, this type of project would improve the communication among employees (Bolton, 1997; Munro-Faure and Munro-Faure, 1992). Ghobadian and Gallear (1997) state that one advantage of small business is that communication and coordination tend to be efficient and effective.

Table 6. Link between the ISO standard and documentation. (+) Positive influence; (o) Very Low influence; (N/A) Not applicable.

Factor	Case evidence				
	A	B		C	D
		1994	2003		
Bureaucracy	o	o	o	N/A	o
ISO direct influence on					
Codification of knowledge	+	+	+	N/A	+
Traceability	+	+	o	N/A	+
Company's procedures	+	+	+	N/A	+
ISO indirect influence on					
Training for new employees	+	+	+	N/A	+
Operational Consistency	+	+	+	N/A	+

All companies reported that their internal communication was good before the implementation or updating of the ISO standard. This was the case for companies A, B and D. Although company C had not implemented ISO at the time of the study, good internal communication systems were evident.

After ISO certification both companies A and B found that the standard encouraged them to increase training sessions and improve internal communication about quality issues. Furthermore, company B also increased the number of quality meetings in an effort to increase communication and knowledge about the quality performance of the company with other employees in key positions.

As company D was implementing the standard at the time of the study, they found that training on quality and 'quality' meetings enhanced the internal communication of the company. At these meetings the progress of their ISO project was discussed. Table 7 illustrates the factors enhancing internal communication across companies after the ISO implementation.

Table 7. Internal communications across companies. (+) Positive influence; (-) No influence; (N/A) Not applicable; (?) Unknown.

Factor	Case evidence				
	A	B		C	D
		1994	2003		
Training	+	?	+	N/A	+
Quality Meetings	-	?	+	N/A	+

Note that there was no evidence available to uncover issues related to internal communication in 1994 in company B. Thus the fourth research proposition that 'ISO does little to enhance internal communication' was not supported.

3.6. Patterns and contrast across cases

The findings suggest that even though the four companies were in the food ingredient business, there were differences in the type of market served, customers and consequently business requirements. However, they all focused on quality improvements and customer satisfaction. Therefore, the customers of companies A, B, C and D influenced the decision whether or not to adopt ISO standards.

Although there were certain similarities between company C and D in terms of employees' expertise and size of the enterprise, the key difference was that company C was a consultancy company and company D a manufacturer. Therefore, the business requirements were different.

Table 8 summarises the benefits achieved by companies A and B after ISO certification. It also illustrates the benefits achieved during ISO implementation and expected benefits for company D.

Table 8: Benefits of the ISO standard across companies A, B and D. (•) Benefits; (☑) Expectation.

Company	A	B		D
Benefits		1994	2003	
Internal benefits				
Accurate documentation	•			•
Traceability	•	•		•
Better performance metrics	•			
Continuous improvements	•			
Sales increase	•			☑
Structure of quality system		•		•
Structure on management		•		
Increase in training			•	•
Increase of quality meetings			•	•
Up to date documentation			•	
Increased efficiency				☑
Develop a competitive advantage	•			☑
Develop a frame of thinking	•	•	•	•
Improved public perception benefits				
Recognition	•	•	•	☑
Credibility	•	•	•	☑
Better and faster response to customers	•			☑

4. Conclusion

Companies A, B and D obtained benefits from their involvement in the ISO accreditation process. While the benefits achieved varied across companies, the common benefits were credibility, recognition, and development of a frame of thinking.

The applicability of the ISO standard depends on many factors such as company characteristics, company customers, markets, competitors and, to a large extent, the implementation process. In this study the ISO standard had practical implications for the three manufacturers, these were removal of informalities in the company, the opportunity to expand sales and develop markets, improved competitiveness, increased quality awareness among employees, faster response to customer queries, business retention and improved operational consistency. While there was evidence to suggest that documentation systems led to rather bureaucratic procedures in the implementation phase, these systems also formalised procedures and were more easily implemented through teamwork.

Communication is very important among employees when a company is engaged in an ISO project. It enables discussion, agreement, and documentation of procedures and manuals within the company. It enhances analysis of actual practices with a view to performance improvements. With efficient communication the development of documentation could take less time and more improvements could be made. The evidence suggests that the process of documentation and internal audits enhance continuous improvements in the company. These generally boost competitive advantage. Such advantage can be explained in terms of quality systems, quality products, right price and services. These led to increased sales and market share. Increasingly, customers focus on supplier processes rather than product and thus less effort is on product testing and more attention is given process auditing. In this way a customer can engage in a more strategic approach to purchasing and the buyer-seller dyad shares more and more management performance information that enhances product quality and cuts cost.

The evidence suggests that when micro/small-scale ingredient supply companies operate in a particular environment they will seek to adopt industry specific systems and standards, and when customers change their systems and requirements they were seen to respond. For example, company A had moved on to BRC and having achieved the ISO standard company B had set about achieving an industry specific system, FEMAS (Feed Material Assurance Scheme). In the particular case of the consultancy company the only benefit of the ISO would have been improvements in the internal management of the company. ISO implementation would not have improved sales nor developed new markets. Rather this company used a rather detailed computerised project management system and this provided a basis for customer service and other measures. This was because, at the time of the study, none of their customers were requesting ISO certification. Thus ISO was not seen as a business requirement. Therefore, it is not only important to adopt a QMS to improve the business, it is also important to adopt a QMS suitable to the business, its customers and general environment. While this study investigated the applicability of one QMS, namely ISO, future research could investigate the factors that influence enterprise awareness of the appropriate systems and their ability to continue to respond to changing requirements.

Notwithstanding the differences between these case studies, they all operate within the same sector and have experience of supplying large-scale companies. Indeed many of the employees/owner-managers in these micro/small-scale ingredient companies have worked within larger companies. It is thought that this background and context influenced their attitude towards ISO and more generally quality management systems. Therefore, it not suggested that all micro/small-scale food companies find such systems useful nor have they all a tendency to use them. In the authors' experience many micro/small-scale food companies producing and marketing niche products in consumer markets are less interested in formalised systems and tend to minimise their use of such systems. Therefore, it would appear that organisational culture influences adoption of quality management systems. Further investigation into the adoption and use of such systems across a wider spectrum of micro/small-scale food companies may provide further insight into the factors that influence company awareness of and attitude to appropriate systems and their ability to continually adapt to changing market and regulatory conditions.

References

Aghaire, A. and K. Popplewell, 1997. Techniques simulation for TQM the unused tool?. The TQM Magazine 9, No. 2, 111-116.

Bolton, A., 1997. Quality management systems for the food industry. A guide to ISO 9001/2, Blackie Academic and Professional.

Cassell, C. and G. Symon, 1995. Qualitative methods in organisational research. A practical guide, Sage.

Curkovic, S. and M. Pagell, 1999. A critical examination of the ability of ISO 9000 certification to lead to a competitive advantage. Journal of Quality Management 4, No. 1, 51-67.

Curran, J. and R.A. Blackburn, 2001. Researching the small enterprise. Sage.

Escaciano, C., E. Fernández and C. Vázquez, 2002. Linking the firm's technological status and ISO 9000 certification: Results of an empirical research. Technovation 22, 509-515.

Ghobadian, A. and D. Gallear, 1996. Total Quality Management in SMEs. Omega The International Journal of Management Science 24, No. 1, 83-106.

Hansen, T., 2001. Quality in the Marketplace: A theoretical and empirical investigation. European Management Journal 19, No. 2, 203-211.

Man, T.W.Y., T. Lau and K.F. Chan, 2002. The competitiveness of small and medium enterprises. A conceptualisation with focus on entrepreneurial competencies. Journal of Business Venturing 17, No. 2, 123-142.

Miles, M.B. and A.M. Huberman, 1994. An expanded sourcebook. Qualitative data analysis. 2nd ed., Sage.

Munro-Faure, E. and M. Munro-Faure, 1992. Implementing TQM. Longman.

Varzakas, T. and D. Jukes, 1997. Globalisation of food quality standards: The impact in Greece. Food Policy 22, 501-514.

Wiklund, H. and P.S. Wiklund, 1999. A collaboration concept for TQM implementation in small and medium sized enterprises. International Journal of Applied Quality Management 2, No. 1, 101-115.

Yin, R.K., 1994. Case study research design and methods. Sage.

Yusof, S. M. and E. Aspinwall, 2000. TQM implementation issues: review and case study. International Journal of Operations & Production Management 20, 634-655.

Company costs and benefits of organic processed food

Cristina Mora and Davide Menozzi
Department of Economics, Università degli Studi di Parma, Italy;crismora@unipr.it

Abstract

The aim of this paper is to present the results of research in progress to evaluate the trend of organic processed food, focusing on the managerial and economic problems that processing industries have to face in this sector, compared to their potential and perceived benefits. In order to achieve this purpose, it has been decided to submit questionnaires to important industries involved in organic food processing to have an overview of the market environment and the reasons behind the industries' choice to produce organic food and on a qualitative and quantitative evaluation of the costs and benefits related to the 'organic food choice'.

Keywords: organic processors, organic chain, organic market

1. The development of organic food products in Italy

It is a well-known fact that the development of agricultural production has been stimulated by the Common Agricultural Policy (CAP). Since the beginning of the 1990s, against a background of production surpluses and the economic and financial problems generated by these surpluses, the European Union decided to offer incentives for the conversion of farms and farmland to organic farming (EC Regulation No 2092/91), offering premiums per hectare as part of the measures accompanying the CAP Reform.

By adopting Council Regulation (EEC) No 2092/91, amended by Council Regulation (EC) No 1804/1999, the European Union was one of the first to set up a policy on organic farming. With this regulation, the Council created a Community framework defining in detail the requirements for agricultural products and foodstuffs bearing a reference to the production methods used in organic farming and foodstuffs.

The trigger was the EC Regulation No 2078/92, which included financial support for crops with the lowest environmental impact, primarily organic farming. The first funds were paid out in 1995, so the boom in organic farming noted from 1996 is therefore quite understandable.

The sharp increase in supply in some parts of the country had an impact on the organic products market, which had until this time been very limited (Table 1).

In Italy the market was slow to take off due to the lack of a dedicated retail structure and appropriate (national) measures aiming to support the autonomous development of product supply chains. A growth in the market was seen only when the distribution network for conventional products became interested also in organic products, driven not so much by a real interest in organics but rather by the serious food crisis of the late 1990s.

The boom in the market did not however manage to ride the wave of the not only national economic crisis following the introduction of the single currency, which was not accompanied

Table 1. Regional distribution of organic farms and firms in 2003 (Ministero delle Politiche Agroalimentari, SINAB (Sistema di Informazione Nazionale sull'Agricoltura Biologica, 2004).

	North Italy	Centre Italy	South	Island	Italy
Farms	9,962	8,509	11,260	12,454	42,185
Processors	2,013	1,002	750	499	4,264
Importers	8	3	1	0	12
Farms/processors	559	681	394	215	1,849
Processors/importers	0	0	0	0	0
Processors/Importers	127	22	6	3	158
Farms/Proc/Imp	1	3	0	1	5
Total certified	12,670	10,220	12,411	13,172	48,473

by suitable market measures. This explanation is linked closely to the high difference in consumer prices between organic products and conventional ones.

While during the initial phase the lack of distribution channels made its mark in terms of lack of intermediate demand for the farmers, during the boom, following the full scale entrance of the large scale retailers on the market, the price level remained high in any case, both due to the position the retail chains give to organic products - precisely due to their reassuring and attractive function that draw in the consumer -, and for the actual management costs for these kinds of products (from the creation of a critical mass of suppliers to the logistics issues involved in the final distribution due to the low turnover).

All the greatest large distribution chains have launched in the recent past their own private label organic line (Table 2).

In 2000 the market share of super- and hypermarkets exceeded the specialised organic shops one (see Figure 1). Therefore, despite the solid assistance at production level, which in practice often placed the production price in line with the prices of conventional agricultural products,

Table 2. Private label organic line.

Name	Organic line brand	No. of organic references
Coop	Bio-logici	258
Esselunga	Esselunga Bio	480
Conad	Da Agricoltura Biologica	80
Carrefour Italia	Scelgo Bio	-
Gruppo Pam	Bio Più	-
Panorama	Da agricoltura biologica	-
Crai	Bio	
Selex	Bio Selex	
Rewe Italia	Si !	
Despar	Biologico	

the low contractual power of the farmers and the low interest in the national food processing industry has led to a slow Italian market, despite the 'attempt' to revitalise it driven by the large scale retailers.

The synergic mechanism foreseen (or at least hoped for) by the Commission, thanks to which the prices reduction through public subsidies offered to organic producers, would be translated into an increase in consumption and thus market growth with the entry (and exit) of new businesses, did not occur. Therefore, while consumers are potentially interested in organic products, they can only get them today at incredibly high prices.

The economic value of the Italian organic market is estimated to be around 750 million Euro in 2003. Over a third of all Italian organic production is exported, mainly to other European countries, but also in the USA and Japan.

2. The role of the food processing industry in the organic market

There has been growth in the number of organic household and processors in Italy over the period 1998 to 2002 (Table 3).

Most of Italian organic processor are in the North of Italy while the farms are more located in the South and the Islands. The moderate presence of the processing industry[1] from the organic market is fairly constant today, if not for a few rare exceptions that are discussed below, both for physiological reasons linked to the most widespread categories of organic products (products with low processing requirements, such as fruit and vegetables) and for reasons of strategic opportunity. These latter were originally bound by the inefficient distribution of the product and then by the presence of too strong competitors: within the retail chains, the presence of private labels saturates the supply for most organic food categories. In fact, today the greatest growth on the overall market is seen by retail private labels (Figure 1).

Food processing industry[2] is still interested in very specific niches, providing these are already specialised in the organic field and produce for already-known brands or for private labels.

Some interesting gaps still remain to be filled, where for various reasons the private labels have not taken hold. Such is the case of fresh milk, where the unopposed national leader is Granarolo, with its Prima Natura Bio brand. In the first type of company, the specialisation is mostly linked to ethical reasons, while in the second case by strategic reasons, closer to those that stimulated the large retailers initially.

Today the situation is very serious, and even the national large scale retail is becoming more strongly oriented towards the more ethical, social and environmental content of organic products, but above all towards their 'GMO free' nature, while orienting the overall communication strategies towards the convenience factor (price). The retail companies working in Italy that have their decision-making centres abroad pay more attention exclusively to price competition. This is

[1] Processors include market operators that preserve, process, package and label organic agricultural produce (including slaughter and butchering of livestock) in order to make them ready for the market.

[2] Organic pioneers like Brio, Ecor, Ki, Almaverde, Alce Nero with interesting turnover to the dairies Scaldasole (a divison of Plada, Heinz Group), Granarolo (the most important italian dairy co-op), some municipal ones (Brescia, Florence, Trento), Monini and Carapelli (olive oil), Galbusera (bakery), Polli (pickles), Fileni (poultry) and Fattorie Novelli (eggs).

Table 3. Evolution of organic processors (unit) (Ministero delle Politiche Agroalimentari, SINAB (Sistema di Informazione Nazionale sull'Agricoltura Biologica, 2004).

	1998	1999	2000	2001	2002	2003
Processors	1,490	2,165	3,006	3,947	4,346	4,264
Holdings	38,616	47,705	53,630	56,440	49,489	44,043
Importers	-	-	97	122	155	175
Farming area[1]	577,475	911,068	1,040,377	1,237,640	1,168,212	1,052,002

[1]The organic farming area is 7% of the total agricultural area.

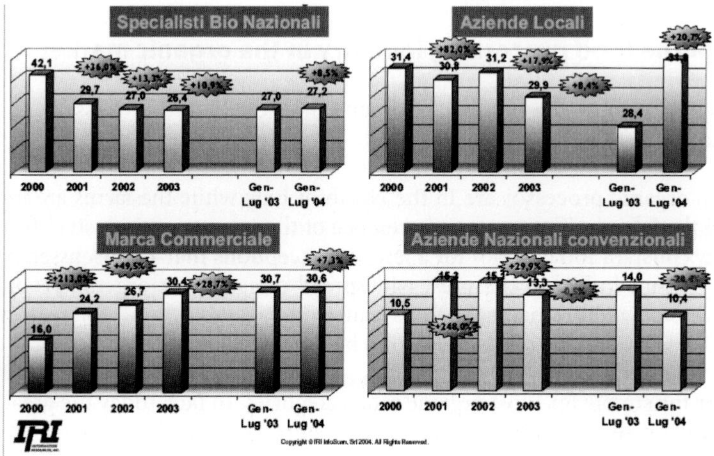

Figure 1. Market share in 2004 (IRI, data).

also due to the recent attention paid by the Commission to the GMOs contamination of organic products, which does not allow for an optimistic future for organic products in countries like Italy, where the public opinion's aversion to GM products is very deep-rooted and sensitive.

3. The guidelines on industrial processing

From different sources it has been widely recognised that the demand for organic agricultural products has grown at a huge rate in the last few years, and this trend is to be welcomed and encouraged. However, organic food products have traditionally reached the demand with a minimum of processing. Nowadays it seems that consumers would also like to see organic products available in processed form and, in principle, all or at least almost all food products should indeed be available in organic versions. This can however raise technical and organisational problems for processors as, for instance, only a few additives are allowed in organic products.

Processors will therefore have to develop new processing methods, where advanced logistical solutions would play an important role, in order to preserve the recognised texture, colour, preservation qualities, etc., of particular products. The information on these new processing and preservation methods owned by a single industry and may act as a technical barrier in the market.

This situation may lead to a sort of monopoly and, as a consequence, to an insufficient provision of organic processed food in the market.

Moreover, processing companies often deal with both non-organic and organic products, thus facing considerable costs in keeping the two processing areas separate, and in storing and transporting the different products. Finally, another problem might occur if the marketing strategies of the two lines were to be in competition with each other.

Currently the EC Regulation No 2092/91 and later integrations allow for the preparation of the products obtained from organic farming methods. However, there are no specific technical standards that regulate the sector, with the exception of the provisions of Annex III, part B and Annex VI, parts A, B and C .

Among the voluntary initiatives, some guidelines have been drawn up by IFOAM (Basic Standards) and various specifications have been drawn up for specific products by the inspection bodies and industry associations.

3.1. Regulatory guidelines

- The preparation of organic products, including preserving, processing, and packaging, must be kept separate from conventional products in both time and space.
- Organic products must be clearly identifiable during the whole processing phase.
- The preparation process of organic products must be risk-free in terms of environmental pollution.
- The processing technologies may be based on physical, chemical and micro-organic methods; the use of processing aids, ingredients and other products which may be used for processing is regulated by EC Regulation No 2092/91, annex VI.
- Product labelling must clearly and distinctly list all the ingredients of organic origin and those of conventional origin.

3.2. Voluntary guidelines

- The preparation of organic products must respect the nutritional and organoleptic characteristics of the ingredients (raw materials).
- The preparation of organic products must aim to reduce the consumption of energy and natural resources.
- The processing technologies, based on physical, chemical and micro-organic methods, must preferably limit the use of additives and other products which may be used for processing.
- All ingredients used in the processing of such foods should preferably be of organic origin.
- The environmental effects of the packaging used must be reduced as far as possible.

3.3. Process selection criteria

An organic agricultural product is a product that, by nature, is much more susceptible to fungal and bacterial disease following harvesting than a conventional one, and therefore the wrong choices made during the processing and/or preservation phase would inevitably lead to the decline in product quality.

The aspects that must be taken into consideration in selecting the most appropriate process or technique can be summarised as follows:

1. Oriented towards processes with low environmental impact.
2. Oriented towards processes that reduce the use of non-renewable natural resources (energy, fossil fuels, water) to a minimum.
3. Excluding processes that require the use of processing aids and other products which may be used for processing or in any case ensuring that their use is minimal (EC Reg No 2092/91, annex VI).
4. Avoiding contact between organic products and non-permitted substances.
5. Choosing contact materials that do not contaminate the product in any way.
6. Drawing up a plan for the control and prevention of health and hygiene risks (Italian Legislative Decree D.Lgs. 155/97).

The choice of preservation process must cover also the product packaging, bearing in mind the following (IFOAM Basic Standards of 30th April 1998):
1. Minimising the environmental effects of the packaging.
2. Eliminating any packaging material that is superfluous or not strictly necessary.
3. Using as far as possible recyclable packaging materials (or materials that can be reintegrated into the system) and which do not pollute the environment.
4. The packaging materials must not contaminate the food (preferably use glass, PET, non-toxic plastic materials, tetra pack, aluminates).
5. Adopt a programme for the reduction of the environmental effects of packaging materials.

3.4. Animal production

Animal production and processing are also regulated by the provisions of the EC Regulation No 2092/91 as modified by EC Regulation No 1804/99, and Italian Ministerial Decrees of 4th August 2000 and of 29th March 2001.
- Transport: During transportation to the slaughterhouse and during slaughter the animals must be handled in such a way as to reduce stress to a minimum.
- Slaughter: Slaughter must only be carried out following stunning by the methods permitted by the applicable national regulations.
- Identification and Traceability of Animal Products: The identification of animals and animal products must be guaranteed throughout the production cycle, as well as the preparation, transportation and sale of the products. Only in this way, the necessary 'traceability' of the product sold on the market can be guaranteed. The guidelines on organic animal product traceability are included in annex II of the Italian Ministerial Decree of 4th August 2000.
- Slaughtering Poultry: setting a minimum age for slaughtering poultry.

4. Certifing bodies and certification costs in Italy

In order to evaluate the certification costs borne by units involved in the preparation (processing, packaging, labelling, sale, import, etc.) of organic products, the tariffs of the inspection and certification bodies operating in Italy were analysed.

At the end of 2004, there are 16 Italian private bodies authorised to carry out the inspections on the national territory of agricultural production activities, as well as the preparation and import of products obtained using organic farming methods[3]. In addition to the national inspection bodies, there are 4 foreign bodies authorised to operate in the Autonomous Province of Bolzano alone.

[3] See the Italian Legislative Decree D.Lgs. No 220 of 17th March 1995, which implements articles 8 and 9 of the EC regulation No 2092/91 concerning organic farming and agrifood processing.

All the bodies analysed adopt annual tariffs for the companies, which vary according to the company turnover certified as organic in the previous year.

In some cases the tariff is calculated as a fixed percentage (varying from 0.3% to 0.5%) of the turnover for organic products; in other cases the percentage is calculated 'in brackets' (with varying percentages from 1% to 0.4% for the first brackets, from 0.5% to 0.05% for higher turnovers). In most cases, minimum and maximum tariffs are established that bind the range of variability of the amounts due. The lower threshold, excluding VAT, goes from a minimum of €220 to a maximum of €1,030, while the maximum tariff goes from a minimum of €7,750 up to €103,291. In the latter case, the great variability that exists among the different bodies is very clear.

One body applies an annual mixed tariff, obtained by adding a fixed tariff to the variable part, which is calculated as a percentage of the turnover. In practice, all the companies involved, whatever their activity (farming, processing, distribution, etc.) are obliged to pay a fixed annual amount (fixed tariff) of €180. In this way, the fixed part of the tariff partially covers the fixed costs of the certification body for the implementation and updating of the system, while the variable part is proportional to the control and certification services provided by the body and the production size of the operator subject to the control and, consequently, to the direct control costs.

All the certification bodies also take into account the type of company subject to control. Generally the processing activities carried out directly by the producing farm pays a lower tariff, as the process control is simpler. In only one case a body applies a higher tariff to producers/processors, who are obliged to pay not only the amount due as farmers calculated according to the type of crop and the certified farmland area, but also the full tariff applied to processors. All the other bodies offer reduced tariffs for producer/processors, through the reduction of the tariff applied to processors (generally 50% less), or through a reduction of the minimum and maximum tariff thresholds.

In one case an entry tariff is required, while generally regional adjustments to the tariffs are foreseen due to the existing specific territorial conditions. Furthermore, in almost all tariffs analysed there is a possibility to reduce the set rates in the event of objective circumstances that reduce the control costs (e.g. the existence of an internal quality control/management system that simplifies the management of the body's tasks, the existence of other certification schemes initiated by the body, favourable conditions applied to producer associations or cooperatives, consortia, etc.).

A further element discriminating the various tariffs is the chemical and biological analysis costs. In half of the cases analysed, these are considered separately from the proposed tariffs and are totally overcharged to the company subject to the control; in the other cases, the analyses that are part of the standard annual surveillance programme are included in the tariffs, while any analyses carried out to verify any extraordinary events (e.g. reinforced control programmes due to non-conformities) are for the company's account. These differences do not however appear to have any repercussions on the proposed tariff levels.

All the bodies set specific tariffs for companies that carry out processing activities on behalf of third parties; in two cases a fixed payment is required (varying from € 103 to € 775) while in one case a minimum tariff is considered, fifty percent of that applied to processors.

To summarise, the great variability of the tariffs applied by the certification bodies to the organic product processing companies makes it particularly difficult to estimate the costs borne by a single company during the certification process. Indeed, the tariffs analysed not only vary in their minimum and maximum references but also, and above all, in their division into fixed and variable parts, in the specific clauses and concessions that differentiate one tariff from another and, last but not least, in the possibility to vary the tariffs applied by the same body in different regions. Figure 2 shows three graphs that represent three examples of tariff application: variable tariff with a lower limit and no upper limit, variable tariff with lower and upper limit, mixed tariff (variable + fixed) with a minimum and maximum limit.

5. The results of the questionnaire

In May 2004, all the firms exhibiting organic products at the CIBUS Fair (Parma, Italy) were contacted by an expert data collector. In total, 110 firms where contacted. Among these, 86 firms agreed to answer to the first questionnaire, either directly or indirectly by returning it by mail.

The companies answering the questionnaire belong to a wide range of sectors within the industry (bread, coffee grinding, frozen foods, fruit and vegetable packaging, from pasta to olive oil, and from milk to fresh cheese). Considering the size of the supply in this niche, the companies answering are certainly representative of the sector, both in terms of the size of overall turnover and the involvement of the most representative companies in some markets and their relations with large scale distribution.

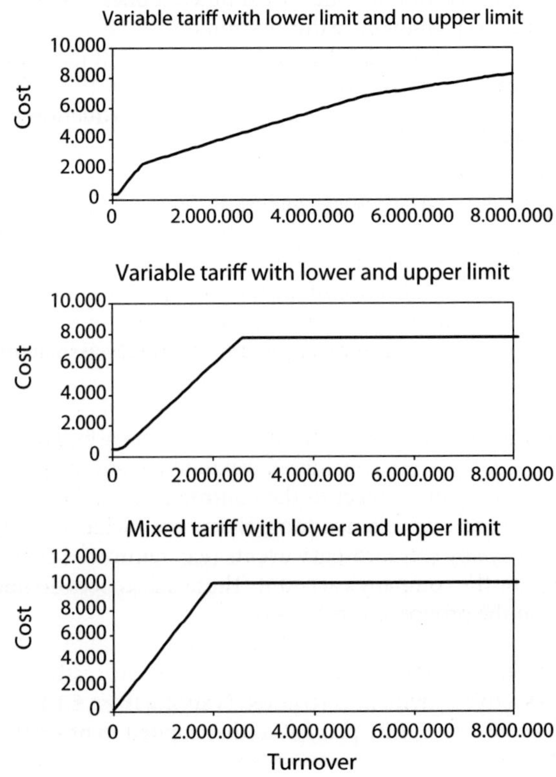

Figure 2. Examples of tariff application (adapted from Inspection Bodies data).

The questionnaire was divided into three parts, with a total of 25 questions, aiming to develop the following issues:

- market;
- technology;
- economics;
- organisation and distribution;
- marketing and information flows;
- policy and normative compliance.

Finally, the companies where asked to fill in a form detailing their structure and activity.

Most of the firms (60%) entered the organic market after 1996 following the phase of rapid growth of the market, and the share of organic products in their total turnover is less than 25%; some large non-specialist industrial brands belong to this group. The others entered the market before the introduction of the EC aid, and are on average smaller with a higher percentage of organic products on the total turnover.

Almost 70% of the examined firms offer both conventional and organic products, and the weight of the latter does not exceed 25% of the gross sales. The majority of these firms started this business for strategic reasons; the others, equally, also for economical and ethical reasons. 80% of the companies supply their products with their own company brand, and most of these also use the EC logo. One fourth sells for private labels, and another 25% sells to other industrial companies without using their own company brand.

5.1. Market issues

This section aimed to verify some defining aspects of organic products. The first two questions aimed to identify the attributes that characterise organic products in the opinion of the companies, and which on the other hand are the attributes perceived by the consumer (Table 4).

Industrial producers feel that organic products respect the environment more than conventional ones, and are therefore healthier and safer also for the consumer. In their opinion, the consumer perceives more the personal aspects of the products and the advantages offered by organic products for the health. This is therefore a correct perception for the Italian market, and an honest attitude by the producers. Animal welfare is not high up on the list, being considered exclusively by those producers with a direct knowledge of the animal production regulations through their milk processing activities.

Table 4. Attributes of organic products compared to conventional ones.

For the producer	For the consumer (in the producer's opinion)
more respect for the environment	healthier
healthier	safer
safer	more respect for the environment
tastier	tastier
more respect for animal welfare	more ethical
more ethical	more respect for animal welfare
more nourishing	more nourishing

Moving on to a question linked to this issue (despite being placed in the next section), concerning organic packaging, most companies feel that also the packaging must have a low environmental impact, even though the consumers do not pay close attention to this, as they are more concerned with the content of the product itself.

A later question aimed to verify market prospects for organic products and integrated farming products, which in the opinion of all the producers are perceived in the same way as organic products.

The market with the greatest potential seems to be that of fresh organic products (fruit and vegetables, milk, meat and eggs); then at the same level processed organic products and typical organic products are considered. This response does not however reflect the actual market condition. Considering however that there are in fact very few producers offering typical organic products (both in the market and the sample of companies interviewed), this may mean that from a first analysis this market may be of interest, although in our opinion there may be a number of technological problems, overlapping of controls and problems linked to cannibalisation between conventional and organic typical products.

The market for fair trade organic products presents excellent prospects, as these enjoy great favour with the consumer. These types of products have very strong ethical characteristics, but do not represent potential competition as in most cases the national and foreign producers are positioned in different markets and at different levels of processing. Generally, however, Italian producers pay much greater attention to the research into raw materials of national origin, as the Italian consumers generally pay much more attention to the origin of the products they consume.

With regard to the market growth prospects, the majority of the companies believe that the size of the market would increase particularly because of the consumers' perception that foods labelled as organic are healthier than conventional ones. This answer must be linked to those given under the 'economic issues' heading.

5.2. Economic issues

In this section only two questions were asked. The first concerns the levers that will lead to the growth in the organic market, and the second aims to verify the opinion of the producers on the origin of the differences in cost between organic and conventional products (Table 5).

All the companies stated that they agreed that both levers are important for the growth of the market, but hope for a growth in the market due to the increase in the consumers' ability to pay higher prices, as they view a possible reduction in prices as very dangerous. However, as can be seen from question 18, 70% of the companies identified the effective level of production costs as the cause of this difference, with the intermediary margins following as an explanation of the price difference (Table 6).

Table 5. Question 17. The growth of the organic market will be mostly linked to:

the consumer's ability to pay
a reduction in prices

Table 6. Question 18. Which factors are mostly responsible for the difference in the final price (multiple answer).

Factor	Percentage of answers
Effective level of production costs	60
Intermediary margins	26
Existence of a low turnover	21
Distribution inefficiencies	14
other: certification and inspection costs, specialist labour costs, no economies of scale	

The issues perceived to be mostly connected to the growth prospects of the companies are the need to provide more information to the consumers, increasing their willingness to pay for organic food, and the need to improve coordination throughout the supply chain by reducing transaction costs.

From the analysis of the question concerning distribution and organisation issues, it is clear that all the companies, with only a few exceptions, are convinced that new synergies along the supply chain could improve the supply conditions of organic goods, both in terms of product quality and the composition of the final price. The companies feel very responsible over the issue of product quality. 70% in fact believe that the consumers rely more on their own brand than on the mark of the certification body, which is indicated on the package by law.

This statement is however partially contradicted by the consumers, who appear only to trust very strong industrial brands; otherwise the certification body mark seems to guarantee greater trust. Linked to this issue is the importance of company communication to make their products known, even though the budgets made available for advertising are very limited.

5.3. Technological issues

To this regard, the operators admit the existence of entry barriers, linked mainly to issues concerning technology and information. The analysis of the economic aspects shows that the specific problems facing a company processing organic foods are the lack of information on the market structure for organic raw materials and their quality, and the lack of trust in a product certification system designed along such lines (Table 7).

Table 7. Entry barriers in order of importance.

Lack of information on raw material producers
Lack of information on raw material characteristics and prices
Lack of information on techniques
Lack of information on regulatory issues
Other: lack of information on competitors
Technological barriers

5.4. Regulatory issues

The final question dealt with public policies the companies would like to see implemented. The first action that all companies wished for is consumer information and education. Following this, a strong demand emerged for the standardisation of certification, inspection and control procedures, as well as production techniques and regulations, at both national and EU level, and above all today the strengthening of the inspection and control system, particularly concerning GM products.

Another key issue perceived by the companies and linked to this is the regulation concerning the existence of different levels of 'organicness' of processed products. 75% of the companies interviewed in fact feel that this generates confusion, even though it is more expensive to supply a 100% organic product, particularly when considering the problems of procuring the raw materials. Small specialist companies also asked for economic support along the whole supply chain, or at least assistance in improving market coordination.

5.5. Cost and benefits of organic choice

The second questionnaire sent by post in January 2005 will be based on some answers given to questions included in the first one. All the companies answering the first questionnaire were contacted (86) by post. Among these, up to now only 15 firms have so far agreed to answer to the questionnaire and have returned it by post.

The questionnaire was divided into several parts, aiming to develop the following issues:

- reasons for the difference in final price between conventional products and organic products;
- initial investments;
- management costs;
- general benefits of organic products over conventional ones;
- cost advantages over competitors.

Finally, the companies were asked to fill in a form to describe their structure and activity. All the companies answering that they have to pay for Initial investments. An example for a medium sized enterprise is proposed in Table 8.

Table 8. Investment cost (size of the enterprise: 11-50 mln. Euro, turnover 2003).

Investment costs	Percentage of total investment costs
Labour	3%
Production plants	5%
Storage	72%
Transport	10%
Information technology	4%
Consultancy	6%

The reasons that keep the prices of organic products high are linked to the higher level of raw materials and other production and logistic costs, the costs of the application of regulated standards (production specifications), certification procedures (compulsory inspection schemes) and a specific labelling scheme, the low rate of turnover, a lack of supply chain organisation, the often 'exaggerated' costs of monitoring the raw materials for both certification bodies and companies themselves (Table 9).

Moreover, the coexistence with GM products and conventional agriculture and the risk of contamination of organic products represent a significant cost that the food industries have to bear (2-5% of management cost in average).

5.6. Prices for organic products

The cost of organic food is on average 30% higher than conventional products, for the processor interviewed, while for the consumer (according to processors answers) this increases to 40-50%. This means that the margin is double that of conventional products. According to the FiBL (2004) premium prices for organic products are still relevant, especially at the retail level for processed imported goods (milk) but because of the small-scale processing plants and inefficiencies in distribution channels, processed food often has very large premium prices.

5.7. Benefits related to market entry

The economic performance of the companies will be investigated in order to quantify the benefits related to market entry. Among the benefits the companies have indicated, we can list the following in Table 10.

Finally, the examined companies have been asked to evaluate their competitive advantage with respect to their competitors (Table 11). Among them, the companies indicated mainly the speed of response to the needs of the market and the qualitative characteristics of their products.

Table 9. Management cost (size of the enterprise: 11-50 mln. Euro, turnover 2003).

Management costs	Percentage of total managements costs
Raw material packaging	40%
Certification	1%
Plant Care	5%
Traceability	4%
Inspection on raw materials	4%
Storage cost	20%
Inspection on final products	3%
HACCP	3%
Inefficiencies	5%
Promotion	10%

Table 10. Benefits.

Increased profit
Strengthening of company brand
Higher product quality
Differentiation with respect to the competition
Relations with the banking system
Reduction of various risks (presence of toxic substances, GM contamination, other)
Acquisition of new markets (national, foreign)
Greater management efficiency, also due to traceability
Internal efficiency (reduction in raw material assessment costs, processing, processing times, greater product quality)
External efficiency (reduction in recalls, reduction in shipping errors, quicker identification of responsibilities)
Increase in product quality

Table 11. In your opinion, do you have cost advantages over your competitors?

Economies of scale (from production to advertising)
Learning
Internal vertical links
Integrations in the supply chain
Differentiation
Speed of response to the needs of the market
Various discretional policies:
Quality characteristics of the product
Quality of service
Investments in quality (research and development, technologies, human resources)
Delivery times
Quality of raw materials
Minimum quality index
Geographic location
Institutional factors
Other?

5.8. Article 18 of the EU Regulation No 178/2002

The costs associated with traceability in compliance with the provisions of EC Regulation No 2092/91, which have in any case required great initial effort, have also assured company compliance with the instructions of the more recent EC Regulation No 178/02, with some minor comments.

For organic farmers and operators, the introduction of the new standard changes very little, as those involved in organic farming were the first to be traced along the whole supply chain: the system in fact controls the whole process, from the seed selected (which must be organic, or approved by ENSE, the Italian national seed selection body, if the seed is not present or not available on the market), to the processed product found on the shelves of the large scale retailer (also passing through private labels). Organic operators have adopted a whole range of obligatory records, including ministerial forms that must obligatorily describe all the processes carried out and the raw materials used (stock book), as well as the receipts used in the case of those companies processing raw materials into packaged products. Organic processing industries have in turn always been able to precisely document where a given product was stored, in which silo, etc.

6. Conclusions

The objective of the research is to evaluate the trend of organic processed food, focusing on the managerial and economic problems that processing industries had to face in this sector, compared to their potential and perceived benefits.

The difficulties inherent in this activity are basically linked to the structure of the market, which is extremely fragmented and where the absence of links between the various phases of the supply chain causes serious problems concerning product and information management.

In despite of the great potential of the product, in Italy organic market have suffered from a lack of attention from industrial operators and institutions, leaving the companies at the mercy of, first of all, community aid and, later, the large retailer's power. The issues perceived to be mostly connected to the growth prospects of the companies are:

- the need to provide more information to the consumers, increasing their willingness to pay for organic food;
- the need to improve coordination throughout the supply chain by reducing transaction costs;
- the need to redistribute the added value throughout the supply chain.

The result is a market in which high prices go hand in hand with a chronic lack of information. In these unhappy economic times, furthermore, the continuing lack of information and excessively high production and distribution costs and prices cannot but cause further problems and crises in this market. Even before we start to tackle the GM issue.

References

Duchateau, K., 2003. Statistics in focus: organic farming in Europe. Theme 8-2/2003, 7.

European Commission, 1999. Agriculture, environment, rural development: facts and figures - challenge for agriculture. Brussels, 262.

European Commission, 2004. European action plan for organic food and farming, Commission Staff Working Document Brussels, 10 June 2004, SEC(2004) 739, Annex to the Communication from the Commission {COM(2004)415 final}.

FiBL, 2004. Research Institute of Organic Agriculture, Country Reports, Italy, 2004 (www.organic-europe.net)

FiBL, 2004. Research Institute of Organic Agriculture, Organic farming in Europe-Statistics per 31.12.2003, 2004 (www.organic-europe.net).

Hansen, J., 2001. Statistics in focus: organic farming in Europe, Theme 2001, 8.

Mastromauro, A., 2004. Conference on organic food and farming. Brussels, 22 january 2004.

Ministero delle Politiche Agroalimentari: www.politicheagricole.it/produzione/agribio/ Italia%202003.pdf.

The effects of certification costs on the success of a PDO/PGI

Giovanni Belletti, Tunia Burgassi, Andrea Marescotti and Silvia Scaramuzzi
Dipartimento di Scienze Economiche, Università di Firenze, Italy; tunia.burgassi@unifi.it

Abstract

To protect and valorise quality agrifood products the European Union provides two quality signs linked to the area of origin of the product: the Protected Denomination of Origin (PDO) and the Protected Geographical Indication (PGI). Through a case study analysis on three Tuscan products the paper shows how PDO/PGI certification costs influence the choice of the actors in using a PDO or a PGI. The paper aims at categorising and qualifying some positive and negative effects of using a PDO/PGI in order to draw a 'cost-benefit balance', so as to evaluate the opportunity of valorisation of an agrifood product through these quality signs.

Keywords: food quality, typical products, PDO/PGI, certification costs, marketing strategies

1. Introduction

In recent times consumers are paying a growing attention to the quality of agrifood products, due to both the greater concern about food safety related aspects, and the need of re-discovering the true values of agriculture and rurality. Agro-food typical products are an important example of these tendencies, as their specific qualities are strictly tied to the territory they come from, including both natural specificities (i.c. animal breed, climate), and human ones (i.e. the knowledge and skills of local producers).

The success of the market of typical products is due to their suitability in responding to consumers needs in terms of genuineness, reply to food mass production and rediscovery of old cultural traditions; this success has pointed out the usefulness of agrifood products territorial origin highlighted in the label to become a strategic tool for differentiation in the agrifood sector. Reg. EEC 2081/92 (now: 510/2006) supported consumers' needs, and provided a strategic tool through the creation of the Protected Designation of Origin (PDO) and Protected Geographical Indication (PGI), two quality signs aiming at pointing out the link between the quality characteristics of an agrifood product and its territorial origin.

PDOs and PGIs are considered an important protection and promotion tool for producers in order to increase their added value and market power, and may represent a means to enhance rural economy and rural development, remunerate locked-in resources, obtain spill-over effects on the local economy and activate or re-enforce rural development.

However, notwithstanding the growing attention and interest on these important 'official' tools, up to now there is no much evidence on costs and benefits deriving to supply chain firms from the use of these quality schemes and more in general on the effects of their application on rural economy and development.

2. Aims and methodology

The theme of costs and benefits coming from the use of PDO/PGI and other certification schemes has already been analysed by some studies, focusing on diverse aspects. Among most recent studies, Agri2000 (2002) and Lazzarin (2004) aim at quantifying the costs to obtain and keep a PDO/PGI operative, Vuystelke *et al.* (2003) highlight the problem of the exclusion of farmers as a consequence of the adoption of quality certification and standardisation schemes; a costs and benefits analysis is carried out by Verhaegen and Van Huylenbroeck (2001), with special reference to the implications of participating in innovative marketing channels.

Within the described general framework, this paper aims at identifying, categorising and qualifying costs and benefits faced by firms when joining a PDO or PGI certification scheme. After giving a description of the application of Reg. EEC 2081/92 in Italy (par.3), our analysis deals with three significant case studies of PDO/PGI products in Tuscany - Italy (par.4). The case study analysis starts from the evaluation of direct certification costs (costs born by the actors of the PDO/PGI supply chain for product certification body services), then focusing on how these costs are shared among the actors of the chain (par.5). The data have been drawn from the rate books of product certification bodies, and from direct surveys to firms involved in the supply chains, to product certification bodies and to representative producers associations and consortia.

Through the evaluation of expected benefits coming from the European Union protection and of other costs related to the PDO/PGI, the paper draws a first cost-benefit balance, which identifies the factors the actors of supply chains have to evaluate in order to join a PDO/PGI scheme (par.6). A comparative analysis on similar products that are marketed with and without a PDO/PGI denomination (where the preliminary, and collective, choice 'to apply for PDO/PGI or not to apply for PDO/PGI' has already been made) has been carried out. The analysis of these aspects, especially regarding to production methods, commercial channels and socio - economic context of the case studies, has been supported by literature and available data deriving from previous research of the authors (Belletti, 2000; 2001; Pacciani *et al.*, 2003; Marescotti, 2000; Marescotti *et al.*, 2004).

3. EC Regulation No 2081/92 and the implementation in Italy

The aim of EC Regulation No 2081/92 is to protect geographical indications and designations of origin for agricultural products and foodstuffs; as the European Commission Directorate has specified: 'The objective of EC Regulation No 2081/92 is the protection of names [...]. Benefits for the producer and the consumer, such as the possibility to individualise products, the protection of consumers from the misuse of names, increased diversification of agricultural output and the development of rural areas, are direct consequences of the regulation' (European Commission, Directorate General for Agriculture, 2004).

Therefore the first objective of EC Regulation No 2081/92 is to protect the name of the agricultural product, and the second one is to develop the context (both social and economic) where the protected product is produced: in this sense the PDO (or PGI) wears the dress of a valorisation tool. From another point of view, the indirect aim of EC Regulation No 2081/92 is to protect the consumer against frauds and, as a product certification, the PDO/PGI may be considered as a quality sign.

To register a product as a PDO or a PGI, EC Regulation No 2081/92 establishes two basic features:

- The PDO/PGI product must 'differentiate' itself with respect to other products thanks to its link with its geographical origin.
- There must be a Code of Rules, where the production area, the characteristics of the PDO/PGI product and of the production process are defined and the designation of an independent Inspection body; anyone who wants to join the PDO or PGI supply chain has to comply with these rules.

EC Regulation No 2081/92 only gives general guidelines for the PDO/PGI systems and gives flexibility to the EU Member States in its application according to their national legislation. This flexibility determines a strong heterogeneity in the PDO/PGI system and consequently prevents the implementation in the EU of an integrated public policy based on sustainability, quality and origin, where PDOs and PGIs could play a major role (Sylvander *et al.*, 2004).

The Code of Rules defines the requisites that the typical product must meet to bear the denomination of origin. The construction of the Code of Rules is left to the negotiation of the actors of the supply chain aiming at enhancing (or preserving) the quality of their product or at protecting the production methods from mass production or imitation, though it is elaborated under the supervision of regional and national administrations and has to be approved by EU institutions in order to guarantee a coherence of the Code of Rules with the 'original' typical product. Nonetheless the Code of Rules can reflect the strategy of the actors of the supply chain with respect to process and product characteristics, hence to the quality of the typical product. This is a very delicate phase of the process of registering a product as a PDO/PGI, for the Code of Rules binds to its prescriptions every subject who takes part to the production process of the PDO/PGI product.

The definition of a Code of Rules needs specific competences and negotiations between the actors of the supply chain and often the intervention of public institutions. Consequently the Code of Rules is the result of a compromise among the actors of the supply chain that must also balance the needs of protection of the product quality with the costs deriving from controls and from regulating the production process with stricter rules (Pacciani *et al.*, 2003; Tregear *et al.*, 2004).

To implement the PDO/PGI system a Control Plan is needed. It is an operational document written by the Inspection body where the whole control system for the respect of the Code of Rules is established. The Control Plan may call for two types of checks: a documentary and inspection verification to prove that the techniques and processes meet the requirements set forth in the regulations, monitoring that each critical activity and phase in the production chain meets the requirements; an analytical verification to demonstrate that the product conforms to the parameters (that can be chemical, physical and sensory) set forth in the regulations.

The structure of the Control Plan strongly depends on the Code of Rules. Even this phase is very delicate because the amount and the distribution of certification costs directly depend on what is established in the Control Plan. The role of the 'quality' of the Control Plan (and in general of the control systems) is widely analysed in literature, see Anania e Nisticò (2003) for an application to the PDO-PGI system.

Our analysis deals with the application of EC Regulation No 2081/92 within the Italian legal (verificare termine inglese) framework, which establishes that in order to have a valid certification it is necessary for the product certification body to be independent, credible and capable of operating competently and responsibly; to work according to a clearly defined, systematic and documented certification procedure; to meet the organisational and operational requirements as set forth in the UNI CEI EN 45011 international standards; to write a 'control plan' that describes in detail the procedure for verifying that the product complies with the regulations. The product certification body can have a public or private nature; in case it is private, the normal situation in Italy, it has to be authorised by the Ministry of Agriculture, which approves also the Control Plan to give guarantee to the consumers.

Another important element of the Italian legal framework, as far as costs are concerned, is that the control costs paid to the certification body must be born by the firms of the PDO-PGI supply chain.

In most cases, in Italy the firms involved in the supply chain of the protected product (agricultural firms, processors, etc.) are represented by a consortium, which is a particular kind of association that can also be officially recognised as representative of the whole category of producers (or processors) if its members produce (or process) at least 2/3 of the production of the category (see Italian D.M. n. 61413 04/12/2000, published in G.U.R.I. n. 9 01/12/2001).

Consortia have been created in reply to the need of the large number of small producers and processors of typical agro-food products for a stronger contractual power and for higher coordination. In other words consortia were born as a possible solution to the fragmentation of the supply chain of most of typical agro-food production systems in Italy. Along years consortia have become more numerous and more important and, in year 2002, the Italian Ministry of Agriculture decided to give them an official 'dress' of representative body of producers and processors.

Until the approval of EC Regulation No 2081/92 consortia were the 'guardians' of the denomination of origin and they directly carried out controls and all other operations connected to the management and control of the denomination. Today consortia are no longer allowed to control the production; nevertheless their role is still fundamental for their members:

- for technical assistance, including the preparation and compilation of the documents for the certification;
- for the negotiating power they have with the Control bodies in order to establish certification fares for the members;
- for their role in collective marketing (and collective investment); often consortia have their own hallmark (which contains the registered denomination of origin) that only their members can use.

Joining consortia is absolutely voluntary for the firms of the supply chain, who in principle are free to certify their products as a PDO or PGI directly with the certification body, without going through consortia.

In the following paragraphs we will analyse three case studies where we highlight the level and the nature of direct certification costs of the PDO or PGI products and draw a qualitative balance between the direct certification and other costs and the possible benefits deriving from the use of the certification.

4. The case study analysis

4.1. The Olio Toscano PGI

The Olio Toscano (Tuscan extra-virgin olive oil) is a product which enjoys great, long-standing 'renown' in Italy and all around the world. The supply of the product is very fragmented, managed by both professional and non-professional agents (olive growers, olive mills, olive pickers, merchants, both small and industrialised mixing and bottling firms), and directed toward heterogeneous marketing channels (from direct sale by the olive growers to supermarket chains); many farmers directly control all the phases of the production process, up to the sale to the final consumer. The Olio Toscano PGI was obtained in 1998 mainly for facing unfair uses of the name 'Tuscan' made also by some large firms.

The Code of practice established the production area in the whole Tuscan region (all the phases of the production process should be made within the boundaries of Tuscany), the allowed olive-tree varieties (the traditional ones), some process specifications that are very important for the quality of the olive-oil (as the picking systems, the time between picking and milling, the maximum yield in olive per tree and in oil), and the sensory and chemical-physical characteristics of the oil to be marketed (at the bottling phase).

The Control Plan sets out a complete traceability of the oil (up to the origin of the olive fruits) and the duties of the firms: some of them are 'una tantum' (olive-tree varieties), others are annual, some of them are based on documentary proofs (with a direct control on a sampling basis) whereas the product characteristics have to be directly verified for each bottling lot by authorised expert panels and chemical laboratories.

The fragmentation of the cultivation and the specificity of marketing channels imply many problems in the agricultural phase. To give an overview on the production level and structure, in 2002/03 the quantity of oil certified as Tuscan PGI was 2,500 tons, produced by 9,900 olive growers, 244 olive-oil mills and 258 bottlers; the certified bottling lots were 358, with a mean of 70 q. per lot but with a very small number of big lots and a lot of small and medium sized lots.

This structural situation implies that all the firms using the PGI ask for the services of the Consortium (Consorzio di tutela dell'Olio di Oliva toscano). The latter supports the process of documentation and traceability of the product and puts at Certification body disposal the results of its activity, thus limiting the direct involvement of the Certification body and hence the associated costs. For this reason the Consortium has built up and manages a website, where each firm of the PGI supply chain has to insert its documents and certifications, that the Certification body can examine.

The Consortium negotiates the certification tariffs with the Certification body; in addition it makes marketing initiatives for the promotion of Tuscan oil around the world. To fulfil this aim the Consortium has a private hallmark that signals the origin of the product under the control of the product certification body.

The activities of the Certification body (inspection and certification) and of the Consortium (support activities) are very strongly bounded, so it's very difficult to separate the costs related to the Consortium services from the Certification body ones. Even if it is possible, up to now no firm has addressed itself directly to the Certification body, and there is no evidence of the tariffs that would be applied in this case (probably they would be much higher than those in column 2 of Table 1). This indicates that the Consortium, as an intermediate organisation, economises some functioning costs of the PDO-PGI system.

The structure of Tuscan Olive Oil direct certification costs (see Table 1) is strongly influenced also by the nature of controls, and in particular by the costs of organoleptic and physical-chemical analyses, that are fixed for each bottling lot of oil. These costs have to be born only by the firms

Table 1. The Olio Toscano PGI: direct certification costs and Consortium costs (Consorzio di tutela dell'Olio di Oliva Toscano).

	PGI certification costs	Consortium costs
Agricultural phase		
Fixed share (per capita)	€ 15.00 una tantum 1st year	€15.00/year +€11,00 una tantum 1st year
Proportional share	-	-
Milling phase		
Fixed share (per capita)	€ 15.00 una tantum 1st year	€15.00/year + €11.00 una tantum 1st year
Proportional share	-	-
Bottling phase		
Fixed share (per bottling lot)	€ 309.87	-
Proportional share (lots ≥ 800 kg)	€0.38 /Kg	

that bottle the oil (olive growers, olive mills, or specialised professional bottlers), which have to pay a minimum fare of €309.87 (+ VAT) per lot: this amount includes all the analyses, control and bureaucratic costs, and the net quota due to the Certification body.

For the lots exceeding 800 kg the bottling firms pay an amount per bottle (variable on the basis of the bottle capacity, i.e. €0.34 for 1 l and €0.26 for 0.75 l); this amount is divided between the Certification body (for inspection activities) and the Consortium (for provision services and marketing activities). All the costs are directly paid by the firms that bottle the oil (farmer or miller that sell the oil, or specialised, and in some cases very large, bottling firms).

As this cost structure penalised small 'bottlers' and in particular farmers in the use of the PGI, during 2002 - 2003 the Consortium decided to cut down both the minimum fee paid by the bottlers (from €309 to €100) and the exceeding weight of the lot for which the producer has to pay a proportional share (lots from 800 kg down to 300 kg): this change allows also (very) small bottlers (especially small farms) to join the PGI scheme.

Table 2 gives an example of the effects of the 'discounted' fares for some amounts of bottled oil; as we can see, the difference between the cost per kilo with and without the Consortium 'discount' is inversely related to the amount of production: the smallest the amount of PGI certified oil, the highest the discount. The rest of the minimum fee due to the Inspection body is covered by the Consortium itself thanks to the proportional part of the tariff.

This means that there's a clear distribution policy aiming at easing the burden of the certification costs to the small producers, which in this way can join the PGI supply chain. This policy is very important also with respect to the price of the certified product, in fact the prices for the Tuscan PGI lots on the gross market normally vary between 600 and 800 Euro per 100 kg. The price for the bottled product is very heterogeneous, but for the 'standard' Tuscan oil normally varies on Italian market between 5 and 10 Euros per bottle of 0.75 l.

Table 2. The Olio Toscano PGI: trend of PGI certification costs (in Euros, bottle capacity 0.75L) (Adapted from Consorzio Olio Toscano data).

Bottling lot (kg)	200	300	400	800	1,000
Per lot					
Without 'discount'	309	309	309	309	378
With 'discount'	100	113	151	309	378
Difference	209	196	158	0	0
Per kg					
Without 'discount'	1.55	1.03	0.77	0.38	0.38
With 'discount'	0.50	0.38	0.38	0.38	0.38
Difference	1.05	0.65	0.39	0.00	0.00

4.2. The 'Vitellone Bianco dell'Appennino Centrale - Chianina' PGI

The Vitellone Bianco dell'Appennino Centrale - Chianina PGI beef is a typical product made of a local breed reared mainly in Tuscany and Umbria in the Chiana Valley (Valdichiana), fed with natural forage and slaughtered in the PGI area (even if the Chianina beef is a PGI, the entire production process must be carried out in the PGI area, which is very wide and includes the whole Umbria and part of Tuscany, Abruzzo, Marche, Campania, Lazio and Emilia - Romagna regions). The Chianina cattle is mainly bred by small farms through closed - loop breeding (Van der Meulen and Ventura, 1994; Marescotti, 2000): there are few firms specialised in fatting calves.

The Control Plan of the Chianina PGI beef divides the production process in breeding, slaughtering and carcass sectioning, plus the marketing phase, and according to the rate books of the product certification body the certification costs are not proportional to the weight of the calf but to the number of animals controlled or carcasses marked, with different costs for breeders, slaughterers and sectioners. The distribution of the certification costs among the Chianina PGI supply chain depends on the services asked to the product certification body, regardless of the phase (Table 3, column 2); in particular, the direct intervention of the product certification body is requested in the breeding phase, which is controlled for the importance of the pureblood and the feeding of the cattle for the quality of the Chianina PGI beef, and in the sectioning phase where the PGI mark is burn in the 18 sections of each carcass, considered a 'critical point' for guaranteeing the origin of the beef.

The control of the Chianina Beef is carried out at documentary level in order to verify the traceability of the product and at inspection level to demonstrate that the product conforms to the parameters established in the Code of Rules. Nevertheless there is not always a connection between the agents of the supply chain that are controlled and the imposition of fees, that may be paid by other agents, as applies to Chianina PGI. The slaughterhouses and the butchers joining the Chianina PGI supply chain have to provide specific documents for the traceability of the carcass and they do not pay any fee to the product certification body, even if their documents are controlled.

It is important to highlight that the PGI certification costs of the Chianina beef are born by those who request the PGI marking of the carcasses to the product certification body, so those breeders

Table 3. Chianina PGI: direct certification costs and consortium costs.

	PGI certification costs	Consortium costs
Breeding phase		
Proportional share	€1.30/head	€19.36/carcass
Fixed share	-	€25.00/year + €25.00 una tantum for inscription
Sectioning Phase		
Proportional share	€19.36/carcass	€ 1.30/carcass
Fixed share	-	-
Marketing phase		
Proportional share	-	-
Fixed share	-	€256.00 una tantum (advertising materials with the PGI logo)

who direct - sell the Chianina PGI beef of their animals have to bear all the certification costs of the supply chain: for controlling their animals alive and for marking the carcasses when they section them for selling.

Most of the breeders joining the Chianina PGI supply chain used to carry the animals to the slaughterhouses, which section them and give them back to the breeders, and even in this case the certification cost for controlling the sectioning phase is born by the breeders themselves: in this sense we can observe that often the whole certification cost is paid by the same actor, for the existence of these 'hybrid' situations, though this cost can be transferred on the other agents downstream the chain, eventually the consumers: we have to consider that a pure-blooded Chianina calf can be paid (if sold for fattening) about €1,600 while a PGI marked carcass can cost €2,000 in average, if sold to butchers.

As in the case of olive oil, behind the whole production process of the Chianina PGI beef there is Consortium, a breeders' Consortium, that is the one who applied for the PGI protection. The Consortium has multiple functions. It not only now manages the registration of the calves to the National Herd Genealogic Book of the Chianina breed, but plays an intermediation role between its members and other actors of the supply chain and local, national and European institutions. In addition it provides its members with many services as promotional activities, management of bureaucratic practices, technical assistance that are included, in terms of costs, in the fee paid to the Consortium itself by the actors of the chain. The Consortium, being representative of a wide number of members has a higher contracting power with the certification body than the individual; furthermore it eases inspection activities by holding every information on the breading joining the Chianina PGI supply chain that may be useful for the control activities. Through its contracting power and the detention and management of significant information the Consortium exerts a catalyst action on certification costs, though as in the case of Tuscan Olive Oil, as all breeders are members of the Consortium, there is no evidence of the tariffs that would be applied to the individual agents addressing themselves directly to the certification body without joining the Consortium.

4.3. The Pecorino Toscano PDO

The Pecorino Toscano is a typical sheep cheese produced in a wide geographical area which includes the whole Tuscany, part of Umbria and Lazio. It has a defined and structured production process which gives a significant mass of product marketed on structured commercial channels.

Before becoming a PDO the Pecorino Toscano was already protected with a national protection system of local production called Denomination of Origin (D.O.). Therefore this sheep cheese has for long been considered as a product needing a protection from misuses and frauds. For the existence of different typologies of pecorino cheese traditionally produced in Tuscany and hence sold as 'Tuscan pecorino cheese', a mark which could include and protect all these heterogeneous products was needed. So, after the D.O. protection, the PDO was obtained, with a not very prescriptive Code of Rules, in order to meet the requirements of all the different typologies of cheese that were produced.

The Control Plan divides the Pecorino Toscano PDO supply chain in four phases: the dairy production phase, the milk collecting phase, the processing phase and the seasoning phase. Every phase of the Pecorino Toscano PDO production process is controlled by the product certification body every year, and the total certification cost is distributed among the whole supply chain.

Also in Pecorino Toscano cheese some of the actors of the supply chain are associated to a Consortium. The Consortium activities concern technical assistance to the members, recording and management of the data related to the milk and dairy production and of the marketing phase of the Pecorino Toscano PDO supply chain, including promotional activities; the data collected are very useful to the product certification body, which contacts the Consortium for the PDO inspection.

In the case of Pecorino Toscano PDO the identification of direct certification costs has been more difficult as only some actors (sheep breeders) pay a fee for certification and a fee for the consortium services (column 2 + column 3 of Table 4), while other actors (dairy processors and seasoners) pay to the Consortium a single fee that includes direct certification costs and consortium services (column 3 of Table 4).

Milk producers (the sheep breeders) and milk collectors pay to the product certification body a yearly fee for its services, independently of the amount of production, while the dairy farms and the seasoners pay an annual fee plus a proportional fee (per form) (Table 4, column 2).

In addition the sheep breeders have to pay to the Consortium an annual fixed share plus an amount per litre of milk destined to become Pecorino Toscano PDO; while both the dairy farms and the seasoners must pay a fixed share per year plus a proportional share per kilogram of Pecorino cheese produced; remark that these payments include the fee to be paid to the product certification body, while milk collectors do not pay any fee to the Consortium, for they are not associated (Table 4, column 3).

As in the other cases analysed, the Pecorino Toscano Consortium plays a fundamental role as an intermediation structure between the agents of the chain and the product certification body. In this case the consortium plays this role even with non members, when collecting the fees that are paid by sheep breeders to the product certification body. In none of the cases the agents of the chain have a direct contact with the certification body.

Table 4. Pecorino Toscano PDO: certification costs and consortium costs per year.

	PDO certification costs	Consortium costs
Milk production phase		
Proportional share	-	€0.001/L
Fixed share	€21.00	€ 00.00
Milk collection phase		
Proportional share	-	-
Fixed share	€620.00	-
Dairy activity phase	(only if not associated)	(including certification costs)
Proportional share	€0.0085/form	€0.10/kg
Fixed share	€300.00	€2,000.00
Seasoning phase	(only if not associated)	(including certification costs)
Proportional share	€0.0085/form	€0.10/kg
Fixed share	€300.00	€1,000.00

In addition the Pecorino Toscano Consortium uses part of the fees collected by its members to give services in terms of documentation to the consortium, administrative support and promotion of the product. Only in this case the survey has made it possible to separate the fee that would be paid by non members (see Table 4 column 2) which is about from 15% to 30% of what is paid when also the Consortium fees are included.

For a complete evaluation of the costs in Table 4 we have to consider that the wholesale average price of the Pecorino Toscano PDO is around €7.50/kg for the 'fresh' typology and €9.00/kg for the seasoned one, but the price is very different depending on the channel, the geographical market considered and the firm marketing strategy.

5. The PDO/PGI direct certification costs

In the cases analysed the certification costs born by firms to sell their productions as PDO or PGI are an important evaluation variable in the decision of any firm whether to use this important valorisation tool. As we have previously discussed, the absolute and relative amount of direct certification costs depends on many elements, among which a major importance can be given on one side to the content of the Code of Rules and (consequently) of the Control plan and, on the other side, to the role of intermediate institutions (in the cases analysed by Consortia).

With regard to the effects of the characteristics of the Code of Rules and the Control Plan on the amount of costs, as we wrote in par.3, these costs depend on the decisions taken by firms mainly before redacting the Code and therefore before applying for the PDO/PGI. The amount of costs is a reply to the product positioning collective strategy, as a result of the negotiating process among the different aims of the actors (firms and institutions) involved (Pacciani *et al.*, 2003).

As concerns the role of intermediate institutions, we mainly refer to the phase after the approval of the PDO/PGI (this paper deals specifically with this phase - see introduction). In this case it is important to analyse the way in which certification costs can be remodulated through the action of these institutions who, in defence of the interests of producers, aim at lowering certification costs and at supporting the action of their member firms supplying them a variety of services.

The remodulation of certification costs can be implemented either vertically, that is among the different phases of the supply chain, or horizontally, that is within the same phase of the production process.

We saw that direct certification costs in some cases are *vertically* spread among the supply chain, in each phase, and in other cases are born by the actors of one or two of these phases: the vertical distribution of direct certification costs reflects the structure of the Code of Rules and of the actors' intentions behind it; in fact the Code of Rules contains the characteristics of the PDO/PGI product which distinguish it from other similar products. The phases of the production process where these characteristics are given to the products represent the 'critical points' needing controls and certification by the inspection body, in order to guarantee that these phases are carried out correctly, according to the Code of Rules; these considerations can explain for example why in the Chianina PGI beef only the breeding and the sectioning phase are directly controlled by the product certification body. Otherwise, as in the Olio Toscano PGI case, we often find that the whole certification cost is born by only one phase (the bottling one) of the supply chain; in this phase the specific analyses and the panel tests on the olive-oil are made, but 'de facto' bottlers pay also the other controls carried out in the other phases of the supply chain. In fact it is in the bottling phase that the larger added - value is created; in addition in this phase there is a small number of firms involved, and hence this allows keeping down functioning (and transaction) costs of certification by concentrating the most important controls but also concentrating the tariff withdrawal.

We can also split direct certification costs in costs proportional to the certified volume produced or sold and in non - variable costs (i.e. annual payment to the product certification body): this division directly influences the horizontal distribution of direct certification costs among small and large producers or processors or sellers and in this sense, influences the adhesion of the actors of small dimension to the PDO or PGI project.

The horizontal distribution - among different agents of the same phase - of certification costs can be strongly influenced by the presence of a Consortium and its strategy: in the Olio Toscano PGI case, the presence of the Consortium made it possible to negotiate a cheaper fixed fee for the analyses and inspections of the product certification body, making in this sense the choice of the PDO/PGI worthwhile even for the smallest actors of the supply chain.

Even the vertical distribution - among some agents of different phases - of certification costs can be influenced by the presence of a Consortium: as we can see in the columns related to the Consortium costs for the three case studies, through the annual or the proportional quota asked to their members, the Consortia can spread the burden of the costs related to certification activity among the actors of the PDO/PGI supply chain: for example the total effective costs (including the Consortium costs) born by the Chianina PGI breeders and sectioners per animal controlled or carcass marked are the same (€20.66, see Table 3).

The analysis carried out has made it possible to point out how the level of certification costs is tied to the design of the Code of Rules and of the Control Plan. Another important element is also the distribution of these costs among the actors of the chain, that can be both, or either, at vertical and horizontal level. The distribution of costs is related to a specific strategy of the Consortium that can be based on an idea of progressive incidence of their weight on the profitability of the activity carried out, but can also be based only on a convenience judgement related to the possibility of making the controls easier and therefore less costly within the bottlenecks of the supply chain.

6. A costs / benefits balance

Although the direct certification costs are one of the most important element considered, the decision taken by the firms whether to use the PDO/PGI scheme depends on a more general balance between costs and benefits.

Among the relevant costs to be considered, besides direct certification costs, there are all the costs firms need to face in order to adapt the whole firm structure, organisation and production process to the contents of the Code of Rule, but more in general to adapt to a 'modern' and codified organisational logic, given that often firms involved in typical products are artificial and small. The size of these costs depends both on how 'strict' a Code has been written when the application for PDO/PGI was submitted to the European Commission, and on the structural and organisation characteristics of the firms potentially involved in the production of the PDO/PGI product.

The structural adjustment of the production process to the Code of Rules and its reorganisation can imply high investment for the firm: for example the Chianina PGI beef Code of Rules establishes strict conditions for exclusive transportation of the livestock (separated from non - destined to PGI livestock) to the slaughterhouses (which must be inside the PGI area); in addition, the lack of slaughterhouses with high capacity obliges breeders to carry the livestock, even if in limited number, to different slaughterhouses, spread in the PGI area; these rules and their implementation often imply investment in truck and costs for transport, which in some cases are very high for the (small) dimension of farms.

Other costs can derive from the use of more expensive raw materials: for example in the case of the Pecorino Toscano PDO the Code of Rules establishes the use of sheep milk of registered breeding of Tuscany instead of cheaper sheep milk with other origins (i.e. from Sardinia) and in addition the milk does not have to be frozen (freezing milk for the cheese production is a common practice); the Chianina PGI breeders have to exclude some food (silage) from the livestock feeding to use a more expensive food which in some cases rises the feeding price up to €0,50/animal per day.

Besides, when firms want to use the PDO/PGI only for a part of the whole production, there may be some organisation costs due to the need of keeping the production lines separate, that may cause some inefficiencies in managing processes and require dedicated assets (storage structures, plants, transport vehicles, etc.), thus increasing the overall costs.

Depending on how strict the contents of the Code of Rules is, another kind of cost may exist for those firms which are forced to adapt the specificity of the firm's product to the standard imposed by the Code. In this case the firm is obliged to make a choice between using the PDO/PGI and its special recipe, when the latter conflicts with the Code. The cost here is given by the renounce to use a differentiation leverage based upon the firm's savoir-faire.

Generally speaking, there are also administrative and bureaucratic costs, connected to filling and keeping registers in order to fit the needs of the Inspection body; other costs that can be considered as 'psychological costs', deriving from the need the entrepreneurs have to adapt to the logic of certification schemes ('being controlled by someone'). We have seen that here the role of Consortia or other kinds of Intermediate Institutions may support firms in lowering their psychological and bureaucratic costs.

All these costs have to be considered in the light of the expected benefits which can derive from the PDO/PGI choice. Among the benefits deriving from joining the PDO/PGI scheme, the interviewed producers firstly consider the premium price which can be gained thanks to the protection against frauds and misuse of the name of the product that the European Commission guarantees: this is the case for example of the Olio Toscano PGI and the Pecorino Toscano PDO and in general of many reputed Italian products, whose production structure is strong enough for the product to be exported, thus being more exposed to frauds and misuse of the name.

The premium price benefit may also come from the use of the PDO/PGI logo on packaging as a quality sign, and includes both the increase of sales and the increase of the price; for example, the Chianina PGI beef increased its sales thanks to the guarantee of the origin during the BSE period, and the higher demand of Chianina cattle resulted in an increase of the price of calves and carcasses. In addition, the PGI sign allowed the Chianina beef to enter new commercial channels, as 'boutiques' and, above all, supermarket chains, which re - evaluated and asked the PGI beef looking for some guarantees for their consumers during the BSE disease. Among the benefits also the positive effect the use of PDO/PGI, even in a very limited way, can exert on the global image of the firms, thus qualifying the whole production, can be considered.

The access to new long - distance and modern marketing channels is a very important effect also for Pecorino cheese and Toscano Oil: the PDO and PGI became a standard of reference for national and international supermarkets chains, that introduced the products in their assortments. This fact has obviously a negative effect for the Tuscan firms that produce similar but not certified products.

Another expected benefit is given by the possibility PDO/PGI products have to obtain EU incentives: the Common Agricultural Policy and the new rural development policy is clearly oriented to support the quality of the agro - food products and both the past reform and the mid term review established incentives for those producers who join a national or EU quality certification system (like PDO, PGI or Organic production); for example, EU incentives, crossing with the BSE disease and the consequent poultry beef sector crisis were a decisive factor in the choice of the PGI tool for Chianina breeders.

Table 5 gives an overlook on the importance of the 'indirect' costs (excluded direct certification costs) and the benefits deriving from the use of PDO/PGI valorisation tool, with reference to a 'representative' situation that hides very heterogeneous situations.

In fact all these categories of costs and benefits are not homogenous and they are vertically and horizontally distributed, that is between the sectors of the supply chain and between different typologies of firms and different territories (for the analysis of these distributional effects see Belletti, 2000; Marescotti, 2003).

The importance of indirect costs and benefits represented in Table 5 derives from a comparative analysis on similar products marketed with a PDO or PGI denomination and without it: all costs and benefits are considered related to the whole supply chain, not divided into single phases.

In the costs and benefits balance there are also wider economic implications within both the supply chain and the whole local economic system. The quantification of these economic implications is very hard, but even their categorisation as costs or benefits may be difficult as it depends on the point of view of the actors considered.

Table 5. Importance of indirect costs and benefits deriving from PDO or PGI tool.

	Chianina PGI	Pecorino Toscano PDO	Olio Toscano PGI
Indirect costs			
Investment for structural adjustment	+	+	+
Raw materials of higher quality	+++	+++	+
Re-organisation of production process	++	+	+
Bureaucratic and psychological costs	+++	+	++
Benefits			
Increase of sales	+++	++	+
Increase of price	++	++	++
Access to new commercial channels	+++	++	+++
EU incentives (quality certification)	+++	+	+

Within the supply-chain the implications deriving from the use of a PDO/PGI can be at structural level (i.e. in the increase of employment), at production level (i.e. induced production of higher quality raw materials for the production processes), or in the profitability of the downstream phases of the chain as in the case of commercial incentives to retailers.

Within the local system, the use of a PDO/PGI, by bounding the income in the PDO/PGI area, by attracting consumers to the area, can exert spill-over effects on other actors of the local system and for the development of activities of other economic sectors as in the case of tourism or handicraft activities. The presence of a PDO/PGI supply chain can also have implications at social level within the whole local system, i.e. the surviving of the traditional production methods and the encouraging of social interactions.

All these implications have to be taken into account not so much by the agents of the supply chain as by the local policy maker whose valorisation and wider aims may be fulfilled or denied through the support of a PDO or PGI valorisation tool.

Acknowledgements

This work has been funded by Azienda Regionale per lo Sviluppo e l'Innovazione in Agricoltura (research 'Prodotti tipici, percezioni di qualità lungo la filiera e possibilità di sviluppo del mercato', Co-ordinator Alessandro Pacciani) and by Ministero dell'Università e della Ricerca, Italy (research 'Agro-food typical products territorial valorisation: problems and opportunities in the use of geographical designations and collective hallmarks', year 2004, co-ordinator Giovanni Belletti).

References

Agri2000 and C. Lazzarin, 2004. I costi per il mantenimento delle denominazioni DOP e IGP. (in press).

Agri2000, 2002. Costi per l'ottenimento e la gestione dei marchi d'origine DOP E IGP: definizione di parametri tecnici ed economici per la valutazione e la razionalizzazione dei costi. (in press).

Anania, G. and R. Nisticò, 2004. Public regulation as a substitute for trust in quality food markets. What if the trust substitute cannot be fully trusted? Journal of Institutional and Theoretical Economics 160, 4.

Belletti, G., 2000. Origin labelled products, reputation, and etherogeneity of firms. In: Sylvander B., D. Barjolle and F. Arfini (eds.), The socio-economics of Origin Labelled Products in Agri-Food Supply Chains. INRA Actes et Communications, n.17-1, pp. 239-260.

Belletti, G., 2001. Le prospettive offerte da Dop e Igp per la valorizzazione degli oli extra-vergini d'oliva. In ARSIA. Olivo e olio: suolo, polline. DOP, pp. 39-110.

European Commission, 2004. Directorate General for Agriculture, Food Quality Policy in the European Union Protection of Geographical Indications, Designations of Origin and Certificates of Specific Character for Agricultural Products and Foodstuffs Working Documents of the Commission Services, Guide to Community Regulations

Marescotti, A. and T. Burgassi, 2004. Problemi ed opportunità derivanti dalla certificazione IGP Vitellone Bianco dell'Appennino centrale - Chinina. Osservatorio agroindustriale per la Toscana, INEA.

Marescotti, A., 2000. Marketing channels, quality hallmarks and the theory of conventions. In Sylvander B., D. Barjolle and F. Arfini (eds.), The socio-economics of origin labelled products in agro-food supply chains: spatial, institutional and co-ordination aspects. INRA, Serie Actes et Communications, 17-2, pp. 103-122.

Marescotti, A., 2003. Typical products and rural development: Who benefits from PDO/PGI recognition? paper prepared for the 83rd EAAE Seminar Food Quality Products in the Advent of the 21st Century: Production, Demand and Public Policy, 4-7 September 2003, Chania, Greece.

Pacciani, A., G. Belletti, A. Marescotti and S. Scaramuzzi, 2003. Strategie di valorizzazione dei prodotti tipici e sviluppo rurale: il ruolo delle denominazioni geografiche. In: Arzeni A., R. Esposti and F. Sotte (eds.), Politiche di sviluppo rurale tra programmazione e valutazione, Franco Angeli.

Raynaud, E., and L. Sauvée, 2000. Common labelling and producer organisations: A transaction cost economics approach. In: Sylvander B., D. Barjolle and F. Arfini (eds.), The socio-economics of origin labelled products in agri-food supply chains. INRA Actes et Communications, 17-2, pp. 133-142.

Sylvander, B. (ed.), 2004. WP 7 final report - synthesis and recommendations, DOLPHINS Concerted Action (Development of Origin Labelled Products: Humanity, Innovation and Sustainability), INRA - UREQUA.

Thiedig, F. and B. Sylvander, 2000. Welcome to the club? An economical approach to geographical indications in the European Union. Agrarwirtschaft 49, Heft 12, 428-437.

Tregear, A., F. Arfini, G. Belletti and A. Marescotti, 2004. The impact of territorial product qualification processes on the rural development potential of small-scale food productions. XI World Congress of Rural Sociology Trondheim, Norway, 25-30 July.

Van der Meulen, H. and F. Ventura, 1994. La costruzione della qualità: produzione commercializzazione e consumo della carne bovina in Umbria. CESAR.

Verhaegen, I. and G. Van Huylenbroeck, 2001. Costs and benefits for farmers participating in innovative marketing channels for quality food products. Journal of Rural Studies 17, 443-456.

Vuylsteke, A., E. Collet, G. Van Huylenbroeck and M. Mormont, 2003. Exclusion of farmers as a consequence of quality certification and standardisation. 83th EAAE Seminar. Food Quality Products in the Advent of the 21st Century: Production, Demand and Public Policy, 4-7 September 2003, Chania, Greece.

Strategic, organisational and human resource management aspects of quality management

Improving quality-related communication in food chains with Quality Function Deployment: the dairy industry

Martina Peupert and Ludwig Theuvsen
University of Goettingen, Department of Agricultural Economics and Rural Development, Platz der Goettinger Sieben 5, 37073 Goettingen, Germany; theuvsen@uni-goettingen.de

Abstract

Quality management in agribusiness is increasingly focusing on quality standards, third-party audits and external certifications. However, quality assurance systems often lack an appropriate methodological underpinning even though such a foundation is paramount for making quality efforts transparent and quality requirements easier to comprehend. Thus, chain-wide collaboration and quality communication is often considered insufficient. In this paper we present Quality Function Deployment as a tool for supporting quality-related analysis, communication and planning processes and systematically integrating quality techniques into the processes of food production and processing along the food chain. The application of Quality Function Deployment is demonstrated in reference to the dairy industry.

Keywords: Quality Function Deployment, food chain, dairy industry, House of Quality, quality communication

1. Introduction

In recent years the design of value chains and chain-wide cooperation has become one of the most vividly discussed topics in agriculture and the food industry. In this discussion different points of view can be identified. Some authors consider, for instance, vertical integration and cooperation paramount for the future competitiveness of food chains. These authors often refer to the United States and Denmark. In both countries vertically integrated hog production systems successfully serve the world market for pork (Windhorst, 2004). Similar statements can be found concerning the beef industry (Fearne, 1998; Hornibrook and Fearne, 2005). In both cases the increasing requirements of large retailers and fast-food companies concerning product quality and traceability are identified as important drivers towards more integrated food supply chains (den Ouden, 1996; Lawrence *et al.*, 2001). On the other hand, 'free entrepreneurs don't need contracts' is still a popular slogan (AgraEurope, 2004) and more vertical cooperation and integration in food chains is strongly rejected by German farmers (Recke *et al.*, 2005).

Another impetus for discussing the organisation of food chains comes from recent developments in quality management in agribusiness. In this sector two different approaches prevail. On the one hand, quality management is incorporated into vertically integrated food chains with close contractual relationships between agriculture and the food industry. In such cases food quality management is organised by a lead company and protected, among other things, by detailed supply, production and marketing contracts. In Germany such chain-wide cooperation is pursued by the PHW Group, which owns the Wiesenhof brand, Germany's best-known brand in the meat sector. The alternative approach relies on certification standards and third-party audits, where quality and quality assurance are guaranteed by firm-specific quality management systems, stage-specific quality requirements and neutral certifications of system members through accredited certification bodies (Spiller, 2004). System participants remain independent and do not enter contractual relationships with upstream or downstream industries. On the contrary, due to the reduction of quality uncertainties, spot market transactions are favored over alternative ways of organising food chains (Schramm and Spiller, 2003).

At downstream levels of food chains, the design of business relationships has also been identified as a source of competitive advantages. Efficient Consumer Response (ECR), for instance, has become a buzzword for attempts to improve the transfer and use of information in food chains, implement more efficient logistic solutions, harmonise marketing concepts and reduce losses due to friction between transaction partners (Seifert, 2001). In recent years, ECR has been considered a major building block of more agile supply chains (Van Hoek *et al.*, 2001).

Irrespective of these and other drivers - resulting, for instance, from new EU legislation on food traceability (Theuvsen and Hollmann-Hespos, 2004 and 2005) - chain-wide cooperation in agribusiness is often considered insufficient and in need of major improvement. Indicators for this shortcoming in food chains are a low intensity of inter-company communication, deeply rooted mistrust between transaction partners, and suppliers' readiness to quickly change transaction partners when expecting short-term price advantages (Traupe, 2002; Gerlach *et al.*, 2004; Recke *et al.*, 2005). Furthermore, methodological shortcomings concerning supplier relationship management can be identified in food chains (Ryder and Fearne, 2003).

Against this background, this paper aims at introducing Quality Function Deployment as a tool and quality technique for supporting intra- and inter-company communication in food chains. The remainder of the paper is organised in two parts: (1) the introduction of Quality Function Deployment and analysis of its previous applications in agribusiness, and (2) a description of its experimental use in the dairy chain. Some final remarks conclude the paper.

2. Quality Function Deployment

2.1. Development and basic ideas of Quality Function Deployment

In 1966 Quality Function Deployment (QFD) was first introduced by Yoji Akao from Japan as a tool for supporting product development processes. The gist of QFD can be given as 'planning and developing the quality functions of a product in accordance with the quality attributes required by customers' (Akao, 1990). QFD is usually considered a process-oriented tool supporting the transformation of customer requirements into quantitative as well as qualitative technical product and process parameters. In companies QFD is typically carried out by interdepartmental teams. It is a systematic analysis, quality planning and communication tool that serves the important function of translating the 'voice of the customer' into the 'voice of the engineer' (Kamiske *et al.*, 1994) in order to guarantee the required quality and the best possible customer orientation throughout the whole product development and production phase (Pfeifer, 2002).

The first successful projects using QFD were carried out at the Mitsubishi Heavy Industries shipyard in Kobe, Japan. In the 1980s the instrument became more and more popular in U.S. industry. But it was not until the 1990s that German companies became interested in QFD (Klein, 1999).

The core element of QFD is the House of Quality (HoQ; see Figure 1). The HoQ is a combination of tabular and graphical delineations and analyses relevant for solving marketing and technical problems (Kamiske *et al.*, 1994). The most typical features of the HoQ are the integration of a number of planning steps and the analysis of interrelationships in a matrix form. Due to this well-structured documentation, it is possible to guarantee a high transparency in the process of selecting and defining product and process characteristics. Furthermore, the HoQ enables consensus-building in the course of product development despite the often conflicting viewpoints expressed by marketing and engineering experts (Akao, 1990).

Figure 1 roughly outlines the major steps of the QFD process. QFD starts with a careful analysis and structuring of customer requirements - the 'WHAT?' at the 'entrance' of the HoQ. These initially generated requirements are subject to a benchmarking process in the right hand part of the HoQ which reflects a product evaluation from a marketing and consumer perspective. Additional columns for such elements as pricing can be added if necessary. The customer requirements are then confronted with product attributes - the 'HOW?' -considered necessary for meeting these requirements.

The relationship matrix in the center of the HoQ represents the core of the analysis. The matrix contains a structured effect analysis in which technical product features are assessed according to the fulfillment of customer-specific quality attributes. The correlation matrix in the 'roof' of the HoQ is devoted to possible trade-offs or harmonies between defined product features. In the

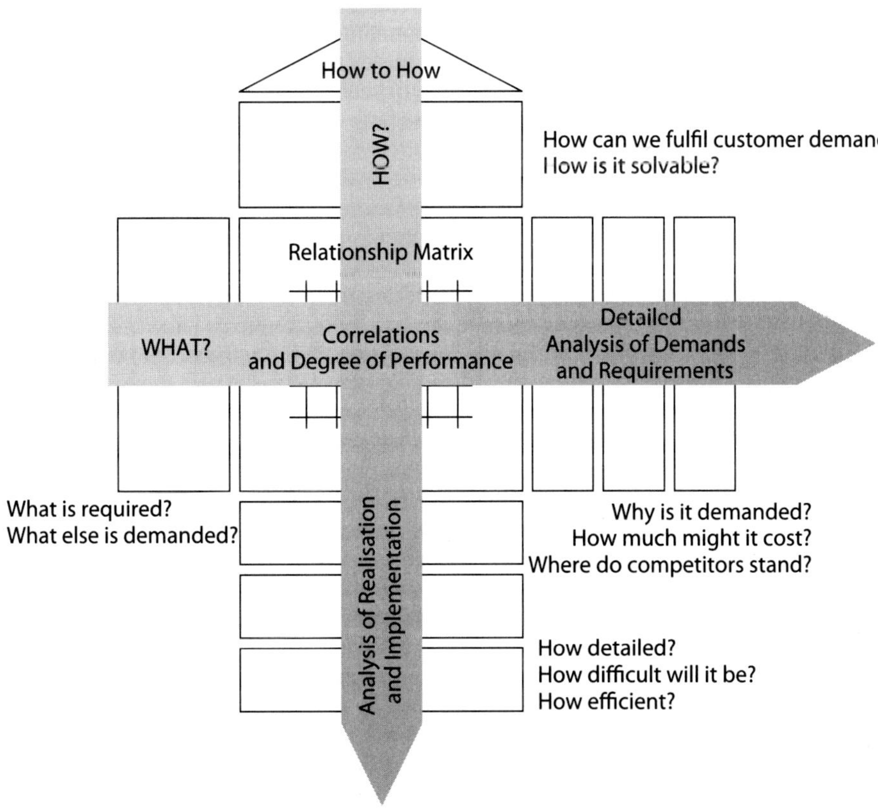

Figure 1. QFD and the House of Quality (Redeker, 2002).

lower part of the HoQ another benchmarking process compares technical product features with the characteristics of technically comparable products that already exist in the market. Other information, such as the difficulty of implementing new ideas and solutions, can also be included in this part of the HoQ. Finally, the central matrix of the HoQ is also analysed in the lower part of the HoQ. The results direct attention towards critical product and process characteristics and, thus, contribute to the development of competitive products and services. This final step takes all prior analyses into account.

The critical points identified in the first HoQ are then subjected to an analogous analysis in another - in this case they represent the entryway of the HoQ, the 'WHAT?' If necessary, this process can be repeated several times, finally resulting in the specification of detailed process guidelines such as job descriptions and job instructions. Thus, in the end, QFD ensures the systematic transformation of customer requirements into product and process attributes (Kamiske *et al.*, 1994; Saatweber, 1997; Klein, 1999).

2.2. Use of Quality Function Deployment in the food industry

QFD was developed in the heavy and machine building industrial sectors, so the agribusiness sector is not its traditional application area. Nevertheless, the German food industry has made initial attempts to utilise the tool in supporting product development processes. But so far, QFD has not proven successful and therefore has not gained much acceptance in the food industry. Despite this, a survey of the German food and stimulant industry revealed that 33% of all companies in the sector are familiar with QFD. Of these companies, about 23% have already applied the tool (Silberer *et al.*, 1999). Most users of QFD in the food industry consider it helpful in improving existing products but have not yet used it to support the search for innovations (Benner *et al.*, 2003).

Respondents in the food and stimulant industry attributed a high transparency to QFD since all decisions are clearly delineated in the HoQ. Furthermore, managers in the food industry extol improved interdepartmental cooperation through the use of QFD. Reasons for rejecting its use are, for instance, its perceived complexity and the resulting time required to implement it. A lack of knowledge about this - at least for many companies in the food industry - new technique and a lack of understanding of the results of analyses in the HoQ are additional obstacles to a successful QFD implementation.

Despite these by and large disappointing results, this paper continues with the description of a successful experiment with the use of QFD in the agri-food sector. Unlike previous applications of this quality technique, it is not presented here as a tool for supporting product development but as a tool for improving intra- and inter-company communication processes in food chains concerning quality-relevant product and process requirements.

3. Experimental use of Quality Function Deployment in the agri-food sector

3.1. From Quality Function Deployment to Quality Food Chain Deployment

The traditional QFD approach has as its goal the translation of customer requirements into product characteristics. The proposed new QFD approach is more process-oriented and translates quality requirements into process characteristics in order to guarantee the quality readiness of the whole food chain. Thus, instead of Quality Function Deployment, QFD may be read as an acronym for Quality Food Chain Deployment.

The idea of a multi-stage QFD process borrowed from the American Supplier Institute is adopted in the quality food chain approach as well. But, whereas the multi-step process usually seeks to transform customer requirements step-by-step into product attributes, process characteristics and process and job instructions on the company level, it is now used to support communication along the chain by forwarding specifications and requirements that have to be met by other companies in the food chain. The first HoQ is usually developed by a chain leader who has the necessary managerial resources and sufficient influence on upstream and downstream industries. The addressees use the information received from the chain leader as input for their own QFD processes. Taking the dairy industry as an example, Figure 2 visualises the sequential development of HoQs in the food chain proposed by the Quality Food Chain Deployment approach.

In the chain-wide approach the HoQ remains the core element of the modified QFD approach although some changes regarding the content of the HoQ are necessary due to the new focus on quality-related communication along the food chain. Since benchmarking procedures and correlation analyses remain very much the same from a technical point of view, the new HoQ looks very similar to the one presented above. Nevertheless, the focus is different. Benchmarking, for instance, no longer deals primarily with technical product characteristics but concerns itself with assessing existing quality-related processes and procedures, the identification of improvement

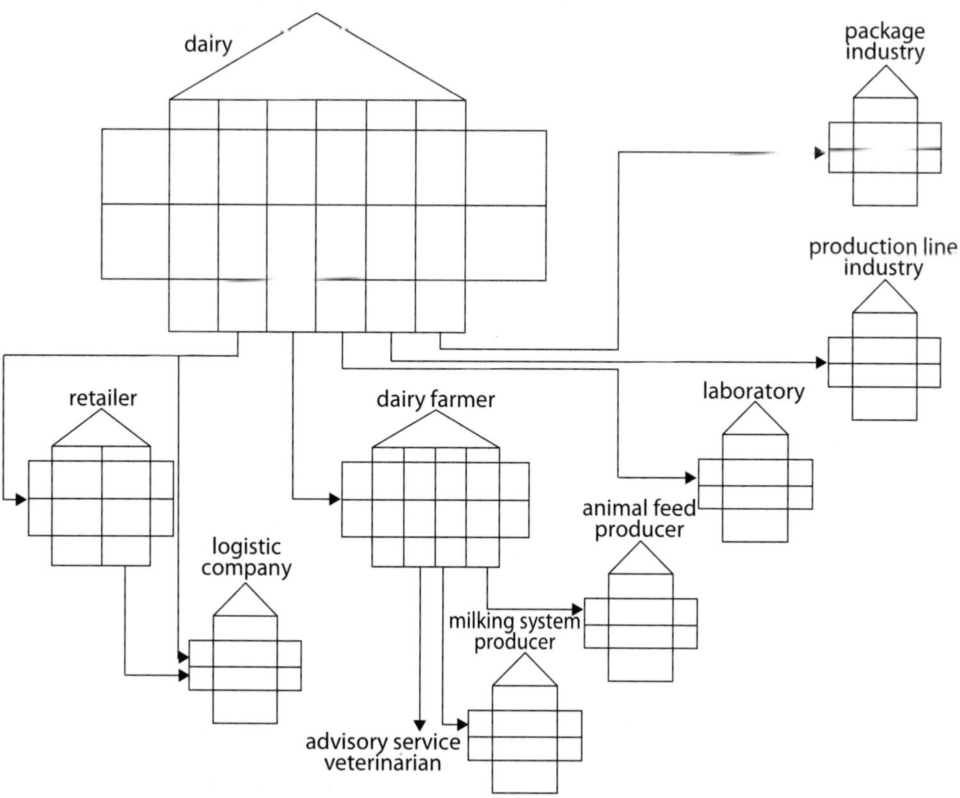

Figure 2. Quality Food Chain Deployment in the dairy industry.

opportunities and targets to be met. Figure 3 illustrates the suggested modifications to the HoQ; a more detailed description can be found in section 3.3.

3.2. Quality Food Chain Deployment in the dairy industry

Now QFD will be exemplarily applied to the dairy chain, which basically consists of dairy farmers, dairies and retailers. Other producers and service providers - animal feed producers, logistic companies, laboratories, the package industry and the production line industry - are also included in the analysis. Future research may also cover marginal parts of the dairy chain, such as extension and consulting services, veterinarians and producers of milking systems. In this paper we refrain from such an extensive definition of the dairy chain. Instead, we start with its central component, i.e. the dairy, which is considered paramount since it functions as a quality driver in the dairy chain and communicates market and production-driven quality requirements to other companies in the food chain.

The dairy chain serves as an example demonstrating the applicability of QFD for supporting chain-wide communication processes concerning quality requirements. For this purpose a

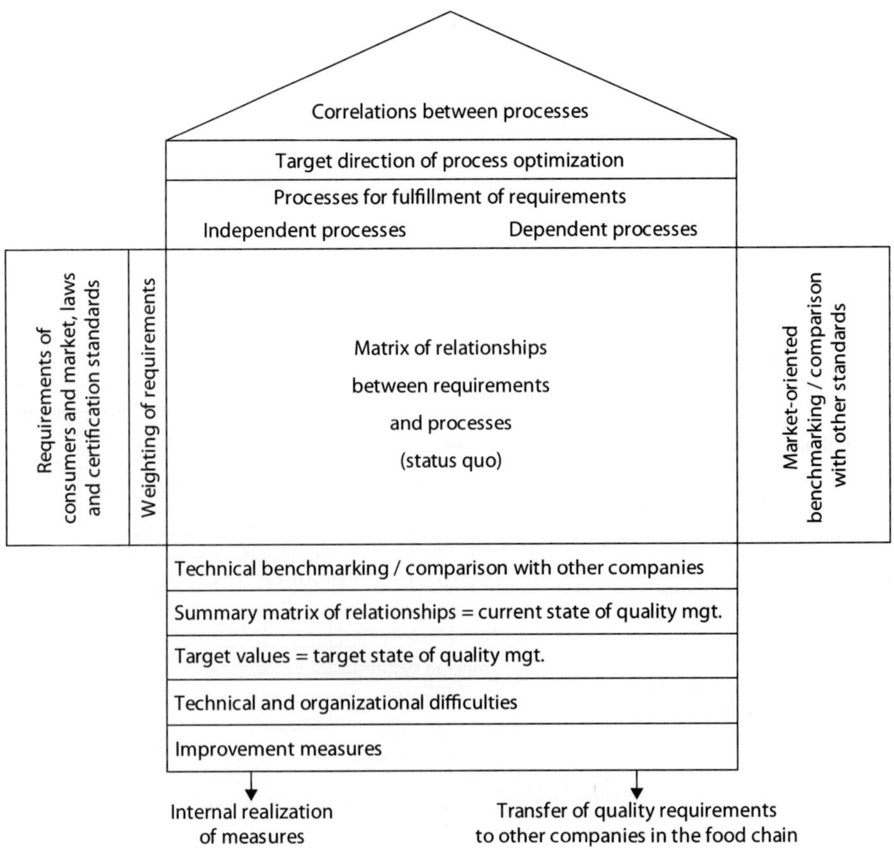

Figure 3. Quality Food Chain Deployment: a modified House of Quality.

group of master students at the Faculty of Agricultural Sciences of Goettingen University was divided up into groups representing different stages of the dairy chain: Group 1 = dairy; group 2 = retail and logistics; group 3 = dairy farmers and animal feed producers; group 4 = laboratory; group 5 = production line industry; group 6 = package industry. Figure 2 has already presented the hierarchical structure of the HoQs resulting from the process described as well as the interdependencies between the different stages of the food chain and the subsequent communication of quality requirements.

Using QFD the aforementioned groups defined and analysed relevant product and process parameters for their particular stages in a given time period. The dairy group was chosen as a starting point since the dairy is considered a 'natural' chain leader or supply chain 'captain' (Goldsmith *et al.*, 2003). For that reason its HoQ is examined in more detail in Figure 4. The following description refers to the development of the HoQ, the subsequent communication of quality requirements to retailers, dairy farmers and laboratories and the development of HoQs at these stages of the dairy chain. The results of the dairy's HoQ are fed into the QFD applications of the other groups. Throughout the process, no communication other than that of requirements stemming from the QFD application took place between the groups. In this way, we sought to gain insight into the likelihood of applying this methodology in the chain-wide context of food production.

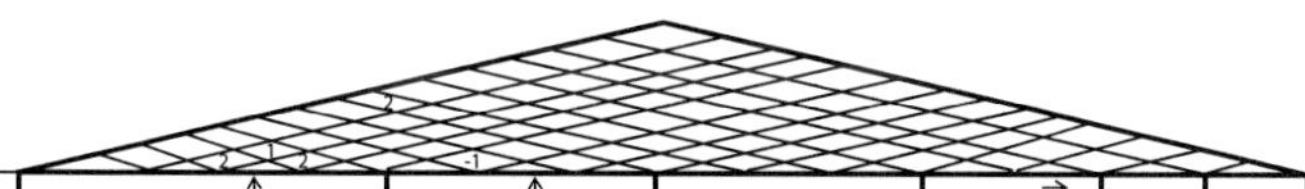

Processes / Requirements		Milk processing				Service			Raw material			Analysis		…	…	Market-oriented benchmarking (1 2 3 4 5)
		Quality management	HACCP	Training	R & D	Handling	Storage/ cooling	Presentation	Selection of raw material	Ingredients	Supply reliability	Analysis of raw material	Documentation of results	…	…	
Freshness	5				5		5		5					…	…	★ 3, ◇ 4
Low fat content	2									5				…	…	★ 3, ◇ 5
Sensory attributes	4								5					…	…	★◇ 5
Appearance	1						5							…	…	◇ 2, ★ 3
Shelf-life	6				5	1	5							…	…	★ 2, ◇ 4
Product standardiz.	4	5			1						5	9		…	…	★ 4, ◇ 5
Absence of hazards	10		9	5		1								…	…	★◇ 4
Hygiene	9		9	5										…	…	★ 4, ◇ 5
Process reliability	8	1		1								5		…	…	★ 3, ◇ 5
Traceability	7	9											5	…	…	★◇ 5
Customer service	5							5						…	…	★ 2, ◇ 4
Difficulties		2	6	3	9	5	8	1	4	9	10	3	1	…	…	
Technical benchmarking (1–5)		★ 3, ◇ 4	★ 4, ◇ 5	★ 2, ◇ 4	★ 2, ◇ 4	★ 1, ◇ 2	★◇ 2	◇ 2, ★ 3	★ 3, ◇ 4	★ 2, ◇ 4	★◇ 4	★ 4, ◇ 5	★ 3, ◇ 5	…	…	Scale benchmarking: 1 - bad, 5 - very good, ★ dairy, ◇ competitor
Analysis of matrix: rank		6	1	4	2	9	8	12	10	3	7	5	11	…	…	
Target values		+ 10	+ 30	+ 50	+ 40	+ 25	+ 40	+ 10	+ 30	+ 40	+ 5	+ 15	+ 5	…	…	
Aims/ sanctions for advancement		Support	Support	Process structuring	Innovative activities	Design of package	< 4 °C	Medium rack level	S-class milk	Cell/microbe count	Daily	Choice of tests	Electronic media	…	…	

Figure 4. Dairy's House of Quality.

3.3. Quality Food Chain Deployment: the dairy's House of Quality

The dairy receives signals from customers and markets regarding the expected quality attributes of its products and processes. The dairy then relates these signals to company processes which can be expected to effectively fulfill the requirements, makes the necessary evaluations, interprets the data and draws conclusions concerning improvements in its own production processes or forwards requirements to other stages of the dairy chain. The quality requirements derived can serve as starting points for goal agreements between the dairy and its customers and suppliers, who in turn take these quality requirements and analyse to what degree their own processes meet the dairy's expectations or to what extent improvements are necessary.

Most of the steps of developing the dairy's HoQ resemble the standard application of the methodology although some modifications are necessary to take inter-company context into account. We present the most popular sequencing; of course, variations in sequence are possible. For reasons of clarity we choose a simple example - the production of commercial durable milk for end-consumer use - and refrain from fully describing all the necessary evaluations in the course of the development of the HoQ.

First, on the left side, customer and market requirements represent the input into the HoQ. This information may stem from customer surveys or analysis of legal requirements and certification standards. In our example we chose freshness, low fat content, sensory attributes/taste, appearance and shelf-life as the most important product characteristics. Process standards to be met refer to product standardisation, absence of production hazards, hygiene, process reliability, traceability and customer service.

In the second step product as well as process requirements receive weights from, for example, 1 to 10. Low numbers represent low importance and vice versa. The weights are necessary for analysing the relationship matrix later on.

The third step takes place on the right side of the HoQ. It already includes an initial benchmarking process in which the dairy's performance concerning product and process attributes is compared to that of competitors. This benchmarking process reveals the degree to which the dairy meets external expectations and where its strengths and weaknesses lie compared to important competitors. If the criteria chosen as input for the HoQ stem from a certification standard, the benchmarking process makes it possible to see where the standard is met, whether any improvements are necessary and whether even a higher standard could be met. On the right side of the HoQ more columns can be added if necessary in order, for example, to highlight customer complaints or deficits criticised by internal or external auditors.

In the next step all processes relevant to fulfilling the requirements identified in the first step are listed in the upper part of the HoQ. Here two kinds of processes can be distinguished. On the one hand, there are independent processes, that is, processes that are only relevant for internal activities and tasks. Typical examples are the maintenance of a quality management system, the implementation of an HACCP system, employee training in hygiene and food handling, and research and development activities. On the other hand, dependent processes have to be analysed, too. These processes are strongly influenced by other companies in the value chain. Retailers, for instance, influence product quality by guaranteeing (or not guaranteeing) a permanent cooling chain and providing specific storage conditions. Dairy farmers determine the quality and availability of the raw product and, thus, strongly influence the preconditions for high-quality products (such as a long shelf-life) and production processes. Laboratories choose more or less reliable and trustworthy testing methods and data transfer procedures and, thus, are also relevant for quality.

The fifth step is to evaluate the extent to which the company's processes meet the product and process requirements identified in the first step. The results are laid down in the relationship matrix in the center of the HoQ. This evaluation is restricted to assessing the status quo in order to allow the identification of necessary improvements later on. It also elucidates whether there is a relationship between processes, on the one hand, and process and product quality attributes, on the other. If such a relationship exists, the evaluation reflects how well the process actually meets the requirements. Numbers in the relationship matrix can represent the degree of achievement (for example, 1 = weak achievement, 5 = medium achievement, 9 = optimal achievement). This procedure can be seen, for example, in the 'quality management' column. Figure 4 shows that the design of the quality management system is very important for product traceability; allotting 9 points means that the system works very well in this respect. Process reliability still needs substantial improvement due to such factors as the insufficient application of advanced quality techniques. Attaching only 1 point highlights this shortcoming and calls for further action. The goal of guaranteeing consistent product quality is partly met; 5 points indicate this but also hint at a need for further improvement. A blank means that a specific process does not influence a product or process attribute. Refraining from entering anything into the relationship matrix in this case improves readability and clearness of the HoQ.

The combination of the previous analyses is carried out in step 6. The values for the actual achievements in the relationship matrix are multiplied by the weights attached to the product and process requirements (step 2) and the values acquired from the market-oriented benchmarking process (step 3). Thus, the value for the field 'freshness'/'R&D' is calculated by multiplying $5 \cdot 5 \cdot 3 = 75$. Then the values are summed up column by column. The total sums are noted in the lower part of the HoQ (summary matrix of relationship). Based on these results the processes can be ranked according to their importance. Before drawing any final conclusions, however, further evaluations and several more steps are necessary.

The seventh step consists of a second benchmarking process. In this case it is done from a technical point of view. In our example, the dairy's process design is compared to that of competing companies in order to identify strengths and weaknesses. Often these evaluations confirm evaluations already noted in the relationship matrix.

Step 8 is devoted to the identification of target values for improvements and, thus, indicates future developments. This can be done in two different ways. First, targeted improvements can be identified in a non-analytical way (for instance, quality management - + 10; employee training - + 50; see Figure 4). But target values can also be laid down in a more analytical and transparent way. To do so the calculation described in step 6 is repeated, but instead of current values derived from the market-oriented benchmarking process the calculation now takes into account the (assumedly) higher values of competitors identified in step 2. Concerning 'freshness' and 'R&D', for instance, the other values have to be multiplied by 4 instead of 3 ($5 \cdot 5 \cdot 4 = 100$; see Figure 4).

The next step on the way to a final decision about the implementation of improvement measures is the assessment of technical and/or organisational difficulties. Here it is often suitable to use a 10-point scale on which higher numbers represent larger technical or organisational problems in realising improvements.

In step 10 (in the roof of the HoQ) a global development trend for each process is defined based on the evaluations in the relationship matrix and the benchmarking processes. The dairy may, for instance, decide to improve, hold or reduce the performance levels in its processes.

Step 11 is dedicated to the determination of interdependencies between the company's processes. The results are noted down in the 'roof' of the HoQ by using symbols or numbers for representing conflicting as well as complementary relationships. Often the intensity of the relationships is also included. In Figure 5, for instance, a strong complementarity is symbolised by '2' whereas a weak trade-off gets a '-1'.

The final step is the most comprehensive and complex one. Based on all previous analyses, measures for improving the dairy's own processes as well as the processes and products of other companies in the food chain are defined as targets. In this way, step 12 also sets the stage for communicating with other partners in the food chain. Here the huge importance of differentiating between different processes, such as milk processing, service, raw material supply and quality analyses, becomes evident. This distinction ensures clarity in the somewhat complex HoQ and allows, for instance, clear differentiation between dependent and independent processes. Milk processing,

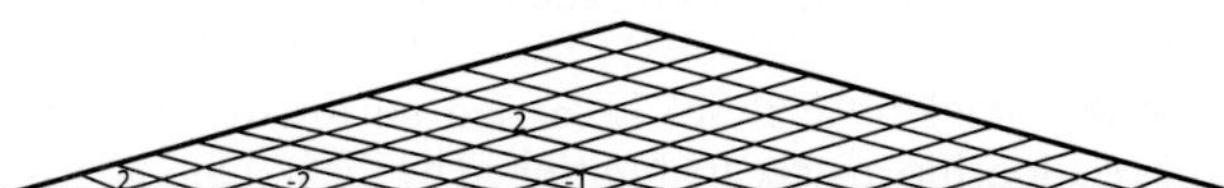

Processes / Requirements		Dairy farmer										Animal feed			Market-oriented benchmarking (1 2 3 4 5)
			↑					↑		→					
		Control of health	Support by veterinarian	Drove management	Vermin control	Variety choice	Livestock husbandry	Milking	Storage/ cooling	Cow planner	Training	Sort of feed	Quality of feed	Homogeneity of feed	
Cell count	6	5	5					5							★ ◇ 4
Microbe count	7		5			5	9	1	9						★ 4, ◇ 5
Daily supply	1			9		5				5			5		★ 2, ◇ 4
Color	2											9		1	★ ◇ 3
Taste	5											5			★ 3, ◇ 4
Odor	3								5			5		1	★ 3, ◇ 4
Hygiene	8			5	5		9				5				★ 4, ◇ 5
Documentation	4									9					★ ◇ 3
Difficulties		4	2	4	6	9	10	10	4	2	5	3	7	8	
Technical benchmarking (1–5)		★ 3, ◇ 4	★ 2, ◇ 5	★ 3, ◇ 4	★ ◇ 3	★ 2, ◇ 4	◇ 3, ★ 4	★ 2, ◇ 4	★ 3, ◇ 5	★ ◇ 4	★ ◇ 3	◇ 3, ★ 4	★ 3, ◇ 4	★ 1, ◇ 3	
Analysis of matrix: rank		13	8	7	2	5	10	12	4	1	3	6	11	9	
Target values		+ 15	+ 35	+ 15	+ 5	+ 25	+ 25	+ 50	+ 15	+ 5	+ 5	+ 15	+ 20	+ 25	
Aims/ improvement measures		Health improvement	Service contract	BSI < 1,5	Biological control	Crossing	Open stable	Milking carousel	Extension of capacity	Support	Husbandry	Without silage	Energy density	Batches	

Scale benchmarking:
1 - bad
5 - very good
★ dairy farmer
◇ +25%-farms

Figure 5. Dairy farmer's House of Quality.

for example, is an internal process in the dairy's production system. Therefore, improvement measures have to be implemented and controlled internally only; they are not communicated to other companies. Service and logistics, on the other hand, require close cooperation with retailers who are responsible for product display and major parts of handling and cooling. Raw material supply and quality analyses require similar approaches. In these cases quality requirements have to be communicated to dairy farmers or laboratories respectively.

Due to space limitations, the lower part of Figure 4 displays only a very limited number of improvement measures and target agreements. The training process, for instance, needs an improved process design and content changes. Storage and cooling need improvements in order to constantly guarantee temperatures below 4 °C to retailers. Concerning raw material supply, the specification of maximum cell and microbe counts can be expected to improve quality but may, of course, also require several additional measures on dairy farms. Laboratories face additional expectations concerning the speed of milk testing so that just-in-time processes can more easily be implemented in the dairy. Furthermore, laboratories are also expected to use only certified testing methods.

3.4. Quality Food Chain Deployment: other Houses of Quality

In the experimental chain-wide application of QFD the other groups were required to refer to the quality requirements of the dairy and to integrate these requirements into their own HoQs. This procedure is considered paramount for improved quality communication in food chains. Analogous to the procedure described for the dairy, the other companies develop their own HoQs, i.e. identify relevant processes, benchmark processes, and identify strengths and weaknesses as well as necessary improvements.

Retailers may, for instance, focus on the technical reliability of their cooling equipment, improved controls for the cooling chain or intensified employee training concerning food handling. Dairy farmers may have to check the status quo of animal health, herd management, livestock husbandry conditions, milking systems and animal feed choice and composition in their HoQ. Finally, laboratories check marking and storage of probes, maintenance of laboratory equipment, choice of testing methods and qualification of staff members. Since the procedures to develop the respective HoQ always remain the same in principle, we refrain from describing them in greater detail. Nevertheless, Figure 5 illustrates this process by presenting a dairy farmer's HoQ.

If the companies in the dairy chain come up with new quality requirements through the application of QFD, then these requirements have to be translated into action internally or communicated to other companies in the food chain. This again reflects the distinction between dependent and independent firm processes.

The resulting HoQ can be used as an instrument for supplier assessment or for controlling the implementation of target agreements. If the dairy's market position allows these activities, an integrated quality chain emerges. In this case it is possible to use the HoQ as a basis for the joint optimisation of the quality of milk and milk products through all partners in the food chain. As such, QFD may become an integral partner of improved supplier relationship management (Gerlach *et al.*, 2004). Furthermore, the HoQ can also support auditing processes since it allows insights into process structures and efforts towards quality improvements.

4. Concluding remarks

The QFD approach presented in this paper was introduced as a way of improving quality-related communication in agribusiness supply chains. Due to this new focus, QFD's central element, i.e. the HoQ, was devoted to the analysis of relationships between quality requirements and process design. This differs clearly from common QFD applications in product development processes, in which requirements and product attributes are related to one another. In our experimental setting, QFD turned out to be a successfully and easily applicable tool for improving chain-wide cooperation. Its systematic application in food chains would provide companies in the agribusiness with a more comprehensive overview of their strengths, weaknesses and needs for improving their own as well as other companies' processes. Furthermore, QFD ensures the consequent translation of customer and market requirements into process design as well as new impulses for product and process innovations. Last but not least, the HoQ can be used for analysis of compliance with certification standards, systematic supplier assessment and thorough argumentation for rising quality standards in the food chain.

But the implementation of QFD may also face impediments. A predominant problem is the huge market power of retailers, whose interest in greater transparency in the supply chain may be limited (Theuvsen, 2004) as may their interest in more systematically supported purchasing decisions and strategies. Nevertheless, since price can also be integrated into the HoQ, retailers may find QFD attractive in the end. Another problem stems from the large number of dairy farmers and their - on average - small size, which limits the application of more advanced quality techniques such as QFD. At the farm level, the integration into the QFD process of extension services or dairy associations serving as bundling organisations for farmers may be a feasible solution.

In summary, the experimental use of QFD proved successful and demonstrated a promising solution for the existing shortcomings in chain-wide cooperation in the agribusiness. Open-mindedness of chain partners and sufficient knowledge about state-of-the-art quality techniques are paramount for successful Quality Food Chain Deployment.

References

Akao, Y., 1990. Quality function deployment: Integrating customer requirements into product design. Cambridge.

Benner, M., A.R. Linnemann, W.M.F. Jongen and P. Folstar, 2003. Quality function deployment (QFD) - Can it be used to develop food products? Food Quality and Preference 14, 327-339.

Den Ouden, M., 1996. Economic modelling of pork production-marketing chains. Wageningen Pers.

Fearne, A., 1998. The evolution of partnerships in the meat supply chain: Insights from the British beef industry. Supply Chain Management 3, 214-231.

Gerlach, S., B. Köhler, A. Spiller and C. Wocken, 2004. Supplier Relationship Management im Agribusiness: Ein Konzept zur Messung der Geschäftsbeziehungsqualität. Working Paper 0406, Institute of Agicultural Economics, Georg-August-University Goettingen.

Goldsmith, P., N. Turan and H. Gow, 2003. Food safety in the meat industry: A regulatory quagmire. International Food and Agribusiness Management Review 6, 1.

Hornibrook, S. and A. Fearne, 2005. Demand driven supply chains: Contractual relationships and the management of perceived risk. Paper presented at 2nd European Forum on Market-Driven Supply Chains, Milano, April 5-6, 2005.

Kamiske, G.F., T.G.C. Hummel, C. Malorny and M. Zoschke, 1994. Quality function deployment - oder das systematische Überbringen der Kundenwünsche. In Marketing ZFP 3, 181-190.

Klein, B., 1999. QFD - Quality function deployment: Konzept, Anwendung und Umsetzung für Produkte und Dienstleistungen.Expert.

Lawrence, J.D., T.C. Schroeder and M.L. Hayenga, 2001. Evolving producer-packer-customer linkages in the beef and pork industries. Review of Agricultural Economics 23, 370-385.

Pfeifer, T., 2002. Quality management: Strategies, methods, techniques. Hanser.

Recke, G., B. Schulze, A. Spiller and L. Theuvsen, 2005. Schweineerzeugung in Nordwestdeutschland: Strukturen, Wettbewerbsfähigkeit, Wertschöpfungspotenziale.Münster-Hiltrup.

Redeker, G., 2002. Ganzheitliches Qualitätsmanagement. Working Paper, Institute of Quality Assurance, University of Hannover.

Ryder, R. and A. Fearne, 2003. Procurement best practices in the food industry: Supplier clustering as a source of strategic competitive advantage. Supply Chain Management 1, 12-16.

Saatweber, J., 1997. Kundenorientierung durch Quality Function Deployment: Systematisches Entwickeln von Produkten und Dienstleistungen. Hanser.

Schramm, M. und A. Spiller, 2003. Farm-Audit und Farm-Advisory-System: Ein Beitrag zur Ökonomie von Qualitätssicherungssystemen. Berichte über Landwirtschaft 81, 165-191.

Seifert, D., 2001. Efficient consumer response: Supply Chain Management (SCM), Category Management (CM) und Collaborative Planning, Forecasting and Replenishment (CPFR) als neue Strategieansätze. 2nd ed., Hamp.

Silberer, G., A. Scharf and M. Nagel, 1999. Quality Function Deployment in der deutschen Nahrungs- und Genussmittelindustrie. Working Paper, Institute of Marketing and Retailing, Georg-August-University Goettingen.

Spiller, A., 2004. Qualitätssicherung in der Wertschöpfungskette - Vor- und Nachteile unterschiedlicher Organisationskonzepte. In Dachverband Agrarforschung (ed.), Lebensmittelqualität und Qualitätssicherungssysteme, Frankfurt/M., 83-96.

Theuvsen, L., 2004. Transparency in netchains as an organisational phenomenon: Exploring the role of interdependencies. Journal on Chain and Network Science 4, 125-138.

Theuvsen, L. and T. Hollmann-Hespos, 2004. Rückverfolgbarkeit von Lebensmitteln: Aktuelle Entwicklungen und Anforderungen an Informationstechnologien. In Integration und Datensicherheit - Anforderungen, Konflikte und Perspektiven, G. Schiefer *et al.* (eds.), Bonn, 49-52.

Theuvsen, L. and T. Hollmann-Hespos, 2005. The Economics of traceability: A model of investments in tracking and tracing systems in agriculture and the food industry. In Proceedings of EFITA and WCCA 2005 Joint Conference, Vila Real, Portugal, July 25-27, 2005.

Traupe, C., 2002. Schlachtschweinevermarktung in Niedersachsen - Stand, Defizite, Entwicklungsmöglichkeiten. Cuvillier.

Van Hoek, R. I., A. Harrison and M. Christopher, 2001. Measuring agile capabilities in the supply chain. International Journal of Operations & Production Management 1/2, 126-147.

Windhorst, H.-W., 2001. Europäische und globale Trends in der Schweinefleischerzeugung und ihre Auswirkungen auf Deutschland. Paper presented at DLG Conference 'Spitzenbetriebe der Schweinemast', Goettingen, November 8, 2001.

Failure Mode and Effect Analysis (FMEA) as a decision support tool within a quality information system in pork production chains

Andreas Gödderz[1], Thomas Schmitz[2], Adriane Mack[1] and Brigitte Petersen[1]
[1]University of Bonn, Institute of Physiology, Biochemistry and Hygiene of Animals, Katzenburgweg 7-9, 53115 Bonn, Germany
[2]PLATO AG; Breite Straße 6-8, 23552 Lübeck, Germany; andreas.goedderz@uni-bonn.de

Abstract

Recent norms of retailer organisations, chain oriented quality programmes, the new ordinance of the German government concerning the hygiene of food (Lebensmittelhygieneverordnung) and of course the EU regulation 178/2002 demand the implementation of self-control and hazard control techniques in terms of Hazard Analysis and Critical Control Point (HACCP) systems of agrofood industry. Such an analysis of course includes a risk analysis. These demands and regulations require a stronger inclusion of the production process in systems of quality assurance. The Failure Mode and Effect Analysis (FMEA) seems to be an appropriate tool to enable animal health services to support farmers to fulfil these requirements. On the level of advisory services a computer aided FMEA tool which includes elements of the HACCP concept is tested. The tool allows to document efforts made to meet the claims of quality assurance and simultaneously provides gathered knowledge in form of a knowledge data base supporting the advisory service to solve concrete problems on farm. The paper describes how to assemble such a system for the *Salmonella* problem in pig farms.

Keywords: FMEA, HACCP, quality assurance, pork production, *Salmonella*

1. Introduction

Recent norms of retailer organisations, chain oriented quality programmes, the new ordinance of the German government concerning the hygiene of food (Lebensmittelhygieneverordnung) and of course the EU regulation 178/2002 demand the implementation of self-control and hazard control techniques in terms of the Hazard Analysis and Critical Control Point (HACCP) system of the agrofood industry. Relating to the supplier chains for meat and meat products this means to include the process of animal production in the process of quality assurance. While in the converting companies lead of experiences in quality methods have been made, there is a lack of those at the level of animal production. The establishment of those chain oriented systems may be an appropriate operational field of the Failure Mode and Effect Analysis (FMEA) in the pork production chain. The conception of this tool of preventive quality assurance is similar to the HAACP system. However the FMEA is wider composed. A combination of the two techniques FMEA and HACCP seems to be promising. To optimise the application of hazard control techniques a method manual which combines those two techniques was developed (Schmitz and Petersen, 2004). Such a 'mixed' FMEA-HACCP concept is provided here.

2. FMEA method

In the range of industrial production the FMEA is an established part of quality management as a tool of preventive quality assurance. The FMEA helps to implement a closed quality control loop by providing gathered expert knowledge. This can be used for planning as well as for executing processes (Pfeifer, 2002).

The FMEA intends to detect potential sources of error and their consequences on quality characteristics as early as possible. So consecutively disturbances can be anticipated (Pfeifer, 2002). Because of this it is essential that the FMEA has to be customer oriented from the first step of production. Therefore all possible consequences of a self-inflicted failure for all succeeding chain members has to be considered.

Three important contents of the FMEA are:
- structured failure analysis including an analysis of the causes and effects;
- risk assessment based on the analysis mentioned before;
- use of the results of risk assessment to carry out an optimisation of process or concept (Edenhofer and Köster, 1991).

This aspects show that the FMEA is in fact part of a HACCP analysis. Within such an analysis the FMEA takes the parts of failure/ hazard analysis, risk assessment and it provides the actions to deal with the revealed failures and hazards.

The knowledge needed to run the FMEA in an efficient way is distributed to many persons. Therefore a team with members from every step of the process of interest should be formed. The team members contribute their knowledge as experts. The discussion in the FMEA team is prepared and presented by an experienced moderator. The moderator has to encourage the other team members e.g. QM representatives and other experts to examine the process of interest very critically. And of course the moderator has to encourage the other members to be self-critical. This in addition to efficient preparation of the FMEA determines the FMEA´s success.

To initiate a FMEA the following steps should be taken. The first step is to fix the analysis´ limits. Then the process is structured and standards are assigned. In step 4 and 5 the failure analysis is done and a FMEA form established. The achieved FMEA´s results are filled in a FMEA form to guarantee documentation as well as systematic and clarity. Step 6, risk assessment, is done by calculating a risk priority number (RPN). The RPN is calculated by using three variables describing the probability of the failure to occur (occurrence, O), the severity (S) of the potential failure mode on the process and the probability to detect the failure (detection, D). Normally an assessment number ranging from 1 (no risk) to 10 (high risk) is used to describe these three variables. To facilitate assessment verbal explanations are assigned to the different values. The RPN is calculated by multiplying the values of the three variables O, S and D. The value of the RPN gives a hint whether optimisation is urgently required. Risk assessment needs a lot of supporting data to be done exactly. Optimisations (step 7) are carried out according to the following principles:
- Strategy amendment to exclude the cause of failure or reduce its severity. This means to restructure the system.
- Increase of the strategy reliability to minimise the occurrence of the failure´s cause.
- More effective detection of failure´s causes.

If the FMEA reveals that optimisation has to be done it has to be defined who has to do which of the recommended actions by when. This is also entered on the FMEA form. After performance of the recommended actions a risk assessment is carried out again. Risk priority numbers which had an effect on the decision are calculated again. A comparison of the two RPNs (previous and improved state) allows a final result assessment and the assessment of the relationship between achievable improvement and utilised effort. The reassessment of the risk after implementation of the recommended actions gives an estimation of the remaining risk of certain failure´s occurrence. Depending on this result the team decides whether the chosen actions were successful or whether additional actions are necessary (Stamatis, 1995).

The application of FMEA software tools proved to be useful in different industrial branches. There are three main advantages:

- The FMEA establishment is systematised.
- The entered FMEA knowledge will be saved onto a knowledge database and can be used again.
- The effort of the establishment is reduced by the optimisation of the teamwork and by the falling back upon information already entered by means of search helps (Schmitz and Petersen, 2001).

3. Adaption of the FMEA concept to farm level

Referring to Noordhuizen and Frankena (1999) a quality-management instrument at farm level should satisfy two basic requirements:

- it should provide the advisory service or the individual farmer with clear and simple procedures for elimination and control of disease risks on the farm;
- it should enable the farmer to prove the execution of these procedures to a third party for herd-health certification and health insurance purposes.

Welz (1994) demonstrated the possibility to adapt the FMEA concept to animal production. In his study he used the FMEA to reveal interferences of product and process quality resulting from animal diseases on farm level.

In the following a FMEA like approach for prevention and reduction of Salmonellosis in pig production is given. In pig production the problem of *Salmonella* is to be considered from two different angels. On the one hand problems in production and economical losses resulting from Salmonellosis during the production period, on the other hand the endangering of human health due to *Salmonella* contaminated pork-products (Waldmann and Plonait, 2001). Steinbach and Hartung (1999) assume circa 20% of human Salmonellosis in Germany to be caused by consumption of *Salmonella* contaminated pork-products. Referring to Van Altrock and co-authors (2000) and Meyer (2004) circa 10% of tested fattening pigs showed a positive test result. This indicates that there is a need for supporting tools to solve this problem.

In literature several possible sources for the introduction of Salmonella in pig producing farms are described. The most important sources are:

- purchase of piglets and gilts (Lo Fo Wong *et al.*, 2004; Berends *et al.*, 1996);
- purchase of feed (Hartung, 2003; Lo Fo Wong *et al.*, 2002);
- biotic and abiotic vectors (Meyer, 2004; Letellier *et al.*, 1999).

Each of these aspects include a lot of different subaspects. Also the transmission of *Salmonella* within a farm is influenced by a lot of factors. The most important are listed in Table 1.

Table 1. Factors with influence on the transmission of Salmonella *within a farm.*

Factor	Reference
hygiene status and farm hygiene	Berends *et al.*, 1996
hygiene lock	Lo Fo Wong *et al.*, 2004
all in and all out	Lo Fo Wong *et al.*, 2004
cleaning and disinfection	Lo Fo Wong *et al.*, 2004
disposal of dead animals	Letellier *et al.*, 1999
farm management	
farm size	Van der Wolf *et al.*, 2001b
pig density in pens	Funk *et al.*, 2001
pen separation	Lo Fo Wong *et al.*, 2004
floor design	Meyer, 2004
manure management	Belœil *et al.*, 2004
feeding system	Lo Fo Wong *et al.*, 2004; Van der Wolf *et al.*, 2001b; Van Schie and Overgoor, 1987
addtion of organic acids to feed or drinking water	Van der Wolf *et al.*, 2001a
number of attending persons	Meyer, 2004
other infections within the herd	Belœil *et al.*, 2004; Wills *et al.*, 2000; Møller, 1998
concomitance of parasitic diseases	Van der Wolf *et al.*, 2001b

Each of these factors is associated with a number of different characteristic values which may even interact. A lot of these factors especially those which refer to the spreading of Salmonella within a herd/farm apply for every step of pig production (farrowing, fattening). So to keep the FMEA clear and to be able to use the gathered knowledge preferably on different types of farm, the following seven system elements were created in Workgroup Computing System SCIO™ FMEA System (PLATO AG, Lübeck):

- production process/ husbandry;
- cleaning and disinfection;
- pest control;
- water;
- feedstuff/feeding;
- hygiene of environment;
- hygiene of staff.

For each of these system elements a FMEA form was established by checking literature for possible risk-factors concerning introduction and spreading of Salmonella associated with the system element of interest. To create the FMEA form the following steps were executed and the following corresponding questions put forward:

- Listing all steps concerning the production process.
- Determination of potential hazards/failures.
  - Which hazard or hygiene failure can be caused by this production step?
  - Which hazards or hygiene failures occurred at this production step in the past?
- Determination of the effects.
  - What are the effects of this hygiene failure on the animals?
  - What are the effects of this hygiene failure on the farm´s *Salmonella* status?

- What are the effects on the next production step?
- What are the effects on the consumer of pork products?

- Search for potential causes for each hazard/failure.
  - Searching for the causes in the surroundings of man, machines, environment, material, method, management or measurement.
- Listing possible actions to avoid the hazard/failure (precautionary and checking actions).
  - What can be done to avoid this hazard/hygiene failure?

The kind of questions indicates that the chosen approach is not just FMEA based but also contains elements of the HACCP concept. Some of the chosen column headings reflect this, too.

The created FMEA form is a table with 22 colums (Figures 1 and 2). In the first column the process of interest is entered. The second column contains the potential hazards or rather the 'hygiene failures'. The potential effects of these failures are listed in the next column. Then an assessment of failures´ severity (S) is done. The next column shows your decision whether this aspect is a controlpoint or not. Then failures´ causes are assembled. Next step is to assess failure´s probability to occur (O). The result of this assessment is entered in the column. In the next column the decision is made whether the current applied control to deal with the failure is precautionary or checking. Then the current control is entered. In the next column an assessment of the probability

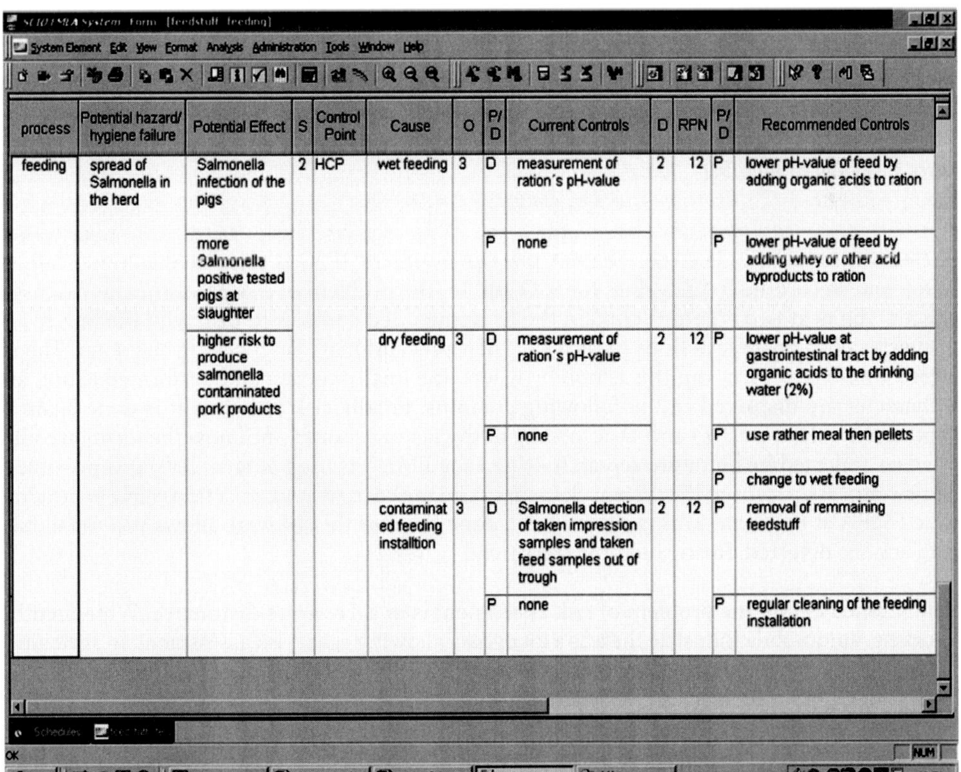

process	Potential hazard/ hygiene failure	Potential Effect	S	Control Point	Cause	O	P/D	Current Controls	D	RPN	P/D	Recommended Controls
feeding	spread of Salmonella in the herd	Salmonella infection of the pigs	2	HCP	wet feeding	3	D	measurement of ration´s pH-value	2	12	P	lower pH-value of feed by adding organic acids to ration
		more Salmonella positive tested pigs at slaughter					P	none			P	lower pH-value of feed by adding whey or other acid byproducts to ration
		higher risk to produce salmonella contaminated pork products			dry feeding	3	D	measurement of ration´s pH-value	2	12	P	lower pH-value at gastrointestinal tract by adding organic acids to the drinking water (2%)
							P	none			P	use rather meal then pellets
											P	change to wet feeding
					contaminated feeding installtion	3	D	Salmonella detection of taken impression samples and taken feed samples out of trough	2	12	P	removal of remmaining feedstuff
							P	none			P	regular cleaning of the feeding installation

Figure 1. Screenshot of the used FMEA form. This screenshot displays only the half FMEA form. The rest is shown in Figure 2.

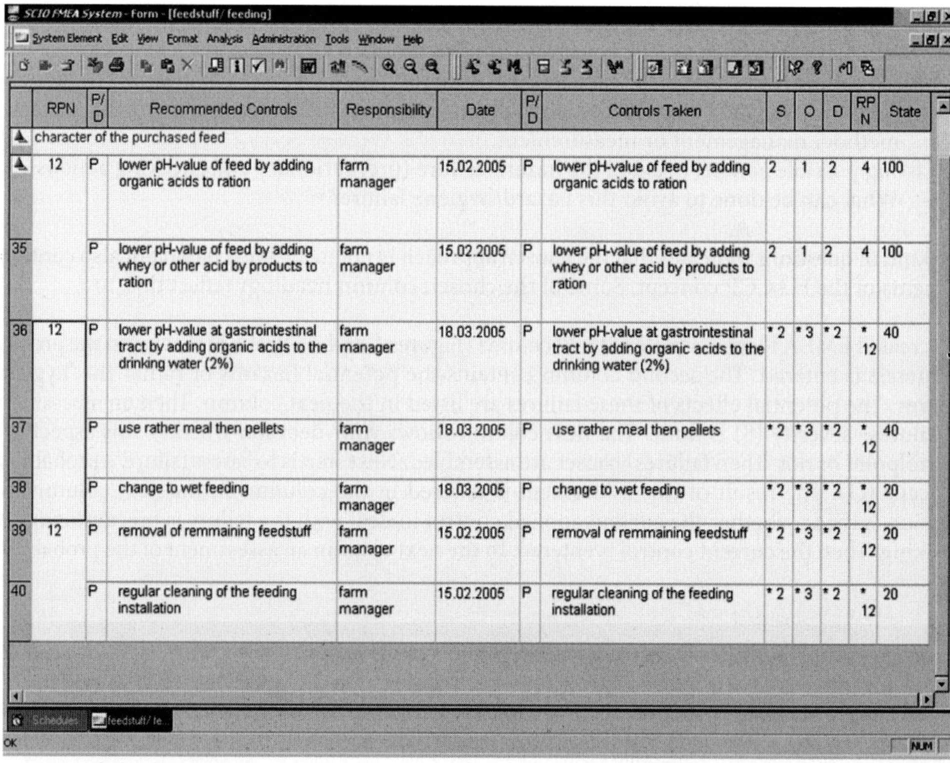

SCIO FMEA System - Form - [feedstuff/ feeding]

	RPN	P/D	Recommended Controls	Responsibility	Date	P/D	Controls Taken	S	O	D	RPN	State
	character of the purchased feed											
	12	P	lower pH-value of feed by adding organic acids to ration	farm manager	15.02.2005	P	lower pH-value of feed by adding organic acids to ration	2	1	2	4	100
35		P	lower pH-value of feed by adding whey or other acid by products to ration	farm manager	15.02.2005	P	lower pH-value of feed by adding whey or other acid by products to ration	2	1	2	4	100
36	12	P	lower pH-value at gastrointestinal tract by adding organic acids to the drinking water (2%)	farm manager	18.03.2005	P	lower pH-value at gastrointestinal tract by adding organic acids to the drinking water (2%)	* 2	* 3	* 2	* 12	40
37		P	use rather meal then pellets	farm manager	18.03.2005	P	use rather meal then pellets	* 2	* 3	* 2	* 12	40
38		P	change to wet feeding	farm manager	18.03.2005	P	change to wet feeding	* 2	* 3	* 2	* 12	20
39	12	P	removal of remmaining feedstuff	farm manager	15.02.2005	P	removal of remmaining feedstuff	* 2	* 3	* 2	* 12	20
40		P	regular cleaning of the feeding installation	farm manager	15.02.2005	P	regular cleaning of the feeding installation	* 2	* 3	* 2	* 12	20

Figure 2. Sequel to Figure 1.

to detect the failure (D) is given. The risk priority number (RPN) is calculated automatically by the software according to the values for S, O and D. The problem of risk assessment is discussed beneath. The next two columns contain the recommended controls to deal with the failure and the specification of these actions in terms of their precautionary or checking character. Then is entered who has to carry out this action by when. The finally chosen and performed action and its character are displayed in the following columns. Finally risk assessment is done again by adapting the values for S, O and D according to the taken actions. Until now the forms are filled with data collected by literature research. When the FMEA is used at farm level it is possible to add new processes steps to the system elements. Also the current applied actions have to be added to the FMEA at the farm. This is simply to be done because the software allows you to fall back on all actions detected during literature research.

As mentioned before the problem of risk assessment is in agriculture distinctive. While in other sectors e.g. automobile industrie there is a lot of data allowing exact risk assessment, in agriculture there is not so many data because of a lack of documentation, until now. But the introduction of quality programmes such as QS (Quality and Security) for example improved documentation and therefore the amount of available data. But still it is very difficult to give an assessment for a failure´s probability to occur, to be detected or its severity. As mentioned above normally an assessment number ranging from 1 to 10 is used to describe these three variables. In small food processing businesses positive lead of experiences were made with assessment numbers only ranging from 1 to 5. In this model for *Salmonella* introduction and spreading assessment numbers ranging from 1 to 3 were determined. Because of the limited numbers of experiences this scale was chosen. If on road tests reveal the need for adjustment it can easily be done by means of the used software tool. The following evaluation patterns were fixed:

To determine the severity of a failure Odd´s ratios (OR) (which are known from literature) for intoduction or spreading of *Salmonella* on a farm were used as an ancillary tool (Table 2). The borders chosen may be adapted by further investigations.

Table 3 shows the evaluation pattern for failure´s or cause of failure´s probability to occur (O). The data for frequency are adopted from a project carried out with small food processing businesses.

In Table 4 the evaluation pattern for failure´s or cause of failure´s probability to be detected (D) is shown. Here as well the frequencies are adopted from the project with small food processing businesses.

Table 2. Evaluation pattern for failures´ severity (S).

Evaluation pattern	OR	Evaluation
High: A cardinal failure occurs which leads to a very fast spread of *Salmonella* within the whole herd. Because of this in all probability a higher percentage of *Salmonella* positive pigs at slaughter is to be expected. This arises the risk of *Salmonella* contamination of pork products while slaughter and will lead to a degradation of farm´s *Salmonella* status.	> 2	3
Medium: The spread of *Salmonella* in the herd is supposable, but may be unique to batches. The percentage of *Salmonella* positive pigs at slaughter may arise.	> 1	2
Low: An influence of the failure on the *Salmonella* situation on the farm is improbable.	≤ 1	1

Table 3. Evaluation pattern for failures´ probability to occur (O).

Evaluation pattern	Frequency	Evaluation
High: The cause of failure´s occurrence is almost inevitable. Failure´s occurrence in a large quantity is very probable.	> 2%	3
Medium: The cause of failure may occur in some cases but the process is controllable.	< 2%	2
Low: Failure´s occurrence is improbable. It was not (it was rarely) detected at similiar processes.	< 0,5%	1

Table 4. Evaluation pattern for failure´s probability to be detected (D).

Evaluation pattern	Frequency	Evaluation
Low: It is almost impossible to detect the failure or the cause of failure. It is a matter of hidden failure.	< 90%	3
Medium: A detection of the failure or of its cause is possible by investigations with ancillary tools like pH-value measurement or bacteriological or serilogical investigations of taken samples.	> 95%	2
High: It is very easy to detect the failure or its cause by visual, manual investigation or by computer supported control (e.g. climate computer). It is a ostensible inspection criterion.	> 98%	1

4. Chain oriented using of the FMEA based knowledge database

The following Figure 3 makes a proposal how to use the FMEA within a chain oriented approach to minimise the risk of *Salmonella* contaminated pork products. The gathered expert knowledge is used to run self-control systems within the chain. By doing this an enhancement of the database and the methods may be achieved.

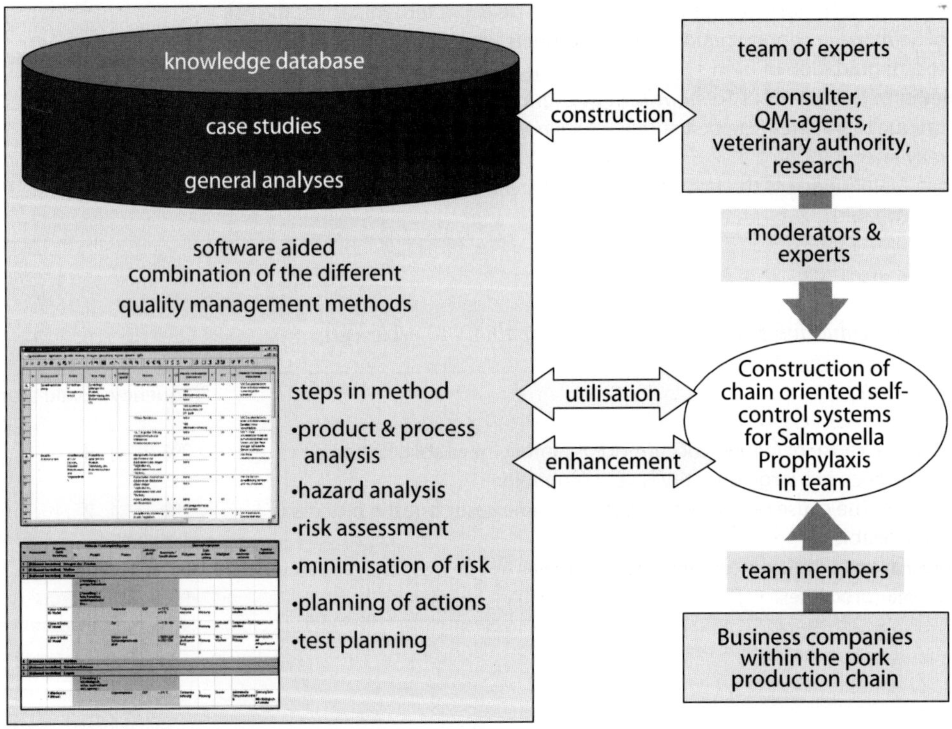

Figure 3. Using the FMEA in a chain oriented approach.

5. Conclusion

A computer-aided FMEA system can be very helpful to run risk analysis within complex processes. The complete documentation of the analysed processes enables to arrange knowledge data bases. This helps to solve concrete problems in a very effective way. Due to the FMEA´s structure those data bases provide clear and simple procedures for the elimination and control of disease risks on farm. Furthermore the FMEA allows to prove the execution of these procedures for health certification and health insurance purposes according to the demands of EU-regulations and distributive trade. Therefore the FMEA seems to be an appropriate tool to support quality information systems in pork production chains. Now the theoretically approach has to be validated by tests with advisory services.

References

Belœil, P.-A., P. Fravalo, C. Fablet, J.-P. Jolly, E. Eveno, Y. Hascoet, C. Chauvin, G. Salvat and F. Madec, 2004. Risk factors for *Salmonella enterica* subsp. *Enterica* shedding by market-age pigs in French farrow-to-finish herds. Preventive Veterinary Medicine 63, 103-120.

Berends, B.R., H.A.P. Urlings, J.M.A. Snijders and F. Van Knapen, 1996. Identification and quantification of risk factors in animal management and transport regarding Salmonella spp. in pigs. International Journal of Food Microbiology 30, 37-53.

Edenhofer, B. and A. Köster, 1991. Systemanalyse: Die Lösung, FMEA optimal nutzen. Qualität und Zuverlässigkeit 36, 699-704.

Funk J.A., P.R. Davies and W. Gebreyes, 2001. Risk factors associated with Salmonella enterica prevalence in three-site swine production systems in North Carolina, USA. Tierärztliche Wochenschrift 114, 335-338.

Hartung, M., 2003. Bericht über die epidemiologische Situation der Zoonosen in Deutschland für 2002. BfR-Hefte, Bundesinstitut für Risikobewertung.

Letellier, A., S. Messier, J. Paré, J. Ménard and S. Quessy, 1999. Distribution of *Salmonella* in swine herds in Québec. Veterinary Microbiology 67, 299-306.

Lo Fo Wong, D.M.A., T. Hald, P.J. Van Der Wolf and M. Swanenburg, 2002. Epidemiology and control measures in pigs and pork. Livestock Production Science 76, 215-222.

Lo Fo Wong, D.M.A., J. Dahl, H. Stege, P.J. Van Der Wolf, L. Leontides, A. van Altrock and B.M. Thorberg, 2004. Herd-level risk factors for subclinical Salmonella infection in European finishing-pig herds. Preventive Veterinary Medicine 62, 253-266.

Meyer, C., 2004. Qualitative und quantitative Risikofaktoren für die Einschleppung und Verbreitung von Salmonellen in unterschiedlichen Produktionsverfahren beim Schwein. PhD Thesis Tierärztliche Hochschule Hannover.

Møller, K. T.K. Jensen, S.E. Jorsal, T.D. Leser and B. Carstensen, 1998. Detection of Lawsonia intercellularis, Serpulina hyodysenteriae, weakly beta-haemolytic intestinal spirochaetes, *Salmonella enterica*, and haemolytic *Escherichia coli* from swine herds with and without diarrhoe among growing pigs. Veterinary Microbiology 62, 59-72.

Noordhuizen, J.P.T.M. and K. Frankena, 1999. Epidemiology and quality assurance: Applications at farm level. Preventive Veterinary Medicine 39, 93-110.

Pfeifer, T., 2002. Quality management: Strategies, methods, techniques. Hanser.

Schmitz, T. and B. Petersen, 2001. Denken in Prozessketten - Transparenz durch IT-gestützte präventive Qualitätstechniken. Die Ernährungsindustrie 5, 56-57.

Schmitz, T. and B. Petersen, 2004. Einsatz softwaregestützter präventiver QM-Methoden in der Beratung von Zulieferketten in der Lebensmittelbranche. In Integration und Datensicherheit - Anforderungen, Konflikte und Perspektiven, Referate der 25. GIL Jahrestagung 08.-09. September 2004 in Bonn. G. Schiefer, P. Wagner, M. Morgenstern, U. Rickert (eds.). Bonn, 2004.

Stahl, P., 1997. Die Qualitätstechnik FMEA als Lerninstrument in Organisationen. Deutscher Universitätsverlag.

Stamatis, D. H., 1995. Failure Mode and Effect Analysis: FMEA from Theory to Execution. ASQC Quality Press.

Steinbach, G. and M. Hartung, 1999. Versuch einer Schätzung des Anteils menschlicher Salmonellenerkrankungen, die auf vom Schwein stammende Salmonellen zurückzuführen sind. Tierärztl. Wschr. 112, 296-300.

Van Altrock, A., A. Schütte and G. Hildebrandt, 2000. Untersuchungsergebnisse aus Deutschland zu dem EU-Projekt '*Salmonella* in pork (Salinpork)' - 1. Mitteilung: Untersuchungen in den Beständen. Tierärztl. Wschr. 113, 191-201.

Van der Wolf, P.J., F.W. van Schie, A.R.W. Elbers, B. Engel, H.M.J.F. van der Heijden, W.A. Hunneman and M.J.M. Tielen, 2001a. Administration of acidified drinking water to finishing pigs in order to prevent *Salmonella* infections. The Veterinary Quarterly 23, 121-125.

Van der Wolf, P.J., W.B. Wolbers, A.R.W. Elbers, H.M.J.F. van der Heijden, J.M.C.C. Koppen, W.A. Hunnemann, F.W. van Schie and M.J.M. Thielen, 2001b. Herd level husbandry factors associated with the serological *Salmonella* prevalence in finishing pig herds in The Netherlands. Veterinary Microbiology 78, 205-219.

Van Schie, F.W. and G.H. Overgoor, 1987. An analysis of the possible effects of different feed upon the excretion of *Salmonella* bacteria in clinically normal groups of fattening pigs. The Veterinary Quarterly 9, 185-188.

Waldmann, K.H. and H. Ploinat, 2001. Erkrankungen der Verdauungsorgane und des Abdomen. In Waldmann, K.H. and M. Wendt (eds.): Lehrbuch der Schweinekrankheiten. 3rd ed., Parey, 335-358.

Welz, M., 1994. Bewertung von Erkrankungen als qualitätshemmende Faktoren mit Hilfe der Fehler-Möglichkeitsanalyse (FMEA) im Rahmen der Erzeugung von Qualitätsfleisch. PhD Thesis University of Bonn.

Wills, R.W., J.T. Gray, P.J. Fedorka-Cray, K.-J. Yoon, S. Ladely and J.J. Zimmermann, 2000. Synergism between porcine reproductive and respiratory syndrome virus (PRRSV) and *Salmonella cholerasuis* in swine. Veterinary Microbiology 71, 177-192.

Quality assurance in agribusiness: evaluating 'Qualität und Sicherheit' and 'QM Milch' from a Total Quality Management perspective

Ludwig Theuvsen and Martina Peupert
University of Goettingen, Department of Agricultural Economics and Rural Development, Platz der Goettinger Sieben 5, 37073 Goettingen, Germany; theuvsen@uni-goettingen.de

Abstract

In the aftermath of several food crises, quality management systems have been widely introduced into the European agribusiness. Nevertheless, many of these systems have been severely criticised due to shortcomings in system design and effectiveness. Therefore, a benchmark is needed which allows to assess the systems implemented so far and provides a starting point for system improvements. Total Quality Management is such a benchmark. In this paper we assess two recently introduced German quality management systems, i.e. Qualität und Sicherheit and QM Milch, from a TQM perspective, highlight their strengths and weaknesses and suggest some improvements.

Keywords: Total Quality Management, Qualität und Sicherheit, QM Milch, benchmarking

1. Quality management systems in the crossfire

Due to several widely recognised food crises in Europe, quality and safety of agricultural and food products have gained much attention during recent years. Many efforts are made to guarantee that food gives consumers no cause for concern and regain consumer trust. The introduction of chain-wide quality management systems is one of the most visible measures in this context. These systems are expected to improve the transparency of food chains (Theuvsen, 2004) and contribute to improved food quality. Most of the recently introduced systems mainly rely on improved self-control, regular third-party audits and a more systematic documentation of procedures and processes.

In Germany many observers and system participants doubt the effectiveness of the newly introduced quality management systems. Jahn *et al.* (2003), for instance, find that German farmers view the quality standards of the German *Qualität und Sicherheit* system in the meat sector as immature and undemanding. Similar concerns have been raised by German dairy farmers participating in the *QM Milch* system (Jahn *et al.*, 2004). Randomly collected feedback from farmers in the Goettingen region conveys a similar impression. Day-to-day operations on farms have remained widely unaffected by Qualität und Sicherheit and business is going on as usual. Thus, fundamental improvements in food quality and safety are doubtful and the question is whether the design of Qualität und Sicherheit and other quality management systems in agriculture and the food industry allows meeting the self-set goals.

Total Quality Management (TQM) evolved as the leading concept in quality management. TQM is most widely spread in the automobile and automobile suppliers, electrical and electronics and machine industries, but in recent years it has also been discussed as a blueprint for the food industry (Beardsell and Dale, 1999; Luning *et al.*, 2002) and has been implemented at least in parts of it. TQM can be considered a comprehensive approach to quality management which puts together the vast experiences of a wide spectrum of industries about how to achieve fundamental improvements in quality and safety. In this paper we consider TQM a benchmark for quality management systems in the agribusiness. Thus, we aim at deriving assessment criteria from TQM and applying these criteria to two recently introduced quality management systems in Germany, i.e. Qualität und Sicherheit in the meat and QM Milch in the dairy sectors. This includes a thorough evaluation of system strengths and weaknesses and the identification of improvement opportunities for both systems. The approach applied in this paper is guided by the idea that half-hearted solutions result in more bureaucracy for farms and firms (Theuvsen, 2005) without considerably reducing the probability of food crises or regaining consumer trust in food quality and safety.

2. Total Quality Management

According to ISO 8402, TQM is 'a management approach of an organisation, centred on quality, based on the participation of all its members and aiming at long-term success through customer satisfaction, and benefits to all members of the organisation and to society.' (Pfeifer 2002: 5). Although the implementation within an organisation is stressed in this definition, a chain-wide approach which includes customers and suppliers in all parts of the value chain is typical of TQM. Thus, it can also be considered suitable for assessing chain-wide quality management systems in the agribusiness as well.

Figure 1 reveals the most important conceptual cornerstones of TQM (Pfeifer, 2002):

- Total: TQM requires a company- or chain-wide implementation of its basic philosophy concerning process, result, people, customer and public orientation.
- Quality: TQM does not only focus on product attributes but prefers a wide definition of quality which also takes into account aspects such as working and systems quality.
- Management: TQM requires leadership, setting of objectives and changes in an organisation's quality philosophy and culture.

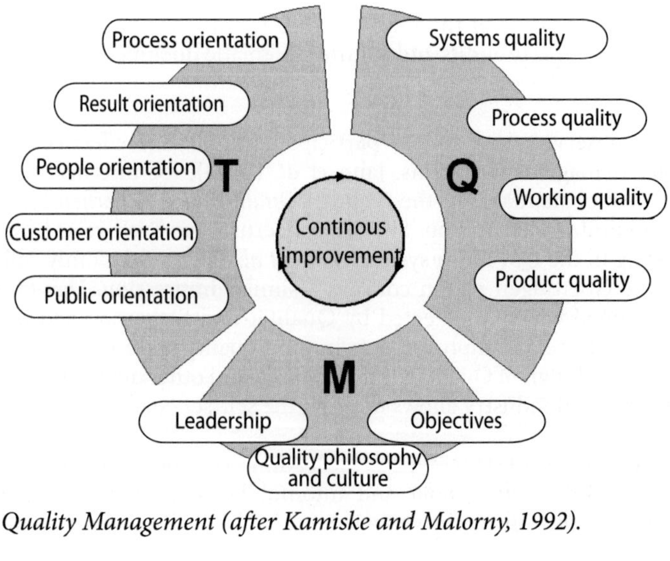

Figure 1. Total Quality Management (after Kamiske and Malorny, 1992).

In the next paragraphs we depict the central elements of TQM in more detail. In doing so, we aim at deriving assessment criteria from the TQM approach which can be used later on for evaluating the Qualität und Sicherheit and QM Milch systems.

2.1. Management

'Quality starts at the top' (Feigenbaum, 1991) is one of TQM's basic principles and highlights top management's pivotal role in quality management. A quality improvement process always depends on top management's motivation and self-commitment (Hackman and Wageman, 1995) and the proper execution of the following tasks:

- Definition of a mission statement: TQM requires top management to integrate an organisation's basic values concerning quality into its mission statement.
- Setting of objectives: TQM considers a balanced and transparent goal system embracing operational quality goals paramount. Furthermore, goal conflicts, for instance between quality and costs, have to be solved (Wagner, 2003).
- Leadership: TQM considers people a decisive success factor and attributes an important role to leadership. Thus, TQM asks top management to
  - start quality programs, implement quality management systems and promote TQM inside and outside the organisation (initiative function);
  - strive for continous improvement, participate in TQM training and meetings and periodically check up management's own performance (role model function); and
  - provide necessary personnel, material and financial resources and enable and support people (service function) (Pfeifer, 2002).

The interplay between the aforementioned aspects is considered paramount for influencing an organisation's quality culture (Abraham *et al.*, 1999). The culture of an organisation is a pattern of shared basic assumptions which are considered valid by its members and from which the organisation derives its most basic sense of core mission, primary tasks or reason to be (Schein, 1992). The pivotal role TQM attributes to top management is motivated by the observation that promoting quality in an organisation or an industry is impossible without influencing the shared basic assumptions. The latter requires an intensive and long-lasting involvement of top management.

From a TQM perspective quality management systems in agriculture and the food industry are evaluated by referring to three questions:

- Does a mission statement exist which defines the industry members' basic values concerning food quality and safety and serves as a guiding principle for all industry members and the development of a quality culture?
- Has a goal system been defined which translates the basic values into operational objectives for everyday management in the food chain and have goal conflicts been solved?
- Does anybody exert chain leadership (initiative, role model and service functions)?

2.2. Broad definition of quality

TQM is based on a broad understanding of quality since focusing solely on product quality is considered insufficient. Instead, a multi-dimensional quality definition is promoted which also takes into account the quality of structures and production processes, working conditions and other factors affecting a company's potential such as human resources, technical equipment and external relationships (system quality; Pfeifer, 2002). What is also important for the coordination of quality improvement activities in a value chain is a shared understanding of what quality basically means in the chain (Kamiske and Malorny, 1992). Thus, a TQM approach to assessing quality management systems in the agribusiness has to ask whether (a) a multi-dimensional approach to defining quality is applied and (b) a shared understanding of what quality basically means exists in the entire food chain.

2.3. Process orientation

Process orientation is a multi-facetted category. The unambiguous definition of process and quality responsibilities is an important aspect due to its strong motivational effects. These effects result from quality-related goal-setting activities (Locke and Latham, 1990) and contingent rewards for attaining quality goals (Wagner, 2003; Zollondz, 2002). Furthermore, process control has to be improved by continuously reviewing the organisation and results of key processes. A third central aspect of process orientation is continous improvement, i.e. a large number of small steps towards improved quality (Pfeifer, 2002). Thus, quality management systems in the agribusiness are expected to (a) clearly define quality responsibilities, (b) improve process control and (c) implement continous improvement in the food chain.

2.4. Result orientation

TQM requires organisations to measure what has been achieved in terms of the overall company goals, the deployment of resources and the fulfilment of needs and expectations of stakeholders (Pfeifer, 2002). Result orientation embraces a wide spectrum of controlling activities such as the use of performance criteria (customer satisfaction, staff satisfaction, image indices and so on). Quality controlling provides input for decision processes, has important behavioral effects (Montes *et al.*, 2003) and should be supported by benchmarking activities (Kaplan and Norton, 1996). From a TQM perspective the ongoing activities for improving quality management in agriculture and the food industry are expected to include controlling activities for analysing and improving (a) the success of all companies in the food chain and (b) the effectiveness of the chain-wide quality management activities.

2.5. People orientation

TQM requires to strictly focus human resource management activities on the organisation's quality goals since human capital is considered not only a cost but also a decisive success factor. Capabilities and motivation of employees are paramount in this respect (Hackman and Wageman, 1995). Furthermore, careful personnel selection, contingent rewards and organisational measures such as quality circles are important aspects. Thus, the criterion for assessing quality management systems in the agribusiness is: In how far do they take the human resource management aspects into account, i.e. personnel selection, improving capabilities through training, motivation through contingent rewards and organisational measures such as quality circles?

2.6. Customer and public orientation

In modern quality management systems quality is defined from the customers' perspective. Customer requirements and expectations are paramount and customer satisfaction is an important overall objective of the organisation. Improving customer relationships and measuring customer satisfaction in market research studies are permanent tasks according to the TQM approach (Pfeifer, 2002). Furthermore, other interest groups' expectations also have to be taken into account. This aspect is especially important in agriculture and the food industry since in modern societies many nongovernmental organisations such as Greenpeace, reform-oriented agricultural politicians and the wider public care about how food is produced. Environment-friendly production, improved animal welfare and freedom from genetically modified organisms are typical examples of stakeholder goals concerning food production and processing (see, e.g. Theuvsen *et al.*, 2005). Company communication has to address customers as well as other constituencies since each stakeholder 'can affect or is affected by the achievement of the organisation's objective' (Freeman, 1984: 25). Thus, TQM-oriented quality management in the agribusiness (a) has to define quality criteria which reflect customer and other stakeholder's requirements and expectations and (b) implement communication procedures with all relevant constituencies.

3. Quality management in the agribusiness

3.1. Qualität und Sicherheit: history and objectives

In 2001 the Qualität und Sicherheit (QS) system was founded by organisations representing the whole value chain in the German meat industry including feed producers, farmers, abattoirs, meat processors and retailers. The QS system is managed by the Bonn-based Qualität und Sicherheit GmbH. This organisation organises neutral controls and sanctions, manages a pool of certified auditing bodies and contracts with system participants. It is also the owner of the QS label and administers its use.

The QS system is a chain-wide quality management approach which integrates all members of the food chain into a standardised certification concept. Its major goals are to guarantee a basic food quality from farm to fork, improve food safety, increase the transparency of food production and processing and regain consumer trust. The QS system is based on a three-tier control system: sytematic self-control by farms and firms, external controls and certification through certified auditing bodies and a control of the external control bodies. The controls are based on the product and process criteria defined by the QS GmbH and laid down in the so-called QS Charta which describes minimum standards for food production and processing. In the meat sector the QS system started in 2001. In 2004 the QS GmbH decided to establish a similar system in the fruits, vegetables and potatoes sectors.

3.2. Integriertes Qualitätsmanagement Milch: history and objectives

Integriertes Qualitätsmanagement Milch (QM Milch) was initiated by the German Farmers' Association, the Raiffeisen Organisation and the Dairy Industry Association. These organisations laid the foundations for a standardised quality management system for milk production, collection and processing. QM Milch mainly documents and integrates formerly scattered requirements concerning quality management in the dairy sector. Similar to the QS system, QM Milch relies on organisational self-controls and neutral external controls.

QM Milch's major goal is the improved safety of production processes by defining quality criteria and documenting product and process qualities. The German dairies integrated QM Milch into their supply contracts with dairy farmers, thus, extending their own quality management system to the farm level. Farmers have to prove the fulfillment of the QM Milch criteria by documenting their processes. Furthermore, conditions on the farms are externally audited. Farmers who do not fully meet the defined criteria become integrated into a compulsory farm advisory system. The feed industry is participating indirectly through confirming the safety of their products.

3.3. Qualität und Sicherheit and QM Milch from a TQM Perspective

In the following paragraphs the QS system and QM Milch are evaluated from a TQM perspective. The assessment is based on the evaluation criteria defined above and publicly available information about both systems (www.q-s.info.de; www. milchwirtschaft.de). When dealing with the QS system we focus on its application in the meat sector.

3.3.1. Management

The management tasks embrace the formulation of a mission statement, the development of a goal system and the the exertion of several leadership functions.

The QS system aims at improved food quality and safety, more transparent food production and processing and regaining consumer trust. Thus, the system has a general mission statement serving as a guiding principle for all system participants. What is missing is a systematic and operational goal system for all system participants derived from the mission statement. Instead, the QS handbook contains a huge number of somewhat unintegrated goals and requirements. Most of these goals can be found where food hygiene is defined by referring to critical temperatures or microbe counts. Many of these goals are taken from laws, norms and industry-specific best practice guidelines. Thus, the QS system mainly restricts itself to systematically documenting existing quality standards and largely refrains from setting new and more demanding ones. This allows most farms and firms to quite easily participate in the system but has also disappointed those stakeholders who expected more, for instance improved animal welfare.

Goal conflicts have been ignored for quite a long time in the QS system. In the very beginning of system implementation, goal complementarities were emphasised instead. The QS GmbH, for instance, argued that the QS system could be integrated into farm management systems and may result in improved animal health and better financial performance. Only just in late 2002, discussions about a lack of price differentiation between QS and non-QS meat as well as a controversy about feeding cheap but risky material revealed so far unsolved goal conflicts.

The QS GmbH organises and manages the QS system but it is not part of the food chain. Its responsibility is mainly restricted to administrative, initiating and coordinating functions. This is most problematic concerning the role model and the service functions since the QS GmbH depends upon the voluntary deployment of human and other resources to the QS system by system participants. Thus, the QS GmbH is not able to independently act as a system and chain leader and to fully exert the necessary leadership functions. Furthermore, companies voluntarily acting as chain leaders are rare in the meat sector. The vast majority of farmers, abattoirs and meat processors are too small and lack market power to perform this task. Large retailers have the necessary power but most of them refrain from actively promoting the QS system. We consider this shortcoming one of the fundamental weaknesses of the QS system since the development of a quality-oriented industry culture is next to impossible without the existence of a chain leader who sets the pace.

Similar to the QS system, QM Milch has a general mission statement. According to this statement, QM Milch aims at a high quality standard in a more and more competitive environment and an improved image by communication with consumers, politicians and retailers. The mission of the QM Milch system was defined by a working group jointly formed by the founders of the system. But due to the unclear responsibility for further developing the core ideas of QM Milch, the lasting impact of the mission statement on system participants is doubtful so far.

The goal system consists of general statements concerning the background and overall objectives of QM Milch: national standardisation of quality management in the dairy sector, improved transparency in the food chain and growing acceptance of quality initiatives. More explicit and more operational goals can be found in the guidelines for farmers and dairies so that all in all QM Milch has a more elaborated goal system than the QS system. Nevertheless, a thorough discussion of goal conflicts is missing so far as well. Just to give an example: Many observers doubt that dairy farmers who do not meet the quality criteria will actually be expelled from deliveries to dairies since the latter are highly dependent on regular deliveries. It is generally expected that at least large milk farmers have enough market power to avoid becoming expelled even if they are not certified by auditing bodies due to serious shortcomings in the production processes. Obviously the goal conflict between the consistent and stringent application of the quality standards and the continuous supply to dairies has yet not been solved completely.

Concerning chain leadership, QM Milch has an advantage over the QS system. Whereas the latter lacks clear leadership, the former is dominated by the dairies which act as quality drivers in milk production and processing and perform the required leadership tasks. The initiative function is carried out by integrating QM Milch into the supply contracts between dairies and milk farmers. The dairies' own quality management systems which have already existed before the introduction of QM Milch act as role models for farmers. As far as continous improvement processes and regular quality controls exist in dairies, this can be interpreted as the exertion of the role model function by which dairies create broad acceptance for QM Milch. Last but not least dairies also perform some service functions by mailing checklists for self-control procedures to their farmers, supporting the implementation of self-control systems on farms and consulting to farmers.

Due to short shelf-life of milk, the dairy sector has been characterised by sophisticated quality management systems for quite a long time. This long-term development is further accelerated by QM Milch and provides a good starting point for the emergence of a quality-oriented industry culture. What is needed at the moment is a more down-to-earth discussion about quality criteria and the overall usefulness of the system. Again, dairies can be expected to play an important role by communicating their point of view throughout the milk chain.

3.3.2. Broad definition of quality

Goals such as freedom from residues and microbes and improved documentation and traceability demonstrate that in the QS system the main focus is on product and process quality attributes. From a TQM perspective, this is a somewhat narrow approach which does not take into account other relevant aspects such as the quality of working conditions and external relationships. Nevertheless, integrating quality criteria into a guideline contributes to a shared understanding of quality attributes throughout the food chain. Improved salmonella inspections, for instance, are extremely important for meat processors. For the first time this goal as well as the current salmonella status are systematically communicated to German farmers.

Product quality of milk is legally defined in the so-called Milchgüte-VO. Thus, the QM Milch guidelines as well as the external audits focus only on process quality. Similar to the QS system, other quality dimensions are not taken into account but the development of a chain-wide shared understanding of what quality means is very much encouraged.

3.3.3. Process orientation

In the QS system, participants are responsible for the correct and complete documentation of production, the necessary self-control activities and keeping to the QS guidelines. The emphasis on personal responsibilities is given expression by the priority of self-control over external controls.

Improved process control in order to safeguard the current quality level is a second aspect of process orientation. Such aspects are very rare in the QS system. Minor exceptions to the rule are the compulsory taking and storage of samples and the regular controls of all system participants.

System participants are obliged to draw their production processes as flow charts, document critical control points, develop measures for controlling these points and analyse processes and the environment. Furthermore, the participants are expected to adapt their self-control and documentation systems to changing process characteristics, correct complaints of external auditors and maintain their communication systems in a good shape. Nevertheless, continous improvement processes are imperfect and incomplete in the QS system. The system is considered too bureaucratic and inflexible and lacking a link to consulting activities.

Stressing the autonomy and responsibility of system participants and, furthermore, making use of the bottleneck function of dairies in the value chain by modifying the milk supply contracts are characteristic traits of QM Milch. The latter allows integrating all farms and all other organisations in the value chain into the system and compels all participants to carefully follow the QM Milch guidelines and use the self-control system.

The dairy industry has very good starting conditions for quality management since dairy products belong to the most extensively tested food products. In order to improve process control, QM Milch mainly refers to legal requirements. This minimises additional administrative burden and costs for farmers and improves acceptance and appreciation by system participants. Furthermore, process control is improved by preferring prevention over checking for faults and ex-post elimination of quality problems. This includes an improved control of production processes.

Through self-controls and self-evaluations farmers are able to identify personal and organisational weaknesses. Neutral external controls are another way to trigger continous improvements of processes. Nevertheless, in the QM Milch system the main focus of continous improvement is on the farm level whereas the other parts of the food chain are left aside.

3.3.4. Result orientation

Result orientation is low and considered a major weakness of the QS system. Self-control and external audits are the only evaluation mechanisms. A systematic evaluation of the success of system participants, benchmarking and the control of the effectiveness of the QS system itself have not been integrated into the system.

Due to the existence of a checklist for self-evaluation which can be applied by farmers, external controls and feedback to farmers by dairies, result orientation is a little bit stronger in the QM Milch system. Nevertheless, it is hard to tell at the moment how the effectiveness of QM Milch is assessed and who is responsible for this task. Since QM Milch is a nationally standardised system, improvements cannot be initiated by single dairies. Instead, a national approach is necessary. In how far this approach works is difficult to judge since QM Milch was developed by a working group. It is unknown so far whether this group is a standing committee and how it will work exactly in the future.

3.3.5. People orientation

A quality-oriented human resource management is only limitedly in existence in the QS system. Careful personnel selection is restricted to system participants with important functions. The QS GmbH checks, for instance, the educational background, training activities, product and industry knowledge and auditing experiences of auditors. Certifying bodies have to be accredited in accordance with EN 45011 and registered by the QS GmbH. Veterinary surgeons have to prove their work permit and a special know-how in their core competence area confirmed in writing by their professional association. Besides that no special attention is given to personnel selection. Instead, this task is completely left to system participants and their selection policies and decisions.

Legally required activities such as hygiene and health trainings are compulsory in the QS system. Auditors are obliged to take part in training sessions about auditing procedures. Outside agriculture, HACCP training and regular information about the QS Charta, are compulsory for employees. Nevertheless, systematic personnel training is not implemented in the QS system and takes only place in particular cases.

System participants are motivated by sanctions in case of infringements of the QS Charta. Punishments are harshest for certifying bodies which are immediately excluded from the QS system in case of a lack of objectivity in audits and obliging audit reports. Positive incentives stem from the design of the auditing schemes. Farms and firms rated QS Standard I are audited less often than those in categories II and III and, thus, save time and money.

Organisational measures such as quality circles or a systematic suggestion scheme are not part of the QS system.

QM Milch is also characterised by a very mixed picture concerning people orientation. The system guidelines do not tell anything about personnel selection except for farm audits which are the exclusive task of trained and accredited auditors. The situation is better concerning training and consulting which are integral parts of QM Milch. Furthermore, farmers and dairies closely cooperate with industry associations and public administrations responsible for improved quality and safety of milk and dairy products.

Similar to the QS system, sanctions in case of infringements of the QM Milch regulation are the major incentive for system participants. The imposed punishments depend on the heaviness of the infringement. Farms and firms which do meet criteria concerning keeping animal registers, use of veterinary drugs or protection of the environment are reported to the relevant public authorities. Infringements of legal requirements are punished according to the sanctions imposed by the law. Positive incentives for quality and safety improvements have yet not been implemented in the QM Milch system.

Although the dairy industry offers favorable organisational conditions due to the existence of producer associations, an organisational infrastructure for exchanging experiences has not been considered when developing the QM Milch guidelines.

3.3.6. Customer and public orientation

The QS system mainly considers legal requirements and common industry standards. Furthermore, the system is characterised by a somewhat narrow definition of quality which emphasises freedom from residues and microbes and improved documentation and traceability. Other quality criteria relevant for consumers' buying decisions such as colour, texture and shelf-life of meat and meat products have not been considered relevant. Systematic market research in order to track consumer expectations and requirements and customer satisfaction is not conducted. Only the ban on animal fat in feed and mechanically recovered meat obtained from the bones of animals in meat products can be attributed to consumer concerns after the rise of the BSE crisis. Meat processors' requirements have guided the thorough improvement of salmonella monitoring in Germany. Apart from the aforementioned exceptions to the rule, the QS system is hardly oriented towards customers.

Public orientation was the guiding principle for including some basic animal welfare aspects into the QS guidelines. But since the QS system does not want to establish above-average standards and deterioate the international competitiveness of German farmers, it refrained from including far-reaching claims of nongovernmental organisations concerning improved husbandry conditions, limited herd size or GMO-free feed.

The QS system is characterised by a comparatively low level of communication with consumers. Thus, the QS label is not very well known by German consumers. Furthermore, the communication strategy of the QS GmbH is somewhat inconsistent. On the one hand, the QS system is designed to guarantee a minimum level of food quality and safety. On the other hand, QS meat is sometimes advertised as above-arage. Empirical studies show that the QS label is a quality signal for weak brands and no-name products but not for strong brands (Enneking, 2004). The latter are rare in the German market for fresh meat but exist in markets for processed meat such as sausages and convenience products.

The situation is somewhat different in the dairy sector. Dairies are often certified after ISO 9001 and are used to ISO's process model and systematic market research. Thus, customer and consumer requirements are usually well-known. Integrating dairy farmers into the QM Milch system allows systematically communication of customer requirements to farmers and improves customer orientation in the dairy sector.

Regarding other stakeholders, the QM Milch guidelines demand, for instance, regaining acceptance of agricultural production in society. Nevertheless, QM Milch mainly sticks to legal requirements in order to avoid additional economic pressure on German dairy farmers. Thus, most requirements concerning feed controls and environmental aspects stem from the relevant laws. Since all these criteria are so-called knock-out criteria for farmers taking part in the system, there is at least symbolic action stressing the high relevance attributed to these aspects.

Although communication of a QM Milch label or something like that has not been planned so far, at least beginnings of an advanced customer and public orientation are recognisable in the QM Milch system.

4. Discussion of results and managerial implications

We can summarise that from a TQM perspective both quality management discussed so far are characterised by strengths and weaknesses (see Figure 2). Both systems fail to fully meet the major TQM principles but QM Milch has at least some advantages over the QS system due to the active chain leadership pursued by dairies.

One might argue that the observed weaknesses are hardly relevant at all since both systems only aim at guaranteeing a minimum standard of food quality and safety. But from a TQM perspective it is doubtful whether even this undemanding goal will be met: '[...] TQM is an all-embracing strategy. [...] TQM cannot be restricted to individual departments in the company; similarly, TQM is impossible when the concept has been embraced only at management level.' (Pfeifer, 2002: 5). Or coined differently: Quality management does not work when some of its central elements are not implemented. When we also take into account that TQM is generally considered a blueprint for successful quality management which summarises state-of-the-art knowledge about quality and safety improvements, the effectiveness of the newly introduced quality management systems seems doubtful. In this case, narrowly focused improvements such as more advanced salmonella monitoring would be easier to implement and cheaper as well.

We know from ISO 9001 that it had a considerable impact on organisations as long as it was predominantly used in industries such as automobiles and automobile suppliers. Due to a long history of quality management, specialised quality management departments and powerful chain leaders in these industries, quality improvements actually took place after certification. But the situation is very much different in small and medium-sized companies which are often certified after ISO 9001 without actively 'living' the idea (Walgenbach, 2000). In those companies, ISO certifications are often hardly more than 'myth and ceremony' (Meyer and Rowan, 1977; Beck and Walgenbach 2003). If the QS and the QM Milch systems which have also been designed for implementation in small and medium-sized farms and firms want to avoid this reproach, improvements of system design are necessary. The following major improvements are paramount:

- Strengthening goal orientation by developing an operational goal system which contains clear statements concerning goal conflicts. This is especially relevant for the QS system which in this respect is characterised by more deficiencies than QM Milch. Goals have an important motivational function, i.e. direct behavior of system participants in everyday operations. Up to now the QS system mainly relies on enumerating duties, for instance concerning documentation, and control through external audits and certification procedures. Since it is very difficult to avoid opportunistic behavior only through controls, generating more intrinsic motivation to act in the best interest of customers and the quality management system is required. This can also be considered a first but important step towards a more quality and safety-oriented industry culture.
- Result orientation is too weak and should be strengthened by a systematic evaluation of strengths and weaknesses of both systems and a benchmarking approach. Benchmarking can be used, for instance, to evaluate procedures in participating farms and firms as well as differences and similarities between certifying bodies. So far the audit market is very intransparent and there is much uncertainty concerning neutrality and quality of external audits (Jahn *et al.*, 2004).
- Both quality management systems have an under-developed people orientation. Thus, an improved quality-oriented human resource management is necessary. More systematic training activities and organisational support by, for instance, chain-wide quality circles or other instruments for supporting a more systematic exchange of knowledge and experiences are paramount for both systems.

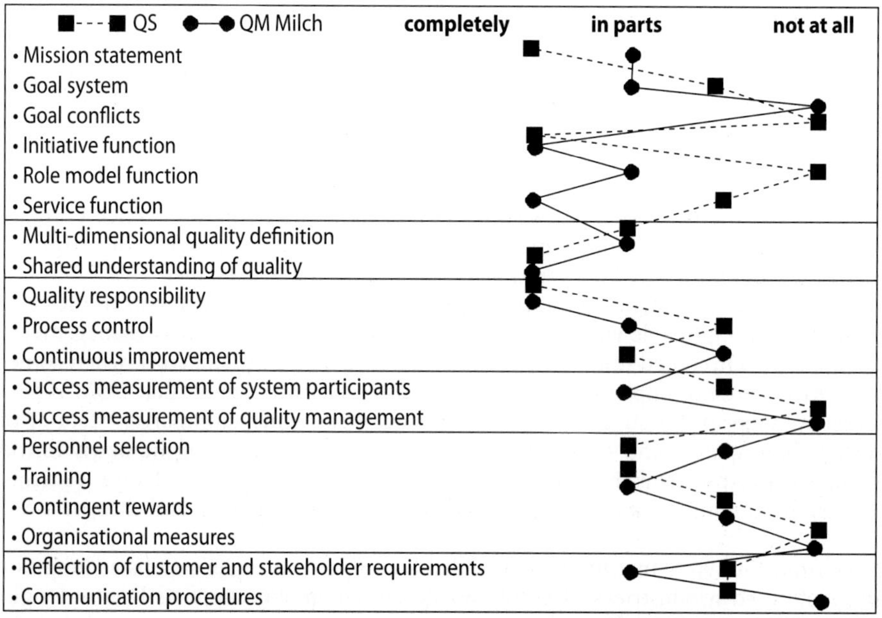

Figure 2. Summary of results.

All these measures have to be implemented by carefully taking into account the large number of small and medium-sized farms and agribusiness firms. Benchmarking, for instance, should be supported as fas as possible by information exchange through the Internet. Quality circles can be organised as regional working groups meeting from time to time to enhance information exchange. Successful examples exist in other industries in which they are organised, for instance, by the German Society for Quality. Industry associations and influential and well-known individuals can also play a more active role in influencing industry culture (Theuvsen, 2004a). Thus, although agriculture and the food industry certainly have structural disadvantages concerning chain-wide quality management, a more ambitious structural and methodological underpinning of the existing certification schemes is not impossible. For the achievement of the self-set goals concerning food quality and safety it is even inevitable.

References

Abraham, M., J. Crawford and T. Fisher, 1999. Key factors predicting effectiveness of cultural change and improved productivity in implementing Total Quality Management. International Journal of Quality & Reliability Management 16, 112-132.

Beardsell, M.L. and B.G. Dale, 1999. The relevance of total quality management in the food supply and distribution industry: A study. British Food Journal 101, 190-200.

Beck, N. and P. Walgenbach, 2003. ISO 9000 and formalization: How organisational contingencies affect organisational responses to institutional forces. Schmalenbach Business Review 55, 294-320.

Enneking, U., 2004. Willingness-to-pay for safety improvements in the German meat sector: The case of the Q&S label. European Review of Agricultural Economics 31, 205-223.

Feigenbaum, A.V., 1991. Total quality control. 3rd ed., McGraw Hill.

Freeman, R.E., 1984. Strategic management: A stakeholder approach. Pitman.

Hackman, J.R. and R. Wageman, 1995. Total quality management: Empirical, conceptual and practical issues. Administrative Science Quarterly 40, 309-342.

Jahn, G., M. Peupert and A. Spiller, 2003. Einstellungen deutscher Landwirte zum QS-System: Ergebnisse einer ersten Sondierungsstudie. Working Paper University of Goettingen.

Jahn *et al.*, 2004. Abschlussbericht zur Evaluationsstudie QM Milch. Working Paper University of Goettingen.

Jahn, G., M. Schramm and A. Spiller, 2004. The quality of certification and audit processes in the food sector. In: Bremmers, H.J. *et al.* (eds.), Dynamics in chains and networks: Proceedings of the 6th International Conference on chain and network management in agribusiness and the food industry, pp. 351-358

Kamiske, G.F. and C. Malorny, 1992. Total Quality Management: Ein bestechendes Führungsmodell mit hohen Anforderungen und großen Chancen. Zeitschrift Führung und Organisation 61, 274-278.

Kaplan, R.S. and D.P. Norton, 1996. The Balanced Scorecard: Translating strategy into action. Harvard Business School Press.

Locke, E.A. and G.P. Latham, 1990. A theory of goal setting and task performance. Prentice Hall.

Luning, P.A., W.J. Marcelis and W.M.F. Jongen, 2002. Food quality management: A techno-managerial approach. Wageningen Academic Publishers.

Meyer, J.W. and B. Rowan, 1977. Institutionalized organisations: Formal structure as myth and ceremony. American Journal of Sociology 83, 340-363.

Montes, J., A.V. Jover and L.M.M. Fernández, 2003. Factors affecting the relationship between Total Quality Management and organisational performance. International Journal of Quality & Reliability Management 20, 189-209.

Pfeifer, T., 2002. Quality management: Strategies, methods, techniques. Hanser.

Schein, E.H., 1992. Organisational culture and leadership. 2nd ed., Jossey Bass.

Theuvsen, L., 2004. Transparency in netchains as an organisational phenomenon: Exploring the role of interdependencies. Journal on Chain and Network Science 4, 125-138.

Theuvsen, L., 2004a. Motivational limits to tracking and tracing: Principal-agent problems in meat production and processing. In: Schiefer, G. and U. Rickert (eds.), Quality assurance, risk management and environmental control in agriculture and supply networks. Vol. A. ILB Press, pp. 223-230.

Theuvsen, L., 2005. Quality assurance in the agrofood sector: An organisational-sociological perspective. In: Odening, M., K. Hagedorn and J. Nagel (eds.), Umwelt- und Produktqualität im Agrarbereich (in press).

Theuvsen, L., H. Brand-Sassen and S. Essmann, 2005. Artgerechte Tierhaltung zwischen Wunsch und Wirtschaftlichkeit: Analyse der Einsatzmöglichkeiten des Target Costing. In: Landwirtschaftliche Rentenbank (ed.), Entwicklungspotenziale ländlicher Räume: Landwirtschaft zwischen Rohstoffproduktion und Management natürlicher Ressourcen, pp. 113-154.

Wagner, K.W., 2003. PQM - Prozessorientiertes Qualitätsmanagement: Leitfaden zur Umsetzung der ISO 9000:2000. 2nd ed., Hanser.

Walgenbach, P., 2000. Die normgerechte Organisation: Eine Studie über die Entstehung, Verbreitung und Nutzung der DIN EN ISO 9000er Normenreihe. Schaeffer-Poeschel.

Zollondz, H.-D., 2002. Grundlagen Qualitätsmanagement: Einführung in die Geschichte, Begriffe, Systeme und Konzepte. Hanser.

Regional quality programs: relevance, objectives and strategies

Oliver Poignée and Gerhard Schiefer
University of Bonn, Chair of Business Management, Organisation and Information Management, Meckenheimer Allee 174, 53115 Bonn, Germany; o.poignee@uni-bonn.de

Abstract

Enterprises in the agro-food sector presently have to deal with a multitude of legislative as well as market-related demands on the assurance of quality, hygiene and safety of their products. Requirements primarily focus on individual companies. However, striving for an efficient realisation, businesses along the network-wide production process have to cooperate. On this note, the article discusses in a first instance, in which dimensions intra- and inter-organisational decisions have to be made in order to establish network-wide quality structures. In a second step, the article presents in which manner enterprises of the agribusiness sector currently make their decisions citing quality programs in the German breadstuffs industry as an example.

Keywords: quality programs, quality chain management, networks, breadstuffs industry

1. Introduction

Companies in the agribusiness are confronted with a wide variety of demands on the assurance of quality and safety of their products and processes. The requirements are uttered by different groupings including legislation, general quality standards, service providers (such as insurers and banks) as well as commercial and private consumers.

In the light of previous food safety problems, the demands all aim at improving the agro-food sector's capacity to guarantee. Reliable granting of warranties requires a closer vertical cooperation of all organisational entities involved in producing, processing, transport and distribution of foodstuffs. Intensified communication and an extensive coordination of quality levels as well as quality support systems (ISO etc.) are pivotal elements for this networking.

The complex requirements for a more coordinated quality production affect both development paths of agrarian markets:

- price and cost-oriented markets for *bulk commodities*; as well as
- quality and high value-oriented markets for *niche products.*

However, when implementing network-wide cooperation, needs for reorganisation are a good deal bigger for businesses in the mass markets. For the most part, they have so far designed their production, distribution and logistic systems for anonymous bulk products. Contrary to this, traditional niche market suppliers are one step ahead in vertical coordination. Quality programs are in place as one form of organisation for these vertical-integrated niche market structures. They make use of their integrated, quality-oriented production and distribution networks to bestow advanced warranties to their customers.

Against the background of dynamically changing general conditions in the agribusiness sector, this paper discusses the status quo of quality programs regarding the organisation of network-wide quality structures. The German breadstuffs industry will be in the spotlight of the article.

2. Network-wide coordination of quality production: complexity of decision making

2.1. From enterprise activity 'quality management' to quality chain management

Quality management is the alignment of all enterprise activities to the best possible fulfilment of customer requirements. These demands are becoming increasingly complex without regards to their source being either commercial or private consumers. A part of customers asks for quality attributes, which cannot be verified at intermediate and final products (origin, specific production methods etc.). Due to a vast absence of control options for these process attributes, customers demand quality warranties to minimise their uncertainties.

The agro-food sector is afflicted with complexities in terms of granting such warranties (e.g. fragmentary information exchange, varying quality characteristics, inconsistent quality regimes). This leads to uncertainties concerning qualitative and quantitative conditions as well as safety of agricultural commodities, which accumulate in the course of networks.

Traditional quality management concepts admittedly reduce this network-view to a sole sequence of customer-supplier relationships. This can result in a suboptimal quality performance of the network as a whole: if every business only tries to fulfil its own objectives and requirements of its direct trading partners at the best, it cannot be assured, that demands of subsequent network nodes is taken into account for process organisation to an adequate degree, too. Instead of linking exclusively enterprise stand-alone systems, new quality concepts demand for a more intensive coordination of production and communication along the whole agribusiness network. Therefore, enterprise-specific quality management systems have to be supplemented with network-wide elements in a next step.

2.2. Decision dimensions of a network-wide coordinated quality production

Multiple fields of science traditionally address the design of interchange relationships between network nodes: New Institutional Economics, Strategic Management, Network Theory, Social Capital Theory, Resource Dependency Theory, System Theory, Game Theory, Supply Chain Management, Information Technology etc. In addition, the increasing focusing on quality aspects asks for a more intensive involvement of quality management in this scientific network. In the light of heterogeneous agricultural structures, coordination of Quality Chain Management measures first of all leads to a complex entrepreneurial scope. The heterogeneity of the agribusiness sector causes a bounteousness of different intra- as well as inter-enterprise requirements on the development of quality structures. This yields large-scale demands on organisational and technological flexibility of such structures.

The University of Bonn accompanies businesses, chains and networks of the agro-food sector through all phases of coordinating quality production in joint projects (further information at www.qm-g.de). One main objective is the development of a reference model as a decision supporting instrument for the implementation of network-wide quality structures.

The above mentioned fields of research particularly focused on the identification of a decision scope for quality-related networking. They provided the base for the definition of decision dimensions, decision variables (Table 1), sub-variables and variable values. Then, the theoretical considerations were reviewed with regard to their practical relevance in several projects with businesses with different structures. At the same time, typical decision archetypes concerning the selection of variable values were determined (even against different commodity groups); the first step of a decision model to be developed.

The analysis of quality programs introduced in the course of this article can be considered as an additional instrument for verifying the decision scope on this note. Moreover, it should answer the question on which variable values were chosen by the analysed programs. Quality programs were selected for the analysis as they traditionally are of several stages of the production chain (cp. chapter 3).

3. Quality programs as a form of vertical-coordinated quality structures

According to Schiefer (2003), two basic types of organisation assuring quality, hygiene and safety of food exist:

- *Coordination* of quality initiatives between different food networks and product lines.
- *Specialisation and concentration* on single food networks and product lines.

Quality programs are representatives of the second alternative. Within well-defined structures, they try to establish self-contained profiles to differ from the mass market in quality and external economies. The higher intensity of vertical-coordinated cooperation assists the programs to deliver warranties to a higher extent in comparison to sectoral, open initiatives. Due to their sector-wide claim, open standards (such as Q&S, BRC, IFS, EurepGAP, GMP+) are afflicted with coordination and control problems concerning the granting of warranties. In addition, open standards normally lack explicit signals, for example risk assessment and risk acceptance criteria (Schiefer, 2003). Therefore, realised improvements in the assurance of quality, hygiene and food safety cannot be communicated to consumers to the desired extent.

Table 1. Dimensions and variables of the decision scope for implementing quality networks.

Decision dimensions	Decision variables	
Strategies	enterprise orientation	strategic focus
Quality management functions	elements	intensity of design
	sequence of integration	intensity of integration
Network organisation	network length	intensity of integration of network participants
	network breadth	network coordination
Process design	strategic processes	commodity flow
	management processes	flow of information
Information technology assistance	technology	organisation of implementation
	organisation of infrastructure	organisation of support

Quality programs tempt to capitalise on these structural and communicative disadvantages of open systems. For that purpose, they mostly support the communication of their activated warranties by a label. However, brand policy of quality programs does not solely lead to an enhancement of their outward-looking market impacts. It develops an inward-looking identification of participants with program objectives and guidelines at the same time. The common brand policy creates a certain 'program culture' in this way (Kagerhuber and Kühl, 2002). This contributes to a reduction of felt disadvantages due to partial abandonment of entrepreneurial autonomy in the course of program participation.

The assurance of promised quality attributes (such as product or production characteristics) is partly implemented by the programs based on quality support systems. Subject to their marketing statements, several programs formulate additional presets on the quality of products, process organisation, process documentation and/or process control. Based on these presets, decisions on selection and implementation intensity of quality coordination activities mentioned in chapter 2.2 have to be made. In which way enterprises of quality programs in the German breadstuffs industry presently accomplish this, is presented in the following chapters.

4. Market analysis of quality programs in the German breadstuffs industry

Scientifically established information on practical implementations of components of the decision scope for quality networks discussed in chapter 2.2 are principally reduced to the meat sector (e.g. Kagerhuber and Kühl, 2002). This is primarily due to past problems in the meat industry in terms of assuring food safety. Therefore, the meat sector leads the way establishing vertical-coordinated quality structures.

However, almost no equivalently detailed analyses were conducted for other branches. On this note, the survey of quality programs in the German breadstuffs industry makes a contribution to generate comparable knowledge about not yet analysed branches.

4.1. The German breadstuffs industry

Complementary to the above discussed requirement complexities and general characteristics of the agribusiness sector, the attributes of the breadstuffs industry as listed below additionally aggravate the network-wide coordination of quality production (Figure 1):

- *Heterogeneous sector structures* with widely ramified commodity flows and complex transformation processes of the raw material breadstuffs.
- Purchase of *grain as a homogeneous bulk product* at international spot markets to produce certain flour qualities.
- *Different quality understanding* of grain at different network levels (farmers: harvesting characteristics, mills: flouring characteristics, bakeries: baking characteristics).
- *Missing knowledge* about the influence of production methods at primary network levels on in- and external qualities of own raw materials.
- Traditional information and communication processes as *'series of disconnects'* (Bouma, 2000).

A couple of the described restraints (e.g. natural variances) do not solely affect the mass market for breadstuffs. Enterprises at the market for quality grain with quality programs being part of this market, also have to deal with these barriers. Their method of solution consists of their integrated, quality-oriented production structures.

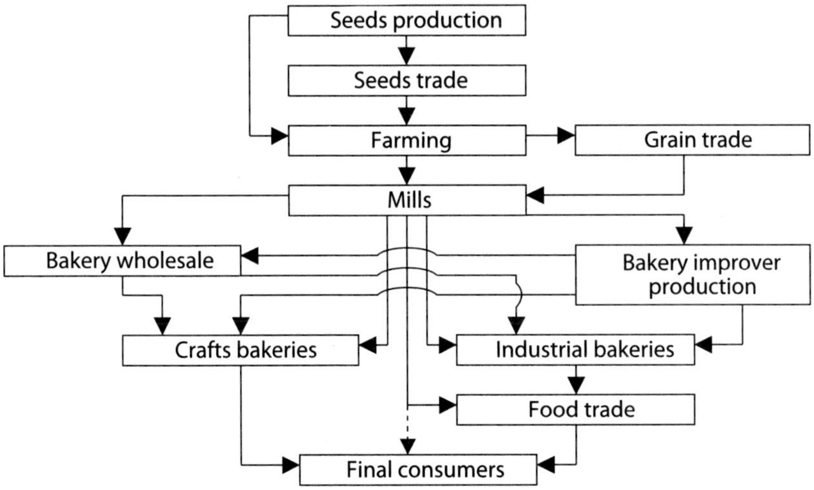

Figure 1. The network for breadstuffs.

Chapter 5 explains which implementation strategies were chosen to support this approach. But first of all, in the following chapter a survey of existing quality programs in the German breadstuffs industry will be presented.

4.2. Main findings of the market analysis

The analysis of the German breadstuffs industry concerning relevance and design of quality programs was divided in two main parts:
- a market analysis; and
- a detailed analysis of selected programs.

The market analysis formed the basis for the whole survey. Within literature and internet enquiries as well as expert interviews, the network for breadstuffs was checked for the general existence of programs. The first market analysis phase included a purely quantitative enquiry for existing quality programs. In a second step, the determined programs were analysed to that effect, if they still were in business and if they met the analysis' prerequisite. The research framework defined quality programs as follows: the programs had to have a network length of at least three levels and they had to operate at the market in a level-overarching manner. Both requirements allow the chain respectively network thought.

By means of the above described analysis phases, a universe of 48 quality programs was identified. Clear south-north as wells as west-east gradients were detected (Table 2).

On the one hand, a reason for this is the more extensive government aid particularly in Bavaria (for three years). On the other hand, the proportionally to the eastern and northern federal states small-sized agriculture and processing industry is an argument for a greater willingness to participate in vertical-coordinated structures. As a result of their small enterprise sizes and their greater number, a lot of businesses are restricted in terms of an active market development.

Table 2. Regional distribution of the identified quality programs (Poignée and Pilz, 2005).

Federal state	Entries	Federal state	Entries
Total	48	Lower Saxony	2
Bavaria	17	Schleswig-Holstein	2
Baden-Wuerttemberg	6	Thuringia	1
North Rhine-Westphalia	6	Berlin/Brandenburg	1
Hesse	4	Mecklenburg-Western Pomerania	1
Rhineland-Palatine	4	n.s.	2
Saxony	2		

Therefore, an enhanced vertical as well as horizontal coordination opens up greater options to bundle capacities and to forge sustainable customer and supplier ties.

A more detailed study of the identified quality programs to their existence and conformity with the chosen analysis approach resulted in the following findings: Five programs were no longer in business at the time of the market analysis. They were not able to bring off sustainable economically profitable production and marketing systems. Higher costs for quality production (documentation, controls etc.) could not be balanced by an adequate price discrimination compared to the spot market. For 13 quality programs, no specific information or contact persons were available even after an intensive enquiry in different media. Therefore, no statements can be formulated for these programs regarding their existence or their conformity with the given analysis approach. Even if the programs still operate at the market, their missing medial availability contrasts an efficient brand policy. Following their own opinion, nine programs did not apply to the definition of quality programs as outlined for the market analysis. Consequently, 21 quality programs remained after finishing market analysis. They formed the universe for the following detailed analysis.

5. Detailed analysis of selected quality programs of the German breadstuffs industry

5.1. Methodology

Of the 21 relevant programs, 16 agreed to take part in a detailed analysis of their system characteristics. The analysis itself was realised via a standardised questionnaire. The questionnaire was subdivided into six main blocks each consisting of 2-6 inferior points. The main blocks were derived from the dimensions of the decision scope in chapter 2.2:

- *General program characteristics:* region, product lines, lot size, *trade radii.*
- *Network structures:* network length, network breadth.
- *Objectives and strategies:* foundation initiative, year of foundation, objectives, strategies to meet these objectives.
- *Network coordination:* distribution of authorities to decide, distribution of costs and benefits, program openness, integration of external institutions in coordination, ways of sanctioning and motivating.
- *Quality Chain Management elements:* implementation of quality design, quality assurance, quality improvement, quality control and tracking & tracing.

- *Design of information exchange:* types of exchanged information, information channels, organisation of information exchange, information technology assistance.

At large, data interpretation was carried out in a qualitative way. In addition, important statements were validated using analyses of correlation.

5.2. Main findings

5.2.1. General program characteristics

The *regional distribution* of the analysed programs shows south-north as wells as west-east gradients according to the results of the market survey. Thus, less programs are seated at typical locations of grain production. *Vice versa*, the identified programs almost solely produce, process and distribute on periphery locations. There, participation in quality programs is often the only option to compensate for economies of scale-based competitive advantages of typical cultivation regions. On this note, all analysed programs offer regionally closed cycles. Merely two programs realise *trade radii* of more than 100 km.

The *product lines* 'grain' and 'flour' are part of all programs. Bakery products and other goods like bread mixes as well as fruits and vegetables supplement the commodity groups. 15 programs are specialised in breadstuffs products; only one program distributes a wide range of goods under regional aspects.

The determination of *qualitative bundle potentials* finishes the illustration of the general program characteristics (Table 3). The bandwidth of processing lot sizes is between 80 and 75,000 t/year. In terms of production quantities, the analysed programs have a stake in the German total breadstuffs comminutions of approximately 2%. The share of program farmers in the total agricultural breadstuffs-sales in Germany is approximately 1%. Broken down to single federal states, program shares between 0.02 and 9% (for agricultural sales) as well as between 0.03 and 23% (for breadstuffs comminutions) were identified. The top values are realised by the biggest program with a comminution of 75,000 t/year. Consequently, this program by itself represents about 60% of total quantities of the analysed market for quality programs.

5.2.2. Network structures

The basis for analysing network structures was the finding of *network lengths and network breadths* of the programs. Typically linear chain structures were identified. The network levels 'farmers' and 'mills' are integral parts of all programs. Both levels have a special interest in cooperating with network partners. Reasons are:

Table 3. Breadstuffs processed in 2002/03 (in t/year; n=16; Poignée and Pilz, 2005).

Quantities	Entries	Quantities	Entries
< 100	1	5,000-10,000	-
100-500	2	10,000-20,000	2
500-1,000	-	> 20,000	1
1,000-5,000	7	n.s.	3

- Qualitative procurement backup with defined market prices and guarantee of market access (by customer as well as supplier loyalty).
- Qualitative as well as hygienic dependency of mills on cultivation, stock-keeping and transport methods at farm level.
- Small-sized structures at both levels.
- Length of the breadstuffs chain, which leads to the fact, that farmers are 'far away' from end-consumers. To have the ability to communicate the excellence of their products and production, farmers depend on downstream levels.

The level 'grain trader' is present in only one program. In the context of their close collaboration, farmers and mills wilfully exclude the trade level to save trade margins. In seven of the analysed programs, the functions of grain traders (bundling of offers, service provider) are taken over by the level 'agricultural association'.

With 14 entries, the level 'crafts bakeries' is also integrated to a significant extent. The ambition to build up continuous guaranty as well as communication chains up to the final consumers invest the integration of the corresponding cut surface with a particular relevance. The level 'others' is integrated in seven programs. Institutions, which play a part in chain coordination (e.g. consortiums or marketing associations), have the highest impact thereby. Typical chain lengths could not be identified in the analysis. They vary between three (7), four (5) and five (4) stages.

Considering that the number of businesses per level solely are snap-shots, the following typical *chain breadths* were constituted:

- 'Farmers' with up to 50 producers to bundle marketable lot sizes.
- 'Mills' with one enterprise and the function of a bottleneck. Mills as important program initiators are generally not interested in cooperating with rival businesses. In case of programs with more than one mill included, the foundation initiative comes from the level 'farmers'. In that cases, participation of mills is reduced to a contract-based milling.
- 'Crafts bakeries' with up to 50 businesses as the case may be with enterprises with up to 50 subsidiaries to sell the produced and processed goods at the market.
- 'Agricultural associations', 'grain traders' and 'others' with one enterprise/institution.

For all analysed chain levels, the identified *enterprise sizes* mirror country-typical market structures. The enterprises' small-size for the most part is one argument for program participation. Only the integrated farmers in Eastern Germany are partially specialised in crop production. The farmers in Southern Germany are typical mixed businesses.

5.2.3. Objectives and strategies

Initiatives to launch the programs primarily come from 'farmers' (9) and 'mills' (8), partially in combination of both levels. In doing so, the focal point of *program foundations* was in the middle of the 1990s. At that time, problems regarding food safety acted as an important catalyst for program founding. Since then, new foundation rate has continuously decreased. Against the background of several program failures during the last years, the market for quality programs in the German breadstuffs industry has contracted at large.

Objectives at program foundation are outlined in Table 4.

Table 4. Program objectives at foundation (n=16; Poignée and Pilz, 2005).

Objectives		Entries
Economic objectives	customer and supplier loyalty, increased creation of value for every participant, competitive advantages, increased marketing options, coverage of market prices and sale quantities, increased sales, differentiation from mass market in the event of loss, maintenance/adding of jobs	14
Environmental protection		9
Regionality		8
Quality objectives	well-defined quality, quality assurance, quality improvement, verification of quality capability	8
Transparency	assurance of customer confidence, verification of product origin, reduction of inner-chain uncertainties, increased risk and crisis management	6
Shaping of opinions		1

In the first instance, economic objectives played a superior role. Due to the increasing legal as well as market requirements, qualitative and transparency-related aspects have constantly gained in importance. *Strategies* to meet these objectives are outlined in Table 5.

In addition to the listed aspects, problems concerning the sale of bigger trade volumes at regional markets also have a great influence on program strategies. Market penetration at traditional regional markets is limited to these sale problems. Against that background, market development seems to be a promising strategy. Yet, practical examples show, that previous attempts were not successful. Price discrimination opposite to bulk flour, which can be realised in the own region, is normally not viable at nationwide markets. Profitability of market development is only given

Table 5. Correlation of identified objectives and strategies (n=16; Poignée and Pilz, 2005).

Objectives	Strategies
Economic objectives	higher market prices, fair price systems, cultivation and supply contracts, trade name, sales promotion, public relation, participating leadership, integration of marketing service providers, up to date producing and processing techniques
Environmental protection	directives regarding production management
Regionality	trade name, marketing, public relation, controls, short itineraries
Quality objectives	supplier audits, documentation, cultivation and supply contracts, directives regarding production management, quality controls, training of employees, latest producing and processing techniques, participating leadership, integration of extension services
Transparency	documentation, labelling, traceability, cultivation and supply contracts, participating leadership, consumer information
Shaping of opinions	consumer information, consumer education

up to definite radii in view of higher distribution and logistic costs. Therefore, most programs focus on an assurance of market access instead off market penetration or development strategies. Covering the status quo becomes a prior objective.

5.2.4. Network coordination

According to Helbig (1997), three basic *organisational models* for network coordination exist: contract-oriented, commanded and participating. The analysed programs feature a combination of these alternatives to allocate the authority to decide. Within the different combinations, contract-based power structures (8) prevail compared to externally commanded (6), internally commanded (6) and participating (4) models. External institutions, which are integrated in network coordination, are mainly extension and control services. Withal, the intensity of organisational involvement of chain participants increases from contract-oriented to commanded and finally participating structures. The type of organisation depends on the network level or enterprise, which has brought the quality initiative into being.

Just like in every cross-company organisational structure, quality programs are also confronted with complexities due to a possible differing of objectives between businesses. Therefore, firm co-operations rely on a general agreement about added values to be reached and their allocation. In principle, quality chains and networks aim for 'win-win-situations' for every enterprise as well as for the chain as a strategic economic entity. In this spirit, 11 of the analysed programs show a proportionate *allocation of costs as well as benefits*. Farmers receive a program additional value (about 5-10% above market quotation), mills obtain higher flour prices from bakeries (about 10% above market price) and bakeries sell their products with a higher market price (about 10-35%). Two programs have no apportionment of costs/benefits, one program separate added values for the farming level.

With respect to the *openness of program organisation*, three general variants were identified: closed (8), semi-open (4) and open systems (3). Closed systems have advantages as to the assurance of their capacity to guarantee. Indeed, in the course of their differentiation from the mass market, qualitative and quantitative shortages could arise due to climatic conditions or plant diseases.

5.2.5. Quality chain management elements

Within the scope of the analysis has been carried out, the implementation of all Quality Chain Management elements (see 5.1) were researched. The following details are restricted to the functions 'quality design', 'quality assurance' and 'tracking & tracing' due to their present relevance for the agribusiness.

In most cases, *quality design* is realised by a combination of licensed quality directives, general quality categories and enterprise-specific presets [normally given by the leading business(es)]. These additional presets result in a higher quality claim compared to the mass market. Quality definitions of the programs basically are: regionality, transparency, customer-specific product quality and environmental protection. Especially in the formation phase, external institutions like extension services were also involved in the process of deciding on product, production process as well as resource specifications.

In terms of *quality assurance*, seven main blocks can be outlined: product-related controls (11), implementation of general quality standards (8), documentation of quality measures (7), process-related controls (6), extension services (4), program directives (3; regarding cultivars, cultivation methods, stock-keeping, product qualities, etc.) and closed program structures (2). Another finding was, that, by the majority, realised quality assurance measures have a solely individual-enterprise focus. Only in the case of extensions services and program directives a certain chain encompassing coordination of quality measures are implemented.

According to this, general quality standards are of a high importance ensuring quality production. The network-level specific distribution of quality standards is depicted in Table 6.

Implementation of quality standards is most widespread at milling- and bakeries-levels. Among other things, this can be deduced from legal requirements, which were obligatory for both but not for upstream levels. Farmers and grain traders were traditionally taken out from legal initiatives (e.g. food hygiene regulations). In the context of new regulations, they meanwhile are an integral part of the whole liability network. This is one reason for an increased adoption of quality systems especially during the last year. Anyway, it must be mentioned, that the identified realisation of quality standards is just a snap-shot. Even since the second half of 2004, the standards 'IFS' (for mills) as well as 'EurepGAP' (for farmers) have gained importance in the German breadstuffs industry. Due to the fact, that program farmers often are not liable to a whole delivery duty and that program mills often market two different product lines (program flour as well as non-program flour), both standards are of a certain importance for the programs, too. Therefore, it can be assumed, that the total number of implemented quality systems has definitely increased in the programs up to now.

Other than the implementation of general quality standards, the development of practical *tracking and tracing* systems henpecks the present discussion in the breadstuffs industry. Table 7 shows implementation alternatives, which were chosen by the analysed programs.

Table 6. Network-level specific quality standards (n=8; Poignée and Pilz, 2005).

Network-level	Quality standards
Farmers	HACCP (2)
Agr. Associations	HACCP (2), ISO 9000 (1)
Grain traders	GMP+(1)
Mills	HACCP (4), GMP 13 (3), ISO 9000 (2), GMP+ (1)
Crafts bakeries	HACCP (4), ISO 9000 (1)
Others	-
No exact assignment possible	HACCP (2), ISO 9000 (2), GMP 13 (1)

Table 7. Implementation of tracking and tracing (n=16; Poignée and Pilz, 2005).

Method	Entries	Method	Entries
Documentation	14	Contracts	1
Controls	13	Chain-wide information system	1
Retain samples	9	IT-based data bank	1
Labelling/proof of origin	6	Quality separation	1
Exchange of documents	3		

Information technology is barely used to allow for an efficient up- and downstream traceability of commodities. Traceability-related data are basically collected by established inner-enterprise forms. To that effect, the detected tracking & tracing methods are mostly limited to isolated and level-specific measures. Chain encompassing crisis management regimes are only realised in few programs.

5.2.6. Design of the network-crossing information exchange

The improvement of efficiency and intensity of cross-chain level information exchange is one central competitive factor for quality production. At the mass market, grain and flour are mostly traded as anonymous bulk commodities. Therefore, flow of information is limited to the exchange of transaction-related data as well as product-related laboratory data. In addition, the researched quality programs include other *infotypes* in their transfer. Trade (10) and product information (8) still are of major interest, but quality requirements (9) as well as process information (9) complete the exchanged information portfolio.

The *organisation of information* exchange is characterised by a high percentage of direct communication (Table 8).

As a result of the predominance of face-to-face communication, *information channels* also exist between businesses, which are not direct trading partners (in 10 cases). The programs also stated, that the *directness of information flow* is up- and downstream. In spite of this integrated information management approach, it must be added, that gatherings and circulars are no exchange types with a high frequency. The regularity of information transfer must be enlarged therefore, especially using standardised information technology.

The advantageousness of *IT-based systems for acquisition, storage and exchange of information* is presently discussed with respect to new legal and market requirements. The researched programs have not adopted these supporting technologies to an appreciable extent yet. Indeed, individual enterprise IT-systems are partially at work. However, these systems stand alone in a chain-wide view. 12 programs have no chain-crossing IT-solution at work, 4 programs go about an according system. Only two systems already have implemented information exchange via an IOIS (Inter-organisational Information System).

Table 8. Organisation of information exchange (n=16; Poignée and Pilz, 2005).

Alternative	Entries	Alternative	Entries
Gatherings	8	Exchange of commercial documents	2
Circulars	6	Exchange of quality documents	1
Communication via traditional media	4	No explicit specification	1
Direct verbal communication	4	n. s.	1

6. Conclusions

One obstacle in the conduction of the analysis was the fact, that the collection of significant data was very difficult. Information regarding existing programs was distributed among only few experts. Therefore, part of the programs could only be detected in separate internet enquiries, oftentimes by secondary sources. On this account, the described market analysis does not claim to be representative. The dynamics of the breadstuffs industry only allow a temporally limited snapshot. In addition, the authors act on the assumption, that several programs could not be identified due to their low name recognition by the interviewed experts as well as to their missing medial presence. For this reason, it cannot be reliably rated, which share the identified 48 programs have in the total amount of quality programs in the German breadstuffs industry. Nevertheless, the researched programs represent a sufficient basis for a detailed analysis of techno-organisational quality program characteristics.

One main result of the analysis is the fact that the number of quality programs in the breadstuffs industry has decreased over the last years. At the same time, a number of programs have increased their market shares. Thereby, the efficient promotion of their core marketing statements was key success factors: regionality, environmental sounded production methods as well as flexibly customised product qualities. In doing so, the programs deliberately differ from the mass market.

However, in the course of new legal requirements as well as sectoral quality initiatives, this deliberate differentiation has become increasingly difficult for all quality programs. From there, programs have to cut their own innovative paths not to completely lose their advantages compared to the mass market concerning the granting of warranties. On the one hand these innovations correspond to organisational aspects. The study could show, that most programs solely realise a low intensity of quality coordination at present. Instead of doing their own businesses under a common brand name in a relatively independent way, increased vertical integration of strategic, management-related as well as operative processes confer necessary unique selling propositions to the programs. In addition, intensified utilisation of information technology (IT) can also result in extra benefits of efficiency. Thereby, IT poses as a major enabler for the techno-organisational linkage of network partners. Withal, enterprises should resort to standards for data acquisition and exchange as far as possible due to the heterogeneity of the agribusiness. Several examples for this are already well-established (e.g. GS 1-systematic) or reside in a determination phase right now (e.g. agroXML).

Apart from these technological data management agreements, standards at process-related level are also in the core of scientific discussion. The breadstuffs industry is afflicted with different complexities in this regard as identified in the framework of the detailed analysis. Indeed, with reference to strategic orientation and chain organisation, wide consistencies could be detected between the researched programs. Therefore, the implementation of strategic management-related as well as operative processes is subject to a broad bandwidth. So, decisions about ideal chain or network compositions can only be reached in individual cases depending on requirements and own objectives to be realised as well as institutional general conditions, resources and restrictions.

This is the point, where the University of Bonn starts with its projects. Accompanied by the theoretical formulation of a stepped decision model, the University supports enterprises, chains and networks in the organisation of efficient quality structures. The pilot project was realised in one of the above mentioned best practice quality programs. The fact that this engagement in vertically integrated quality networks is linked with the creation of added values, can also be documented by a set of projects with businesses, which traditionally operated at the mass market for breadstuffs. As driving forces at this juncture, the network stages 'farmers' and 'agricultural associations' can be identified. In view of coming competitors (e.g. Black Sea abutters), the development of network-wide quality structures shall complicate market access for third parties in terms of non-tariff trade barriers. Traceability and transparency are major elements of this strategy to assure own market stakes. This quality initiatives will only be partially accompanied by a trade name. However medium-term, the market for vertical-coordinated production of quality breadstuffs will totally gain in importance due to these developing networks with their high product quantities.

References

Bouma, J., 2000. Value Chains: A strategy for the Alberta Agri-Food Sector. In: Trienekens, J.H. and P. J. P. Zuurbier (eds.), Chain Management in Agribusiness and the Food Industry. Wageningen Pers, pp. 333-340.

Helbig, R., 1997. Organisational alternatives for Chain Encompassing Quality Management in agrifood production. In: Schiefer, G. and R. Helbig (eds.), Quality Management and Process Improvement for competitive advantage in agriculture and food. ILB, pp. 35-42.

Kagerhuber, M. and R. Kühl, 2002. Unterschiede und Gemeinsamkeiten vertikaler Prozessorganisationen im Bereich der integrierten tierischen Produktion und Vermarktung-Ein europäischer Vergleich erfolgreicher Konzepte. In: Landwirtschaftliche Rentenbank (ed.), Lebensmittelsicherheit und Produkthaftung - Neuere Entwicklungen in der integrierten Produktion und Vermarktung tierischer Erzeugnisse. Landwirtschaftliche Rentenbank, pp. 8-24.

Poignée, O. and C. Pilz, 2005. Abgestimmte Qualitätsproduktion über die Kette - Konzeption und praktische Umsetzung der deutschen Brotgetreidewirtschaft. Report B-05/1, ILB.

Schiefer, G., 2003. Vom Unternehmenskonzept 'Qualitätsmanagement' zur Sektorinitiative 'Qualitätssicherung' - Entwicklung, Situation und Perspektiven. Report B-03/4, ILB.

Development of a process oriented management system based on a data bank: an approach to successfully fulfil quality requirements in food industry

Hartmut F. Binner and Christiane Jansen
Dr. Binner Consulting & Software, Schützenallee 1, 30519 Hannover, Germany; info@cim-house.de

Abstract

The increasing demands for the fulfilment of norm and legislative requirements in food industry call for effective and open management systems. The paper reflects on projects presenting practical experience with an approach which stresses the participation of employees in the design of a management system. A database used as the main tool allows for the transparent visualisation of all of the company's processes and provides management documentation to the employees by using a minimum amount of paper. It is shown that the new management system does not only fulfil external demands but also creates positive effects within the company.

Keywords: process orientation, integrated management, participation of employees, data bank, transparency

1. Background

Companies in the food industry face a steadily increasing number and extent of requirements expressed in different standards. Starting in the middle of 2003 all major retailers in Germany asked the producers of their private brands for a certification according to the German International Food Standard (IFS). IFS is supposed to contribute to a higher food safety level and defines very exact requirements for the introduction and maintenance of a Hazard Analysis and Critical Control Points (HACCP) system, for the plant's surroundings as well as for a comprehensive quality management system based on the international standard DIN EN ISO 9001. In addition, companies often have to present further certifications such as those according to the Standard of the British Retail Consortium (BRC) or for an environmental management system following DIN EN ISO 14001. Further requirements demanded by law, such as traceability, or directly by the customers enhance the pressure created by a continuously changing market.

Companies have to fulfil the task of integrating all these different requirements in their management system and to ensure its implementation and continuous improvement. An indispensable prerequisite is to achieve the acceptance of the system among the employees so it can work without requiring too much extra time and money. In light of the rapid development of the market, every management system has to be open for the easy integration of new requirements and standards.

2. Approach

External advice and support by data processors are helpful measures for efficiently coping with this comprehensive task and for rapidly accomplishing the desired goals. Dr. Binner Consulting & Software has developed an approach which uses the process management software SYCAT® as a tool and has proven its practicability in numerous projects. This approach is part of a holistic and integrated procedure of organisational development planning. Its basic principle is the process orientation which is also demanded by various standards. The procedure presented here follows a model of four phases when (re-)structuring the management system (see Figure 1).

The main task during phase one is developing the process model, i.e. the identification of the company's processes. Many companies face great difficulties defining and managing processes across the traditional boundaries of departments. For too many years they had stressed thinking and working within the departments, each one trying to achieve success on its own. But this often leads to mutual blaming of failures, e.g. when sales and distribution sell too many different products which due to capacity restrictions cannot be turned out by the production department. So, processes have to ensure that interfaces between departments are described as clearly as possible following the objectives of the company as a whole.

At this point it already becomes clear that the definition of a process asks for inputs from all different parties involved. In order to define the company's processes it needs a change of point of view: Starting with the customer and taking the company as a whole as the acting unit one strictly follows the value-added chain of the product. In addition processes of management and

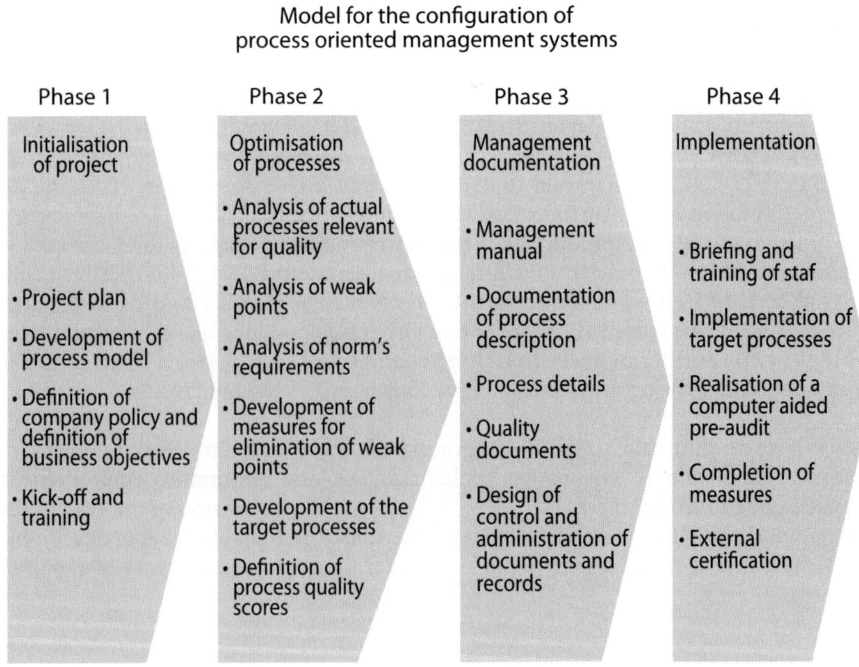

Figure 1. Model for the configuration of process oriented management systems.

of support are identified. These often comprise a high potential for improvement. An example of a food company's process model is given in Figure 2.

Another fundamental step during this phase is the determination of the company's policy and of adequate annual business objectives. They serve as guidelines for the system and are used in the kick-off meeting when all people involved in the project are trained. In phase two processes are optimised. First the actual status of each process is described during a workshop with the employees. Here two factors are essential for further success:

First it is important to include representatives from all departments involved in the process. It may also be necessary to invite the persons assigned to special tasks, such as for example, hygiene, occupational health and safety, environment and others. This helps to ensure that all aspects of the process are dealt with and that the different points of view of the employees are presented and discussed during the workshop.

Secondly, it is important to visualise cooperation between the different departments involved as well as the flow of information, may it be written or oral. Visualisation is the clue for understanding, acceptance and deployment of procedures. Figure 3 shows a standard process visualised using the process management software SYCAT.

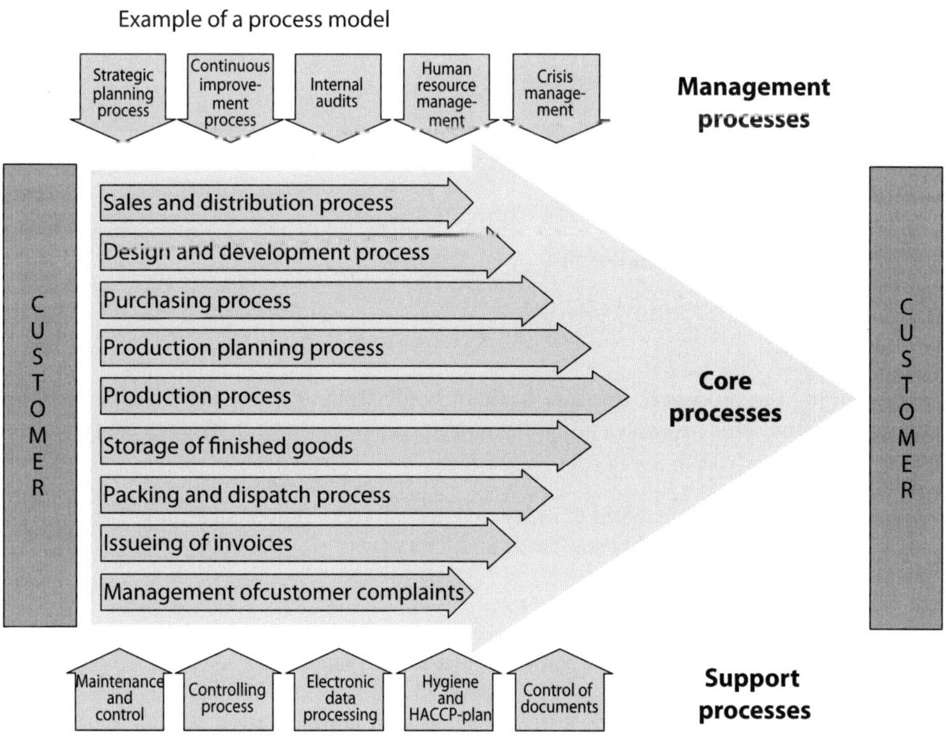

Figure 2. Process model of a company in the food industry.

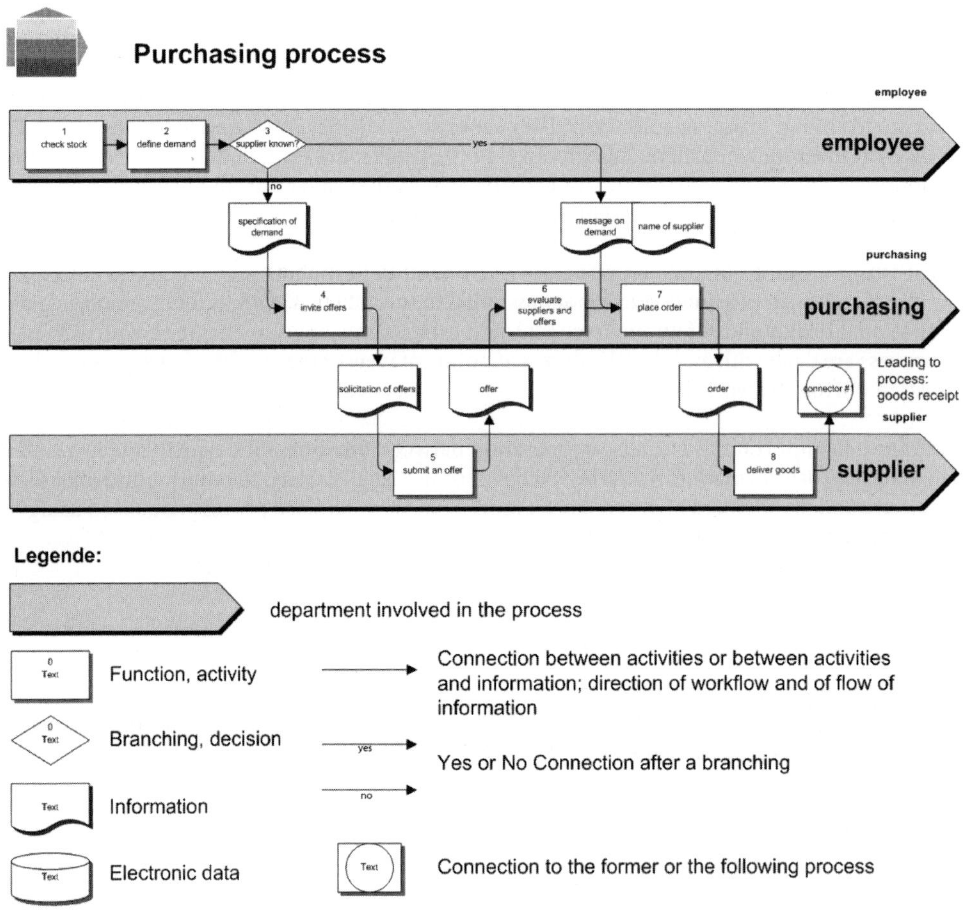

Figure 3. Process visualised with SYCAT.

By documenting the process in the data bank and publishing it in the intranet the results are visible to everyone. This promotes a high level of transparency and helps avoiding the spread of rumours.

Subsequently processes are analysed in order to detect all weak points regarding:
- the requirements of the quality standards and the laws;
- the needs of employees; and
- the strategies of executives.

Many of those weak points already become obvious when visualising and discussing the process in detail. Together, the departments involved develop and agree upon measures for elimination of the weak points. In all workshops a tight and systematic moderation is the prerequisite for considerable success. Weak points and measures are localised exactly within the processes and are put down in the data bank for follow-up.

In addition to standardised procedures, a company in the food industry has to establish safe and hygienic surroundings for food production. Continuous rounds in the production area with checklists derived from the norms supply further weak points which are also recorded in the data bank. Likewise, the HACCP team may find further weak points and necessary measures when analysing the processes.

For all measures, responsibilities and deadlines are pinpointed. Due to the steady participation of all the interested parties in the elaboration of the new procedures, motivation, acceptance and compliance among staff are very high. A forceful monitoring of the measures is built up, and also supported by the data bank. This way the project coordinator is able to supervise a large number of changes and to always document the actual status in the data bank (see Figure 4). An automatic mailing function supports him in the task of reminding his colleagues of their duties.

The defined measures serve as guidelines for the elaboration of the target processes which meet the standards' requirements. The data bank offers two portrayals: the first one is the graphic already shown in Figure 3. This visualisation is complemented using colours and symbols, e.g. for activities important for traceability, functions where further documents are attached or which are described in more detail in a subprocess. The other portrayal is a description of the process flow in a table. Here, the user finds detailed information about the task to carry out, work steps, forms and other documents to apply, software systems to use and others (see Figure 5).

SYCAT®

Plan of Measures acc. to Graphic No

Purchasing process

Function Weak Spots:

No Function / **No Weak Spot**	**Function Area** / **Due-Date**
1 check stock	**employee**

Weak Spot: ID: 1 Priority: 2

1 often it is forgotten to check stock of lubricating oil;
consequence: machines cannot be maintained adequately; high risk of machine breakdown and losses in output
reason: frequent changes in maintenance personal

Solution(s):

the check of stock of lubricating oil is carried out by the production manager

Measures:

change table of responsibilities accordingly; remind new person in charge for a month; check on the effectiveness of measure in 3 months

Responsible: Smith (jsm) Date is overdue! **2/1/2005**

Remark to the Slipped Date:	Slipped Date:	New date:
Worst case has become true: the change in responsibility has to be realized immediately!!	3/16/2005	2/1/2005

Figure 4. Report of plan of measures from the data bank.

SYCAT®

Process Flow Description

Purchasing process

No	Function	Responsibility	
		E executing R responsible V is involved I is informed	← Input → Output

Process: Purchasing process

No	Function	Responsibility	
☐ 1 Note: cont. at Fct. 2	**check stock** notice of demand for main stock material is automatically created by the storage management system but has to be verified by the person in charge **Used Informations:** (Int-008) Table of responsibilities for stock	**FA/E: employee**	
☐ 2 cont. at Fct. 3	**define demand** **Used Informations:** (F-001) Form for specification of goods	**FA/E: employee**	
◇ 3 Yes (90%) = 7 No (10%) = 4	**supplier known?**	**FA/E: employee** → [Yes] message on demand → [Yes] name of supplier → [No] specification of demand	

Figure 5. Process flow description complements the process' graphic description.

In order to enable process managers to continuously improve the new processes in the future, key performance indicators are defined for each process and they are also documented in the data bank. These indicators have to be derived from the overall business objectives. The Balanced Scorecard is a valid tool for this purpose.

When all decisions are made about the target processes the new system has to be documented completely. In phase three new documents are developed and integrated into the process documentation. If not yet established, the process of control and administration of documents is designed. By using a workflow function based on the data bank it is guaranteed that a document is always edited, revised and released by the person responsible and that users only have access to the most current version.

Via intranet users have access to the entire system as well as to single documents. They are automatically notified via e-mail when documents have been changed. This way all updates are promptly spread without unnecessary use of paper. All necessary documents are available at all times and at all work stations with access to the intranet.

The management documentation in the intranet is generated from the data bank and contains the following information:

- The company's structure in the process model.
- A graphic presentation of all processes with links to other processes, showing responsibilities, tasks, the flow of information and material and decisions to be made.

- A map of the sequence and interaction of processes.
- Description of the process flow including the documentation of:
  - flow of information (input/output);
  - responsibilities;
  - work steps;
  - utilised processing systems;
  - special features such as comments regarding traceability;
  - Critical Control Points;
  - links to attached documents (such as checklists, instructions, forms, standard letters); they can be opened by the user directly from the process description.
- Characteristics of the process including:
  - input and output, purpose, scope, process ownership;
  - documents and information generally valid for the whole process;
  - statistical data (number of interfaces, involved departments).
- Key performance indicators.
- Plans of measures with deadlines and responsibilities.
- Further features.

In phase four, staff training and implementation of the new processes are followed by a pre-audit. Again appropriate measures are derived from its results. This audit, as well as all following internal audits are managed with support of the data bank, thus reducing the time necessary for the preparation of audits and for the edition of audit reports. The realisation of the measures defined in the pre-audit complete preparations for the external certification audit.

3. Benefits

Of course, a successful certification is the main confirmation for proving that the development of the new system has been worthwhile. But a project of this dimension shall create further direct benefits within the company. The following advantages can be achieved by applying the above mentioned approach and data base:

- High transparency of the whole management system for all employees as a result of the detailed process description.
- Responsibilities are clearly defined across department boundaries.
- Processes are optimised aiming at the general company's outcome, e.g. by reducing time, losses, errors, risks and other costs.
- Processes are accepted by the staff as a result of their participation in their definition.
- Practical and easily applicable management system.
- Manageable and efficient implementation of improvement by a systematic monitoring of measures.
- Motivated employees.
- Less time required for updates using the data bank.
- Topicality of the management documentation in the intranet.
- Great reduction of paperwork as a result of electronic update and distribution of documents.
- Electronic archives of former released versions of documents.
- Support for incorporating new employees.
- Easier preparation and implementation of standardised internal audits.
- Base for new management tasks such as:
  - the implementation of a work flow management, integrating the different applications used by the company;

- creating specifications for the selection of new applications;
- the integration of further management systems instead of creating new manuals.

The organisation of successful workshops, with a systematic elimination of weak points - seen from the different points of view - is a good practice to institutionalise the process of continuous improvement by showing its results to both, the management and the employees. This way motivation is high to keep the system working.

Essential for the success of this approach are the participation of the employees, the transparent visualisation of the processes, a data base with extensive process documentation as well as an effective project management.

References

Binner, H.F., 1993. Strategie des General-Management. Ausweg aus der Krise. Springer.

Binner, H.F., 1996. Umfassende Unternehmensqualität. Ein Leitfaden zum Qualitätsmanagement. Springer.

Binner, H.F., 1997. Integriertes Organisations- und Prozessmanagement. REFA-Fachbuchreihe Unternehmensentwicklung. Hanser.

Binner, H.F., 2002. Prozessorientierte TQM-Umsetzung. 2. Auflage. Hanser.

Binner, H.F., 2004. Handbuch der prozessorientierten Arbeitsorganisation. Methoden und Werkzeuge zur Umsetzung. Hanser.

Network management for dairy productivity and quality in Ethiopia

Marijke D'Haese[1], Gian-Nicola Francesconi[1] and Ruerd Ruben[2]
[1]Development Economics Group, Wageningen University, Hollandseweg 1, 6706 KN Wageningen, The Netherlands; marijke.dhaese@wur.nl
[2]Radboud University Nijmegen, Centre for International Development Issues Nijmegen (CIDIN), P.O. Box 9104, 6500 HE Nijmegen, The Netherlands

Abstract

Market access for Ethiopian small scale dairy producers is highly constrained by a set of transaction costs. Farmers create associations for marketing milk in an attempt to mitigate these costs. The aim of this paper is to investigate how these networks have an effect on the productivity and the quality of the milk produced and traded. We present unique data linking farm characteristics, production records and lab-results across association members and non-members. Main outcome of the study is that membership of the association seems to contribute to productivity, yet, its impact on quality and bacterial contamination is limited.

Keywords: dairy production, milk quality, agricultural development, Ethiopia

1. Introduction

Milk provides essential amino-acids and micro-nutrients particularly needed in marginal diets that are based on staple grains and root crops. Recent calculations expect the demand for food of animal origin in developing countries to double over the next 20 years as a result of population growth, urbanisation and income growth. Delgado *et al.* (1999) estimated that the demand for dairy products in sub-Sahara Africa could grow with 3.8-4% from reference year 1993 to 2020. Similarly, it is expected that Ethiopian dairy market will expand. Ethiopia shows great opportunities for dairy production. The country can rely on the largest livestock population in Africa, estimated at 29 million cattle, 24 million sheep and goats and 18 million camels. The Ethiopian Highlands (covering two-thirds of the country) provide the necessary conditions to support the use of high yielding dairy cows including temperate climate, relatively disease free environment and substantial opportunities for feed production (Ahmed *et al.*, 2004). A majority of dairy cows is kept on family farms that are often small-scale. Yet, given the global and regional demand and the local production potential, dairy could become an important source of income for smallholders.

For many however, marketing the milk remains problematic, due to the limited production scale and the consequent capillary dispersion over the territory, the difficult access to information, communication and transportation throughout the country and the highly perishable nature of dairy products. The latter is of particular importance in Ethiopia as electrical power and cooling facilities are limited and unreliable. As a result marketing and transaction costs are high. The horisontal integration of farmers for collecting and marketing the milk is a potential catalyst for mitigating these costs. This paper reports on networks between farmers around Debre Zeit Town in Ethiopia. The Ada'a-Liben Woreda dairy and dairy products marketing association started informally in

1997 with 34 members and received its legal certificate by 2000, the year in which it started milk collection and marketing activities. In 2003 over 540 members supplied to the association. The association thus contributes to the livelihood of many households[1] (Tegene, 2003).

In the paper, we discuss how the association impacts on production and quality of milk. Our findings show that the association has contributed to changes in farm management, resulting in higher production and productivity levels. We could however not find any link between membership of the association and the quality characteristics such as milk composition and bacterial contamination.

The paper is organised as follows. Section 2 describes the major characteristics of dairy sector in Ethiopia. Section 3 gives some insight in how literature explains the emergence of networks to mitigate transaction costs and puts forward the two research questions of this paper. Section 4 then gives a short overview of the survey and presents general characteristics of the farmers interviewed. This is followed by a description of the main results of the analysis, which are discussed in section 6. Section 7 concludes with some critical issues and challenges for future research in the Ethiopian dairy sector.

2. Dairy sector in Ethiopia

Ethiopia is one of the poorest countries in the world. The World Bank (2004) estimates that 45% of the population is living below the minimum basic needs at a GNP per capita of 100 US$ in 2002. Agriculture is a key economic sector, its value added accounting for more than 40% of the GDP. Within the agricultural sector, livestock and livestock products contribute for 40%. The non-monetary contribution of livestock to the households, i.e. draft power fuel, manure and transportation should be added to this figure. Ethiopia counts the largest livestock population in Africa (Redda, 2001; Ahmed *et al.*, 2004). The increase in milk production has benefited much from policy reforms starting from 1993. Along with population growth and technological intervention, policy reforms contributed to an estimate growth rate of milk production of 3% compared to 1.8% in the period of 1975-1992. Most milk is sold as raw milk (50 million litres in 2002) or processed in butter or cheese (285 million litres). Raw milk is not often bought but taken from own production (home consumption accounts to 370 million litres). Cheese, or ayib, accounts for 9% of the total milk produced (Ahmed *et al.*, 2004). Demand is geographically differentiated in the sense that households in lowlands daily consume milk, while the highlands milk is mainly consumed by children and elderly.

Ahmed *et al.* (2004) distinguish three types of dairy farming. First, the smallholder dairy households keep a majority of the cattle in Ethiopia accounting for 97% of all milk produced and 75% of commercial milk production. They own mainly indigenous cows of low-productivity native zebu breed, producing between 400 to 680 kg milk/cow per lactation period. Secondly, state farms that have been privatised or are in the process of privatisation, use high-graded animals. The third group are peri-urban or urban households, mainly keeping crossbred animals in the central highland plateaus. These cows can produce 1,120-2,500 litres per lactation period.

The dairy cooperatives and associations provide an access of small farms to the formal milk market as shown in Figure 1. On the other hand, the informal market consists of farmers delivering to

[1] ILRI calculated that in the Kenyan formal dairy industry 0.2 to 0.4 jobs were created per 100 litres of milk handle daily, while the same amount creates employment for 2 to 3 informal traders in Kenya and Tanzania (ILRI, 2004).

their immediate neighbourhood, local and regional traders, shops, restaurants, bars and hotels, whereby only 5% is marketed as liquid milk. The small peri-urban dairy households can deliver milk to one of the 30 cooperatives/associations or to informal urban traders. The quality control of milk and dairy products is limited. The deliveries at the cooperative/association are reviewed with an alcohol test and a gravity test to check on basic food safety and possible fraud. Finally, rural farmers sell surplus production of raw milk on the rural market as well as butter and cheese. Prior to 1991 the state-owned DDE (Dairy Development Enterprise) controlled the cold chain and pasteurised-packed milk production (Holloway *et al.*, 2000). Since the policy reforms of the early 90s private farms and dairies entered the market (Figure 1).

3. Importance for network management

We are particularly interested in how network activities could influence the productivity and quality of the milk production. Definitions of networks in management and sociological literature are multiple and are focussing on both individual and organisational levels. The following definition of networks therefore seems to fit with the association analysed, namely 'long-term arrangements among distinct, but related for-profit organisations that allow those firms in them to gain and sustain competitive advantages *vis-à-vis* their competitors outside the network' (Jarillo, 1988), whereby each farm is an individual firm characterised by its manager, location and farm endowments and capital. These networks form part of a supply chain, or as Dyer (1996) puts it: 'interfirm specialisation with firms specialising in a small range of products and services, embedded in a complex chain of input-output relations with other firms'. The core business of the dairy association is to become part of the complex chain through organising the collection and marketing of the milk and thereby reduce transaction costs.

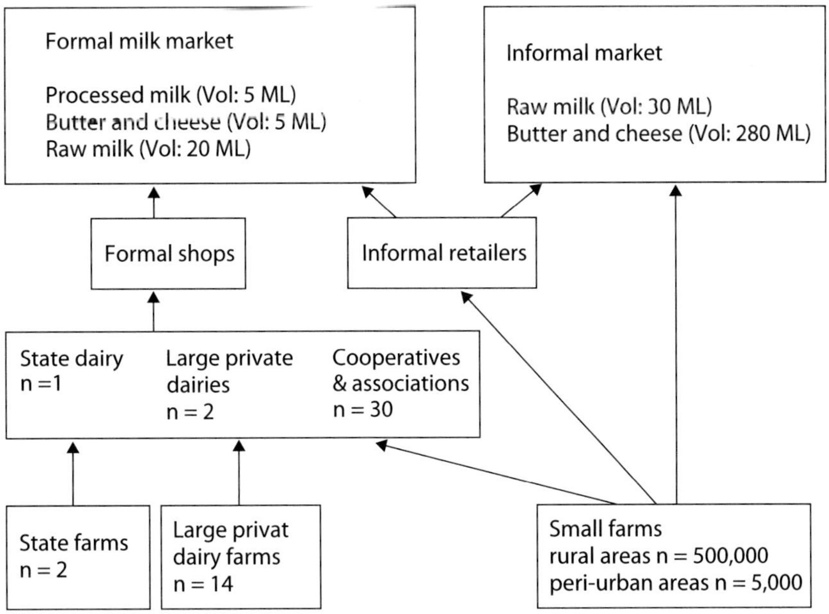

Figure 1. Schematic overview of dairy marketing channels (volumes of production and trade in millions of litres (ML) and number of producers (n)) (Ahmed et al.*, 2004).*

In New Institutional Economic terms, the dairy association can be seen as a hybrid organisation, which is characterised by inter-firm collaboration by individual firms holding property rights whereby joint decisions are made (Williamson, 1996; see Ménard, 2004 for more details). Ménard (2004) proposes some 'regularities' in the structure of hybrids, namely: (1) importance of pooled resources, (2) relational contracting (characteristic for hybrids is that contracts regulate pooling of activities and resources amongst partners who are also operating in unconnected transactions), and (3) their relation to competition (this is because, while cooperating on some issues, partners within the hybrid organisation are each other's competitors and as a group they compete against other firms and organisations).

Following Williamson, joining into a hybrid organisations may become more profitable than selling individually on a spot market when transaction costs are high (Williamson (1985) puts forward the 'discrete alignment principle' by which agents are believed to adopt the mode of organisation that fits the best with the characteristics of attributes at stake; see also Williamson, (1991), Williamson (2003) and Ménard (2004)). As already mentioned in the introduction, smallholder dairy farmers in East Africa face high transaction costs (Omiti and Staal, 1996). According to Staal *et al.* (1997) transaction costs in dairy are particularly high due to (1) the perishable nature of raw milk; (2) its high water content making it a bulky commodity; and (3) the variability of milk composition and quality. High transaction and marketing costs arise as milk has to be transported quickly to the consumer or processor, its bulkiness limits the amount a farmer can transport without vehicle and monitoring its quality and composition is costly and may lead to high levels of rejection. In other words, specialisation in milk production and marketing includes need for asset specificity. This explains the emergence of hybrid organisation such as dairy association in mitigating the transaction costs (see Staal *et al.* (1997) and Holloway *et al.* (2000) for a detailed description of these costs). Yet, for the sustainability of the dairy association, the increased market opportunities could lead to increased milk production and productivity of acceptable quality. Previous studies on the dairy sector in East Africa have mainly focussed on the amount of milk marketed and its resulting potential for income generation (Staal *et al.*, 1997; Holloway *et al.*, 2000). The quality of dairy products has not been an issue of much research. This is regardless of its importance for the health of its consumers and the high rejection rates of milk at processing points. Furthermore, a major part of the milk market is informal and thus outside the formal cold-chain including pasteurisation and packaging with no quality control at all. Two main research questions arise: (1) can the association influence production and productivity on the farms of its members?; and (2) has the association an effect on quality levels of the milk at farm gate in terms of its composition and possible bacterial contamination?

The influence on production and productivity are considered a result of external economies or passive collective efficiency resulting from collective action within the association because its core business is the collection and marketing of milk (McCormick, 1999). The next section presents the models used and explains how the data was collected.

4. Data and methodology

A random sample was taken of 70 small dairy farms. Due to missing records, not all cases could be used in the various regressions. Besides general household information, farm, farm infrastructure, dairy production records and milk-handling characteristics were assessed. All participants where selected from the urban and peri-urban and rural areas of Debre Zeit that is a village 50 km south of Addis Ababa.

Members of the Ada'a-Liben Woreda Dairy and Dairy Products Marketing Association were randomly selected from the list handed by the associations, while non-members were randomly selected on the census of the local unit of the Agriculture ministry. From each farm a sample of raw milk was collected and analysed using standard microbiological and chemical sampling procedures at farm gate. This enabled us to build a data set combining farm and household data with records on fat content and test results on total bacterial count (TBC) and acidity level.

The farmers in our sample have between one or two cows, with an average milk production of 4.5 litres per cow per day (Table 1). The average production on farm is less than 11 litres per day. The average age of the farmers in the sample is relatively high, compared to the life expectancy at birth of 42.1 year (World Bank estimates for 2002). The average monthly expenditure on feed on the farm is equivalent to 8.9 Euro/cow (exchange rate in February, 2005).

Table 1. General characteristics of members and non-members of the association.

		Non-member (n=35) (0)		Member (n=35) (1)		t- stat
		Mean	Std. Dev.	Mean	Std. Dev.	
Personal characteristics						
Age	years	46.03	16.40	49.97	11.63	-1.160
Family size	members	5.74	2.44	6.46	2.59	-1.186
Education level	scale[a]	2.06	1.33	3.74	2.05	-4.086**
Information source	scale[b]	3.88	0.58	3.57	0.85	1.804*
Farm characteristics						
Cows	number	1.46	0.61	2.11	1.39	-2.563**
Cattle	number	4.28	3.03	4.06	2.51	0.343
Feed	Birr/cattle/month	86.15	72.87	112.9	69.06	-1.577
Expenditure on concentrates[c]	Birr/cattle/month	45.45	49.62	53.60	42.01	-0.742
Expenditure on roughage[d]	Birr/cattle/month	40.69	42.60	59.31	60.98	-1.480
Concentrates[e]	kg/day	6.63	6.13	8.79	5.57	-1.534
Roughage[f]	kg/day	10.34	8.70	14.56	14.04	-1.504
Income from dairy	% tot. hh. income	17.14	34.94	29.71	37.49	-1.451
Time milk preserved	hours	52.28	164.37	24.28	26.96	0.995
Location						
Distance to market	km	5.66	4.05	1.25	1.32	6.114**
Performance indicators						
Milk production of farm	l/day	4.06	2.96	15.04	10.10	-6.169**
Milk productivity	l/cow/day	3.05	2.63	6.70	1.81	-6.384**
Fat content	%	4.69	1.14	4.18	0.79	2.119**
Acidity level	%	0.21	0.02	0.21	0.02	0.630

[a]Scale: 1= illiterate to 7 = bachelor.
[b]Scale: 1= extension to 4 = parents.
[c,d]Refers to expenditure on concentrates and roughage respectively on farm divided by the number of livestock.
[e,f]Refers to amount of concentrates and roughage fed to the cows on farm divided by the number of farms.

Dairy is not the main income source for most of the households. Farms are located on average 3.42 km away from the market, which is usually accessed on foot. Only two farmers in the sample transport their milk by bike and another two by car. Not all milk is sold as part of the milk is consumed within the household. Some households in the survey produce butter (n=21; 29%) and some ayib or cheese (n=23; 31.9%) for own consumption or the local market. The number of farms on which yoghurt is made is insignificant (n=2). Women are not only responsible for the production of these dairy products, but they are also in charge of dairy in general, including milking, selling milk and dairy products and management.

We could not prove the non-equality of the age and family size of the respondent, yet, scores on the scales of average education level and information sources are different, with members of the association having enjoyed more education and rely less on parents as information source (Table 1). The average number of cows is not equal for members and non-members, with a higher number of cows on farms member of the association. With regard to location, farms of members of the association are relatively closer to the market than non-members of the association.

The next sections discuss the results of two-stage least square (2SLS) regressions of productivity of milk, acidity levels and fat content on personal and farm characteristics, location and membership of the association. The dummy for membership was instrumented because we suspect it otherwise be correlated with the error term. The productivity of milk is defined as the amount of milk produced on farm divided by the number of cows. The regression also focussed on several personal and farm characteristics that are expected to significantly influence milk production and its quality. The TBC count revealed a binary distribution, with some farmers having major problems with bacterial contamination, and others less. We therefore divided the sample over two groups, and created a new TBC-related dummy variable. A probit test was then performed on this new variable, comparing the influence of farm characteristics and association membership on the probability of showing a high TBC count.

5. Results

5.1. Characteristics of membership of the association

The Ada'a-Liben Woreda Dairy and Dairy Products Marketing Association counted in 2003 about 540 members, with in total 2,500 dairy animal, supplying 175,000 litres of milk per month over 8 collection centres. While its core-business is the collection and marketing of milk, the association provided training in dairy cattle management and in milk handling and processing at the ILRI station in Debre Zeit. The association furthermore acquired a cream separator and butter churn to make butter, cottage cheese and cheese. Members pay a fee, and in return they are paid dividends at given intervals. They are paid twice a month for milk supplied (Tegegne, 2003).

Table 2 shows that the majority of the farmers member of the association consider their homestead as urban (this study does not make a geographical distinction between rural, peri-urban and urban, yet the variable location refers to how farmers perceive their home-environment). The farmers member in our survey all keep crossbred-type cows, while two-third of the non-members have Zebu-type cows.

The results of the probit analysis on factors influencing the probability of being member of the association are shown in Table 3. It seems that age is important, thus the older the farmer, the higher the probability of membership of the association. Education is significant: the higher the education level, the higher the probability of membership. As explained above, many of

Table 2. Distribution of farmer over association membership, location perceived by respondents and breed of cows kept (number of farmers).

	Non-member (n=35)	Members (n=35)	Total (n=70)
Location			
Urban (1)	12	30	42
Peri-urban (2)	6	1	7
Rural (3)	17	4	21
Breed			
Zebu-type (0)	23	0	23
Cross-bred - type (1)	12	35	47

Table 3. Probit analysis of membership counts using Huber/White/sandwich estimator of variance ((1) member n= 35; (0) non-member n=35).

		Coef.	Std. Err.	Z	P>[z]
Personal characteristics					
Age	years	0.036	0.017	2.15	0.031
Sex of respondent	male= 1	-0.058	0.399	-0.15	0.884
Family size	number	0.064	0.069	0.92	0.359
Education level	scale[a]	0.427	0.114	3.74	0.000
Income from outside agriculture	yes = 1	-0.075	0.443	-0.17	0.865
Location					
Location	scale[b]	-0.781	0.186	-4.18	0.000
Constant		-2.042	1.022	-2.00	0.046

Wald chi^2: 34.88; Prob > Chi^2: 0.000; Pseudo R^2: 0.3825;
Log likelihood: -29.96
Percentage correctly classified: 81.43%
[a]Scale: 1=low to 7= high.
[b]Scale: 1= urban, 2 = peri-urban, 3 = rural; based on answers of farmers in the survey.

the members consider their home as peri-urban or urban. This is not surprising because the collection centre is in the urban centre. Sex, family size and whether or not the farmer has an income outside agriculture do not influence membership probability.

All farmers member indicated that they changed their farming system since their participation in the association, namely: feed input increased (n=29 of 35 members in the sample), more veterinary services (n=19), general improvement of health of the cows (n=24), better breed (n=25), increased in herd size (n=21), increased yield (n=22), better market access (n=32), increase in income (n=29) and increase in hygiene on farm (n=34). However, farmers are not satisfied with the extension services of the association, for 26 of them it is considered poor. farmers who are not member indicate poor yield as a major reason why they do not join an association (n=25; 71% of non-members), with the membership fee as second reason (n=6; 17% of non-members).

Farmers member of the association recognise the importance of quality yet, almost a third of the farmers member did not know which tests are used to monitor the quality of their milk (n=12 for members, n=31 for non-members). They point out the fluctuation of the milk price as their main problem (n=17), followed by the prices of feed (n=11) and quality control (n=6).

5.2. Milk production and productivity

While crossbred cows are more common on the urban farms, the rural farmers in our analysis keep indigenous, Zebu, animals which have low production performance. The average cow lactation yield is 5,25l (of lactation of 239d) of which 2,86l is suckled by the calf (Ketema and Tshehay, 1995). Similarly, we found a difference between milk output per cow for members (6.707 l per cow per day) and non-members (3.05l per cow per day) (Table 1).

Table 4 gives the results of the 2SLS model on milk production. The instrumented variable of membership of the association is positive and significant. This confirms the results of the t-test, namely the non-equality of milk production on farms of members of the association compared to those who are not. Other significant variables are: (a) number of men in the household (negative sign) and (b) age of the farmers (in concave relationship). Table 4 also shows the results of a regression of milk output per cow on farm characteristics. These results again confirm the importance of the association for milk productivity on the farm. Other significant variables include the number of men in the household, age of the farmer (in concave relationship), and the amount of concentrates fed to the dairy cows.

Table 4. Regression of milk output and productivity using two-stage least square model with robust standard errors.

		Milk yield per day (ln) (l per day)		Milk productivity (ln) (l per cow per day)	
Number of observations		66		61	
R-squared		0.62		0.74	
		Coef.	t-stat	Coef.	t-stat
Association					
Member	yes=1	1.185	4.70**	0.774	4.50**
Personal characteristics					
Age	years	0.083	1.68*	0.085	3.17**
Age square	years^2	-0.001	-1.68*	-0.001	-3.15**
Gender (women)	number	-0.062	-0.91	0.048	0.247
Gender (men)	number	-0.122	-1.19*	-0.081	-2.63**
Farm characteristics					
Breed	crossbred =1	0.271	1.21	0.199	1.14
Concentrates	kg/day	0.0362	1.67	0.031	2.42**
Roughage	kg/day	0.002	0.22	-0.002	-0.47
Dairy cows	number			-0.044	-1.37
Location					
Distance to market	km	-0.019	-0.52	0.006	0.43
Constant		-0.69	-0.76	-1.051	-2.00

5.3. Milk quality

5.3.1. Milk composition: fat content

The fat content of the milk samples measured fall between 2.5% and 7% with the highest frequency at 4.5%. We find an higher average fat content for farmers who are non-member of the association (mean: 4.69%) compared to members (mean: 4.18%). A higher fat content is preferred as it contributes to the amount of butter and cheese which can be produced from one litre of milk.

Table 5 gives the results of the 2SLS model on fat content of the milk samples. The dummy on membership has been instrumented with similar variables as the regression on production and productivity. From the t-test results, we would have expected a negative sign of the association dummy, yet this could not confirmed in the 2SLS-regression. The regression results show a significant and positive sign of the number of men in the household and a negative sign of the breed. That milk of Zebu cows has a higher fat content than crossbred cows is confirmed by e.g. Musangi (1971; in FAO (1990)) who found on average 5.5% fat in milk in East-African Zebu's. Normal fat content of milk in temperate zones ranges between 3.5% and 6%.

Table 5. Modelling fat content and acidity of milk samples using two-stage least square regression with robust standard errors (%).

		Fat content (%)		Acidity level	
Number of observations		63	67		
R-squared		0.23	0.20		
		Coef.	t-stat	Coef.	t-stat
Association					
Member	yes=1	-0.063	-0.14	0.008	0.83
Personal characteristics					
Age	years			0.000	0.75
Gender (women)	number			-0.002	-0.83
Gender (men)	number	0.152	1.95*	0.000	0.17
Information	parents = 1	0.175	0.55	0.005	0.50
Farm characteristics					
Breed	crossbred =1	-0.942	-1.75*	-0.021	-2.01*
Concentrates	kg/day	-0.002	-0.06	0.003	3.42**
Roughage	kg/day	-0.009	-0.75	-0.000	-0.42
Cattle	number	0.042	0.89	0.004	1.00
Cattle	numb. squared			-0.000	-1.20
Location					
Distance to market	km	-0.054	-0.74	0.001	0.81
Constant		4.70	6.50**	0.183	9.79**

5.3.2. Acidity levels

Milk and dairy products are highly perishable. Hygiene levels on farms in Ethiopia are considered unsatisfactory due to poor housing, inadequate ways of milking and the type of milking vessels used (FAO, 1990). This was confirmed in our survey, 87.3% of farmers did not test the first strip of milk during milking; there were still farmers who did not clean the udder before milking (n=11; 15%) and sometimes did not dry it after washing (n=24; 33.8%). Moreover, 68% of the respondents (and 83% of the non-members) said not to know how to observe mastitis. The lack of hygiene on farms can result in bacterial contamination. In addition, a majority of the farmers does not have a fridge (only 17 farmers, all members had a fridge) to keep the milk fresh.

Acidity in milk results from bacteria fermentation activity. The more bacteria in the milk, the more acid it is. Acid milk means either that the milk is old, i.e. not fresh, or fresh but highly contaminated with bacteria. Normal acidity range is between 0.14% and 0.17%. Milk samples included in the study showed high levels of acidity, all equal to 0.17% or above. The regression of acidity levels in Table 5 shows the significance of amount of concentrates the cows get per day and the breed. The coefficient of breed is significant and negative. This indicates that our model associates crossbred cows with lower acidity levels. The coefficient of concentrates the cows are fed per day has a positive sign, indicating higher acidity levels related to more concentrates. High levels of concentrates (in particular if concentrates exceed 40% of feed), or if its quality is poor, might affect the cows metabolism in the rumen, resulting in ketosis. Ketosis may furthermore increase risk of mastitis and thus milk contamination. We need more research to confirm the influence of concentrates on the acidity levels.

5.3.3. Total Bacterial Count (TBC)

Table 6 shows the results of a probit analysis on the TBC. Factors contributing to a higher probability on bacterial contamination include: (a) small herd livestock, (b) higher yield per cow, (c) high distance to the collection centre or market and (d) the source of information on farming (this goes from formal to informal, thus highest score is given for experience through parents). The more remote farms thus seem to have more problems of milk contamination. The model suggests that farms with higher productivity show a higher bacterial contamination. Furthermore, the small herds have a higher probability towards bacterial contamination. This can be due to low quality inputs that are available to the smaller farms. Farms with more cattle endowment are the ones better off, thus having more access to, for example, feed of better quality and veterinary care. Also the source of information seems to be important, with households farming along traditions learned from their parents showing a higher probability on high bacterial contamination.

Other factors as age and size of household are not significant. The model is significant, yet not convincingly presenting the whole picture. We suspect that other factors are important to explain the occurrence of high bacterial counts, yet we lack the necessary quantitative data on milking and milk handling. Other factors not influencing a probability on high TBC include education, expenditures on concentrates and roughage, and location.

Table 6. Probit analysis of TBC using Huber/White/sandwich estimator of variance ((1) high counts n= 35; (0) low counts n=29).

		Coeff.	Std. Err.	Z	P>[z]
Personal characteristics					
Age	years	-0.005	0.013	-0.39	0.694
Family size	number	-0.03	0.069	-0.43	0.664
Farm characteristics					
Cattle	number	-0.137	0.685	-2.00	0.045
Milk per cow	litre/cow/day	0.164	0.071	2.30	0.021
Source of information	scale 1 to 4	0.450	0.258	1.74	0.081
Location					
Distance to market	km	0.133	0.054	2.44	0.026
Constant		-1.77	1.373	-1.29	0.197

Wald chi^2: 12.68; Prob > Chi^2: 0.0484; Pseudo R^2: 0.1626
Log likelihood: -36.912
Percentage correctly classified: 68.75%

6. Discussion and conclusions

The results show the effect of the association on production and productivity levels. This may be due to innovation in the farming system (breed and farm size) induced by the minimisation of transaction costs related to the search and screening of a trading partner and exchange negotiation. Yet we did not find any evidence that membership of the association is conducive towards milk quality upgrading.

We find that farmers mainly sell on local markets, and to neighbours, friends or just at farm-gate. Furthermore, it concerns mostly the sales of surplus production as the household consumes part of the milk. What is not sold can be processed into butter or cheese. In this case transaction costs of selling the milk are low. Farmers do not bear much risk, little negotiation is involved, no long term contracting and limited transport costs. The buyers of the milk generally trust the farmer on the basis of close social ties or past experience. In their current livelihood strategies rural farmers do not have to invest in specific assets. As shown above, it concerns small numbers of Zebu cows, with limited additional feeding and farm infrastructure and a minimum of monitoring and control of milk quality. Quality is not mentioned after control, it is not measured so that it is also not rewarded in a better price.

Other households living closer to urban centres have more opportunities for selling the dairy produce. Yet entering the more formal processing chain presents higher requirements in terms of quantity and quality. Income opportunities of selling to a cooperative or association may be better, but as quality is checked, rejection rates are higher. As a consequence smallholders face the need for more asset specific investments and more risk. The resulting increase in transaction costs can be too high for accessing the market individually, making the association an interesting alternative. Instead of trading on a spot market, pooling efforts in a hybrid organisation structure becomes profitable (see section 3).

The association structure, as analysed in this paper, principally serves as a collection point where milk is sold in bulk. This benefits the farmers in terms of reduction in transaction and marketing costs and it increases the bargaining power of the farmers as a group (Staal *et al.*, 1997; Holloway *et al.*, 2000). The association, or more general the proximity of the market, gives the farmers an incentive to invest in their livestock, buying more cows of a better breed, and thereby generates positive externalities of co-specialisation. Yet, problems of quality standards persist. Although quality control at the association is limited to an alcohol test and a specific gravity test, rejection rates remain significant, which are immediate income losses for the farmers. The association seems not to have an impact on the monitoring on-farm towards better quality and safety of the milk. The lack of quality of the milk could also become problematic for the performance of the association in the supply chain. To secure its selling contracts, the association needs to increase quality control at collecting points and monitoring at the farms. This is likely to increase transaction costs even more.

Furthermore, the success of the association depends largely on how well the co-op is securing itself against free-riding (if one farmer adds water to its milk or its quality is not adequate and not checked sufficiently at delivery, it can spoil the whole tank); and investment horizon problems (this is that the farmers as a group should think in investing on a longer term, and not just once because if the NGO or ILRI stops its aid to the co-op it might not 'survive') (see Sykuta and Cook (2001) for more insights in this discussion). Much will then depend on how much the farmers trust each other within the association, and how the association acts as third-party control mechanism.

High levels of bacterial contamination are a matter of concern for food safety. Consumers are entitled to safe and fresh milk of acceptable quality levels. The results of our study show that milk collection and collective action within the milk association are an incentive to increase number of cows and invest in better breeds. This seems to pay off in terms of relative higher levels of production as well as productivity. However, impact on quality is limited. Bacterial contamination is high, as shown by high TBC counts and high acidity levels. In other words, networking is conducive to the quantity of produce, yet transition towards increased quality appears to be more difficult. Farmers' interest in quality needs to be promoted along with increased monitoring and control of quality management at farm-level.

The association intends to expand its activities to the rural communities. We thereby believe that the association has a particular role to play in extending and improving its information campaigns in order to increase quality standards of its current and new members. This could result in a decrease in rejection rates of milk at collection point or at processor level, and thereby increase the farmers' income levels. Moreover, it strengthens the position of the association in the supply chain.

Further research is needed to analyse the difference in information acquired by farmers and to find out how this has influenced the willingness to participate in the association and its possible impact on farming system and milk handling. This should help to formulate better strategies for future network management. With regard to quality management within the associations and cooperatives, more analysis is needed to assess how transaction costs are influenced by quality standards set while decreased by hybrid institutional arrangements.

References

Ahmed, M.A.M., S. Ehui and Y. Assefa, 2004. Dairy development in Ethiopia. EOTD Discussion Paper no. 123. International Food Policy Research Institute, Washington D.C.

Delgado, C., R. Rosegrant, H. Steinfeld, S. Ehui and C. Courbois, 1999. Livestock to 2020: The next food revolution. Food, Agriculture and the Environment. Discussion Paper 28. IFPRI, FAO and ILRI.

Dyer, J.H., 1996. Specialized supplier networks as a source of competitive advantage: evidence from the auto industry. Strategic Management Journal 17, 271-291.

FAO, 1990. The Technology of traditional milk products in developing countries. FAO, Rome, Animal Production and Health Papers n°85.

Holloway, G., C. Nicholson, C. Delgado, S. Staal and S. Ehui, 2000. Agroindustrialization through institutional innovation. Transaction costs, associations and milk-market development in the east-African highlands. Agricultural Economics 23, 279-288.

ILRI, 2004. Making the most of human capital in the tropics. Options in dairy processing and marketing. www.ilri.org.

Jarillo, J.C., 1988. On strategic networks. Strategic Management Journal 9, 31-41.

Ketema, H. and R. Tsehay, 1995. Dairy production system in Ethiopia. In: FAO. Strategies for market orientation of small-scale milk producers and their organisations. Proceedings of a workshop held at Morogoro, Tanzania, 20-24 March 1995. (www.fao.org, accessed February 2005).

McCormick, D., 1999. African enterprise clusters and industrialization: Theory and reality. World Development 27, 1531-1551.

Ménard, C., 2004. The economics of hybrid organisations. Journal of Institutional and Theoretical Economics 160, 345-376.

Omiti, J. and S. Staal (eds.), 1996. The Ugandan dairy subsector: A rapid appraisal. ILRI/NARO/MAAIF Research Report, Nairobi.

Redda, T., 2001. Small-scale milk marketing and processing in Ethiopia. In Proceedings of the South-South Workshop on Smallholder Dairy Production and Marketing - Constraints and Opportunities. March 12-16, Anand, India.

Staal, S., C. Delgado and C. Nicholson, 1997. Smallholder dairying under transactions costs in East Africa. World Development 25, 779-794.

Sykuta, M.E. and M.L. Cook, 2001. A new institutional economics approach to contracts and associations. American Journal of Agricultural Economics 83, 1273-1279.

Tegegne, A., 2003. Urban and peri-urban dairy development in Ethiopia: Lessons from the Ada'a-Liben Woreda Dairy and Dairy Products Marketing Association, Ethiopia. Paper presented at Regional Symposium on Situation Analysis and Prevention of Famine in the Amhara Region, May 2003, Bahir Dar, Ethiopia.

Williamson, O.E., 1985. The economics of capitalism. The Free Press-McMillan.

Williamson, O.E., 1991. Comparative economic organisation: The analysis of discrete structural alternatives. Administrative Science Quarterly 36, 269-296.

Williamson, O.E., 1996. The mechanisms of governance. Oxford University Press.

Williamson, O.E., 2003. Transaction cost economics and agriculture: An excursion. Presented at the 80th EAAE Seminar New Policies and Institutions for European Agriculture, September 2003, Ghent.

World Bank, 2005. www.worldbank.org, accessed February 2005.

Organisational process alternatives for process simulation in agri-food processing and trade: the case of the grain chain

Thomas Hannus, Volker Jahn, Oliver Poignée and Gerhard Schiefer
University of Bonn, Chair for Business Management, Organisation and Information Management, Meckenheimer Allee 174, 53115 Bonn, Germany; th.hannus@uni-bonn.de

Abstract

The agri-food chain faces multiple new demands on efficiency of food production, food quality and food safety. These demands cause changes in the management of traditional processes in the agri-food chain. As little knowledge on processes related management activities in terms of cost structure, organisational flexibility and control opportunities is available, this paper discusses principles with respect to the demand of traceability for the grain chain. In a first step new objectives as well as corresponding control opportunities are identified. As to check for their feasibility, organisational flexibilities are discussed as well as efficiency issues. Results are then compared to findings from case studies within the grain chain.

Keywords: process models, process control, grain chain

1. Introduction

Organisation and quality of processes in the agri-food chain have become core issues for the efficiency of food production and the quality and safety of food (Schiefer, 2002). A multitude of quality programs or quality management programs focus on elements of process organisation or process control. Process improvements need to rely on process related feedback information that could support process management activities. However, despite the importance of process related management activities, there is little scientific knowledge on cost structures, efficiencies, and organisational flexibility or control opportunities.

The goal of this paper is to give some insights into the organisational elements which constitute process organisation and introduce ranges of organisational flexibility. Thereby the paper aims at giving first insights on control opportunities and their impact on the production process. Control opportunities as well as their impact are discussed with respect to efficient traceability.

The paper will start off with a short introduction of the grain network as a whole and a focus on the breadstuffs industry. Second, different approaches of viewing the networks production and trade processes are outlined and basic organisational elements are introduced. In the following chapter, process objectives and control requirements are identified with regard to the organisational elements of the processes. Based on the identified control requirements, approaches to control organisation for tracking and tracing optimisation are illustrated. The introduced issues are underlined by findings of case studies, which were conducted with businesses of the German grain sector over the last few years. The paper concludes with a summary of recommendations possible at this point of knowledge as well as a proposal for further research.

2. The grain chain scenario

In the following chapter the grain chain is described emphasising the breadstuffs chain. This chain can be described as a horizontal and vertical network which is tightly interwoven with other branches of the sector (see Figure 1). Key players in the breadstuffs chain are farmers, grain traders, mills and bakeries. Various linkages to related industries exist on all stages of the chain. Farmers interact with the seed industry as well as other suppliers of operating resources. Grain traders often function as sales organisations in this set up. Moreover they are customers for grain, function as storage facilities for farmers or provide brokerage services. Products are sold to the food industry at the levels farm, grain trader and mill. Connected branches are all food industries producing on the basis of grain, such as starch and malt industries. Food industry on the other hand supplies processed grain products (e.g. malt) for mills and bakeries. All stages of the breadstuffs chain sell products and byproducts to the feedstuffs industry. Therefore the feed mix industry is the main interface to livestock farming. (for a more detailed description of the sector see Poignée, 2003).

The grain sector is a complex network with multiple options of production and trade scenarios. Complexity is not only due to the illustrated diversity of trade linkages and process scenarios, but existing cycles of recursive trade constellations (e.g. farmers and grain traders).

Another characteristic trait of the sector is the variety of enterprise structures between and within different levels of the chain. Structural changes affected the branches with varying intensity. The mill industry for example, is characterised by a strong aggregation of production capacities within few companies and a large number of small and medium sized enterprise still at the market (e.g. VDM, 2003).

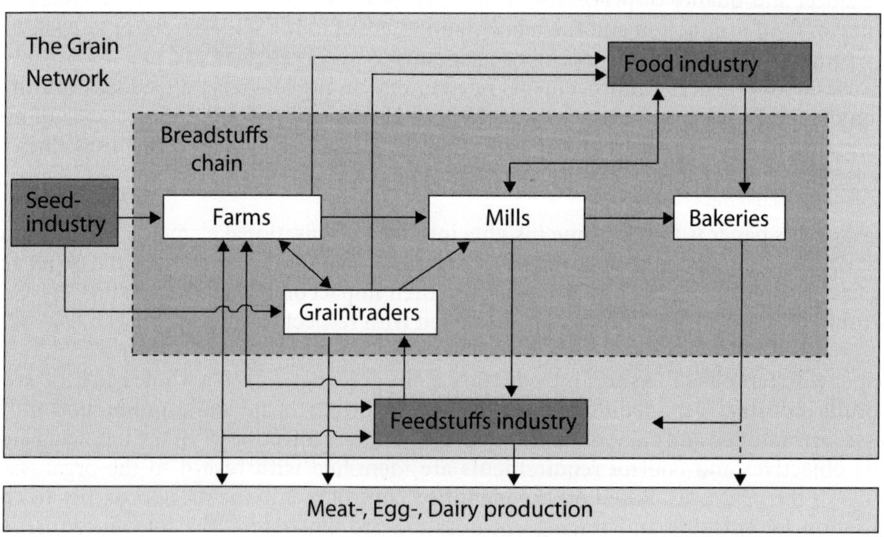

Figure 1. The Grain Network (Poignée et al.*, 2005).*

3. The process view

In the following section, models for process representation are discussed against the background of traceability. In order to model traceability in the grain chain, relevant organisational elements need to be identified in the first place.

From a single enterprise point of view, basic organisational elements can be outlined as depicted in Figure 2. These are delivery, receiving inspection at the company, internal processing, pre-delivery inspections and quality assurance and delivery of the product to the subsequent enterprise (view 1) (adapted form the SCOR model; Supply-Chain Council, 2004). This representation resembles the current demand on transparency as enforced by implementation of the EU Regulation 178/2002 (EU, 2004). However the described differentiation of organisational elements does not enable a focused traceability: As the enterprise itself is treated as a black box, no linkage between incoming and outgoing goods can be established.

In order to enable a design for efficient traceability, the internal processes of enterprises must be integrated into the organisational model (view 2). One approach to do so is the identification and linkage of transport, storage- and production batches (BLL, 2001). Batches, in this context, are defined as clearly separable entities of products or intermediates and need to be identified every stage of the production and trade process. Traceability then is established by linking batches to one another if they are connected by trade and production processes. Following these design principles a model can be established which can map the chain view as well as the internal enterprise view (Hannus *et al.*, 2003).

A further detailing of process modelling by integrating process internals can results in a better focus of traceability. Consequentially the next step of refinement is the modelling of the product flow within the individual process steps of trade and processing (view 3 - not depicted in Figure 2). Research on this topic is conducted in the field food technology. As for the grain chain dealing with bulk commodities, this involves modelling of grain flow through silos, for example. However, only few of these results are realised in day to day business of the grain chain. This is mostly due to the high complexity of these simulation models which require multiple input parameters in order to provide satisfying results. Therefore the following chapters will discuss view 1 and view 2 in comparison and especially view 2 in greater detail.

Figure 3 illustrates the above mentioned consequences of different process model designs on product traceability for view 1 and view 2. Tracing of production runs through a typical four

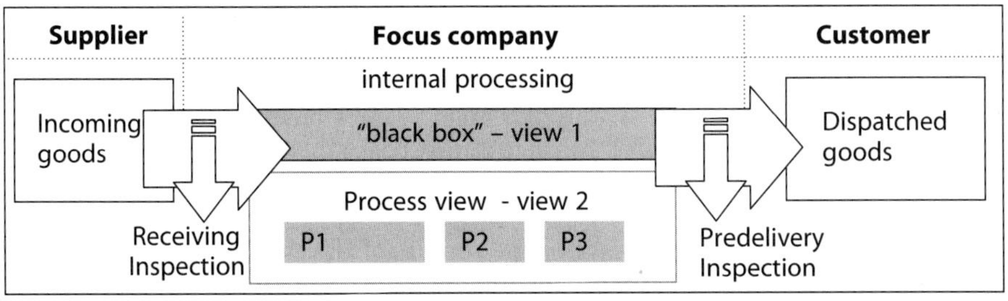

Figure 2. The process view.

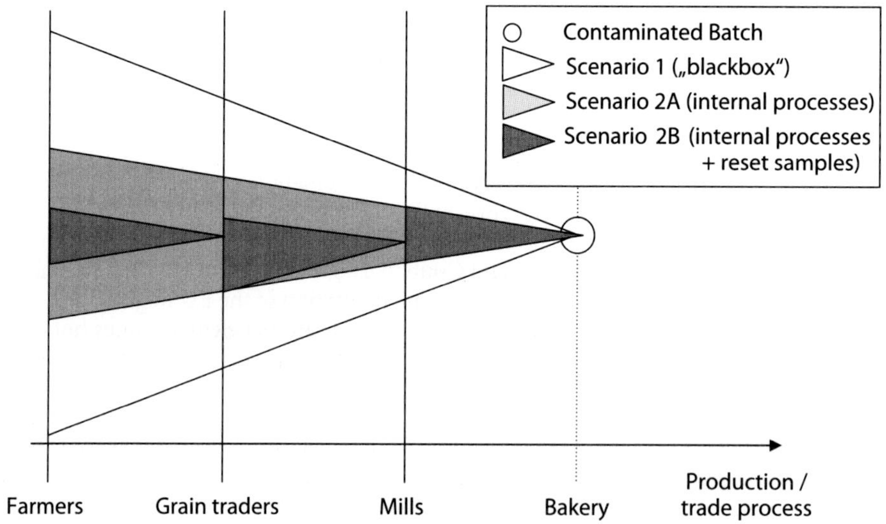

Figure 3. Tracing scenarios.

stepped breadstuffs chain. The tracing process starts with one batch at the bakery level and identifies potentially involved batches in the upstream chain.

Scenario one - resembling view 1 - shows the number of batches, which can be identified as part of the production process if no internal traceability is implemented. Scenario 2A illustrates the setting, once traceability is implemented on the basis of view 2 (internal process modelling). The number of potentially involved batches declines as a relationship between incoming and outgoing goods can be established. Scenario 2B illustrates the effect of drawing reset samples at supply delivery or product dispatch. Reset samples enable the identification of contaminated batches on each level, thus reinitialising the tracing process with one or at least very few batches.

The described example shows, how changes in the process set up in terms of establishing control opportunities will affect traceability. The following chapters will therefore illustrate where these controls can be implemented and how they affect traceability in greater detail.

4. Process objectives and control requirements

Based on the structuring of the production and trade processes as outlined in chapter 3, a number of objectives can be derived for traditional process control (view 1) as well as for process control with the focus on traceability (view 2). Whereas part of the control opportunities are identical, both for traditional and traceability view, others only are implemented when the focus is set on enabling traceability. Table 1 illustrates the interrelation of organisational elements to objectives of view 1 and view 2 with regard to corresponding control opportunities.

A number of objectives gain importance in the context of traceability. They can be summed up under the heading of an increased focus on key parameters of supply chain design and reorganisation of internal production processes.

Table 1. Objectives traditional and traceability focused process control.

Organisational elements	Objectives traditional (view 1)	Objectives new (view 2)	Control opportunities	Change
Supply	large homogenous batches	large homogenous batches	number / size of delivery batches	no change
	-	short upstream process chain	length of upstream chain	new
Internal processing	few large storages	small storages	number of storage facilities	increased number of storages
	-	small batches	number of storage batches	new
	continuous processing	discontinuous processing; small batches	number of production batches	new
	-	few blending points	number of blending points in process	new
	continuous blending	few repetitions	repetitions of blending	decreased number of blending processes
Sampling	-	reset samples of every incoming and dispatched delivery	sample management	new

Objectives of supply chain design focus on the length of the upstream chain as well as the number of division and blending processes. These are to be kept minimal, as they cause the number of involved batches to accumulate along the chain.

Objectives concerning the reorganisation of internal processes revolve around the need to change from a continuous to a discontinuous, batch oriented storage and production process. This contradicts established principles of bulk commodity production, especially in storage and milling processes, where lower plant utilisation is considered a major cost driver. However as of now, this is the only commonly accepted way of modelling traceability in a way, which represents actual business practices. Case studies show, that the necessary change of paradigms - discontinuous vs. continuous storage and processing - is also realised in day to day business by adapting process management accordingly.

The introduction or intensifying of reset sampling both at the reception and issuing of goods is one last objective that enables efficient traceability as illustrated in the scenario 2B of Figure 3. From an enterprises point of view, reset samples not only safeguard in case of a contamination, but are also a way of identifying the minimal number of potentially involved batches at each level of the tracing chain.

Summing up, three main statements can be formulated on the issue of objectives for production and trade processes. First, a number of new objectives need to be taken into account. Second, only one of the objectives, the size of delivery batches, is alike both for view 1 and view 2. Last, a number of objectives of view 2 contradict those of traditional set ups.

5. Control organisation for tracking and tracing optimisation

In this chapter, based on the objectives identified, control opportunities are discussed in terms of supporting traceability. Control opportunities are therefore analysed in a twofold way: First to the flexibility, which the process design typically allows in the domain of a specific control opportunity and second to its impact on traceability.

5.1. Enterprises ability to influence controls

Most controls within the process design can only be influenced in a limited way (Table 2). Reasons for this differ among control options:

- Number and size of delivery batches are impaired by structural limitations. Internal and external structures are oftentimes presets of the enterprises or legal requirements. For example the infrastructure for delivery (e.g. ship, train, truck) or the agricultural structure in a specific area might hinder the delivery of large, homogenous batches. Relevant legal presets could be maximum weights of transport units. Structural parameters can only be changed at high expenses or not at all and therefore only have limited flexibility.
- The length of the upstream process chain can be influenced by the deliverers, however within certain limitations. Limitations are set by the need to blend in order to obtain certain quality parameters (e.g. moisture due to storage requirements) at upstream levels.
- Number and size of storage facilities can be adjusted at high costs and therefore only at a mid to long term range.

Table 2. Flexibility and impact of control opportunities.

Control opportunities	Influence on control	Impact of the control
Number and size of delivery batches	low	less batches involved in the process
Length of the upstream process chain	medium / high	less blending of batches in the upstream chain
Number and capacity of storage facilities	low / medium	simplified identification due to separation in storage
Number of storage batches	low/medium	less batches involved in the production process
Number of production batches	partially restricted	less batches involved in the production process
Number of blending points in the process	low	less batches involved in the production process
Repetitions of blending	low / medium	less batches involved in the production process
Sample management	high	facilitates identification of batches due to certain parameters

- The number of storage facilities leads to consequences regarding the size of storage batches. Constricted warehouse capacity limits the ability of empting storages, thereby raising the number of involved upstream batches through increased mixing in the storage process.
- Problems concerning batch blending in the upstream chain are mirrored in the internal chain of an enterprise, storage or production likewise. Products are combined in order to achieve certain parameters due to market necessities or the set-up of the production line. Therefore this control opportunity also has limited flexibility.
- Reset samples management is easily introduced, allowing high managerial flexibility for utilising this control opportunity.

5.2. Impacts of controls

All control options in storage and production processes cause an accumulated effect on the number of involved upstream batches. Therefore the effect of mixing batches further downstream tends to results in a stronger accumulation of involved batches.

Drawing of reset samples has a major influence on the number of batches potential involved in the upstream production. By means of reset samples the source batch for a contamination can be identified at every level of the production chain. As tracing is reinitialised with specific batches at each stage where reset samples are drawn, the overall number of involved batches declines.

6. Lessons from case studies

During the last few years the University of Bonn has conducted a series of case studies with enterprises and quality programms in the German grain sector. In these studies, a number of practical approaches already employed in the businesses were identified:

- Reset sampling gained importance due to changed legal settings (e.g. changes in laws on product liability) and because of traceability considerations. During the course of the last few years an intensified discussion on legal validity of reset samples, methods of drawing representative samples, storage times as well as storage requirements have populated the relevant technical press, signalling that businesses ascribe high importance to this issue.
- Changes in storage processes include:
  - Increased seperation at the reception of grain: Reception and internal transportation facilities tend to be bottlenecks during the harvest. These may cause unfavourable storage strategies e.g. division of one delivery batch to two or more storagebatches in order to optimise internal transportation. Grain traders and mills in the case studies have invested in order to overcome these problems.
  - Division of storage facilities whenever possible at low cost e.g. subdivision of boxes in warehouses to allow for better identity preservation.
  - Emptying of silos on a regular basis to minimise the number of involved upstream batches.
  - Reduction of blendings by employing planned storage strategies during the harvest.
- Changes in the processing take place only slowly and unwillingly. The changes oftentimes involve the acquisition of new know-how and technology, as well as the overcoming of traditions.

As awareness of the problem of traceability only developed during the last few years, findings of case studies reflect only short term action. Nevertheless, as no detailed knowledge on e.g. cost structures, risk perceptions and efficiencies is available, the enterprises implementations can be looked upon as expert's choices and thereby serve as one indicator for efficiency of adaptations in process management.

7. Conclusion

This paper mainly focuses on techno-organisational issues, setting a basis for further research in order to provide valid decision support. Doing so, a number of aspects need to be considered. Exemplarily two points - cost structures and process objectives - are mentioned here:

7.1. Ambiguous cost structure

First, cost factors which arise in the course of reorganising the process chain have to be identified. This might pose difficulties when the reorganisation of trade processes leads to changes in transaction costs, as the determination of these is inflicted with problems (Picot, 1982).

Determination of processing costs is complicated in those cases, where costs and benefits of process reorganisation are distributed across the border of businesses e.g. when limiting the blending of product batches.

Second, some cost as well as benefits arise only with the necessity to track and trace e.g. due to a product contamination. Therefore the cost benefit evaluation is strongly influenced by the individual entrepreneur's perception of the probability of such events.

7.2. Objectives taken into account

The paper solely discusses traceability as the new objective. However, others need to be taken into account, e.g. quality- and risk management. When considering those, new control options as well as contradicting control objectives will need to be considered.

Still a number of recommendations for the set-up of chain wide process controls can be made based to the current findings. These recommendations generally focus on the organisation of the product flow as well as reset sample management. Both focal points have already gained importance in day to day practice of the grain chain.

Within the grain chain there is an apparent necessity for decision support. However the question of ideal set ups can, as of now, only be answered on a case to case bases. More flexible support can be delivered on the bases of the above mentioned model for describing processes. When designing a decision model on this basis, other relevant objectives and perceptions of enterprises as well as cost benefit structures will need to be taken into account.

References

Bund für Lebensmittelrecht und Lebensmittelkunde (BLL), 2001. Leitfaden Rückverfolgbarkeit - Die Organisation der Rückverfolgbarkeit von Produkten in der Lebensmittelkette. BLL.

European Commission, 2004. Food Law, Implementation Guidance Document, http://europa.eu.int/comm/food/food/foodlaw/guidance/guidance_rev_7_en.pdf. (status February 2005).

Hannus, Th., O. Poignée and G. Schiefer, 2003. The implementation of a web based supply chain information system - experiences with a regional quality program. In: Harnos, Z., M. Herdon and B.T. Wiwczaroski (eds.), Information technology for a better agri-food sector, environment and rural living. Proceedings of the Second European EFITA Conference. Vol. II, pp. 594-600.

Picot, A., 1982. Transaktionskostenansatz in der Organisationstheorie: Stand der Diskussion und Aussagewert. Die Betriebswirtschaft 2, 267-284.

Poignée, O., 2003. Qualitätskommunikation in der Getreidewirtschaft. Report B-03/1, ILB.

Poignée, O., Th. Hannus, V. Jahn and G. Schiefer, 2005. QM-G: Organisatorisch-technische Alternativen für Rückverfolgbarkeit und Qualitätssicherung in Unternehmen der Getreidewirtschaft. In: Schiefer, G., Rückverfolgbarkeit und Qualitätsmanagement in der Getreide und Futtermittelwirtschaft, pp. 21-43.

Schiefer, G., 2002. Environmental control of process improvement and process efficiency in supply chain management - the case of the meat chain. International Journal of Production Economics: Manufacturing Systems, Strategy & Design 78, 197-206.

Supply-Chain Council, 2004. Supply-Chain Operations Reference-model - Overview Version 6.1. http://www.supply-chain.org/SCOR/SCOR_Overview_6.1.pdf (status April 2005).

Verband deutscher Mühlen (VDM), 2003. Annual Report 2002/2003.

Maintaining organic integrity: tackling fraud in organics

Jochen Neuendorff and Ulrich Fischer
Gesellschaft für Ressourcenschutz, Prinzenstrasse 4, 37073 Goettingen, Germany;
jochen.neuendorff@gfrs.de

Abstract

Within the framework of the project, a systematic gap analysis was made of the inspection system and the inspection procedures under Regulation (EEC) No. 2092/91 on organic production of agricultural products and indications referring thereto on agricultural products and foodstuffs as amended. Proposals were identified for proactive risk management on the basis of this analysis. They comprise both the introduction of risk-orientated inspection elements (e.g. identification and management of 'organic critical points' in the production process) and the introduction of control mechanisms throughout the production chain (e.g. cross checks by the inspection bodies).

Keywords: organic food, risk management, certification

1. Introduction

The basic principles of organic farming and the processing of organic foodstuffs are regulated across the world by laws and regulations.

Below are mentioned three of the most important and precedent-setting norms: the Council Regulation (EEC) No 2092/91 of 24 June 1991 on organic production of agricultural products and indications referring thereto on agricultural products and foodstuffs (EEC-Regulation No. 2092/91) has been in effect since 1993; in Japan there has been a special section on organic farming in the Japanese Agricultural Standards (JAS) since April 2001 and in the United States the dispositions of the 'Organic Foods Production Act' and the 'Final Rule' of the National Organic Program (NOP) have been in effect since October 2002.

Apart from these legal requirements, suppliers of organic products, depending on the final market destination, eventually must fulfill additional private standard requirements of organic farmers associations.

The legal and private production rules for organic farming have in common that the organic nature of a product is defined by its production process. The compliance with the organic production standards is verified by private certification bodies, which are approved and supervised by state authorities, through inspection visits, document reviews and laboratory tests.

The implementation of the EEC Regulation in the European Union has been in course for many years now. In Germany, under the framework of the 'Federal Program for Organic Farming', a systematic gap analysis of the inspection system stipulated by the EEC Regulation was done. Proposals for improved risk management resulted from this analysis. The results of the study are presented below.

2. EEC-Regulation No. 2092/91 on organic production

Food safety in the EU is regulated by different EU regulations based on the EEC Regulation 178/2001. These regulations apply to all food- and feedstuff, regardless of the production method.

If the agricultural products carry certain, legally protected indications on their labels, the requirements of further EU regulations may apply as well. EEC Regulation No. 2092/ on organic farming is one of those regulations.

Consequently, Article 5 regulating the labeling of organic products is the central element in the EEC-Regulation No. 2092/91 (Rathke *et al.*, 2002; Schmidt and Haccius, 1994, 1998). Only if agricultural products are labeled with references to organic farming, the regulation applies. Article 5 also refers to the production rules and the inspection system.

The production rules guarantee a process of farming particularly environment-friendly, under which the use of synthetic chemical fertilisers and pesticides is prohibited. Appropriate animal husbandry is a primordial requirement for organic animal production. When organic products are processed, only a very limited number of conventional ingredients and processing aids are permitted.

Article 11 of the Regulation describes the requirements for importing organic products from non-EU countries, referred to as third countries.

The inspection system is described in Articles 8 and 9 of the Regulation. The Member States can choose between having inspection authorities apply the inspection system, or having private inspection bodies do it; or a combination of both. Private inspection bodies must be authorised and supervised by competent authorities. A purely state-administered system is in place in only two EU-Member States (Denmark and Finland). The majority of the EU-Member States have opted for a system incorporating private inspection bodies subject to state supervision. These private inspection bodies are obliged, under Article 9 (11) of the EEC-Regulation No. 2092/91, to meet the requirements of the quality management norm EN 45011 and to 'certify' in accordance with that norm the products covered under that EEC-Regulation No. 2092/91, whenever the requirements of the EEC-Regulation No. 2092/91 are met.

The entities responsible for the 'inspection system' in the case of the EEC-Regulation No. 2092/91, in contrast to the EurepGAP certification system, are not private companies but rather State entities (Jahn *et al.*, 2003). The verification system is implemented by a Public-Private Partnership, in which the advantages of private certification bodies, in terms of their efficiency, and the advantages of the public system, for example, the authority to issue and enforce sanctions, are combined.

Appendix III of the EEC-Regulation No. 2092/91 contains detailed 'minimum inspection requirements' for agricultural operations as well as processing companies, importers and companies processing animal feedstuffs. Its weakness in focussing on particular risks has been criticised in the past (Baummann, 2001; Darnhofer and Vogl, 2002; Heinonen, 2001).

In Germany the application of the inspection system set out in the EEC-Regulation No. 2092/91 is regulated by the German law for the implementation of EEC Regulation No. 2092/91.

While the Federal Ministry of Consumer Protection, Food and Agriculture represents German interests in Brussels, within Germany the Federal Agency for Agriculture and Food is the entity responsible for authorising private inspection bodies and making decisions on importer's applications for permission to import organic products originating in third countries (Article 11 (6) of the EEC-Regulation No. 2092/91.)

Farms and companies subject to the inspection system, according to the EEC-Regulation No. 2092/91, are registered by the competent authorities of the 16 Bundesländer (regional governments in Germany, similar to provinces or states). They also supervise the 21 private certification bodies currently operating in the country (as of March 2005) and are responsible for issuing major sanctions.

3. Study approach

In the initial phase of the study, the existing documentation was reviewed in order to identify the critical points.

This was followed by an examination of the statistics concerning sanctions imposed between 1999 and 2002, that the Federal Agency for Agriculture and Food (BLE), notified the EU authorities about. Additional questionnaires were mailed out to the competent authorities in the different Bundesländer and to the inspections bodies requesting their statistics of sanctions issued and asking for the presumed critical points. Furthermore, representatives of selected certification bodies, the competent authorities and the organic farmers' associations were interviewed by telephone. As well, inspectors from different inspection bodies were asked to provide information about their perceptions regarding the existing critical points.

The opinion of the suppliers of organic products concerning the efficiency of the inspection system was evaluated in three workshops (one each for cereals, fruit and vegetables, and meat). Based on the survey results, proposals for an improved risk management were developed.

4. Results

The statistics from the BLE indicate that from 1999 to 2002 the certification bodies carried out at least one scheduled inspection per year of the companies which are subject to the inspection system according to EEC-Regulation No. 2092/91. Additional unannounced inspection visits were carried out at a rate between 10 and 19 percent of the companies. The frequency of such unannounced inspections was lowest for farms, medium for processing plants, and highest for importers.

The surveys sent to the competent authorities of the Bundesländer were returned duly filled out by 8 of the 16 competent authorities. Upon evaluation of the statistics regarding the sanctions issued, it was immediately noticed that in the years in the period in question in the case of processing plants and importers, one of the most significant sanctions - removal of the indications referring to organic farming from the label of the lot affected - was ordered in only 4 cases, whereas in the case of organic farms, the same sanction was imposed in 118 cases. It was also interesting to see that the number of times this penalty was imposed increased sharply when the very complex regulations regarding organic animal production went into force in the middle of the year 2000. A more detailed analysis of the sanctions practices of five German certification bodies selected for that purpose confirmed these findings.

One might conclude from these data that organic integrity is endangered by ever more frequent infractions on farm level, which requires, therefore, more intensive inspections. This observation, however, falls short in capturing the real situation.

In the few cases of significant frauds involving organic foodstuffs, detected between 1999 and 2002 in Europe, and having a significant impact on the market, processing companies and importers were the main players. In Germany, frauds were detected in products such as cereals, legumes and oilseeds. Cases in neighboring Member States involved products such as meat, wine and rice. In terms of products imported from third countries, there were irregularities involving coffee, oilseeds and raisins. One thing common to all of the cases investigated was the labeling of conventional products as organic. The declaration of conventional goods as organic was not detected by the inspection bodies responsible for the infringing companies, but rather at later stages of the marketing chain.

The cheaters took advantage of some gaps in the inspection system. Until July 1, 2005, trading companies were exempt from the inspection system. Therefore, it was not possible for the inspection bodies to exercise any kind of supervision over these companies (wholesalers, brokers, etc.). The guarantee of organic quality from one end of the chain to the other had gaps and the cheaters were able to introduce conventional product lots.

As well, the falsification of documents proving organic quality, such as certificates, bills of lading, invoices and certificates of origin, was quite common. In particular, the cheaters were favored by the fact that the organic certificates granted to companies by the inspection bodies are not sufficiently standardised and are frequently provided via fax machines (i.e. copies, rather than originals). This led to the situation in which the authenticity of certificates presented could not be verified quickly and efficiently.

The step in which conventional lots is introduced into the 'organic system' is a decisive one. Once this is achieved, the conventional products declared as organic can be distributed without significant obstacles through the different levels of the market. Therefore, it is extremely important that inspection bodies verify the entry of merchandise into the organic system and the first buyer's guarantee of its origins and that the inspection bodies 'cross-check' between them the data about the merchandise received and the data from the suppliers about its commercialisation.

In supra-regional cases or in cases in which the cheaters operate through various Member States, neither the competent authorities nor the certification bodies took action quickly enough, nor did they coordinate their actions.

The results of the evaluations presented above were corroborated and complemented by information gathered in the interviews.

All those surveyed were in agreement that the inspection system stipulated in the EEC-Regulation No. 2092/91 has had good results and allows the reality of every holding or company to be clearly seen. On the other hand, inspection bodies still might improve the efficiency of their inspection through focusing on 'critical control points', especially in larger processing companies and larger importers. Until now, the EEC-Regulation No. 2092/91 requires an equally intensive evaluation of all the different aspects that are subject to norms applicable to all companies. According to those surveyed, this does not reflect the potential for real risks, which vary according to the kind of company. For example, the importance attributed to the danger of commingling cereals in an old installation is different from the danger that exists in a new storage facility.

As well, the operations procedures of the inspection bodies concerning the kind of inspections to be performed (announced and unannounced inspection visits), their frequency and scheduling, were criticised. The perception was expressed that the significance of the size of the company and its importance in the market are not considered as important as they should be.

Whereas the competent authorities see that there is some potential for improving the documentation from the inspection bodies regarding infractions committed, the inspection bodies and the suppliers of organic products complain that the interpretations of the dispositions of the EEC-Regulation No. 2092/91 by the competent authorities are not always clear, nor do they take into account the practical reality of the operators' situations, nor are they consistent.

All those surveyed agreed that not enough cross-checking is done by inspection bodies of the certificates themselves and of the flow of organic goods between different operators.

5. Conclusions

Only through appropriate, consistent rigorous practices on the part of all of the suppliers of organic products (producers, processors, importers and distributors) the organic quality of foods can be assured efficiently from one end of the grower-to-consumer chain to the other. In order to achieve this kind of reliability, the contents of the EEC-Regulation No. 2092/91 must be widely accepted among the actors and be considered appropriate and useful.

5.1. Further development of the production rules according to Regulation (EEC) No. 2092/91

The initial version of the EEC-Regulation No. 2092/91, published on June 24, 1991, referred only to organic plant production and the processing of foodstuffs composed of primarily plant ingredients. Animal production was not covered. Fairly general provisions for organic plant production were established (e.g. 'appropriate crop rotation'), and some 'knock-out criteria' stipulated (e.g. non- use of mineral fertilisers and synthetic pesticides)

By December 2004, the EEC-Regulation No. 2092/91 had been complemented and modified by 57 amendments and supplementary regulations. The result is an exhaustive legislative text. Currently, as well as organic plant products, the legislation also regulates organic animal production, organic beekeeping, the processing of organic agricultural and animal products and the production of organic feedstuffs. The production rules for some kinds of activities such as organic animal production are very precise, and for that reason, very detailed and long (with approximately 100 pages of regulatory text) and for the non-expert difficult to understand. This leads to an inadequate understanding of the requirements of the EEC-Regulation No. 2092/91 by the suppliers of organic products.

Some of the provisions cannot be implemented in all EU-Member States (e.g. purchase of 100-percent-organic seeds, stabling conditions for animals and their access to open air). For that reason, a number of exceptional authorisations had to be created, thus generating a lot of relatively unproductive administrative work for the inspection bodies and the competent authorities of the Bundesländer.

The current approach in the EU of having a highly detailed regulation that guarantees flexible application in the different Member States by providing for a large number of possible exceptional authorisations is not well-adopted to the reality of the suppliers of organic products and requires a lot of resources for administrative purposes. It would be more appropriate to place greater emphasis on the need for the principle of subsidiarity. The feasibility of application of any proposed future legal provision should be evaluated at an early stage.

5.2. Further development of the EEC Regulation's inspection system

To maximise public acceptance of the EEC Regulation's inspection system, it is extremely important that its implementation be in clear accordance with the four principal objectives of the EEC-Regulation No. 2092/91 (protection of the consumer, fair competition, transparency in the marketplace, encouragement of organic farming). Its application must be straightforward. That is way to make the best possible use of human resources, especially for deciding which infractions are most relevant and, therefore, consider the interests of the consumer as much as possible. When inspecting the holdings and companies, the amount of time spent must be flexible, and be oriented to goals and risks. Sanctions must satisfy, on one hand, the principle of proportionality, but on the other, also be effective and strongly discourage violations.

5.3. Organic Critical Points (OCP) as the basis for ensuring organic process quality

For large specialised holdings and for processing companies and import companies it makes sense to prepare diagrams of the flow of the product(s) and identify the 'critical points' on the holding or company. The critical points are those points in the handling of the product where its organic quality can be put at risk. For that reason the company must take measures to ensure that the potential risks are reduced or eliminated. The implementation of these measures must be monitored and documented. In the case of non-compliance, corrective measures must be established. It is appropriate to identify only a small number of key critical points, at which errors that cannot be tolerated may arise.

The term 'not tolerable' in organic farm production is applicable to, for example, the use of prohibited substances. In a processing company errors that cannot be tolerated include, for example, using conventional ingredients or employing processing aids such as ionising radiation in the processing of organic foods. In situations where parallel production is taking place, inadequate cleaning measures do not prevent commingling of organic with conventional products and could also not be tolerated. In an import company, traces of pesticides in agricultural products being imported from third countries are not acceptable.

In order to identify the critical points during the process, every stage of the process must be analysed (e.g. the use of certain inputs, the reception and storage of organic raw materials, ingredients and processing aids, cleaning in the case of parallel production) in order to verify if it is possible that an error that cannot be tolerated could occur there. If so, the supplier must then analyse whether precautionary measures can be taken to prevent such errors. If precautionary measures can be taken, then that is a critical point.

The precautionary measures at the critical points must be designed in such a way that they ensure that the organic integrity is subjected to the least possible risk during the production process. In the case of parallel production in processing operations, the precautionary measure would be an adequate cleansing that has been demonstrated to be effective; in the case of the importers, the measure to be taken would be the introduction of an internal plan to take samples to be tested for possible traces of pesticides.

The implementation of the precautionary measures shall be adequately documented. When the inspection is done, the inspection body should verify with particular emphasis the critical points.

This system guarantees that the internal quality assurance is oriented to the risks associated with the particular type of production. It is a preventive approach. In this regard, intensive training of the employees is also important. The employees who identify with the 'organic objective', who are involved in the process and assume responsibility for it, are more careful, because they understand the sense of 'quality assurance' and do not feel that it is an additional burden, impractical, bothersome, a mere formality and bureaucratic.

5.4. Classification of the risks as the basis for external inspection

Irregularities that can compromise the integrity of organic products can occur at any step in the chain, from farm production to sale to the final consumer. The risk of irregularities increases with:

- increasingly similar demands of quality for conventional and organic products;
- more sophisticated production technology required for organic production;
- shorter shelf-life of the product, or greater turnover of products;
- a growing price differential between conventional and organic products;
- greater availability and/or easier handling of prohibited technologies and ingredients (in situations of parallel production); and
- decreasing liability of the supplier.

Inspections carried out by inspection bodies should assign the agricultural holdings and companies inspected to risk categories. The following classification criteria can be applied: the structure and complexity of the company; production of risk products; the risk of commingling; the company's importance in the market; the risk of liability; previous sanctions imposed and the existence of an internal quality assurance system. The kind and frequency of inspections to be carried out by the certification body can be deduced from the risk category.

Thus, for example, the inspection bodies might carry out the annual inspection of the entire unit as an unannounced inspection. This would be appropriate in those companies in which the turnover of perishable products is high (e.g. certain specialised holdings such as those of laying hens or large-scale mushroom production, certain processing companies such as catering services, restaurants and cafeterias, or importers of organic fruits and vegetables). Other operations such as grassland farms can continue to be inspected with prior notice.

The frequency of the inspections carried out by the certification bodies should also be set based on the risk category. For example, it makes sense to inspect an organic fruit and vegetable wholesaler with a significant share of the market several times a year. As part of the inspections, special emphasis should be placed on the critical points. To complement this, spot check inspections can be done to monitor certain higher-risk steps in the process (e.g. sufficient separation of the lots in the case of parallel production) and to verify compliance with the conditions imposed by the inspection body.

5.5. Cross-Checks

Cross checks at different levels of the production chain make it possible to verify certificates and the flow of goods. They provide an opportunity to verify the data prepared by the inspected company so that monitoring outside of the 'company logic' can be done.

To prevent the falsification of certificates, certificates issued by inspection bodies to operators in Germany, Luxemburg and Austria can be verified through the BioC.info database (http://www.bioc.info) (see Figure 1).

The inspection bodies export from their internal database the data concerning their operators to BioC. The data are transmitted through an encrypted connection to the BioC server. After verification that the dataset is transmitted correctly, the data for the respective inspection body are unlocked and made available immediately.

Participation in BioC is currently still voluntary. The authenticity of the certificates of inspection bodies that do not participate in this system can be verified with the help of cross checks between inspection bodies by using forms, which have been provided in different languages during the project.

Within the framework of the current system, the inspectors calculate the flow of organic products for the operator inspected. The intention is to prevent the holdings from selling more organic products than they can produce on their own crop areas and to prevent the processing companies and importers from selling more organic products than would be normally produced with the amount of organic raw materials and ingredients they have bought.

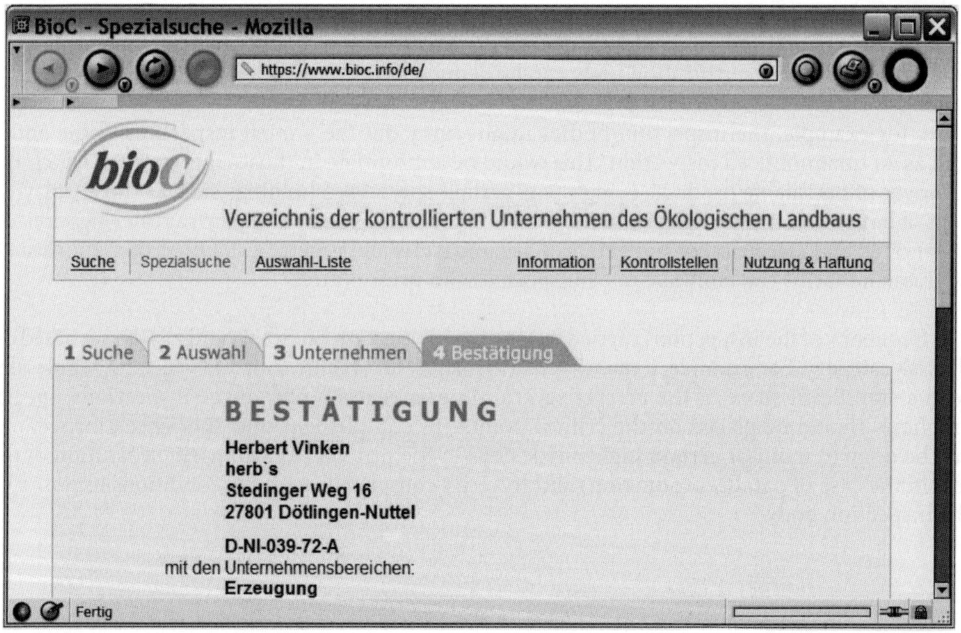

Figure 1. BioC internet-database.

Additional cross-checks can be done by sharing information gathered during the inspection process with other inspection bodies. In this system, the inspection body certifying the purchaser of an organic lot consults the inspection body certifying the supplier, on a random basis, to verify if it is plausible that a certain consignment has come from that particular supplier, and if the data on the documentation from both the purchaser and the supplier are identical.

With the exception of urgent cases, the request for cross check is filed in the supplier's file. At the next inspection, the data in the documents of the supplier and the buyer are compared, and the origin and organic quality of the lot sold are verified. Once the verification has been done, the information must be provided to the certification body making the request.

References

Baumann, P., 2001. Securing against fraud. In: Proceedings of the European Conference Organic Food and Farming. Copenhagen, May 10-11, 2001.

Darnhofer, I. and C.R. Vogl, 2002. Certification and accreditation of organics in Austria: Implementation, strengths and weaknesses. In: Ecolabels and the greening of the food market. November 7-9th, Boston.

Heinonen, S., 2001. The role of legislation. In: Proceedings of the European Conference Organic Food and Farming. Copenhagen, May 10-11, 2001.

Jahn, G., M. Peupert and A. Spiller, 2003. Auf dem Weg in eine Zertifizierungsgesellschaft? Eine Typologie der Zertifizierungssysteme in der Lebensmittelwirtschaft. In: Budde, H.-J., R.A.E. Müller and U. Birkner (eds.), Mobile Information. Chancen für die Agrarwirtschaft. Göttingen, pp. 61-64.

Rathke, K.-D., B. Weitbrecht and H.-J. Kopp, 2002. Ökologischer Landbau und Bioprodukte. Der Ökolandbau in Recht und Praxis. Beck.

Schmidt, H. and M. Haccius, 1994. EG-Verordnung 'Ökologischer Landbau', Müller.

Consumers' expectations of quality assurance and management systems

Food quality signals and customer confusion: is there a relevance for the marketing of food products?

Rainer Kühl and Wiebke Schulz
Department of Agricultural Economics, Justus Liebig University of Giessen, Senckenbergstr. 3, 35390 Giessen, Germany; Rainer.Kuehl@agrar.uni-giessen.de

Abstract

Customer confusion about the reliability of quality information can cause sub-optimal buying decisions. In a 2 month panel survey consumers shopping pattern and behaviour were evaluated for different product groups of frozen food subject to quality signals and the complexity of information. For 20% of the panel participants complex information, confusion about similar brands and the great variety of brands are responsible for an inconsistent buying behaviour. Socio-demographic differentiation of consumers show that quality information transfer is distorted and that customer confusion is prevalent and can be an important problem for the successful implementation of quality labels.

Keywords: food quality signals, customer confusion, marketing, food products

1. Introduction

Consumers' perceptions of food safety and risk have changed consumer behaviour. It is said that customers ask for more information on the procedures of food production and processing and they demand for enhanced transparency. Food suppliers try to fulfil these requirements by implementation of traceability techniques and by providing quality assurance schemes throughout the whole supply chain. The introduction and communication of quality labels and brands should support these activities. An overall aim is to regain lost consumer confidence and to establish loyal relationships. But, the growing content of different quality dimensions that are linked to a product has risen the complexity of a food product. A number of arguments indicate that increased availability of product and quality information can lead to a situation of information overloading. Increasing market in-transparency with the likely consequence of sub-optimal buying decisions could occur. This brief description raises some interesting research questions and asks for theoretical as well as empirical analyses.

The paper will focus on the following questions:

- Are customers irritated by the variety of brands of food products?
- Are they confused by the similarity of different brands?
- If so, how can these confused customers be characterised?
- If so, do the existing quality labels irritate them even more or could they help prevent confusion?
- Can confused customers discriminate between different quality signals or do they show similar patterns of confusion reduction as with food brands?

The paper is structured into three sections. First follows a brief review of the literature on the theoretical foundation of consumer confusion and its impact on variety seeking behaviour and consumer loyalty. The second section will present the data (of a panel analysis), the empirical findings, and the methodological approach used in the analysis. Finally, results and conclusions are given.

2. Theoretical background of customer confusion and development of the hypothesis

Customer confusion can be seen as a conscious or unconscious defection from the processing of information and making of decisions. This can cause sub-optimal buying decisions (Wiedmann *et al.*, 2001). Research on customer confusion has been conducted for more than 20 years and has so far focussed on four main topics.

Chernev (2003), Grünewald (2002), and Iyengar and Lepper (2000) analysed consumers` attitude towards the growing number of different brand varieties. They determined an optimal level of food variety up to which customers appreciate product differentiation. The theses were tested with food products such as chocolate and marmalade. Mitchell and Papavassiliou (1999) and Rafiq andCollins (1996) research is on the perceived similarity of brands, which especially occurs between the original and the me-too-product. They identify some form of confusion among customers. As a result of this confusion, an unconscious purchase of a brand different from the expected or even a disinterest towards food brands may occur. Another aspect of customer confusion has been analysed by Wiedmann *et al.* (2001) and Turnbull *et al.* (2000). They took the complexity and overload of information supplied by brands. Up to now, empirical studies concerning this topic have been conducted on the market of mobile phones, but not in the food sector. Here, several products, e.g. functional food and so-called 'wellness products', have complex characteristics which are difficult for customers to understand. Inaccurate information, like 'healthy' or 'low-fat', could create confusion. The research of Rudolph and Schweizer (2003) is being conducted on the link between the lack of shopping time and customer confusion. They identify that there is an unwillingness to spend time looking closely at every possible option before buying a product. No empirical studies concerning the lack of time while purchasing food products have so far been carried out. Most of the recent research, however, has focused on testing one to two reasons of customer confusion. To the authors knowledge no former study has been conducted including the analysis of all four potential reasons.

Prior studies show that approx. 10-40% of all customers can be regarded as confused. Confusion may have a number of negative outcomes: confused customers show diverge reactions, e.g. switching to another brand or through reduction strategies such as postponing their buying, delegating the purchase to another person or even refusing to buy any product. Another strategy that is often seen is reduction of alternatives by buying the best-known brand or the cheapest product. Furthermore, studies prove small levels of brand loyalty and satisfaction of confused customers (Meyvis and Janiszewski, 2002; Walsch and Hennig-Thurau, 2002; Foxman *et al.*, 1990). To be aware that brand and quality signal proliferation can create customer confusion as well as to recognise the causes of confusion is important to food marketers' strategies.

3. Data and research methodology

The theoretical framework was empirically tested as part of a two-month panel study in spring 2004. More than 300 participants (90% rate of return) took notes of their frozen food shopping pattern and behaviour. Frozen food products show a great variety of brands and products and therefore offer a great spectrum of methods of analysis. This great variety moreover results in a great stress of competition. Packed food products need to be sold through their own packaging, so these products especially need clear attributes to avoid customer confusion.

A recent ACNielsen (2004) research has documented that frozen food belong to the food products with the highest growth rates. Across all 59 countries included in the study, the aggregated sales value grew by 4%. Frozen fish/seafood, meat, fruit, pizza and potato products show growth rates from 4% up to 9%. These facts confirm that frozen food products form an interesting product group both for scientists and marketers. In addition, the study identified three key trends that seem to have driven the growth: a focus on health, the demand for convenience as well as the growing importance of private label. Frozen food can be related to two of these trends: vegetables, fish and fruits serve the trend towards healthy food, whereas frozen complete meals, ice-cream, potato products and pizza meet the desire for convenience food.

The participants answered a ten-page questionnaire containing several questions about their opinions and thoughts on food and quality brands. These questions were asked to determine the impact of customer confusion in three dimensions: the over-variety of food brands, the similarity of food brands, and the complexity of information given by food brands.

The participants were selected by quotas. To the authors knowledge no former study has been conducted including both the analysis of customer confusion and the actual food-purchasing pattern over a longer period of time. Therefore this study may offer new results as it both analyses the food buying behaviour over two months and analyses different potential causes of confusion.

Multivariate analysis was used to assess the characteristics of highly confused customer groups. Factors being studied were the social background, the degree of brand loyalty and novelty seeking, the opinion towards and the usage of loyalty programs as well as the ability to give and accept opinions and advice concerning food products. The panel data was analysed in the four most purchased product groups: frozen pizza, frozen vegetable, frozen fish and ice cream.

4. Empirical findings

In this chapter, all participants will firstly be classified according to their degree of confusion. Secondly, the analysis of possible reasons and motives for confusion will follow. In addition, these results will be compared with the theoretical background. In this context the participants will be divided into clusters with similar characteristics. Thirdly, the ascertained groups (especially the 'confused' participants) will be analysed in consideration of socio-demographic differences. Finally, customer confusion will be connected with other empirically tested buying attitudes and behaviour, e.g. brand loyalty and variety seeking, in order to find further explanations for customer confusion.

4.1. Classification of the participants according to their degree of confusion

One part of the questionnaire was designed to examine the degree of customer confusion. Thus, the participants were asked for their degree of agreement/disagreement with nine statements. For all items a six point scale was used. Statements were e.g. 'I perceive the range of different brands as too complex' and 'While shopping, it is difficult for me to distinguish between the different brands, because the products look alike'. First the mean value was assessed for every participant, which was then rated on a scale between zero and four. The partition was conducted as followed:

0.00 - 1.00: no acceptance → no confusion
1.01 - 2.00: slight acceptance → slight confusion
2.01 - 3.00: acceptance → confusion
3.01 - 4.00: great acceptance → great confusion.

As can be seen in Figure 1, 13% of participants did not agree with the statements and further 51% only agreed with them to a certain extent. In total, 64% of participants can be considered as customers with no or only slight confusion. In contrast, 36% of participants generally accepted the statements. These people can therefore be considered as being slightly or greatly confused.

In the following step, the degree of confusion was compared with socio-demographic characteristics. To examine this issue, univariate analyses of variance were conducted considering all four groups. Significant differences (smaller than 0.05) were found concerning age, education as well as employment. The degree of confusion increases with age. With an increasing level of confusion, the level of education decreases. In addition, the findings also show that confusion is related to the employment factor. While employed participants often show no or slight confusion, unemployed participants tend to be confused.

4.2. Analysis of the motives of customer confusion

To better understand the phenomenon of customer confusion, it is important to examine the motives for confusion. On a closer inspection of the nine aforementioned statements, they can be divided into three categories which represent three assumed motives for customer confusion. The questionnaire was designed in a way that the participants were not able to distinguish between these categories. The three motives and their empirical importance are explained in this chapter, beginning with the least important.

Firstly, the perceived similarity of brands is the least important reason for customer confusion. The mean value of the three statements is 1.28, which can be interpreted as 'slight confusion' (between 1.01 and 2.00). This indicates that the participants hardly have difficulties in distinguishing between quite similar brands.

Secondly, the great variety of brands is a motive for customer confusion, although this motive still has quite a low mean value (1.74). As it is slightly higher than the mean value of the first motive,

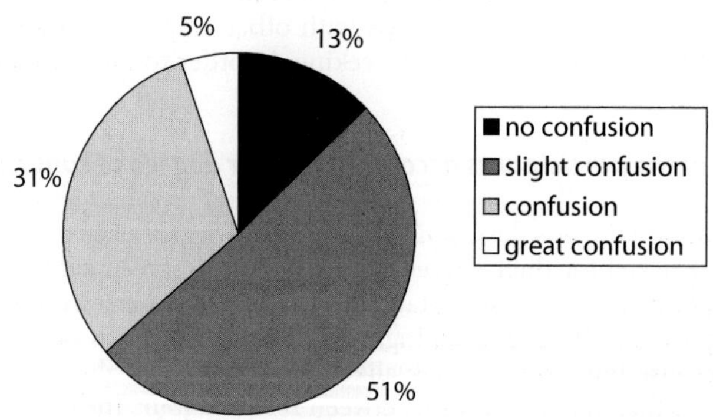

Figure 1. Classification of the participants according to their degree of confusion.

customers can seemingly be slightly confused by the brand variety. On average, the participants do not perceive the variety of brands as too complex. In addition to that, they neither prefer smaller assortments nor find it easier to choose a brand when being confronted with a large variety of brands.

Finally, the complexity and overload of information supplied by brands is the most important motive for customer confusion. The participants tend to agree to the statements (mean value of 2.58), and are therefore confused by the complexity of information. They are confronted with ingredients and product attributes (e.g. probiotic, native, pasteurised) that are rather unknown to them. Furthermore, they would appreciate packages showing only important information for the sake of clearness.

In addition to the presentation of average values, the motives of confusion can be shown individually for each participant. The central aim is to identify sub-groups (clusters). The members of each group should be as homogenous as possible in their opinions, and so represent a specific customer style. Between the groups, a big heterogeneity is desired. Therefore an agglomerative, hierarchical cluster analysis was run for all 302 participants, using the Ward method as fusion algorithm.

As a result, three clusters emerged during the first analysis: 31% of participants show average characteristics, 51% are less confused than the average and 18% are highly confused. However, the analyses of correlation show a strong correlation between the variables variety and similarity (0.548**). In addition to this, the variety correlates quite strongly with the complexity (0.491**). The results from these analyses suggest that it is advisable to eliminate the variable variety. Another cluster analysis was run with the two remaining variables similarity and complexity.

The elbow criterion suggests three to four clusters. When four clusters are considered, the analysis of variance (ANOVA) shows no significant results concerning the first two clusters. Therefore the option with three clusters is analysed in detail. Here both variables show significant differences in variance in all three clusters. The clusters range in size from 58 to 183 participants. These three clusters are presented graphically in Figure 2.

As can be seen from Figure 2, 19% of participants (cluster B) are not at all confused about the similarity of brands. This group only shows slight confusion about the complexity and overload of information. Therefore this group can be called the 'non-confused customers'.

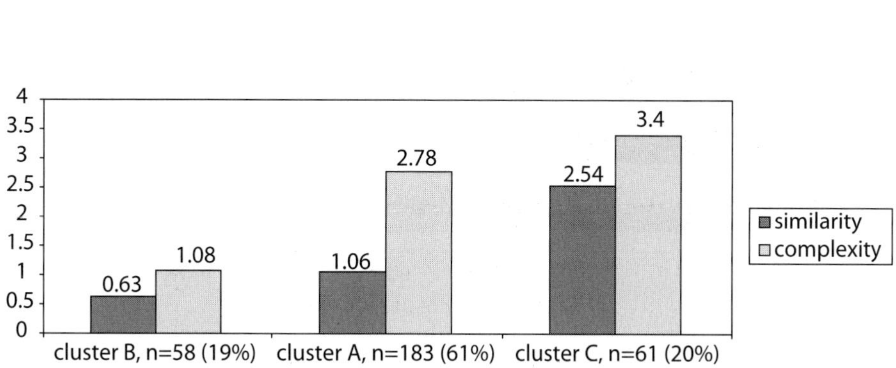

Figure 2. Three clusters similarity and complexity as motives for confusion.

The largest group of participants (61%, cluster A) consists of customers who are slightly confused about the similarity of brands and quite confused about complex information. They can be named the 'slightly confused customers'.

In contrast, the last group (cluster C, 20%), can be called the 'confused customers'. They perceive brands as similar and are strongly confused about complex information. All three clusters show higher values in the variable complexity than in similarity.

These results require further research on the group of confused customers. As confused customers represent a substantial proportion of customers (20%), they should not be neglected. The following chapter will present a more detailed examination of confused customers.

4.3. Socio-demographic characteristics of confusion

In a next step, these three clusters were compared regarding socio-demographic characteristics. To examine this issue, univariate analyses of variance were conducted. Here again, there were significant differences (smaller than 0.05) concerning age, education and employment. The average age is:
- non-confused customers: 40 years;
- slightly confused customers: 45 years;
- confused customers: 55 years.

This trend can also be seen in the next following bar chart (Figure 3). The degree of confusion increases with age.

At the same time the degree of confusion decreases with the quality of academic education (see Figure 4). While non-confused customers often have a college or university degree, confused customers often learnt a practical job through an apprenticeship.

Confusion differs as far as the employment is concerned. While employed participants often show no or slight confusion, unemployed participants tend to be confused (Figure 5).

In fact, however, there were no significant differences between the genders, the number of household members, and the levels of income and school education.

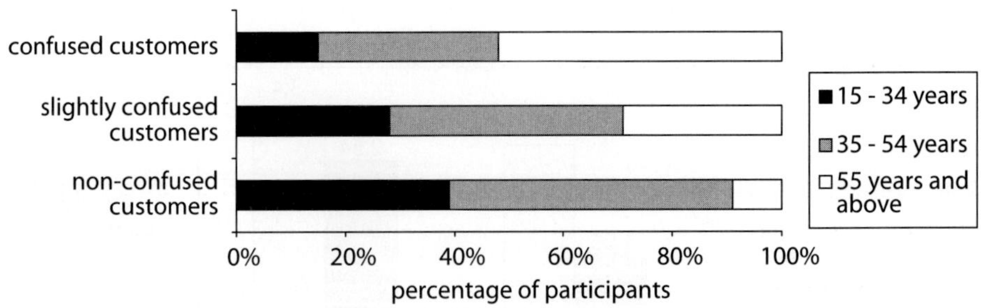

Figure 3. Confusion in three clusters according to the age (3 groups).

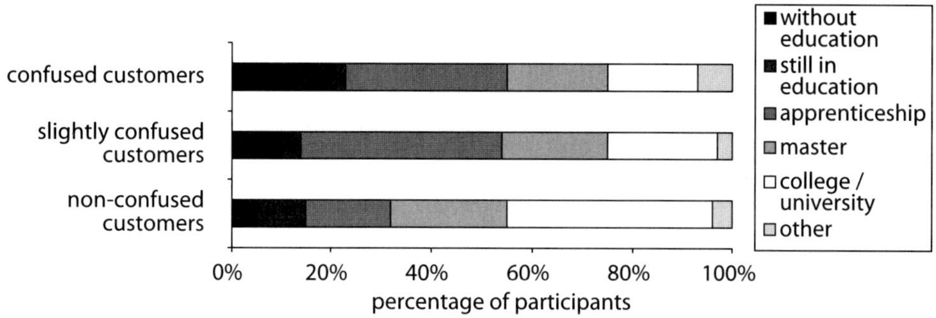

Figure 4. Confusion in three clusters according to the education.

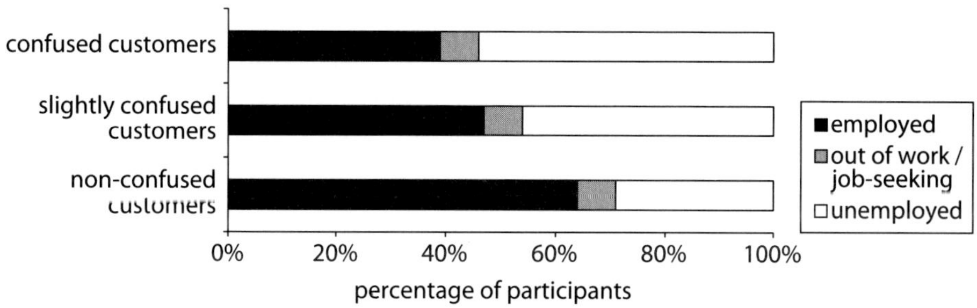

Figure 5. Confusion in three clusters according to the employment.

4.4. Buying attitudes and behaviours of confused customers

In a final analysis step the clusters were compared regarding other buying attitudes and behaviour. Again, analyses of variance were conducted to determine the impact of these factors.

First of all, a possible lack of time was analysed. It has been theorised that a lack of time can cause customer confusion (Rudolph and Schweizer 2003). This variable, too, was scaled between zero and four. Consistent with expectations, the analysis of variance shows significant differences between the three clusters. For both non-confused and slightly confused customers a lack of time is not a major criteria when buying food, whereas confused customers are lacking time during shopping. These participants do not compare all the alternatives before deciding which brand to buy.

Secondly, it was of interest to determine the connection between the attitude towards loyalty programs and customer confusion. Walsch and Hennig-Thurau (2002) observed that customer confusion can raise loyalty. In contrast, other research assume a decline in loyalty (Foxman *et al.*, 1990). For this purpose the customers were asked to answer several questions about loyalty clubs, loyalty cards, coupons and other forms of loyalty programs. Again, the results of this variable were transformed into a scale from zero to four. The results clearly confirm significant differences between the three clusters. Both non-confused and slightly confused customers do not approve of loyalty programs. In contrast, confused customers show slightly better attitudes towards these programs. Surprisingly, no specific difference within the usage of loyalty programs could be found on the basis of variance analysis. A summary of the results is shown in Figure 6.

Thirdly, the degree of passing on references was analysed. Non-confused customer pass on more references than confused customers do. Yet, these differences are too small to be significant. The degree to which references are being accepted from other people and marketing activities shows no relation to confusion as variance analysis suggests no significant results.

All further analyses are related to specific food products. The panel data from spring 2004 tested four product groups: frozen pizza, frozen vegetable, frozen fish and ice cream. The satisfaction among customers with the frozen food they bought was analysed. Confused customers are found to be less satisfied when buying products (Iyengar and Lepper, 2000; Walsch and Hennig-Thurau, 2002). Each participant was asked to rate his or her individual satisfaction with each product purchased. Interestingly, none of the four product classes showed significant results. The results indicate that both confused and non-confused customers display rather similar patterns of satisfaction.

The degree of customer loyalty was connected with customer confusion. Here, similar results occurred, i.e. confused as well as non-confused customers can be loyal customers. Regarding the degree of novelty seeking, both confused and not confused customers seek new products in a rather similar way. Table 1 will summarise the previous presented results.

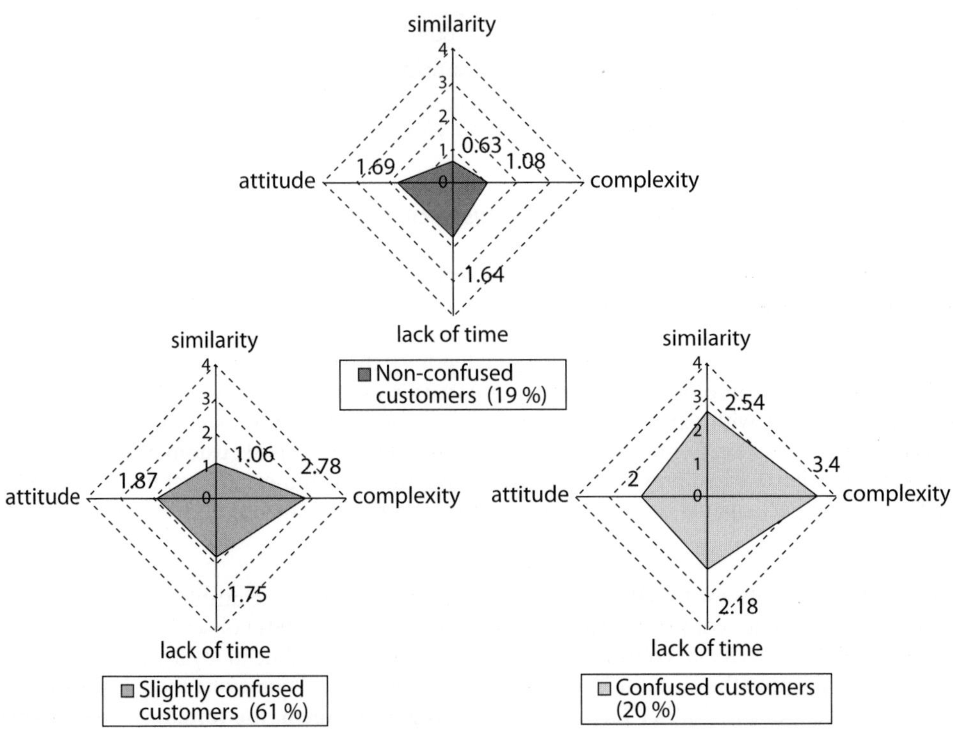

Figure 6. Customer confusion (similarity, complexity) - lack of time - attitude towards loyalty programs.

Table 1. Buying attitudes and behaviours of confused customers.

	Non-confused customers	Slightly confused customers	Confused customers
Lack of time (0.001**)	1.64	1.75	2.18
Attitude towards loyalty programs (0.012**)	1.69	1.87	2.00
Usage of loyalty programs	no significant differences		
Passing on references	rises with confusion, but not significantly		
Accepting references	no significant differences		
Satisfaction	no significant differences		
Loyalty	no significant differences		
Novelty seeking	no significant differences		

In abstract, 20% of participants can be regarded as being confused. They particularly show confusion about complex information. Besides, they also show confusion about similar brands and the great variety of brands. This group is characterised by high unemployment (often housewives and pensioners) and low education. Their average age is 55 years. In comparison to other customer groups, they allow the least amount of time for shopping. They furthermore show the highest attitude towards loyalty programs. However, some assumptions could not be verified: no significant differences could be analysed between confused and non-confused customers concerning loyalty, novelty seeking and satisfaction attitudes.

5. Conclusions for quality management systems and marketing

The results reveal that confusion towards food brands exists and therefore the question arises to which extent this confusion also influences the customer reaction towards quality labels. Possible causes of customer confusion could be the large number of quality labels and the amount of information. In this case good quality management systems could decrease confusion, but several aspects are to be considered. Information complexity in particular should be eliminated as far as possible to avoid confusion.

Further studies could address the problem of the appropriate amount of quality labels to avoid confusion due to an over-variety. Although the similarity of brands does not seem to be the main reason for customer confusion, the creators of quality labels should nevertheless put importance on a clear and unique communication of their labels.

References

ACNielsen, 2004. What's hot around the globe - insights on growth in food & beverages 2004. www.acnielsen.de [09.02.2005].

Chernev, A., 2003. Product assortment and individual decision processes. Journal of Personality and Social Psychology 85, 151-162.

Foxman, E.R., D.D. Muehling and P.W. Berger, 1990. An investigation of factors contributing to consumer brand confusion. The Journal of Consumer Affairs 24, 170-189.

Grünewald, S., 2002. Der ALDI-Boom.www.rheingold-online.de/db/download/ph_dn1_2092002124145156.pdf (27.05.2003), 1-7.

Iyengar, S.S. and M.R. Lepper, 2000. When choice is demotivating: Can one desire too much of a good thing? Journal of Personality and Social Psychology 81, 995-1006.

Meyvis, T. and C. Janiszewski, 2002. Consumers' beliefs about product benefits: The effect of obviously irrelevant product information. Journal of Consumer Research 28, 618-635.

Mitchell, V.W. and V. Papavassiliou, 1999. Marketing causes and implications of consumer confusion. Journal of Product & Brand Management 8, 319-339.

Rafiq, M. and R. Collins, 1996. Lookalikes and customer confusion in the grocery sector: An exploratory survey. The International Review of Retail, Distribution and Consumer Research 6, 329-350.

Rudolph, T. and M. Schweizer, 2003. Kunden wieder zu Käufern machen. Harvard Business Manager 25, 23-33.

Turnbull, P.W., S. Leek and G. Ying, 2000. Customer confusion: The mobile phone market. Journal of Marketing Management 16, 143-163.

Walsch, G. and T. Hennig-Thurau, 2002. Wenn Konsumenten verwirrt sind. Empirische Analyse der Wirkungen eines vernachlässigten Konstruktes. Marketing: Zeitschrift für Forschung und Praxis 24, 95-109.

Wiedmann, K.-P., G. Walsch and A. Klee, 2001. Konsumentenverwirrtheit: Konstrukt und marketingpolitische Implikationen. Marketing: Zeitschrift für Forschung und Praxis 23, 83-99.

Brands as quality signals in the meat market: lessons from the poultry sector

Maren Lüth and Achim Spiller
University of Goettingen, Department of Agricultural Economics and Rural Development, Platz der Goettinger Sieben 5, 37073 Goettingen, Germany; a.spiller@agr.uni-goettingen.de

Abstract

In recent years the German meat market has been shaken by several crises and quality problems. Compared to other countries such as Denmark or the Netherlands the share of branded meat from integrated production chains is very low and most fresh meat is sold unlabelled. This study analyses the advantages of brands from an information economic perspective and emphasises branding as an important quality assurance and signaling tool. As brand extensions offer the opportunity to introduce brands to new markets at much lower costs we examine the brand transfer from the poultry to the red meat market taking Wiesenhof, the German brand leader for chicken and poultry, as an example.

Keywords: branding, quality signal, brand extension, meat market

1. Quality problems in the meat market

The German meat market has lately had to face severe crises and fluctuations in demand. Hardly any other branch in the food industry has had to cope with so many food scandals. BSE, MKS, salmonellae, dioxin and nitrofen have effectively lowered the consumer's trust in animal food and have reinforced the long-term trend towards lower meat consumption (Hansen and Thiele, 2005). This is especially true for the red meat sector while the demand trend for poultry is on the whole positive.

In the last few years, several food quality and safety initiatives have been established to control the traceability of the meat and the fodder. With the introduction of the QS-label in Germany the whole value chain was integrated into one certification system for the first time. The mainly production-oriented features aim at regaining and strengthening consumer trust in meat. However, the cooperative labelling approach is limited by financial restrictions and free rider effects. A successful certification label can work as a basic guarantee in the market only if it is well recognised and trusted.

On the other hand, the importance of brands as quality signals and customer relationship instruments has completely been neglected by the meat industry. Particularly in terms of safety aspects consumers often do not know whether the meat is of good or poor quality and are not willing to pay for a better quality they cannot identify. Information economic theory considers brands as suitable quality signals to bypass imperfect and asymmetric information by guaranteeing high and consistent quality standards. In the German meat market there has been only one serious attempt to establish a classical brand: Wiesenhof belonging to the Paul-Heinz-Wesjohann-group (PHW) in the poultry sector. In the red meat market, in contrast, only unbranded meat, low-priced private labels or quality production programmes are offered which mainly refer to producers' requirements and have not yet reached a high level of consumer awareness (Von Alvensleben, 1997).

The following study deals with the relevance of brands as quality signals and discusses the possibility of brand transfers to introduce brands in the red meat market at low costs. To analyse this procedure in particular and the potential of meat brands in general an empirical analysis has been conducted on the basis of the brand Wiesenhof. It is the German leader in the market for poultry and poultry products.

2. Branding in the fresh meat sector

2.1. Brands as quality signals

Access to information is an elementary condition for the functioning of markets. Indeed, manufacturers and retailers are inevitably better informed about the quality of their products than individual consumers (Henson and Traill, 1993). Imperfect and asymmetric information is a typical example of market failure and may lead to problems especially in food markets (Jahn *et al.*, 2005; Latvala and Kola, 2003). Akerlof argued that pour quality will prevail over high quality products if there are no signaling instruments in the market which ensure credible information. Without any quality signaling, high quality products will not be able to achieve a price premium and will finally disappear. Markets will only offer pour standards and as a result break down (Akerlof, 1970). Information economic research has often stressed the relevance of signals in markets with high information asymmetry in order to reduce it. From an information economic point of view, goods can be categorised according to different types of quality attributes: Search attributes whose quality can be controlled before purchase, experience attributes whose characteristics can only be determined after purchase and credence attributes which cannot be assessed either before or after purchase by consumers but only by reliable third parties (see Table 1) (Darby and Karni, 1973; Nelson, 1970). Depending on the kind of attribute, the costs of gathering information about the quality of the product increase from search to credence attributes. Food safety can either be categorised as experience attribute, e.g. in the case of microbiological contamination which results in illness, or as credence attribute, e. g. in the case of country of origin. It is largely treated as credence attribute because the appraisal of product quality is not practicable for individual consumers (Caswell and Mojduszka, 1996: 1250). In food markets, where unobservable product quality has gained rising importance, reliable quality signals, such as brands, prices, warranties, therefore, are fundamental instruments in consumer marketing.

In the following discussion, brands will be treated in greater detail as they are considered to be important drivers of food quality. Brands enhance product recognition and serve as communication instruments. As quality signals, brands are especially useful to communicate and guarantee a high level of unobservable product quality as brand producers make several investments to build

Table 1. Typology of attributes and quality signals.

Typology of attributes	Search attributes	Experience attributes	Credence attributes
Suitable quality signals	packaging, design, appearance	brand, price, reputation	brands, guarantees, test labels

brand equity, i.e. advertising, public relations, product design or packaging. These large sums of fixed costs may turn into irreversibly sunk costs in the case of a food scandal or image loss. Therefore, the producers share a vital self-interest and make great efforts to guarantee credible claims about unobservable quality. Brands take on the function of information surrogates and often are an important part of a company's capital and have to be protected from any damage. For the meat sector, brands might help to keep up high standards in food safety and, consequently, to diminish food scandals. This has empirically been proven in the German market for cold cuts, where strong brands such as Herta, Gutfried, Zimbo or Rügenwalder only faced slight declines in demand or even achieved growth rates during the BSE-crises (Katterbach, 2001: 54).

2.2. Wiesenhof: an example of a quality-oriented marketing strategy in the meat sector

The PHW-group in Rechterfeld, which is one of the outstanding companies in the German agricultural sector, has been dedicated to branding since the 1950s. Their traditional chicken brand Wiesenhof marks the most important business segment of the firm with an annual sales volume of 695 million €. Altogether, the PHW-group covers 48 % of German poultry production, i.e. 200 million birds per year. The company is the market leader in this segment by a wide margin and maintains a 29% share of the market based on the national chicken consumption (Lebensmittelzeitung, 2004).

As a confirmation of the quality leadership, the PHW-group is in charge of an integrated production chain for their brand Wiesenhof. The company early started on to certify their agricultural fattening units and was the first poultry producer to introduce the traceability concept in 1995. It documents the origin of each animal and includes the production steps parent herds, hatcheries, fodder mills, raising farms, slaughter-houses and processors in one company ('5-D-Quality'). All 700 contract farmers receive their young birds and the fodder exclusively from the integrated production stages. The company turned down animal protein and abandoned antibiotic feeding for production at an early stage. The fodder component, soja, is not genetically modified. An outstanding role is Wiesenhof's monitoring of salmonellae. Production of fodder in its own mills reduced the salmonellae infection to only one to two percent of cases. Moreover, all Wiesenhof chicken and other poultry have been QS-certified since October 2002.

The Wiesenhof example shows the importance of brands for the quality segment. Brands are promotors of innovations. The high marketing costs force the producers to be involved with an excellent and sustainable quality policy so as not to endanger the brand value and the economic survival of the firm.

2.3. Problems of branding in the meat market

While in many product groups of the food industry, famous brand producers have built up important standings and reach price premiums for their products, there are very few established brands in the fresh meat market. Many industry insiders even doubt that branding is possible in the red meat sector. The following aspects are often mentioned (see Table 2).

Although the above mentioned technical arguments present large challenges for establishing a meat brand there are several arguments for branding in this sector, such as the dynamic growth of the self-service segment, new packaging technologies and the increasing techniques to standardise meat quality by genetic engineering and fodder (Balling, 1990). The high perceived buying risks with meat (see also Von Alvensleben, 1997) - in contrast to the opinion of many members of the meat industry - are an important reason for branding.

Table 2. Arguments from practitioners against branding in the fresh meat market (based on discussions with industry experts).

Argument	Reason
Unpacked product	because of the predominance of sales over the counter branding is neither necessary nor possible
Natural product	the meat quality cannot be standardised adequately enough for the needs of a brand
High quality risks	crises such as BSE affect all firms, also the brand producers
Customers as co-producers	the lack of cooking skills of many consumers leads to disappointing taste experiences which rubs off negatively on the brand
Lack of financial power	the low profitability of the industry rules out expensive advertising campaigns

The cross reference to the co-producer role of the consumers is also not convincing since for other product groups, which also require a special level of consumer competence (e.g. coffee, tea or wine), successful brands can be found. It can be ascertained that knowledge on how to prepare certain foods is decreasing in various parts of the population, but the method of preparation has little influence on the perceived quality (Grunert, 1998: 108). Furthermore, the trend towards convenience meat products will gradually reduce consumers' influence on the taste experience of a meal.

The strongest remaining argument against branding is the low financial background in the meat industry. In the introduction period of a new brand, investments often account for up to 25% of the expected annual sales (Spiller, 2000). In the face of these high financial barriers, it may be wiser to extend an existing label to the pork and beef sector by brand extension. The following chapter discusses the theoretical background of brand extensions.

3. Brand extension strategy

Brand extension, i.e. the use of established brand names to launch new products, is one of the most frequently employed branding strategies. For fast moving consumer goods, often more than 85% of new product introductions are brand extensions (Völckner and Sattler, 2004: 2). Brand extensions are considered profitable because brands that are already known and recognised are generally assumed to require lower new-product introduction expenses such as advertising or promotion costs. These benefits are mainly due to the transfer of the awareness and association of the parent brand to the new product. In the long run, positive spill-over-effects from the transfer product to the parent brand are expected (Aaker, 1990: 49; Balachander and Ghose, 2003: 4). Risks might be seen in the possible weakening of the brand image or in bad will-effects in the case of quality problems (Sattler, 2001b: 147).

Nevertheless, the decision for a brand transfer is not without risk. Several studies show that the transfer potential of a brand mainly depends on three influence factors (Broniarczyk and Alba, 1994; Caspar, 2002; Esch, 2002):

- Strength of the parent brand: A high level of brand recognition and a positive brand image are important requirements for a successful brand transfer. The brand image should not too

strongly be characterised by product specific elements, but by emotional or abstract items (Esch and Fuchs, 1999: 685). It is more difficult to transfer a brand which is strongly related to a special product.
- Perceived quality of the parent brand: The perceived quality of the parent brand turns out to be the most important buying reason for many consumers. It can be expected that consumers assign their quality associations from the parent brand to the transfer product to reduce their risk of mispurchase. A higher perceived quality of the parent brand therefore increases the chances for a successful brand transfer (Agarwal and Rao, 1992; Bottomley and Doyle, 1996; Boush *et al.*, 1987; Smith and Park, 1992; Sunde and Brodie, 1993).
- Perceived fit of parent brand and new product category (transfer fit): The chances for a successful brand extension increase with higher perceived fits of associations between the parent and the transfer brand. The fit may refer to product-related attributes, similar consumer situations or consumer types as well as complementary usage of parent and transfer brand (Aaker and Keller, 1990; Boush and Loken, 1991; Bridges, 1992; Sattler *et al.*, 2003).

The main objective of the following analysis was testing the potential of brand transfers in the meat sector. In our survey the case of Wiesenhof served as an example of parent brand for the red meat sector. The three success factors described above were analysed via direct statements. Furthermore, the brand transfer potential of Wiesenhof was surveyed by the decompositional method of Conjoint Analysis. Respondents were not asked for their brand preferences directly but were asked to rank different products (Green and Srinivasan, 1990). An advantage of the Conjoint Analysis is the indirect calculation of the willingness to pay which reduces the bias towards overstated price sensitiveness and understated brand awareness. Conjoint Analysis have been used for questions of brand transfer before and can be considered as suitable and valid (Green *et al.*, 1993; Rangaswamy *et al.*, 1993).

4. Survey design

With the lack of strong brands in the red meat market an important quality signal is absent. The biggest problem against branding is the financial weakness of this segment. Nevertheless, the example Wiesenhof from the poultry market shows the potential for meat brands in Germany. As brand transfers are a cheaper possibility to introduce brands to a market we analyse if the the brand Wiesenhof can be assigned to the red meat market. Our case study addresses the brand transfer from Wiesenhof poultry to the new hypothetical product Wiesenhof pork.

In January 2004, a consumer survey was conducted in retail stores of the German retailer Edeka in Lower Saxony. Edeka is the leading German retailer by food sales volume and, as generally known, a quality-oriented company. The retail group has more than 10 years experience of the management of their premium private meat label 'Gutfleisch'. Therefore, the area and target group of the sample were selected with regard to the topic brand leadership in the meat market. Altogether, 177 customers were interviewed, of which 67% were women. Compared to the German population as a whole younger people between 20 and 30 years were overrepresented and older people over 70 years underrepresented. Also people with higher education and income as well as households with more members constituted a higher proportion than average. All in all, the survey is a convenient sample.

In the first part of the survey a Conjoint Analysis was conducted in which the respondents were shown several meat offers which they had to rank by preference. The product attributes were systematically varied and combined to several meat products. This procedure allows the calculation of importance of the different product attributes in the buying process (Backhaus *et al.*, 2003: 567). Furthermore, by the calculation of price equivalents the monetary value of different alternatives, in this case especially of the transfer product, can be determined. The second part of the study consisted of questions on brand recognition and image of Wiesenhof, on the quality of the mother brand and the transfer product as well as on the perceived fit of the two product groups. These results will be presented at the beginning of the next chapter.

5. Results of the study

5.1. Strength of the parent brand, perceived quality and transfer fit

In 2003, the advertising costs for the brand Wiesenhof accounted for almost 8 million € (Stryck, 2004). In our survey, the brand was indirectly recalled by 12.4% of respondents which is the highest value achieved with meat brands followed by the quality private label Gutfleisch of the retail group Edeka (11.9%). On the other hand, the direct brand recall is much higher (93.8%) than that for Gutfleisch (57.6%). Compared to fresh meat, brands are much more common in the market for cold cuts, but in this field the recongnition for Wiesenhof is much lower than in the meat sector.

Asked for spontaneous associations towards the brand Wiesenhof 65% of respondents mentioned 'chicken/poultry' and 14% meat/sausages. Positive associations like 'without chemistry, untreated, good quality, from controlled production' were named by 8.5% of those questioned and 'meadow, farm, organic, from the countryside' by 3.4%. Negative associations, such as 'factory farming, hen cages or *Salmonellae*' were mentioned by 6% of the respondents. Finally only 4% of respondents remembered a connection between Wiesenhof and current advertisements or sponsored TV shows. In summary, most of the respondents associated the brand with the product category poultry but not with any abstract associations. This strong focus on the product could constrain the transfer potential.

The image values for Wiesenhof on a semantic differential show positive ratings. The products are considered as pleasant, light, tasty and appetising, but not as sporty or unique. The price of Wiesenhof is perceived as middle of the market.

In a further step we asked about the quality of the parent brand and how it fits the potential new product group. More than half of the respondents (53.3%) rated the quality of the previous Wiesenhof products as good. The similarity between Wiesenhof and the analysed transfer product was balanced. 46.9% of the respondents had a positive and 38.4% a negative attitude towards a brand transfer to the meat market (μ: -0.05 on a scale from -2 to +2). The quality of the new fictitious Wiesenhof product was always rated positive (μ: 0.54).

The interim findings show a high brand awareness of Wiesenhof and positive characteristics such as high-quality, credibility and liking. A problem for the brand transfer to new product categories is the low emotional level of the brand and the strong focus on poultry. Accordingly, the transfer to new product categories is considered sceptically.

5.2. Brand value on the basis of Conjoint Analysis

The conjoint design consists of the product attributes brand, price and origin. The detailed conjoint design with all attributes and characteristics is shown in Table 3.

Table 3. Product attributes and characteristics of the Conjoint Analysis.

Product attributes of pork cutlet	Attribute characteristics
Brand	Wiesenhof
	Gutfleisch
	Gut&Günstig
Price	4.99 € / kg
	6.49 € / kg
	7.99 € / kg
Guarantee of quality	without guarantee of quality
	with '5-D-Guarantee of quality and origin'

The choice of the three brands for the conjoint design can be ascribed to the aim of the study. The transfer potential of the quality brand Wiesenhof is analysed in comparison to the low-priced private label Gut&Günstig. Gut&Günstig is a label which is positioned in the low price segment and not promoted in the media. Such labels constitute a suitable reference basis because of their nationwide distribution level and their low brand equity. The marketing literature has already documented this procedure (Ailawadi *et al.*, 2003).

The high priced private label Gutfleisch is a good example of the efforts of large food retailers to build up meat brands. Gutfleisch was first introduced in Northern Germany before it was extended on a national level. The brand has a long tradition in the survey region and, therefore, mirrors the competition situation that a new brand on the market would face. The particular strength of the brand Wiesenhof is the well communicated guarantee of origin and traceability which accounts for the special quality image of the company in the poultry market.

By means of the statistical program SPSS the attribute characteristics were systematically combined to an orthogonal design of 9 stimulus and 2 holdout cards. These 11 picture cards had to be ranked by the respondents. The calculation of the part worth utilities took place on the basis of the additive model of Conjoint Analysis (Backhaus *et al.*, 2003: 558; Skiera and Gensler, 2002: 204):

$$y = \mu + \beta_A + \beta_B + \beta_C$$

The model is defined as y = constant term which mirrors the average rank relating to all assigned (metric) rank values, A = 3 attribute characteristics, B = 3 attribute characteristics and C = 2 attribute characteristics. For the price a negative linear relationship was assigned, for the other variables a discrete one. Table 4 shows the importance of the single product attributes.

When buying pork most consumers rank the price as the most important criterium (40%). The brand and the guarantee of quality each account for 30% and are, therefore, a little less relevant. The main reason for the predominance of the price aspect could be the lack of powerful brands in the meat sector so that consumers have not developed strong brand sensitiveness in this product category so far.

Table 4. Relative importance of product attributes.

Product attribute	Relative importance
Price	40.11
Brand	29.97
Guarantee of quality	29.92

Internal validity of estimation: Pearson's R = 0.964, Kendall's Tau = 0.833.
Predictive validity of estimation: Kendall's Tau = 1.0.

In comparison, branding and labelling seem to play a similar important role in the meat market. In a situation in which no company has invested in brand equity so far, this result demonstrates the potential signaling effect of brand management.

Detailed information about the assessment of the transfer product is shown in Table 5. The part worth utilities demonstrate the preferences for the single brands and the differences in between.

The results in the pork segment clearly show the higher preferences for the quality private label whereas the low-priced private label Gut&Günstig achieves lower rankings compared to the two quality brands.

The difference between the low-priced private label and the two quality brands can be transferred to price equivalents which reflect the brand value (Brockhoff and Sattler, 1996: 218):

$$PE = (TW_1 - TW_0)\frac{dP}{dTW_P}$$

with:
PE: price equivalent;
TW_1: part-worth utility for brand 1;
TW_0: part-worth utility for brand 0;
dP: difference of prices;
dTW_P: difference of part-worth utilities of prices.

Table 5. Part-worth utilities of the product attribute brand.

Characteristics of the product attribute brand	Part-worth utilities of the attribute characteristics
Wiesenhof	-0.0584
Quality private label	0.3691
Low-priced private label	-0.3107

Internal validity of estimation: Pearson's R = 0.964, Kendall's Tau = 0.833.
Predictive validity of estimation: Kendall's Tau = 1.0.

With the aid of these price equivalents, the consumer's willingness to pay for a special brand compared to an unknown brand c. p. can be determined (Brockhoff and Sattler, 1996: 218; Sattler, 1991: 200). In our case the low-priced private label Gut&Günstig was taken as a reference basis because of its low brand value. It is positioned in the low price sector, is not promoted by any advertising and strongly corresponds to a generic article.

Compared to the low-priced brand, the quality private label Gutfleisch has a brand value of its disposal of 76 cents (= average additional willingness to pay). Wiesenhof pork can achieve an additional willingness to pay of 28 cents and therefore clearly lags behind. Nevertheless, the willingness to pay for the quality private label Gutfleisch lags behind its effective additional charge in the store. This result shows that even regular Edeka customers do not recognise the company's private label although it is associated with large investments on the producer's side. The reason could be the lack of advertising for the private label as it is only promoted by sales promotion.

In summary, the brand equity for the existing pork brand Gutfleisch has been low so far. One important reason could be the lack of strong, competing brands. Exactly for this reason, a brand transfer to the pork market seems to be much more worthwhile than to other product categories.

5.3. Prices and market shares on the basis of Conjoint Analysis

In order to predict the share of preference that a real or hypothetical product stimulus is likely to capture in a special market scenario, simulations can be run on the basis of the preference data obtained in the Conjoint Analysis. In our scenario, only the price for Wiesenhof pork was varied while the other prices stayed fixed and referred to real market prices (quality private label with origin denomination for 6.99 € and the low-priced private label without origin denomination for 5.99 €) (see Table 6). For Wiesenhof pork the guarantee of origin was assumed to be the same as for Wiesenhof poultry products.

For the choice simulator, different specifications are possible for simulation of consumer choices. We chose the BTL (Bradley-Terry-Luce) model which computes the probability of choosing a profile as the most preferred by dividing the profile's utility by the sum of all the simulation total utilities (Hair *et al.*, 1995; Green and Krieger, 1988). With increasing benefit of a stimulus its likelihood to be chosen rises, even if the respondent does not necessarily prefer the product with the highest overall utility.

Table 6. Market shares of the pork brands in percent with varying prices of Wiesenhof.

Varying price of Wiesenhof in €	Market share of Wiesenhof	Market share of Gutfleisch (6.99€)	Market share of Gut&Günstig (5.99€)
5.99	48.87	34.18	16.95
6.49	38.98	35.59	25.42
6.99	23.45	40.11	36.44
7.49	22.88	40.68	36.44

When estimating the market shares it is important to check whether brand preference is correlated to purchasing intensity, i. e. in the meat market a regular pork buyer could, for example, be a heavy user of private labels. In this case the market shares based on the conjoint results should be corrected by a weighting factor (Hensel-Börner and Sattler, 2000). In our study a significant relationship between brand preferences and buying intensity could not be verified and a weighting did not have to be conducted. The following table displays the hypothetical market shares of the transfer product Wiesenhof pork at different price levels in Edeka-stores.

At a starting price of 5.99€ Wiesenhof reaches the highest market share of 49% and the low-priced private label Gut&Günstig the lowest of 17%. Up to the price of 6.99€ the market shares of Wiesenhof show a strongly declining trend. Remarkably, it is the brand Gut&Günstig which can profit best from the losses of Wiesenhof and can better enlarge its market share than the quality private label Gutfleisch. It seems as if the Wiesenhof buyers belong to a price-sensitive group of customers. This result confirms the findings of the image analysis where Wiesenhof is recognised as traditional brand in the middle price segment. The market coverage of Gutfleisch is quite stable with a variation of ca. 6.5%

A largely inelastic progression of market shares can be seen in the price area between 6.99€ and 7.49€. This brand awareness is possibly connected with a high degree of quality insecurity of many consumers who infer a higher quality of those products from higher prices. Actually, 25% of respondents show an atypical buying behaviour in raising their demand at higher prices. An analysis without these customers results in lower market shares for Wiesenhof while the low-priced private label reaches a higher level. The brand name Wiesenhof, therefore, functions as a quality signal for those consumers who are especially looking for safe products.

6. Conclusion

Information economic research argues that brands can act as important drivers of quality improvements in the meat sector. In comparison to generic labelling strategies, which have mainly been favoured, information economic literature emphasises the advantages of branding strategies. Brands are characterised by high communication spending which can be regarded as sunk costs leading to a larger degree of recognition and of self-commitment. The producers, therefore, share a personal economic interest in protecting their brand value which can be taken as a guarantee of high quality standards. Generic labels on the other hand are often characterised by small advertising spending and a low degree of brand recognition. As the arising sunk costs from labelling are much smaller than those from branding the lower degree of self-commitment might enforce free rider behaviour and does not contribute to strengthen and improve the production chain. Brand leaders are interested in improving their quality standards and differentiate in the market by intrinsic motivation. Labelling systems on the other hand are based on extrinsic quality motivations. The producers in these systems have no incentives to produce quality that exceeds the standard level which is controlled by external certification institutions.

As can be seen from the results in our study, branded meat reaches almost two third of market share while the low priced private label always gets the lowest proportions of consumer preferences. Given the choice customers do not always prefer the cheapest offer but trust in branded meat even more. This market potential is actually not used to advantage by the producers. The overall total market share of meat brands lies below 5% in Germany. Agricultural policy has mainly focused on regional labelling strategies such as PDO, driven by hidden intentions to protect farmers' interests and assure subsidies. As a result this approach has led many firms in the meat industry to favour labelling strategies instead of developing their own brands. The chances of branding for the quality segment in general should not be neglected.

The German meat industry will only overcome its quality problems with new attempts to establish premium segments. Branding could, therefore, be a more useful and successful approach than generic labelling in the past.

References

Aaker, D.A., 1990. Brand extensions. The good, the bad, and the ugly. Sloan Management Review 31, 47-56.

Aaker, D.A. and K.L. Keller, 1990. Consumer evaluations of brand extensions. Journal of Marketing 54, 27-41.

Agarwal, M.K. and V.R. Rao, 1992. An empirical comparison of consumer-based measures of brand strength. Working Paper, Cornell University.

Ailawadi, K.L., D.R. Lehmann and S.A. Neslin, 2003. Revenue premium as an outcome measure of brand equity. Journal of Marketing 67, 1-17.

Akerlof, G.A., 1970. The market for 'Lemons': Quality uncertainty and the market mechanism. Quarterly Journal of Economics 84, 488-500.

Backhaus, K., B. Erichson, W. Plinke and R. Weiber, 2003. Multivariate Analysemethoden: Eine anwendungsorientierte Einführung. Springer.

Balachander, S. and S. Ghose, 2003. Reciprocal spillover effects: A strategic benefit of brand extensions. Journal of Marketing 67, 4-13.

Balling, R., 1990. Marketing-Konzeption für einen Markenartikel Rindfleisch. Agrarwirtschaft Special Issue 125, Frankfurt/Main.

Bottomley, P.A. and J.R. Doyle, 1996. The formation of attitudes towards brand extensions: Testing and generalising Aaker and Keller's model. International Journal of Research in Marketing 13, 365-377.

Boush, D.M. and B. Loken, 1991. A process-tracing study of brand extension evaluation. Journal of Marketing Research 28, 16-28.

Boush, D., S. Shipp, B. Loken, E. Gencturk, S. Crockett, E. Kennedy, B. Minshall, D. Misurell, L. Rochford and J. Strobel, 1987. Affect generalization to similar and dissimilar brand extensions. Psychology and Marketing 4, 225 237.

Bridges, S., 1992. A schema unification model of brand extensions. Working Paper Report No. 92-123, Wake Forest University, Cambridge, Massachusetts.

Brockhoff, K. and H. Sattler, 1996. Schwartauer Werke: Markenwert und Qualitätszeichen. In E. Dichtl (ed.), Markterfolg mit Marken, C.H. Beck, 207-224.

Broniarczyk, S.M. and J.W. Alba, 1994. The importance of the brand in brand extension. Journal of Marketing Research 31, 214-228.

Caspar, M., 2002. Markenausdehnungsstrategien. In: H. Meffert, C. Burmann and M. Koers (eds.), Markenmanagement: Grundfragen der identitätsorientierten Markenführung mit Best Practice-Fallstudien. Gabler, pp. 233-259.

Caswell, J.A. and E.M. Mojduszka, 1996. Using informational labelling to influence the market for quality in food products. American Journal of Agricultural Economics 78, 1248-1253.

Darby, M. and E. Karni, 1973. Free competition and the optimal amount of fraud. Journal of Law and Economics 16, 67-88.

Esch, F.-R., 2002. Markenprofilierung und Markentransfer. In: S. Albers and A. Herrmann (eds.), Handbuch Produktmanagement: Strategieentwicklung - Produktplanung - Organisation - Kontrolle. Gabler, pp. 189-218.

Esch, F.-R. and M. Fuchs, 1999. Konzeption und Umsetzung von Markenerweiterungen. In: F.-R. Esch (ed.), Moderne Markenführung: Grundlagen - innovative Ansätze - praktische Umsetzungen. Gabler, pp. 669-703.

Green, P.E. and A.M. Krieger, 1988. Choice rules and sensitivity analysis in conjoint simulators. Journal of the Academy of Marketing Sciences 16, 114-127.

Green, P.E. and V. Srinivasan, 1990. Conjoint-analysis in marketing: New developments with implications for reserach and practice. Journal of Marketing 54, 3-19.

Green, P.E., A.M. Krieger and M.K. Agarwal, 1993. A cross validation test of four models for quantifying multiattribute preferences. Marketing Letters 4, 369-380.

Grunert, K.G., 1998, Qualitätserwartung und Qualitätserlebnis bei Lebensmitteln. In: M. Neuner and L.A. Reisch (eds.), Konsumperspektiven: Verhaltensaspekte und Infrastruktur. Duncker & Humblot, pp. 101-111.

Hair, J.F., R.E. Anderson, R.L. Tatham and W.C. Black, 1995. Multivariate data analysis with readings; 4th ed., Prentice Hall.

Hansen, A. and H.D. Thiele, 2005. Die Märkte für Vieh und Fleisch. Agrarwirtschaft 54, 49-68.

Hensel-Börner, S. and H. Sattler, 2000. Ein empirischer Validitätsvergleich zwischen der Customized Computerized Conjoint Analysis (CCC), der Adaptive Conjoint Analysis (ACA) und Self-Explicated-Verfahren. Zeitschrift für Betriebswirtschaft 70, 705-727.

Henson, S. and B. Traill, 1993. The demand for food safety. Food Policy 18, 152-162.

Jahn, G., M. Schramm and A. Spiller, 2005. The reliability of certification: Quality labels as a consumer policy tool. Journal of Consumer Policy 28, 53-73.

Katterbach, W., 2001. BSE und die Folgen. Impulse 18, 50-55.

Latvala, T. and J. Kola, 2003. Impact of information on the demand for credence characteristics. International Food and Agribusiness Management Review 5, URL: http://www.ifama.org/nonmember/OpenIFAMR/Articles/v5i2/kola.PDF, date: 01.02.2005.

Lebensmittelzeitung, 2004. Firmenprofile Industrie: PHW Gruppe, URL: http://www.lz-net.de/, date: 01.03.2004.

Nelson, P., 1970. Information and consumer behaviour. Journal of Political Economy 78, 311-329.

Rangaswamy, A., R.R. Burke and T.A. Oliva, 1993. Brand equity and the extendibility of brand names. International Journal of Research in Marketing 10, 61-75.

Sattler, H., 1991. Herkunfts- und Gütezeichen im Kaufentscheidungsprozess: Die Conjoint-Analyse als Instrument der Bedeutungsmessung. Metzler und Poeschel.

Sattler, H., 2001b. Brand-Stretching: Chancen und Risiken. In: R. Köhler (ed.), Erfolgsfaktor Marke: Neue Strategien des Markenmanagements. Vahlen, pp. 141-149.

Sattler, H., F. Völckner and G. Zatloukal, 2003. Erfolgsfaktoren von Markentransfers - Eine empirische Analyse für kurzlebige Konsumgüter. Marketing ZFP 25, 147-168.

Skiera, B. and S. Gensler, 2002. Berechnung von Nutzenfunktionen und Marktsimulationen mit Hilfe der Conjoint-Analyse (Teil 1). Wirtschaftswissenschaftliches Studium 5, 200-206.

Smith, D. C. and C.W. Park, 1992. The effects of brand extensions on market share and advertising efficiency. Journal of Marketing Research 29, 296-313.

Spiller, A., 2000. Erfolgschancen mittelständischer Hersteller als Handelsmarkenspezialisten: Eine institutionenökonomische Analyse. In: J.-A. Meyer (ed.), Jahrbuch der KMU-Forschung 2000: Marketing in kleineren und mittleren Unternehmen. München, pp. 391-412.

Stryck, I., 2004. Produktentwicklung und Markenaufbau im Geflügelfleischsektor am Beispiel von Wiesenhof. Vortrag auf der Sommerakademie des Forschungs- und Studienzentrums für Veredelungswirtschaft Weser-Ems der Georg-August-Universität Göttingen am 23.09.2004 in Vechta.

Sunde, L. and R.J. Brodie, 1993. Consumer evaluations of brand extensions: Further empirical results. International Journal of Research in Marketing 10, 47-53.

Völckner, F. and H. Sattler, 2004. Markentransfererfolgsanalysen bei kurzlebigen Konsumgütern unter Berücksichtigung von Konsumentenheterogenität. Hamburg.

Von Alvensleben, R., 1997. Consumer attitudes and behaviour on the meat market in Germany. In: R. von Alvensleben, S. von Cramon-Taubadel, A. Röhr, U. Scheper and K. Schleyerbach (eds.), Problems of meat marketing. Arbeitsbericht Nr. 2 des Lehrstuhls für Agrarmarketing, Kiel, pp. 15-18.

Food quality from a consumer's perspective

Maike Bruhn and Carola Grebitus
University of Kiel, Department of Agricultural Economics, Olshausenstraße 40, 24098 Kiel, Germany; cgrebit@agric-econ.uni-kiel.de

Abstract

Consumers' quality judgement depends on their knowledge concerning certain quality attributes. Against this background this paper presents the results of two qualitative methods for measuring consumers' cognitive structures. On this account we apply a free-association technique and a variation of concept mapping adequate for the research object quality of pork. Our analysis shows that consumers' cognitive structures are quite complex and heterogeneous. Product characteristics dominate consumers' knowledge concerning the quality of pork. To predict the eating quality consumers prefer intrinsic quality cues. In contrast to that extrinsic quality cues are used to predict the origin of the given product.

Keywords: food quality, cognitive structure, memory schemata, free-association technique, concept mapping

1. Introduction

Dealing with consumers' quality perspective is very important with regard to consumer-oriented communication strategies. From consumers' perspective food quality is a perceptual and an evaluative construct which is relative to person, place of purchase and purchase situation (Cardello, 1995). Against this background we take consumers' cognitive processes into account to describe food quality from a consumer's perspective.

Our paper presents preliminary results from a consumer survey which took place in Germany in the year 2004. We intend to uncover consumers' internal information concerning the quality of pork. This approach contributes to actual research in the field of consumer-oriented approaches regarding food quality (e.g. Steenkamp, 1990; Grunert *et al.*, 1996, 2004; Becker, 1999; Grunert, 2001; Scholderer *et al.*, 2004).

The structure of the paper is as follows: it starts with a brief description of the theoretical background of our study. This section shows how memory is organised and its role within information-processing. The paper then presents the results of two qualitative methods for measuring knowledge, namely a free-association technique and a variation of concept mapping. We chose a qualitative approach as we attempted to access consumers' individual level of understanding. This is not always possible with more standardised approaches (Marschall, 1997). We finish our paper with some conclusions and further research prospects.

2. Consumers' quality judgement: theoretical framework

The concept of food quality adopted in this paper is totally consumer-oriented. This means that quality is defined by the end user, the consumer. Consumer-oriented quality is the subjective quality perception of the user. In this context the term perceived quality stresses that quality judgements are dependent on the perceptions, needs and goals of the consumers. Consequently, consumer-oriented quality can differ for the same product between users. Thus, it has to be measured at individual level (Steenkamp, 1990; Grunert *et al.*, 1996; Henson, 2000). The importance of this discussion due to consumer-oriented communication strategies is outlined by Henson (2000): 'The translation of the quality and safety perceptions of consumers into physical product and process attributes requires knowledge and understanding of the overall quality/safety evaluation of consumers, how this evaluation relates to intrinsic, extrinsic and cost cues, and how these cues relate to physical product and process attributes'. In this regard intrinsic quality cues refer to physical characteristics of the product such as colour, form or odour. Extrinsic quality cues refer to everything else for example price, brand, distribution outlet, etc. (Kaas, 1994; Tolle, 1994; Bech *et al.*, 2001). Especially in the field of unprocessed foods such as meat, fresh fruits and vegetables consumers may have difficulties to evalute their quality because of a lack of central quality cues such as brands (Bech-Larsen and Grunert, 2001; Bredahl, 2003).

In other words, from consumers' perspective specific quality dimensions have to be in evidence and comprehensible (Von Alvensleben and Scheper, 1997; Meyer, 2000). If any information is to influence consumers' decision-making it must be processed by their cognitive system. Reduced to its essence consumer decision-making involves three important cognitive processes:

1. Consumers must interpret relevant information in the environment to create personal meanings or knowledge.
2. Consumers must integrate this knowledge to evaluate products or possible actions and choose among alternatives behaviours.
3. Consumers must retrieve product knowledge from memory to use in integration and interpretation processes (Peter and Olson, 1996).

Olson and Reynolds (1983) believe that particularly consumers' cognitive structures of knowledge in memory are a key factor in developing a useful understanding of consumer behaviour. Grunert (1981) points out evaluating food quality consumers prefer attributes which correspond to concepts already stored in memory. Consequently consumers' quality judgement highly depends on their knowledge concerning certain quality attributes.

The schema theory and the spreading activation network model deliver insight into consumers' cognitive structure and its role within decision-making. Any information that an individual is exposed to is organised in memory through schemata, to give meaning to stimuli and to enable interpretation and comprehension (Erasmus *et al.*, 2002). A schema for an object or event is comprised of a set of attributes or slots (Jonassen *et al.*, 1993). For instance, most of us have a well-developed schema for food quality that includes attributes such as price, colour or seal of approval.

The spreading activation network model helps to understand how consumers' knowledge is organised in memory. In this model the nodes in the network represent concepts such as attributes and product characteristics, while the links that connect the nodes represent the type and the strength of the association between the concepts. Consumers' general knowledge structures which contain mostly episodic and semantic general knowledge are called schemas. Schemas can be activated in decision-making situations, and they can influence consumers' cognitive processes. That means activating a node results in a flow of activation through the links from that node to all connected nodes and when the activation level surpasses the threshold level, retrieval of the concept will occur. The greater the strength of a link the more activation passes through it to the connected node (Collins and Loftus, 1975; Anderson, 1983; Peter and Olson, 1996; Cowley and Mitchell, 2003).

Such memory schemata may allow inference to be made regarding product attributes not presented in the immediate task environment. The interrelationships between concepts were previously acquired/learned via classical and/or instrumental learning principles (Olson, 1978). In processing a given cue from a task environment, one might develop beliefs about other aspects of the task stimulus not represented by environmental cues (Fishbein and Azjen, 1975 cited in Olson, 1978). Such inferential beliefs are often formed without much conscious analytical thinking, apparently as a function of one's past experiences. Against this background, the schemata concept can be seen as a theoretical representation of the cognitive structure created by past experiences (Olson, 1978; Peter and Olson, 1996). Results from different empirical studies demonstrate some inferences consumers made evaluating food quality respectively the quality of meat (Table 1).

To sum up, consumers' quality judgement highly depends on their knowledge concerning certain quality attributes or product characteristics. That means, information, fitting into a schema, is processed faster. It eases product evaluation and will be remembered more easily (Kroeber-Riel and Weinberg, 2003).

3. Design of the study and empirical results

Methodologically consumers' cognitive structure can be surveyed by means of associative methods (Olson and Muderrisoglu, 1979; Olson and Reynolds, 1983; Bonato, 1990; Jonassen *et al.*, 1993; Joiner, 1999; Holzmann and Wührer, 2000).

In spring 2004 we accomplished a consumer survey with a total of 260 respondents (random sample). The data were collected in Kiel, Germany. To capture consumers' knowledge structures we applied two different associative methods: a free-association technique and a variation of concept mapping. Those techniques are open and non-directive, which increase the likelihood that the full variety of cognitive associations in a person's memory structure may be evoked (Olson and Muderrisoglu, 1979; Joiner, 1998). They also allow more information to be obtained

Table 1. Inferential beliefs concerning food quality.

Inferences	Author(s)
Muscle colour, at the point of purchase, is an indicator of freshness and anticipated palatability for consumers.	Brewer, 2002 cited in Livingston *et al.*, 2004
The colour of the fat and the colour of the meat were used as indicators of the freshness of the meat.	Bredahl and Poulsen, 2002 cited in Scholderer *et al.*, 2004
Pieces of meat that were cut in unequal sizes or meat that was poorly trimmed were considered to be of low quality.	Bredahl and Poulsen, 2002 cited in Scholderer *et al.*, 2004
A European research project uncovers the following relations: Origin and colour are in general the most helpful cues to predict eating quality of fresh meat. Origin is not only an indicator for eating quality but as well for the safety of meat.	Becker, 2001
Qualitative studies from Denmark suggest that consumers make a whole range of positive inferences from the label 'organic', and these refer not only to concern for the environment and health, but also to animal welfare and even better taste.	Bech-Larsen and Grunert, 2001; Bredahl and Poulsen, 2002 cited in Scholderer *et al.*, 2004

from those consumers who have more elaborate knowledge structures for the object in question (Joiner, 1998).

3.1. Free-association technique

An appropriate method to survey consumers' knowledge structures are free-associations (Olson and Muderrisoglu, 1979; Olson and Reynolds, 1983; Bonato, 1990; Jonassen *et al.*, 1993; Holzmann and Wührer, 2000). The test persons had to describe freely in their own words, what came into their minds thinking of the quality of pork. By means of this technique we try to activate a particular structure of stored memory respectively to elicit consumers' most salient beliefs (Olson and Muderrisoglu, 1979).

Each interviewee mentioned on the average 2.8 concepts. The number of concepts per test person ranged from zero to eleven. To analyse the large number of associatives (n=738) we conducted a content analysis (Lamnek, 1995; Mayring, 2002). We developed six categories that capture the essential aspects of most of the thoughts expressed by the test persons. Table 2 shows our six categories and their most important contents.

47% of the answers are related to product characteristics. Pork of a good quality has to taste good. The colour of meat and the share of fat are important quality cues. Generally consumers look for meat with low fat (Bredahl and Poulsen cited in Scholderer *et al.*, 2004). In this context, Verbeke (1999) points out that fat has become of major issue of health concern for the consumer. Another 30% deals with process attributes such as animal husbandry or country of origin. As meat clearly dominates food safety debates consumers are increasingly interested in production

Table 2. Most important contents of the knowledge elements (n= 738).

Knowledge element	Contents	Share of nominations
Product characteristics	fat content flavour water content freshness colour health value	47%
Process characteristics	animal husbandry feeding country of origin	30%
Meals/preparation	types of dishes methods of cooking	7%
Product environment	butcher discounter shop counter	6%
Specific signals	price seal of approval packaging	6%
External events	swine fever food safety issues	4%

and processing methods (Verbeke, 1999; Gellynck and Verbeke, 2001). The other knowledge elements are present to a lesser extent in the consumer's mind.

But with this illustration (Table 2) we have not been able to reveal whether there are interrelations between single categories or single concepts. It is possible that there is a relation between 'process characteristics', e.g. animal husbandry and 'specific signals', e.g. Demeter, but we have not been able to prove this by now. However, these interrelationships are important, because process characteristics are credence quality attributes. This means consumers cannot ascertain the quality before during or after purchase which may lead to uncertainty concerning the quality of the product (Darby and Karni, 1973; Becker, 1999; Gierl and Stumpp, 2000; Bech *et al.*, 2001). Tolle (1994) points out that in situations of quality uncertainty consumers use extrinsic quality cues predominantly. But compared to the knowledge element 'process characteristics' as it is the category of the credence quality attributes 'country of origin' and 'animal husbandry' the knowledge element 'specific signals' is of little importance. Therefore, we assume that due to consumers' inferential belief formation other product attributes might act as quality signals. One might think of the following relationships: the higher confidence in weekly markets compared to the discounter or supermarket may lead to a more positive perception of animal husbandry. With regard to the category 'external events' we assume relationships with every other category. For example, a food scandal could result in putting all product characteristics in a bad light. This effect is the so-called halo-effect (Peter and Olson, 1996; Kroeber-Riel and Weinberg, 2003).

Against this background the analysis of the interrelations between the individual concepts is our next step. On this account we used a variation of concept mapping.

3.2. Variation of concept mapping

Concept mapping is a qualitative methodology which can be useful for accessing consumers' associations (Joiner, 1998). It is a technique for representing knowledge in graphs. Knowledge graphs are networks of concepts. These networks consist of nodes and links. The nodes represent concepts and the links represent the relation between concepts (Tergan, 1986; Jonassen *et al.*, 1993). The technique assumes that information is displayed in such a way that its arrangement on the page reveals something about the relationships inherent in the information represented in a person's mind. It is assumed that, once a structure is activated, a person should be able to report the majority of the structure content (Joiner, 1998).

We interviewed 69 consumers due to the research object 'quality of pork'. Furthermore, we conducted three other surveys with 69 respondents each to reveal the cognitive structures concerning the key concepts 'pork', 'quality of potatoes' and 'potatoes'. Results are not available yet. We used the paper and pencil technique of concept mapping. To elicit an individual network different steps are needed (Table 3).

An example of an individual semantic network can be seen below (Figure 1). Our test person mentioned altogether ten concepts, seven of the predetermined ones and three individual concepts (dashed boxes). Five concepts are directly related to the key concept 'quality of pork'. Dashed lines indicate missing evaluations. Figure 1 shows that the evaluations of the linkages as well as the evaluations of the concepts are rather incomplete. The evaluation process (Table 3: step 4) was far too difficult for most respondents. A simplification of this process will be one challenge for future research. However, despite these problems our analysis provides some interesting results, which will be presented in the following.

Table 3. Steps to be taken to elicit semantic networks.

Step 1	Each test person got a sheet of paper with the key concept 'quality of pork' in its centre.
Step 2	The test persons had to write down every concept they had in mind due to the key concept. To ease the requirements the interviewees received eight additional concepts. These concepts were freshness, leanness, marbling, butcher, price, seal of approval, taste and country of origin. They were derived from different empirical studies which had taken place between 1982 and 2003 (for a review see Engelage, 2002; Lüddeke, 2002; Frey, 2003). These eight concepts could be used but their use was not required. Moreover, the respondents were asked to add individual concepts which they associated with the 'quality of pork'.
Step 3	Accordingly, the test persons had to think of possible linkages between the different concepts.
Step 4	Finally, the test persons had to evaluate the significance of every single concept and the strength of the interrelations. On this account, the respondents received three predetermined signs: X for less important, O for regular and !! for very important. By means of these signs, the test persons had to evaluate the importance of their concepts with respect to the 'quality of pork'. Afterwards, they had to evaluate the linkages between the concepts as to whether they are weak (X), regular (O) or strong (!!). Furthermore, the test persons evaluated the concepts themselves as to whether these were positive (+) or negative (-).

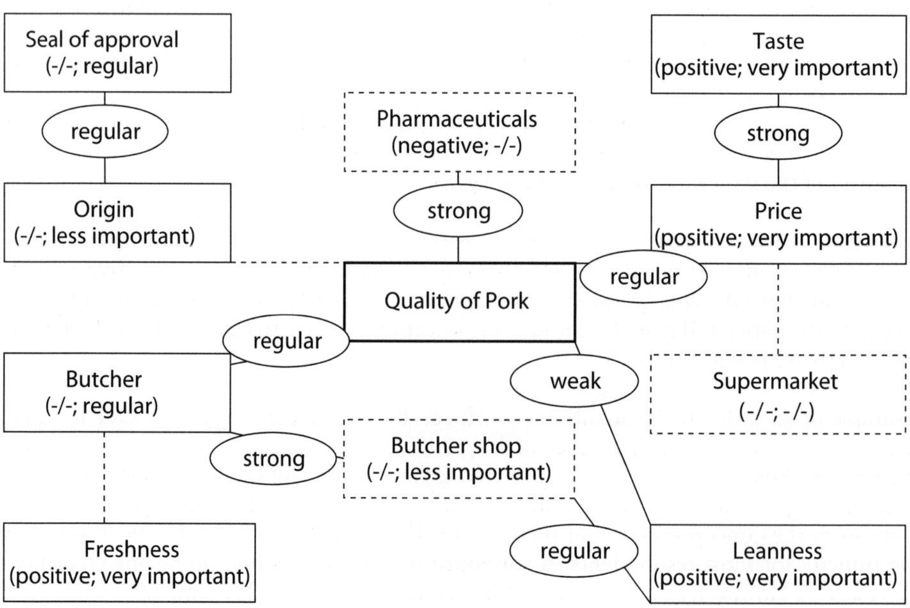

Figure 1. An individual semantic network.

The technique of concept mapping led to massive amounts of data. The number of concepts per test person ranged from nine to 29. The participants replied with 370 different concepts. Of these, eight were from the predetermined concepts, which means 69 test persons wrote down 362 new/individual concepts. In sum, the participants replied with a total of 888 concepts, 379 concepts of the predetermined ones and 509 of the individual concepts. On the average each respondent mentioned 12.9 concepts, 5.5 of the predetermined ones and 7.4 of the individual ones (Table 4). The results clarify the heterogeneity and complexity of the quality concept from a consumer's perspective.

We aggregated the individual knowledge networks by means of Ucinet 6.0 to show the linkages between the single concentrated concepts (Borgatti *et al.*, 2002). Figure 2 depicts the interrelations between the key concept 'quality of pork' and the Top 15-concepts as well as the linkages between the Top 15-concepts. Top 15-concepts are concepts mentioned by at least 10% of all respondents.

Table 4. Number of nominations.

	Number of different concepts	Number of concepts	Average number of concepts per test person
Predetermined concepts	8	379	5.5
Individual concepts	362	509	7.4
Sum	370	888	12.9

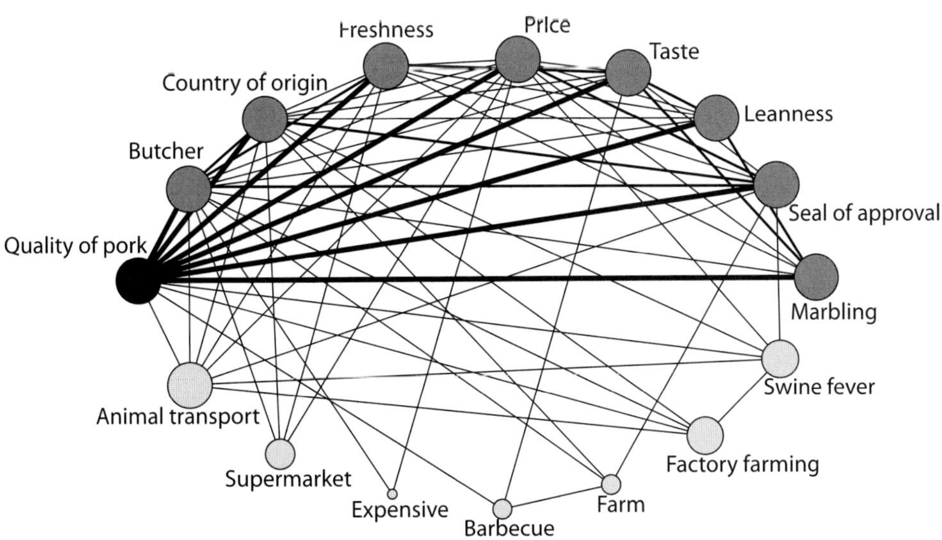

Figure 2. Aggregated conceptual network of the Top 15-concepts (n=69).

The thickness of the connecting lines indicates the frequency test persons mentioned linkages between different concepts. Every given concept is highly related to the key concept 'quality of pork', which is self-evident as its selection based on previous empirical research. It is interesting to note that the first individual concepts mentioned by the test persons were negative ones, namely swine fever and factory farming. Furthermore, Figure 2 shows for example strong interrelations between the credence quality attribute 'country of origin' and the quality cues 'butcher' and 'seal of approval'. In this context, Table 5 presents the relevance of different linkages in percentage.

As Figure 2 and Table 5 are very complex we categorised the given concepts by means of the economics of information approaches: quality dimensions are commonly categorised into search, experience and credence dimensions (Darby und Karni, 1973). Search characteristics respectively visible quality cues may be both physical characteristics of the product (intrinsic quality cues) and other characteristics such as seal of approval or price (extrinsic quality cues) (Bech *et al.*, 2001). Figure 3 shows the most important linkages between the predetermined concepts which have been mentioned by at least ten percent of the test persons.

Table 5 and Figure 3 show that the credence quality dimension 'country of origin' is highly related to the extrinsic quality cues 'seal of approval' (34.8%) and 'butcher' (33.3%). That means about one third of the test persons may use these quality cues to evaluate the origin of a given product. In contrast to that the experience quality dimension 'taste' is highly related to the intrinsic quality cues 'leanness' (27.5%) and 'marbling' (21.5%). To predict the eating quality of pork the test persons rely on intrinsic quality cues mainly.

Table 5. Relevance of different linkages in percentage (n=69).

	Quality of pork	Seal of approval	Country of origin	Butcher	Price	Taste	Leanness	Freshness	Marbling	Swine fever	Factory farming	Farm	Barbecue	Expensive	Supermarket	Animal transport
Quality of pork	-	61	72	78	71	70	67	70	61	10	7	0	6	0	0	6
Seal of approval	61	-	35	17	16	3	0	4	0	1	0	1	0	0	0	1
Country of origin	72	35	-	33	12	6	3	14	0	4	7	6	0	0	1	1
Butcher	78	17	33	-	16	6	1	29	3	0	1	3	0	3	1	1
Price	71	16	12	16	-	6	6	10	3	1	0	0	0	12	7	1
Taste	70	3	6	6	6	-	28	29	22	0	0	0	6	0	0	0
Leanness	67	0	3	1	6	28	-	7	25	0	0	0	0	0	0	0
Freshness	70	4	14	29	10	29	7	-	4	0	0	0	0	0	1	1
Marbling	61	0	0	3	3	22	25	4	-	0	0	0	0	0	0	0
Swine fever	10	1	4	0	1	0	0	0	0	-	3	0	0	0	0	1
Factory farming	7	0	7	1	0	0	0	0	0	3	-	0	0	0	0	3
Farm	0	1	6	3	0	0	0	0	0	0	0	-	1	0	0	0
Barbecue	6	0	0	0	0	6	0	0	0	0	0	1	-	0	0	0
Expensive	0	0	0	3	12	0	0	0	0	0	0	0	0	-	0	0
Supermarket	0	0	1	1	7	0	0	1	0	0	0	0	0	0	-	0
Animal transport	6	1	1	1	1	0	0	1	0	1	3	0	0	0	0	-

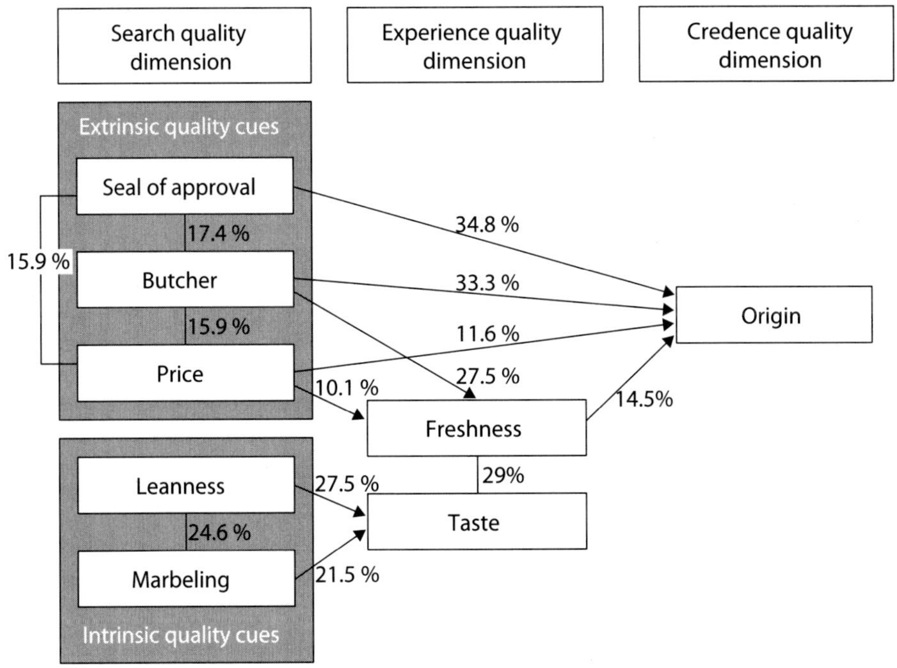

Figure 3. Most important linkages in percentage (n=69).

4. Conclusions and future prospects

The issue of food quality has come to the forefront in Europe and North America (Steenkamp, 1990). Individual cognitive structures are an important part in regard of consumers' quality judgement. The literature favours the view of memory being organised in the form of an associative network. According to this associative network concept, memory consists of a series of nodes (representing concepts) and links, which represents associations between nodes (Olsen, 1978; Engel *et al.*, 1995). Such structure of knowledge which contains mostly episodic and semantic general knowledge is called a schema (Peter and Olson, 1996)

Against this background, the aim of the present paper was to analyse food quality from a consumers' perspective by uncovering consumers' knowledge structures concerning the quality of pork. By means of associative techniques we attempted to access consumers' individual level of understanding. The results of the free-association technique showed that product characteristics followed by process characteristics dominate consumers' mind due to the research object. Then, we analysed the interrelations between individual concepts by means of a variation of concept mapping. Concept mapping enables one to understand the relationship between ideas or concepts by creating a visual map of connections. Identifying such relations is one step forward in developing consumer-oriented communication strategies. To develop effective strategies marketers have to provide consumers with information which they understand and they seek for. Especially in the case of credence quality attributes consumers rely on information they get from producer, retailer etc. without being able to verify them. In view with our case study, research within this field is important since empirical studies by Grunert and co-workers (2004) have shown that customers have difficulties in assessing the quality of meat. With the aid of an aggregated conceptual network we could show that the credence quality dimension 'country of origin' is highly related to the extrinsic quality cues 'seal of approval' and 'butcher'. In contrast to that consumers may use predominantly intrinsic quality cues to predict the eating quality of pork.

In our opinion concept mapping is a suitable technique to analyse consumers' knowledge structure. Because of complex instructions the paper and pencil technique puts high demands on the interviewers as well as on the interviewees. We have already mentioned above that a simplification of the evaluation process will be one challenge for future research. To improve this method we will have to carry out more consumer surveys. In doing so, we will thus have to change the key concept and have to perform unsupported surveys. Moreover, we have developed some software to carry out online investigations with concept mapping.

Besides this, further research has to concentrate on the question how much and what kind of information will be retrieved by consumers in purchase situation. It is possible that information is stored in memory but will not or cannot be retrieved in purchase situation due to a lack of activation. On that account, we accomplished a second survey at different retailers in January 2005.

References

Anderson, J.R., 1983. A spreading activation theory of memory. Journal of Verbal Learning and Verbal Behavior 22, 261-295.

Bech, A., K.G. Grunert, L. Bredahl, H.J. Juhl and C.S. Poulsen, 2001. Consumers' quality perception. In: Frewer, L., E. Risvik and H. Schifferstein (eds.), Food, people and society - A European perspective of consumers' food choices. Springer, pp. 97-113.

Bech-Larsen, T. and K.G. Grunert, 2001. Konsumentenentscheidungen bei Vertrauenseigenschaften: Eine Untersuchung am Beispiel des Kaufes von ökologischen Lebensmitteln in Deutschland und Dänemark. Marketing-ZFP 23, 3, 188-197.

Becker, T., 1999. The economics of food quality standards. Proceedings of the second interdisciplinary workshop on standardization research, University of the Federal Armed Forces, Hamburg, 24-27 May 1999.

Becker, T., 2001. Food safety and health - consumer perceptions of meat quality. Programme abstracts - Presentation of the speakers: Food Chain 2001: Safe - Sustainable - Ethical, 14-16 March 2001 Uppsala, Sweden: 56.

Bonato, M., 1990. Wissensstrukturierung mittels Struktur-Lege-Technik. Eine graphentheoretische Analyse von Wissensnetzen. Peter Lang.

Borgatti, S.P., M.G. Everett and L.C. Freemann, 2002. Ucinet for windows: Software for social network analysis. Havard: Analytic Technologies.

Bredahl, L., 2003. Cue utilisation and quality perception with regard to branded beef. Food Quality and Preference 15, 65-75.

Cardello, A.V., 1995. Food quality: relativity, context and consumer expectations. Food Quality and Preference 6, 163-170.

Collins, A.M. and E.F. Loftus, 1975. A spreading-activation theory of semantic processing. Psychological Review 82, 407-428.

Cowley, E. and A.A. Mitchell, 2003. The moderating effect of product knowledge on the learning and organisation of product information. Journal of Consumer Research 30, 443-454.

Darby, M.R. and E. Karni, 1973. Free competition and the optimal amount of fraud. The Journal of Law and Ecomomics 16, 67-88.

Engel, J.E., R.D. Blackwell and R.W Miniard, 1995. Consumer Behavior. 8th ed., The Dryden Press.

Engelage, A., 2002. Qualitätswahrnehmung bei Lebensmitteln: Das Verbraucherbild in Rechtssprechung und Wissenschaft, PhD Thesis, Berlin

Erasmus, A.C., E. Boshoff and G.G. Rousseau, 2002. The potential of using script theory in consumer behaviour research. Journal of Family Ecology and Consumer Sciences 30, 1-9.

Frey, I., 2003. Qualitätswahrnehmung und Zahlungsbereitschaft für Produktqualität bei Nahrungsmitteln, Master Thesis.

Gellynck, X. and W. Verbeke, 2001. Consumer perception of traceability in the meat chain. Agrarwirtschaft 50, 368-374.

Gierl, H. and S. Stumpp, 1999. Diffusion und Informationswert von Qualitätssignalen. Zeitschrift für Betriebswirtschaft 11, 1241-1268.

Grunert, K.G., 1981. Explaining consumer attribute selection: A semantic network approach. Paper prepared for the Annual Meeting of the European Academy for Advanced Research in Marketing, Copenhagen, March 25-27.

Grunert, K.G., 2001. Perception of food quality - expectations, inferences and process characteristics. Food Chain 2001: Safe - Sustainable - Ethical: Programme abstracts, presentation of speakers, 14-16 March, Uppsala, Sweden, pp. 29-32.

Grunert, K.G., A. Baadsgaard, H. Harvig Larsen and T.K. Madsen, 1996. Market orientation in food and agriculture. Kluwer Academic Publishers.

Grunert, K.G., L. Bredahl and K. Brunso, 2004. Consumer perception of meat quality and implications for product development in the meat sector - a review. Meat Science 66, 259-272.

Henson, S., 2000. The process of food quality belief formation from a consumer perspective. In: Becker, T. (ed.), Quality policy and consumer behaviour in the European Union. Wissenschaftsverlag Vauk, pp. 73-89.

Holzmann, T. and G.H. Wührer, 2000. Kognitive Landkarten zur Entscheidungsstützung im Markenmanagement. In: Foscht, T., G. Jungwirth and P. Schnedlitz, (eds.), Zukunftsperspektive für das Handelsmanagement. Deutscher Fachverlag, pp. 429-443.

Joiner, C., 1998. Concept mapping in marketing: A research tool for uncovering consumers' structural knowledge. Advances in Consumer research 25, 311-317.

Jonassen, D.H., K. Beissner and M. Yacci, 1993. Structural knowledge: Techniques for representing, conveying and acquiring structural knowledge. Lawrence Erlbaum Associates.

Kaas, K.P., 1994. Ansätze einer institutionenökonomischen Theorie des Konsumentenverhaltens. Forschungsgruppe Konsum und Verhalten (ed.), Konsumentenforschung. Vahlen, pp. 245-260.

Kroeber-Riel, W. and P. Weinberg, 2003. Konsumentenverhalten. 8th edition, Vahlen.

Lamnek, S., 1995. Qualitative Sozialforschung, Band 2 Methoden und Techniken. 3rd ed., Beltz Psychologie Verlags Union.

Livingston, M., M.S. Brewer, J. Killifer, B. Bidner and F. McKeith, 2004. Shelf life characteristics of enhanced modified atmosphere packaged pork. Meat Science 68, 115-122.

Lüddecke, K., 2002. Verbraucherwahrnehmung der Lebensmittelqualität und -sicherheit. Ergebnisse einer Verbraucherbefragung, Master Thesis, Kiel.

Marshall, D., 1997. Flexible options: An overview of qualitative research methods and their relevance to food choice. Food for the consumer - measurements of consumers' attitudes: Workshop: Qualitative methods in food quality research, 13-24.

Meyer, R., 2000. Potenziale für eine verbesserte Verbraucherinformation - Endbericht zum TA-Projekt 'Entwicklungstendenzen bei Nahrungsmittelangebot und -nachfrage und ihre Folgen, Arbeitsbericht Nr. 69, Büro für Technikfolgen-Abschätzung beim Deutschen Bundestag (TAB).

Olson, J.C., 1978. Internal belief formation in the cue utilization process. Advances in Consumer Research 5, 706-713.

Olson, J.C. and A. Muderrisoglu, 1979. The stability of responses obtained by free elicitation: Implications for measuring attribute salience and memory structure. Advances in Consumer Research 6, 269-275

Olson, J.C. and T.J. Reynolds, 1983. Understanding consumers' cognitive structure: Implications for adverstising strategy. In Percy, L, and A. G. Woodside (eds.), Advertising and consumer psychology. Lexington Books, pp. 77-90.

Peter, J.P. and J.C. Olson, 1996. Consumer behavior and marketing strategy. 4th ed., Irwin.

Scholderer, J., N.A. Nielsen, L. Bredahl, C. Claudi-Magnussen and G. Lindahl, 2004. Organic pork: Consumer quality perception. Project Paper no 02/04, URL.: http://130.226.203.239/pub/mapp/pp/pp0204.pdf (18.10.2004)

Steenkamp, J.-B.E.M., 1990. Conceptual model of the quality perception process. Journal of Business Research 21, 309-333.

Tergan, S.-O., 1986. Modelle der Wissensrepräsentation als Grundlage qualitativer Wissensdiagnostik. Westdeutscher Verlag.

Tolle, E., 1994. Informationsökonomische Erkenntnisse für das Marketing bei Qualitätsunsicherheit der Konsumenten. Zeitschrift für betriebswirtschaftliche Forschung 46, 675-682.

Verbeke, W., 1999. Factors influencing the consumer decision-making process towards meat. PhD Thesis, University Gent.

Von Alvensleben, R. and U. Scheper, 1997. The decline of the meat image in Germany - Theoretical considerations. In: Von Alvensleben, R., S. von Cramon-Taubadel, A. Röhr, U. Scheper and K. Schleyerbach (eds.), Problems of meat marketing. Department of Agricultural Economics, Agricultural Marketing, working paper 2. Kiel, pp. 1-7.

Perceived safety of organic and regional food from a perspective of uncertain consumers

Rainer Franz[1], Ulrich Enneking[1] and Richard Balling[2]
[1]Technical University of Munich, Agricultural and Food Marketing, Alte Akademie 14, 85350 Freising-Weihenstephan, Germany; franz@wzw.tum.de
[2]Bavarian Ministry of Agricultural and Forestry, Germany

Abstract

Due to the raising uncertainty of consumers concerning food safety organic and regional food are supported by several governments in order to regain consumer confidence. A face-to-face study was conducted in Bavaria with the aim to examine how several labels indicating both regional and organic food production influence the purchase decisions in three consumer segments which are characterised by a different degree of uncertainty. Whether or not uncertainty leads to purchases of organic or regional food depends on the product category evaluated. The a priori assumption that higher uncertainty generally corresponds to a higher demand for organic or regional produce has to be rejected.

Keywords: food safety, uncertainty, eco-labelling, logit analysis, regional food

1. Regaining consumer confidence by means of regional and organic food

Europe has faced several food crises in recent years. Especially in Germany there has been a great debate on consumers' 'uncertainty[1]', which has become a key word in food safety discussions dealing with attitudes, e.g. loss of confidence, and behaviour of consumers. As a consequence, particularly after the BSE crises and the collapse of the beef market in several European countries, changes of administrative structures have taken place, the foundation of the European Food Safety Authority in 2002 being the most prominent example (BYRNE, 2001).

However, not only administrative structures have changed, but also consumer protection has more and more being set on the political agenda. For instance, there have been considerable efforts in building up systems to enable traceability all over the food chain in order to trace back the origin of food and food ingredients. Along with this, a new label indicating food safety, the QS-label, has been introduced in Germany.

Apart from these efforts to improve objective food safety, politicians are interested in enhancing perceived safety as a subjective judgement of the consumer. In Germany and in Bavaria, the official targets of supporting regional and organic food production have been extended to the aim of regaining consumer confidence (Künast, 2002; Stoiber, 2001). Literature on uncertainty suggests that uncertain consumers favour regional and organic food (e.g. Balling, 2000). Notwithstanding

[1] The German term 'Verunsicherung' can hardly be translated into English, the term 'uncertainty' seems to be the best solution. Nevertheless, it is an unusual term in the international literature, where the terms 'risk' and 'risk perception' are more popular. However, these terms do not exactly fit the meaning of the German term 'Verunsicherung'.

the general acceptance of this assumption, empirical evidence that is based not only on measures of attitude but rather on real consumer behaviour is still lacking. This paper presents a study which permits deeper insights into the potential of regional and organic food to regain consumer confidence and thus to improve perceived food safety.

This article shows results from a consumer survey conducted in Bavaria in July 2004. A specific aim of this survey was to examine the influence of several labels indicating both regional and organic food production on the purchase decision in different product categories. By measuring this influence in consumer segments which differ in their perceptions of food safety it could be analysed whether or not uncertainty enhances the probability to choose regional or organic food items.

2. The nature of consumers' uncertainty with regard to food safety

Considering the relevant literature, e.g. Von Alvensleben (1998), Willlhöft (2001), Bergmann (2000) and structuring the causes of uncertainty, two different spheres should be distinguished as it is pointed out in Figure 1. First, there are factors which refer to changes in social and cultural conditions due to a long-term civilisation process. These factors are not considered direct triggers of uncertainty, but they rather represent a framework of underlying mechanisms which are linked to the modern industrial society and form a favourable 'medium' for uncertainty to grow. As Figure 1 shows, one of these factors is alienation from the food production process meaning that the more people grow up in big cities the more they are lacking experience with their environment in general and the food production process in particular. This leads to decreasing knowledge of how food is produced and processed. In a similar vein it can be argued that people in modern societies like Germany face an oversupply of food. They do not longer worry about whether or not they will get enough to eat but they spend more time taking care of food safety (Von Alvensleben, 1998). Another factor of this kind is the diminishing trust in institutions. As Lipset and Schneider (1983) point out, confidence in social and political institutions has decreased in the last 30 years. These three aspects - alienation from the food production process, oversupply of food in industrial societies and diminishing trust in institutions - are examples of the first 'sphere' of uncertainty and they describe general developments in industrial societies.

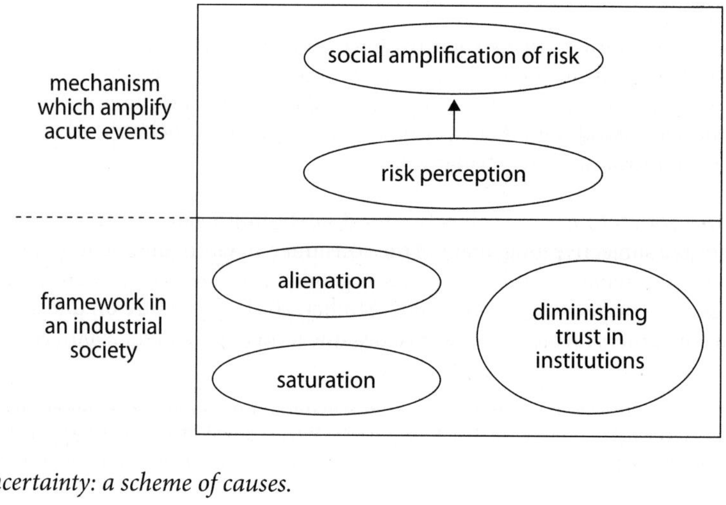

Figure 1. Uncertainty: a scheme of causes.

Second, there are psychological mechanisms concerning risk perception on an individual level (Sparks and Shepherd, 1994) and amplification mechanisms on both the individual and societal level (Kasperson *et al.*, 2003) that form and intensify acute uncertainty. Risk perception on an individual level is an important research field in psychology and focuses on how lay people perceive hazards. The judgement of a hazard by lay people is critically influenced by their judgement of several characteristics of the hazard, which are called 'risk characteristics'. Risk characteristics describe whether a risk is controllable, voluntary, whether it is natural or made by humans, whether its costs are distributed equally etc. Hazards are more acceptable to lay people if they are controllable, voluntary etc. The term 'social amplification of risk' (Kasperson *et al.*, 2003), as it is mentioned in Figure 1, describes another important aspect of uncertainty in this second sphere. This 'theory postulates that the social and economic impacts of an adverse event are determined not only by the direct biological and physical consequences of the event, but by the interaction of powerful psychological, cultural, social, and institutional processes that amplify or attenuate public response to the event' (Burns *et al.*, 1993). Both the information itself and the reactions of individuals receiving this information represent stages of information flow where mechanisms of amplification or attenuation might occur. A result of this amplification mechanism is that people judge hazards more severe than they really are.

The main result of a literature review is that uncertainty is an individual and therefore subjective psychological construct and operationalisation of this construct is still missing. The subjectivity of uncertainty asks for specific consumer research. It is a basic necessity to find out, what factors contribute to uncertainty from a consumer's point of view and not only from an expert's perspective.

3. Measurement of uncertainty

For the purpose of the study described in this paper it was necessary to measure consumer uncertainty regarding food safety. We were able to use an 'uncertainty scale' of three items which we developed in an unpublished study before. This was possible because both studies were part of a large project funded by the Bavarian Ministry of Agriculture and Forestry. With this 'uncertainty scale' consumer segments could be identified which differ in terms of their perceived uncertainty. A short description of how the scale was developed is given in Figure 2.

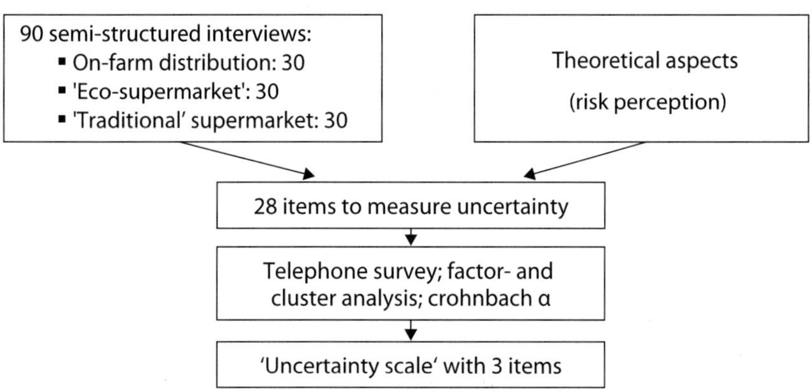

Figure 2. Development of an 'uncertainty scale'.

As it is shown in Figure 2 we measured the construct 'uncertainty' by generating a total of 28 items which we derived from (1) a qualitative study with semi-structured interviews and (2) from the literature focussing on risk perception of lay people. In the former 90 shoppers were interviewed; 30 persons in each of the following distribution channels: an on-farm shop, a supermarket specialized on organic food and a traditional supermarket. The 28 items were used to measure uncertainty by means of a representative survey in Bavaria (n = 814) taking place in June 2003 (Franz, 2004: 100ff.).

The data were first analysed using factor analysis. Nine dimensions of uncertainty emerged out of the principal component analysis (varimax rotation) with 'scandals' being the first and therefore most important factor and 'alienation' from the food production process the least important factor. However, results are not discussed in detail because in this paper factor analysis is just conceived as an intermediate step to construct the uncertainty scale.

However, before doing so we performed a cluster analysis to verify whether it is justified to treat uncertainty as a simplified, uni-dimensional construct. A solution with four clusters was chosen taking into account both statistical criteria and criteria as regards content. Results of the cluster analysis can be summarised as follows: people who worry most about 'scandals' (cluster 1) do so about the other eight factors of uncertainty. The same pattern was observed for respondents who worry second-most about 'scandals' (cluster 2): they also worry second-most about all the other eight factors. The same observation was made for the remaining two clusters. Hence, the resulting clusters represent different degrees of intensity of perceived uncertainty.

That is why construction of a one-dimensional uncertainty scale is justified. In order to construct this scale we analysed all 28 items by means of reliability analysis using Crohnbach's α. Items were eliminated one by one as long as this led to an improvement. The best score for Crohnbach's α was reached with 21 items. Due to the fact that 21 items can not be considered practical in a large scale survey, we eliminated one by one those items with the lowest reduction of Crohnbach's α until we came up with a scale consisting of three items. These items are listed in Table 1.

As a result of reliability analysis using Crohnbach's α, the items in table 1 are the best items to measure uncertainty when a highly reduced item battery of no more than three items is desired. With two of the remaining three items the attitude toward food scandals was measured. This was an appropriate result taking into account that scandals turned out to be the most important factor extracted in the principal component analysis. Crohnbach's α of this 3-item scale is 0.627 which

Table 1. Items to measure the construct 'uncertainty'.

Uncertainty	Crohnbach's α (previous telephone survey)	Crohnbach's α (survey discussed here)
Item 1: BSE has shown that more food scandals will occur in the future.	α = 0.694	α = 0.627
Item 2: I am concerned about the consequences that food ingredients have on human health.	α = 0.694	α = 0.627
Item 3: The food scandals in recent years have scared me.	α = 0.694	α = 0.627

is above the required level of 0.6 (Malhotra, 1993: 308). Therefore, our scale can be considered appropriate to measure uncertainty. As can be seen in table 1 Crohnbach's α is similar in the previous telephone survey (0.694) providing another indication of the reliability of the scale.

4. Design of the study

A consumer survey was conducted in Bavaria in July 2004 with the aim to examine whether subjective consumer uncertainty regarding food safety influences the purchase decision in a way that regional and/or organic food is preferred over conventional food. This was tested across several product categories.

The survey took place in four Bavarian cities (Munich, Augsburg, Nürnberg and Freising). Face-to-face interviews were realised with respondents which were contacted in the city centres in front of shopping malls. Telephone interviews were not adequate because interviewees had to be introduced into their choice task with the aid of product photos. However, due to financial constraints no random-route sample could be drawn resulting in a non-representative sample. Finally, a total of 1,676 interviews were realised.

Table 2 shows the design of the study. Interviewees were shown a picture of a product, where three variables were systematically varied, viz. a label indicating regional production, an eco-label and the product price. The levels of these attributes are also displayed in Table 2. There were two levels representing organic production: (a) the official German eco-label and (b) the official Bavarian eco-label. A third level gave no indication of organic production. Four different labels were used

Table 2. Design of the choice-experiment.

	Name of variable	Levels
Dependent variable		
Binary choice task of the respondents: Would you buy this product in a separate supermarket area specialised on ecological and regional food or would you rather choose a conventional product	choice	buy not buy
Independent variables		
Experimental variables, which were systematically varied across respondents	regional label	on-farm distribution own region Bavaria Germany
	organic label	no organic production German eco-label Bavarian eco-label
	price	low high
Classification variable		
Variable to separate three sub-samples each representing a different degree of uncertainty	degree of uncertainty	low middle high

to represent regional production: (a) on-farm distribution; (b) products from the own region; (c) products from Bavaria; and (d) products from Germany. No brand names were displayed in the pictures. Each label was offered at two different price levels which have been determined in a pretest. There were two reasons to choose these attribute levels. First, we wanted to use real labels that do exist in the market place. Second, the labels used can be interpreted as different stages of the attributes 'regional' and 'organic'. On-farm distribution represents the closest contact between consumer and producer because they meet each other directly. When food is produced in a consumer's own region, a close relation between producer and consumer does still exist but it is not as close as in the case of on-farm distribution. In a similar vein the eco-labels were chosen: the Bavarian eco-label indicates higher production standards than the German eco-label.

As to the experimental design, a total design was realised meaning that all possible combinations of regional and organic labels were shown to the respondents. However, only one regional and only one organic label was shown at a time. Hence, 24 different combinations of labels and prices were realised. Eight different products were used in the study, but only those five products are presented in this paper where a significant effect of at least one label was found.

People were told to imagine that these products can be purchased in a separate area within a conventional supermarket. Each respondent had to indicate, whether or not he would buy a product shown on the picture (dependent variable). A dichotomous choice had to be made and therefore binary logit regression analysis was used to analyse the data. As the choice task was repeated for eight product categories eight logit models were specified. Enneking (2004) points out that a binary choice experiment can reveal similar results compared to a choice-experiment with more than two alternatives. In the present study simple binary choice tasks were employed in order to minimise the cognitive burden of respondents who were already confronted with eight choice tasks.

In order to take the impact of uncertainty into account, the sample was split into three sub samples each containing the same number of respondents. This 'uncertainty segmentation' was performed on the basis of the uncertainty scale introduced above.

5. Empirical results

Table 3 summarises the socio-demographics of the sample. Most of the respondents lived in households with one (27.7%) or two persons (32%). A household size with three persons or more was registered less often. Only 19.9% of the respondents indicated that they had children younger than 14. Due to the fact that the survey was realised in university cities younger people were overrepresented in the sample. Respondents in the group up to 29 years formed 36% of the sample. Furthermore, most respondents were highly educated with 30.4% having some sort of university degree. Finally, 58.8% of the respondents were female.

In order to give an efficient overview of our findings not all results of each single logit model are displayed. In fact the empirical results of the binary logit models are summarised in Table 4. It shows whether the impact of the respective label on the likelihood to purchase the respective product is weakly significant (0.1%-level), significant (0.5%-level) or highly significant (0.01%-level).

The a priori assumption was that purchases of regionally and organically labelled produce increase with growing uncertainty of consumers. To confirm this assumption, the label's influence should highest in the sub sample with a high degree of uncertainty and lowest in the segment with low

Table 3. Sample statistics (n = 1,676).

Variable	Levels	Frequency (%)
Size of household (number of persons living permanently in the household)	1	27.7
	2	32.0
	3	17.9
	4 and more	22.4
Children under 14 years living in the house	yes	19.9
	no	80.1
Age	up to 29 years	36.0
	30 to 39 years	21.6
	40 to 49 years	15.4
	50 to 59 years	11.6
	60 years and older	15.3
Education	Hauptschule	13.8
	Realschule	26.2
	Abitur	24.8
	University	30.4
	other	4.8
Gender	female	58.8
	male	41.2

uncertainty. As the results in Table 4 show this is the case for none of the products evaluated. Hence, the a priori assumption has to be rejected.

Nevertheless some interesting issues have to be discussed. In three product categories - roast beef, fresh milk and eggs - a similar pattern of behaviour was observed: there was a significant influence of one or two labels in the sub sample with a medium degree of uncertainty, but no significant effects in those sub samples with low or high degree of uncertainty. In this product category the relation between uncertainty and the preference for regional or organic food is non-linear. The question arises, why highly uncertain consumers do not opt for organic or regional food whereas somewhat uncertain consumers do. Our empirical results provide no clear answer, so that further research should be dedicated to this phenomenon. One assumption is that people with a high degree of uncertainty do not trust a label on a product offered in a conventional supermarket, but rather choose a different distributor to purchase their food, like organic speciality shops or farmers' stores.

However, this type of relationship was not detected in the case of cheese ('Emmentaler') and potatoes. Obviously, the relation between uncertainty and the purchase of organic or regional food depends on the product category evaluated. Furthermore, label design and label awareness might also have some impact on choice what has to be taken into account when interpreting the findings. In particular the latter might be an important aspect. Both the Bavarian and the German eco-label are quite popular because they either exist since a long time (Bavarian label) or they are promoted intensively (German label). Although the Bavarian label had to be changed due to a decision of the EU, Bavarian consumers are used to the existence of a label indicating the region of origin 'Bavaria'. The situation is very different with regard dot the label 'on-farm distribution'

Table 4. Influence of regional and organic labels on the simulated purchase decision in consumer groups with different degrees of uncertainty.

Products	Degree of uncertainty	Label				
		Origin			Organic	
		On-farm distribution	Region	Bavaria	German Eco-Label	Bavarian Eco-Label
Roast beef	low					
	medium				+	
	high					
Fresh milk	low					
	medium		++			++
	high					
Eggs	low					
	medium			+++		
	high					
Cheese 'Emmentaler'	low					++
	medium					
	high			+++		+++
Potatoes	low				++	++
	medium					+++
	high					

+ significant at the 5%-level; ++ significant at the 1%-level; +++ significant at the 0.1%-level.

which can usually not be found in conventional supermarkets. Therefore it might be rarely known to consumers who do not buy at farmers' shops. A fictitious label like the label 'region' used in this study is completely unknown. Following this argument the influence of the own region purchase argument might be underestimated in this study.

As to the label 'on-farm distribution' another point seems to be important. As already explained, respondents were told to imagine that the products shown on the pictures were available in a separated area in a supermarket. The question arises whether a label on a product is able to communicate on-farm distribution in a supermarket context, where no farmer is present to sell his produce. If this was the case, a label 'on-farm distribution' might not be appropriate to sell regional products in a supermarket.

In the light of potential product specific and label specific influences this research should be continued in order to obtain a more profound view on the phenomenon of uncertainty.

References

Von Alvensleben, R., 1998. Risikowahrnehmung des Verbrauchers: Woraus resultiert die Verunsicherung? BLL-Schriftenreihe, No. 127, 28-43.

Balling, R., 2000. Ergebnisse von Verbraucherbefragungen zur Bedeutung der regionalen Herkunft bei Nahrungsmitteln. In: Agrarspectrum, Regionale Vermarktungssysteme in der Land-, Ernährungs- und Forstwirtschaft, No. 30.

Bergmann, K., 2000. Der verunsicherte Verbraucher, Springer.

Burns, W., P. Slovic, R. Kasperson, J. Kasperson, O. Renn and S. Emani, 1993. Incorporating structural models into research on the social amplification of risk: Implications for theory construction and decision making. Risk Analysis 13, No. 6, 611-623.

Byrne, D., 2001. Risk analysis - Risk communication: The decision maker's dilemma. Joint WHO/FAO/OIE Conference on BSE and its Risks, Paris, 13. June 2001. Source: http://europa.eu.int/comm/dgs/health_consumer/library/speeches/speech105_en.html

Enneking, U., 2004. Willingness-to-pay for safety improvements in the German meat sector: The case of the Q&S label. In European Review of Agricultural Economics 31, 205-223.

Franz, R., 2004. Bestimmungsgründe der Verbraucherverunsicherung und Ansatzpunkte zur Vertrauensbildung durch ökologisch und regional erzeugte Lebensmittel. PhD Thesis TU München-Weihenstephan.

Kasperson, J., R. Kasperson, N. Pidgeon and P. Slovic, 2003. The social amplification of risk: Assessing fifteen years of research and theory. In: Pidgeon, N., R. Kasperson and P. Slovic (eds.). The social amplification of risk. Cambridge, pp. 13-46.

Künast, R., 2002. Die Zukunft der Landwirtschaft mit oder ohne Gentechnik? Berlin. http://www.verbraucherministerin.de/reden/ministerin/2002-07-01-landwirtschaft-ohne-gentechnik.pdf

Lipset, S.M. and W. Schneider, 1983. The decline of confidence in American institutions. Political Science Quarterly 98, 379-402.

Malhotra, N., 1993. Marketing research - an applied orientation. Prentice-Hall.

Sparks, P. and R. Shepherd, 1994. Public perceptions of the potential hazards associated with food production and food consumption. Risk analysis 14, 799-806.

Stoiber, E., 2004. Der Staat handelt und hilft. Verbraucherinitiative Bayern 2001/2002. http://www.bayern.de/Politik/Initiativen/Verbraucher

Willhöft, C., 2001. Verbraucherverunsicherung - eine ausweglose Situation? Vortrag vom 20. Juni 2001, Bundesforschungsanstalt für Ernährung, Karlsruhe.

Consumers' view and awareness of genetically modified food

Antje Wirthgen and Anja Faße
University of Hannover, Institute of Economics in Horticulture, Faculty of Economics and Management, Königsworther Platz 1, 30167 Hannover, Germany; a.wirthgen@gmx.net

Abstract

Since April 2004 European consumers should have the possibility to check, if the offered food is genetically modified. Currently there is not much genetically modified food on the German market, as retailers often refuse these products. This is probably due to previous studies, emphasising that the majority of consumers reject genetically modified food. Against this background the central question of this study is: do German consumers really reject genetically modified food? The findings show that 60% of the interviewees refuse genetically modified food even at lower prices.

Keywords: consumer awareness, genetically modified food

1. Introduction

Several studies show that genetically modified food is very controversially discussed and by European consumers mostly refused (Gaskell, 2000; Eurobarometer, 2000). Since April 2004 European consumers should have the possibility to choose and to check, if the offered food is genetically modified (Werner *et al.*, 2004; Spelsberg, 2004). Genetically modified food (with contamination higher than 0.9%) has to be labelled (Europäische Kommission, 2003; Wissenschaftlerkreis Grüne Gentechnik, 2004). Before that date genetically modified labelling was voluntary and rarely used. Currently there is not much genetically modified food on the German market (Merkur online, assessed 31.03.2005), as German retailers very often refuse these products. This is probably due to previous studies (e.g. GfK), emphasising that roughly 50% of German consumers oppose genetically modified food (Oeko-Test, 2005; Merkur online, assessed 31.03.2005). Against this background the central question of this study is: do German consumers really reject genetically modified food? Is their uncertainty reflected in their purchase behaviour? It is aimed to show, if consumers really search for labelling of genetically modified food and how consumers view genetically modified food compared to conventionally produced food. As differences in the view and awareness of organic purchasers and consumers, who do not buy organic food, are expected, the results are compared for these two consumer segments.

2. Theoretical and methodological background

The underlying theoretical concept for this research is a partial structural consumer behaviour model explaining consumers' purchase decision based on the product perception and resulting product image (see Figure 1: SOR-model (Stimuli-Organism-Reaction)). The product image can be seen as a bundle of single product attributes (multiattribute) with subjective different weightings (e.g. Kroeber-Riel and Weinberg, 1996; Trommsdorff, 1993). Consumers will only consider a specific set of - for them - most important attributes. Generally known and shown in diverse studies is the importance of the attribute 'genetic modification' for the food purchase (Wirthgen *et al.*, 1999). But what do consumers associate with this attribute especially in the

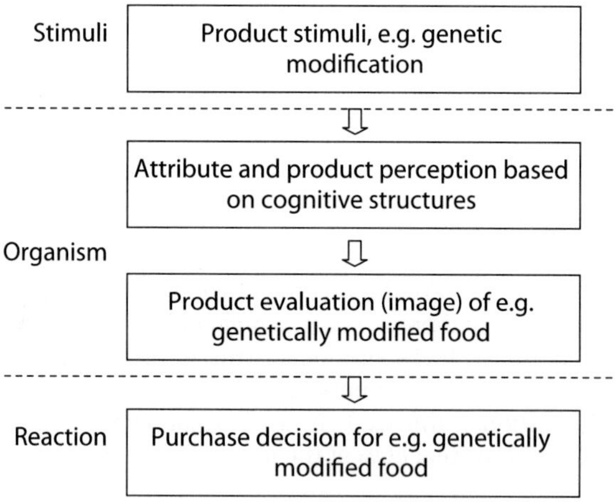

Figure 1. Partial S-O-R purchase decision model for genetically modified food (adapted from Kroeber-Riel and Weinberg, 1996 and Trommsdorff, 1993).

context of food and what is their knowledge, the cognitive structure behind their impression (scheme theory; Kroeber-Riel and Weinberg, 1996)? Consumer perception and knowledge are tied closely (ibid.). The cognitive structure can be presented in terms of semantic nets, showing smaller sense-units of e.g. associations (Kroeber-Riel and Weinberg, 1996).

Against that theoretical background a consumer research was conducted in Hanover in August and September 2004. The survey is based on standardised, face-to-face consumer interviews concentrated on purchasers (n=500 see appendix). Different association tests (e.g. word and serial association, sentence completion test) were applied to examine the perception and cognitive structure influencing the perception of genetic engineering in general and genetic modified food in particular. In this context a multitude of food products were checked in an unaided and aided recall. Furthermore a semantic differential complemented by a rating of selected purchase criteria helped to measure the image of genetic modified food compared to conventional food and the relevance of genetic modification as purchase criterion. Further questions concentrated on the food purchase decision, examining the care and attention to genetic modification, the way of recognising not genetic modified food and finally the willingness to buy depending on price influence.

3. Preliminary results and discussion

With regard to the theoretical concept, the presentation of the results starts with consumers' perception of genetic engineering. At first it was basically asked for the feelings consumers have, thinking of genetic engineering in general. It was aimed to draw a first conclusion of their emotional access to the topic, without any information in advance. The findings show - similar to other studies - that the majority (60%) of the interviewees has a more negative feeling towards genetic engineering in general. Merely 5% of the interviewees indicated a positive feeling, 25% are undecided and 10% did not answer that question. Comparing the group of organic purchasers with non-purchasers the picture gets even more evident, nearly 80% of the organic purchasers have no good feelings towards genetic engineering (compare Figure 2).

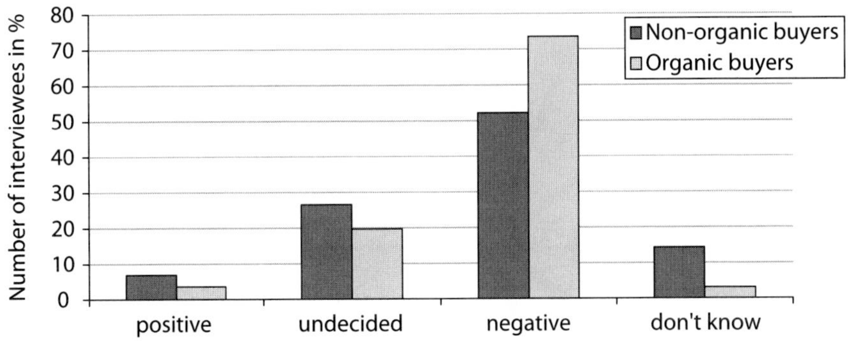

Figure 2. Feelings towards genetic engineering. Question: You know, how it is, you hear a specific word and you get directly a positive or negative feeling. How is that feeling, when you think of genetic engineering?

Deeper consumer insights are given by the results of the association tests. The first tool - a sentence completion test - starting with 'If there would not be any genetic engineering…' presents a slightly more positive picture than the other association tests. Roughly 50% of the statements present arguments contra genetic engineering, unexpected 30% pro and about 20% are neutral. This is surprising considering that only 5% of the interviewees showed a positive basic attitude towards genetic engineering. Quite a lot of people with a negative basic attitude however, seem to think that genetic engineering is necessary and provides essential advantages regarding e.g. medicine, world hunger and plant protection. The arguments contra genetic engineering reflect mainly fear and insecurity especially concerning food.

These arguments pro and contra genetic engineering are confirmed by the findings of the serial association test based on the scheme theory, that was undertaken in two steps. To begin with the first and supposing main association with genetic engineering, a content-analytic categorisation referring to Mayring (2002) was conducted (see Figure 3). In the next step, the second association, that refers to the first term, provides deeper insights which are presented in a semantic net(work), reflecting underlying cognitive structures (see Figure 4). The findings of the quantitative content analysis show clearly that the mainly negative attitudes towards genetic engineering are based on associations like cloning and manipulation as well as the application of genetic engineering in the food sector. The small number of interviewees with positive feelings towards genetic engineering see especially opportunities for progress in medicine and planting (production) (see Figure 3).

Regarding the whole semantic net including the first and the second associations, the cognitive connections between the different association categories become clear (see Figure 4). The main points shall be clarified at three examples: As already be seen in Figure 3 lots of interviewees relate with genetic engineering cloning and manipulation as well as specific foods like maize and soy. The associations cloning and manipulation, go back to the ethical conflict on the one hand and are connected with risks on the other hand (see Figure 4). The terms cloning and manipulation stand for cloned human beings, animal testing and the interference in hereditary material. Regarding specific foods, Figure 4 presents that they relate with specific foods insecurity, consumer protection and politics. Furthermore insecurity seems to be based on a lack of knowledge and presumable health risks. This findings strengthen the hypothesis of Bredahl (1998) and House *et al.* (2004) stating that at least subjective knowledge would increase consumers' acceptance of genetically

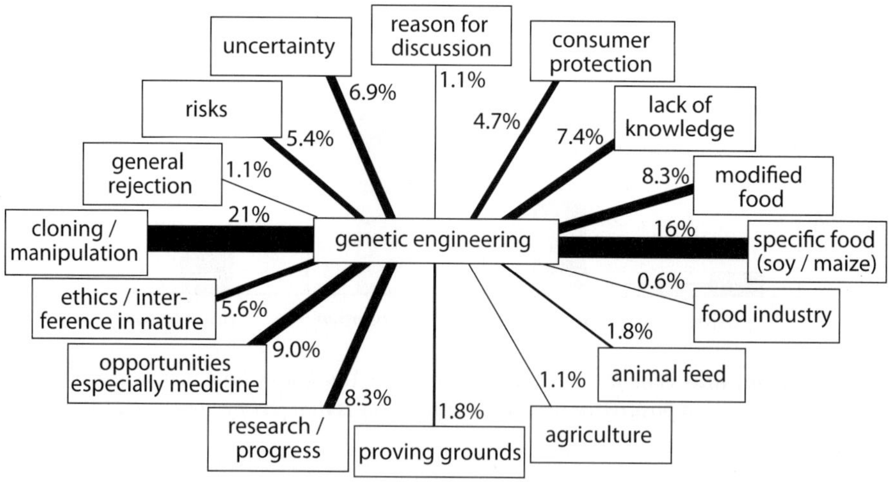

Figure 3. First Associations with genetic engineering. The strength of the lines represents the percentage part of interviewees sharing these associations. Question: What do you associate with the term genetic engineering in general? What comes spontaneously up to your mind? (Only the first association was noted).

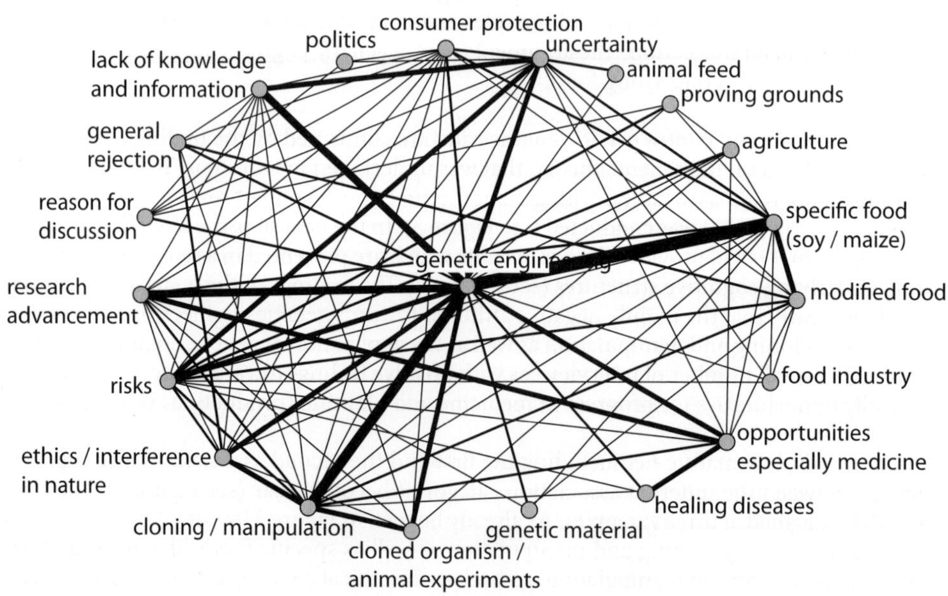

Figure 4. Semantic Net based on serial association: cognitive association structures concerning genetic engineering. The strength of the lines represents the percentage part of interviewees sharing these associations. Question 1: What do you associate with the term genetic engineering in general? What comes spontaneously up to your mind? (Only first association). Question 2: And what do you think of, when you hear … (Second association)?

modified food. Comparing the most negative associations with the most positive associations, the third example regards research and progress. Associations with research and progress are mainly concentrated on opportunities for medicine and plant production. The opportunities for medicine are mostly seen in healing, but also in embryo research and the perfect human being[1]. The opportunities for plant production are traceable by the connection to agriculture, including for instance reduced pesticide use. The environmental benefits of genetically modified food are also stated by Caulder (1998) and Grunert *et al.* (2001).

In the following, the main associations focused on genetic engineering in the context of food or agriculture - so-called green genetic engineering - are pointed out as this paper focuses on consumer's food purchase behaviour (see Figure 5). Compared to figure 3 the food focus becomes evident, although it has to be considered that for this association test several indications were noted. The most often indicated association is the category of specific food products like maize, soy, tomatoes etc. named by more than 80% of the interviewees. An overview over the most named food products gives Figure 6.

The second most often named association, counting still for 50% of the interviewees, is now clearly manipulation and cloning. The next following association of 40% of the interviewees is agriculture, summing up mainly agriculture in general, but also uncontrolled distribution and animal feeding. The next three association categories show the fear and uncertainty, that concerns especially the application of genetic engineering in the food sector: Uncertainty, lack of information, health risks like allergies and resistance against antibiotics as well as artificial modified food. They all account for about 14-17% of the interviewees. The results of about all together 50% of the interviewees fearing genetic modification of food products gets confirmed by Kuckartz and Rheingans-Heintze (2004) who did a representative consumer research about the environmental consciousness in Germany. The other aspects like ethic and the main advantages

[1] Both aspects are included in the associations of cloning and manipulation.

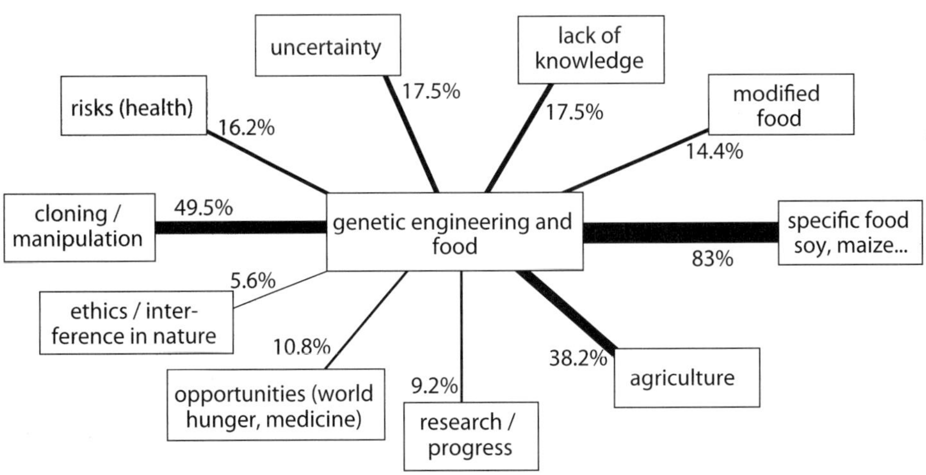

Figure 5. Associations with genetic engineering focusing food. The strength of the lines represents the percentage part of interviewees sharing this associations (several associations are considered). Question: What do you think of genetic engineering in the context of food?

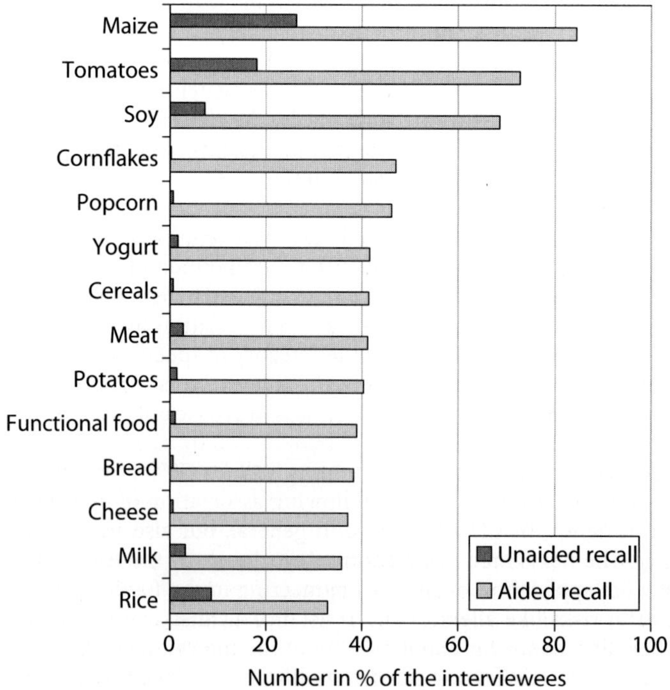

Figure 6. Unaided and aided recall of associations with genetic engineering and specific food products. Soy products are presented by a mean-value, the standard deviation is relative low. Products with an aided recall < 30% were skipped (chocolate, eggs, jelly babies, baby food, noodles). Question: Part 1 - What do you think of genetic engineering in the context of food? Part 2 - And what about these products - do you associate these products with genetic engineering? What about milk? Etc. (only food products that were not named unaided were asked in this aided recall).

of genetic engineering got in Figure 5 similar scores like in Figure 3. However, in the context of food, some interviewees mentioned additionally the opportunity to satisfy the growing world food demand. This was only stated by a small number of interviewees which confirms the results of the Food Standards Agency in 2003.

Aiming at examining how far the objections towards genetic engineering are reflected by consumers' food purchase behaviour, the next Figure 6 shows which specific food products are the most connected ones with genetic engineering (see Figure 6).

Considerable is in particular, that all selected food products are at least for 25% of the interviewees in the aided recall associated with genetic engineering; for maize these are even 85% and for tomatoes and soy products about 70% of the interviewees. Compared to the unaided recall, the same products score for the main associations. Furthermore it is remarkable that cornflakes and popcorn, as two obvious maize products, are nearly only named in the aided recall and thereby only of roughly 45% of the interviewees. This confirms the hypotheses that quite a lot of consumers do not realise that the well known and with genetic modification associated products are often also included in further processed food.

The association analysis showed already that quite some interviewees refuse genetically modified food as being artificial, risky and unhealthy. The following figure presents the image of genetically modified food compared to conventionally produced food measured by a semantic differential (Figure 7).

Genetically modified food seems to be in particular significantly less natural, less ethical tenable, less safe, more risky, less healthy as well as more harmful to the environment and less appropriate to the species. The interviewees did distinguish less, but still significant (t-test at 5% significant level) in the look, price, storability, control and taste of genetically modified and not modified food. Comparing the view of organic buyers and non-organic buyers, overall, non-organic buyers regard both genetically modified and conventional food a little better than organic buyers (see Figure 7). The distinguishing factor was the purchase of organic food in the last two weeks.

The next question is, how much weight the distinguishing attributes for the food purchase. Table 1 assesses the relative importance of selected attributes as purchase criterion for the food purchase. This question was examined at the example of maize and cornflakes, giving the opportunity to compare the importance of genetically modification for maize as one product with strong association with genetic modification and cornflakes as the processed product with less association (compare Figure 6). Like expected, taste is the most important criterion in both cases. For these two products 'no genetic modification' scores on the second place and thus a little higher than usual, higher than product freshness. This might be due to these two sensitive products as well as to the sensitisation during the interview. Comparing the ratings for maize and cornflakes, 'no genetic modification' is a little more important for maize than for cornflakes. However, overall most of the selected attributes score a little higher for the purchase of maize than of cornflakes.

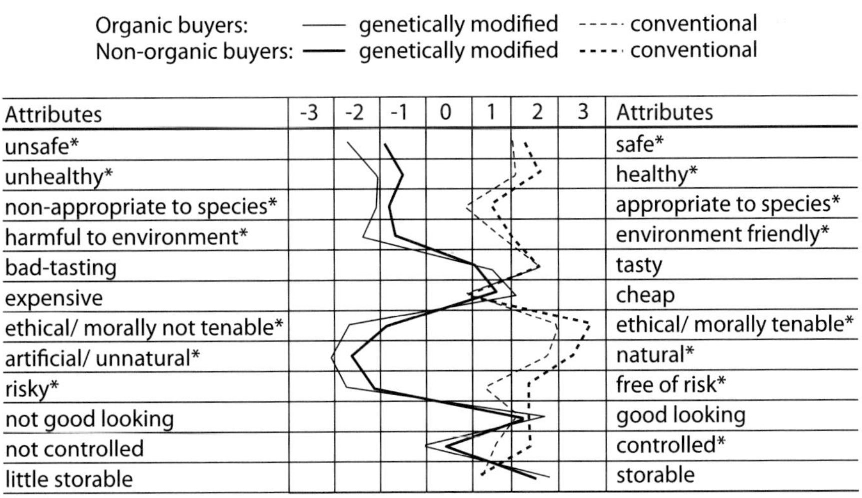

Figure 7. Image of genetically modified food compared to conventional food. Question: I name some attributes and you tell me please, how far - in your opinion - these attributes match conventional not genetically modified food. And now, the same again for genetically modified food, please.
**Significant differences between means of organic and non-organic purchasers at 5% significant level (t-test).*

Table 1. Relative importance of genetic modification as purchase criterion.

Purchase criteria	Maize in tins (means)	Cornflakes (means)
Taste*	1.35	1.26
Free of genetic engineering*	1.85	1.94
Freshness	1.90	1.96
Price*	2.34	2.50
Health*	2.48	2.67
Signet of control*	2.63	2.83
Organic production*	2.93	3.14
Regional production*	3.33	3.62

Scaling: 1=very important, 2=important, 3=neither nor, 4=unimportant, 5=very unimportant
* = significant differences between the means of maize and cornflakes at 5% significant level (t-test).
Question: How important are the following criteria for your purchase of maize in tins? ... And how important are these criteria for your purchase of cornflakes?

Like expected, genetic modification is more important than organic or regional production, even than the price. The importance of the regional aspect differs highly depending on the region. There is one example in Germany, the region Reutlingen, where the regional aspect scores higher than genetic modification (Evangelischer Pressedienst, 2005). But this is an exception, usually other studies confirm the ranking in Table 1 (e.g. Wirthgen *et al.*, 1999). Comparing organic buyers and non-organic buyers, in particular the price falls back, the rest of the ranking stays the same.

Against the background that genetic modification is one of the most important purchase criteria and maize is the most associated food product concerning genetic modification, it is no wonder that maize is also the leading product in the list of food products, of which consumers pay attention when purchasing food (see Table 2). In this context, the processed maize products cornflakes and popcorn rate quite a lot behind maize - similar to the ranking of the association test (compare Figure 6). Surprising - on the other hand - is that 50% and more of the interviewees stated to pay attention to genetic modification purchasing meat and milk products. Considerable are also the relatively high ratings of milk, baby food and noodles in the unaided recall compared to the aided recall.

However, regarding the matter that consumers seem to be uncertain and consider genetic modification as one of the most important purchase criteria, the overall percentage part of interviewees paying attention to genetic modification when purchasing food is low - in particular considering the unaided recall (see Table 2). Only 25% of the interviewees stated in the unaided recall that they pay attention when purchasing maize and nearly 20% when buying tomatoes. Most other products are mentioned by less than 10% of the interviewees. But this is not, because consumers only distrust maize and tomatoes. The right column of table 2 shows that nearly none of the interviewees knows food products, of which they are sure, that these products are produced without genetic modification. Eggs followed by milk seem to be the most trustable products. Although, in the unaided recall only about 5% of the interviewees stated to be sure that these products are free of genetic modification.

Table 2. Genetic modification and consumers' care with specific food products. Question 1: For which food products do you care about genetic modification? Question 2: For which food products you are sure, that they are not genetically modified?

Food	Pay attention (% of the interviewees)		Sure genetically modified free (% of the interviewees)
	aided recall	unaided recall	unaided recall
Maize	65.79	25.43	0.21
Tomatoes	61.40	18.42	1.04
Meat	56.14	14.91	2.92
Yoghurt	56.14	11.40	1.04
Soy	53.51	7.31	0.34
Cornflakes	52.63	8.77	0.21
Milk	50.00	13.16	3.96
Bread	50.00	7.02	1.25
Cereals	48.25	7.89	0.21
Potatoes	48.25	7.89	1.25
Eggs	42.98	7.89	5.00
Chocolate	42.11	3.51	0.83
Cheese	40.35	5.26	1.67
Functional food	37.72	1.75	0.63
Popcorn	35.08	1.75	0.21
Rice	35.08	0.88	1.04
Jelly babies	32.45	0.00	0.42
Baby food	30.70	7.89	2.71
Noodles	30.70	4.39	0.83

Up to now, the findings showed that consumers associate with genetic modification in the food sector mostly fears, uncertainty and health risks. They are especially sensitive with specific food products like maize, tomatoes and soy products. Genetically modified food has generally a negative image and the attribute genetic modification is assessed as one of the most important purchase criteria. Nearly 70% of the interviewees state, that they pay attention purchasing maize while only about 50% care when purchasing cornflakes or soy products. No considerable number of interviewees feel sure about any food product to be free of genetic modification. But how do they pay attention, if they care? How do people recognise not genetically modified food?

In this context only 35% of the interviewees stated they would at least for some products really look for a kind of labeling. If consumers try to figure out if the offered food is genetically modified they mainly look for the ingredient list or for specific labels. Roughly 15% rely on organic food or - but these are only 2% - trust in specific outlets. Like expected the latter consumer group contents of significantly more organic buyers. Further 3% are sure, there is no genetically modified food on the German market and 17% seem to rely on the obligation for labeling and believe, not marked products are free of genetic modification. Only 15% of the interviewees did not yet try to figure out genetically modified food and another 20% do not know how to cope with that problem (see Figure 8).

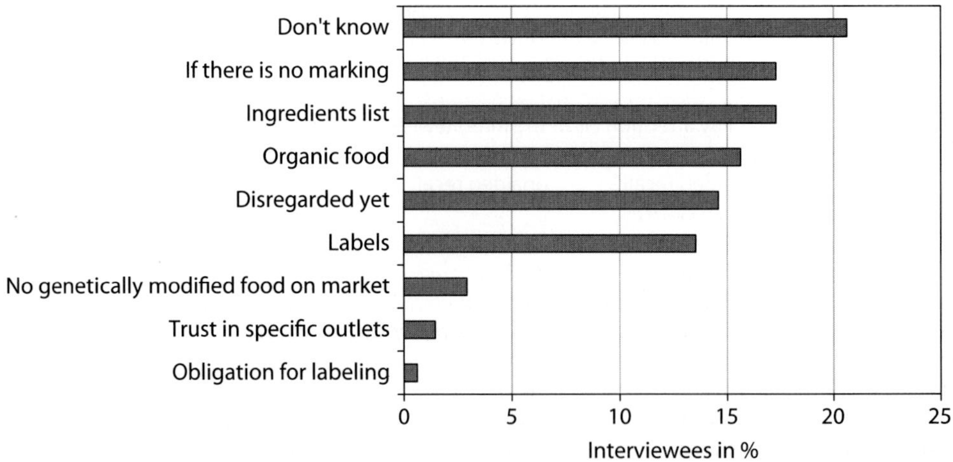

Figure 8. Recognition of not genetically modified food. Question: How do recognise not genetically modified food?

The last and main purchase question, if consumers would buy genetically modified food, in case it is labeled, was asked by means of statements differentiating the buying probability and considering a possible price influence. The results show that more than 60% of the interviewees oppose genetically modified food even at lower prices (see Table 3). On the other hand the price context revealed that unexpected 3% of the interviewees would even buy genetically modified food in case these products are more expensive than conventional food. Another roughly 35% of the interviewees represent market potential depending on the price relation of genetic modified and not modified food. This number includes 14% of the interviewees who would potentially buy GM food in case of lower prices than conventional food. The other 21% of the interviewees could be potential buyers of genetically modified food, in case conventional ones get more expensive.

Table 3. Willingness to buy genetically modified food.

Would you buy genetically modified food, in case these products are labeled?	% of interviewees confirming the statement
Yes in any case, even if they were more expensive than conventional products.	3.2
Yes, but only if they are cheaper.	5.7
Yes probably, if they are cheaper.	2.5
Yes may be, if they are cheaper.	5.9
Yes but only, if conventionally produced ones get more expensive.	5.7
Yes probably, if conventionally produced ones get more expensive.	2.3
Yes may be, if conventionally produced ones get more expensive.	13.1
No, I would never buy genetically modified food.	61.5

Every interviewee could choose only one statement.

Compared to Kuckartz and Rheingans-Heintze (2004) the general tendency of these figures seems realistic. Regarding the results of Gath and Von Alvensleben (1998) with only 12% market potential, the rejection of genetically modified food found in this contribution seems to be low. The difference is probably due to the price impact considered in the underlying study. Overall, all survey figures concerning willingness to buy and thereby potential social wished behaviour have to be interpreted carefully.

4. Summary and conclusions

Since April 2004 German consumers should have the possibility to check if the offered food is genetically modified. Currently there is not much genetically modified food on the German market, as retailers often refuse these products. This is probably due to previous studies, emphasising that the majority of consumers reject genetically modified food. Against this background the central question of this study is: Do German consumers really reject genetically modified food? The conducted consumer research was based on a partial structural consumer behaviour model explaining consumers' purchase decision based on the product perception and resulting product image. The research was undertaken by means of a standardised questionnaire and analysed by qualitative and quantitative methods. The main focus lied on a deep perception analysis showing that consumers associate with genetic engineering in the food sector specific food products and thereby health risks and uncertainty, amongst others due to a lack of information They are especially sensitive with specific food products like maize, tomatoes and soy products. Currently there are significant differences between the care about these products and the related processed food, like Cornflakes or Popcorn. But overall, no considerable number of interviewees feels sure about any food product to be free of genetic modification. Genetically modified food has generally a negative image and the attribute genetic modification is assessed as one of the most important purchase criteria. Last but not least the findings show that more than 60% of the interviewees refuse genetically modified food even at lower prices. Nearly 40% represent market potential, but mostly only in case genetically modified food is offered at lower prices than not modified food.

Overall, drawing conclusions, the rejection of GM food is strong, holds for the majority even at lower prices for conventional food. Considering the high uncertainty, in the near future, in Germany a growing demand for genetically modified food should not be expected.

References

Bredahl, L., 1998. Consumers' cognitions with regard to genetically modified foods: Results of a qualitative study in four countries. MAPP working paper, No. 59, Aarhus, Center of Research and on Customer Relations in the Food Sector, Denmark.

Caulder, J., 1998. Agricultural biotechnology and public perceptions. AgBioForum 1, 38-39.

Eurobarometer, 2000. The Europeans and biotechnology. Eurobarometer 52.1 http://europa.eu.int/comm/research/pdf/eurobarometer-en.pdf.

Europäische Kommission, 2003. Verordnung (EG) Nr. 1830/2003 des Europäischen Parlaments und Rates vom 22. September 2003 über die Rückverfolgbarkeit und Kennzeichnung von genetisch veränderten Organismen und über die Rückverfolgbarkeit von aus genetisch veränderten Organismen hergestellten Lebensmitteln und Futtermitteln sowie zur Änderung der Richtlinie 2001/18/EG.

Evangelischer Pressedienst, 2005. Umfrage zeigt: Keine Liebe zur Gentechnik. Internet News of 1st of February www.epd.de.

Food Standards Agency, 2003. Consumer views of GM food. London. Food Standard Agency (eds.).

Gaskell, G., 2000. Agricultural biotechnology and public attitudes in the European Union. AgBioForum 3, No. 2&3, 87-96.

Gath, M. and R. von Alvensleben, 1998. The potential effects of labeling GM foods on the consumer decisions: Preliminary results of conjoint measurement experiments in Germany. Air-Cat 5th Plenary meeting: Effective Communication and GM Foods. Vol. 4, No. 3, As, Norwegen, 18-28.

Grunert, K.G., L. Lahteenmaki, N.A. Niellsen, J. B. Poulsen, O. Ueland and A. Astrom, 2001. Consumer perception of food products involving genetic modification: Results from a qualitative study in four nordic countries. Food quality and preference 12, 527-542.

House, L., J. Lusk, S. Jaeger, B. W. Traill, M. Moore, C. Valli, B. Morrow and W.M.S. Yee, 2004. Objective and subjective knowledge: Impacts on consumer demand for genetically modified foods in the United States and the European Union. AgBioForum 7, 113-123.

Mayring, P., 2002. Einführung in die Qualitative Sozialforschung. Beltz.

Merkur online, 2005. Gentechnik erreicht Supermärkte: Nachweis fällt oft schwer. www.merkur.de.

Kroeber-Riel, W. and P. Weinberg, 1996. Konsumentenverhalten.6th ed., Vahlen.

Kukartz, U. and A. Rheingans-Heintze, 2004. Umweltbewusstsein in Deutschland 2004. In Bundesministerium für Umwelt, Naturschutz und Reaktorsicherheit (BMU) (eds.), Köllen Druck.

Oeko-Test Verlag GmbH, 2005. Gentechnik in Lebensmitteln: Das darf doch nicht die Wahl sein. Oeko-Test April 2005.

Spelsberg, G., 2004. Kennzeichnung gentechnisch veränderter Lebensmittel. Transgen Wissenschaftskommission Aachen, 1-2.

Trommsdorff, V., 1993. Konsumentenverhalten. 2nd ed., Kohlhammer.

Werner, A., B. Kniel and U. Berg, 2004. Die neue Gentechnik-Kennzeichnung und Rückverfolgbarkeit - Sind diese Regelungen in der Praxis anwendbar? Deutsche Lebensmittel-Rundschau 100, No. 5, 1-12.

Wirthgen, B., H. Kuhnert, M. Altmann, J. Osterloh and A. Wirthgen, 1999. Die regionale Herkunft von Lebensmitteln und ihre Bedeutung für die Einkaufsentscheidung der Verbraucher. In Bundesministerium für Ernährung, Landwirtschaft und Forsten (BMELF) (eds.), Berichte über Landwirtschaft 77, 243-261.

Wissenschaftlerkreis Grüne Gentechnik (WGG) e.V., 2004. Mehr Durchblick für den Verbraucher!? Was sich durch die neue Kennzeichnungsverordnung ändert, 1-2.

Appendix 1. Survey statistics

Table A1. Sample Characteristics (n=500).

		total % of interviewees
Age (quota)	18-34 years	30
	35-60 years	40
	>60 years	30
Sex (quota)	female	70
	male	30
Education	school without matriculation	11
	matriculation	8
	apprenticeship	52
	study degree (e.g., B.Sc., M.Sc.)	29
Average household size	2.2 (mean)	

Information releases concerning food scares and their impact on the aggregate demand: using artificial consumer societies

Volker Saggau
Christian-Albrechts-University Kiel, Department of Agricultural Economics, Olshausenstrasse 40, 24098 Kiel, Germany; vsaggau@agric-econ.uni-kiel.de

Abstract

The motivation of this paper is lying on the questions 'How does a food scare and information of a food scare influence the buying decision of one single agent and the changes in the aggregate demand?' and 'How can we evaluate effective risk communication strategies?'. Since we investigate an artificial society of consumers which have their own decision functions, we can observe how new information could influence the behaviour of each consumer and more interesting the aggregate changes in the demand by creating a population of agents. This multi-agent simulation can be used to investigate how different information releases and decision functions influence the aggregate demand, i.e. the benefits of a risk communication strategy can be evaluated with this multi agent approach.

Keywords: Multiagent simulation, Consumer response, Information diffusion, Risk communication.

1. Introduction

The relation between trust, information and demand is a research field, which has been investigated in many different ways. Our approach is to use an agent based simulation model. We want to investigate how heterogeneous interacting agents process information and change attitude and trust by communicating with related agents.

We concentrate on two kinds of behaviour. One is to get information from the networks in which the agent is present. The agent processes this information. The other is to see how related agents behave in terms of deciding on consuming a food item or not and to take this behaviour into account for the own decision.

Our intention is to measure how different information strategies influence the aggregate demand. For this purpose we use a multi-agent method in order to follow a bottom up approach where each agent acts individually. The interaction between the agents leads to an emergence of an aggregate demand that comes from the bottom up. Each agent follows its internal updating and decision algorithms so that on the aggregate level the demand changes according to the outcomes of the interaction and its related updating processes. After the communication phases the aggregation of the outcomes of each agent shows the result of the information strategy that was selected.

2. The model

The modelling starts with the construction of an economy which consists of an initial population of agents. There are primary agents like consumers and there are auxiliary agents which often represent different social and model environmental objects, e.g. media agents.

First the initial state of the economy has to be specified, i.e. the agents will be equipped with their initial attributes. These attributes are type characteristics, internal behavioural norms and internal information about themselves and other agents like the network connections.

The development of the economy happens in time by interaction and updating of internal states of the agents, i.e. updating the trust value for a specific good or supplier. The simulation model adequately reflects the development of trust respectively distrust within a population of consumer agents after a food safety incident happened.

2.1. The networks and the information sources

The model consists of several networks which serve as information sources for the agents (Saggau and Patelli, 2004: 2f). We have decided to implement decentralised and centralised networks. Both types of networks serve as information sources for the agents. The decentralised networks are social networks which again are several networks like the demographic network or friend's networks. These networks are endogenous, i.e. the information processing is endogenous whereas the centralised networks are exogenous. They spread the information from one single information source via the population of agents.

An initial population of consumer agents found the basis for the simulation. They are connected to the networks. They communicate information about the safety of food and their trust regarding this food item.

We differentiate two respectively three kinds of social networks. Each network serves as an information source for the agents. The networks are differentiated by unique identifiers so that the information sources are well known to the agents.

In the demographic network each agent of the population finds its place in a family structure. The demographic network consists of three generations: the grandparent generation, the parent generation and the children generation. The agents are consistently assigned to this network, i.e. the family relations are kept, and there are no inconsistent states (see Figure 1).

The demographic network can be varied by changing three parameters: the population size, the average number of offspring in the second generation and the average number of offspring in the third generation.

The population size determines the size of the adjacency matrix which will be used as the connections matrix. This connections matrix is quadratic and the size corresponds to the number of agents in the population. This matrix will initially be filled with zeros. The zeros change to one when the generations will be upset and only if the connections are determined. During this phase the initial parameters for the agents will be set.

The size of the first generation results from the quotient of the population divided by the product of the average number of the 2nd and 3rd generation. The size of the second generation is the

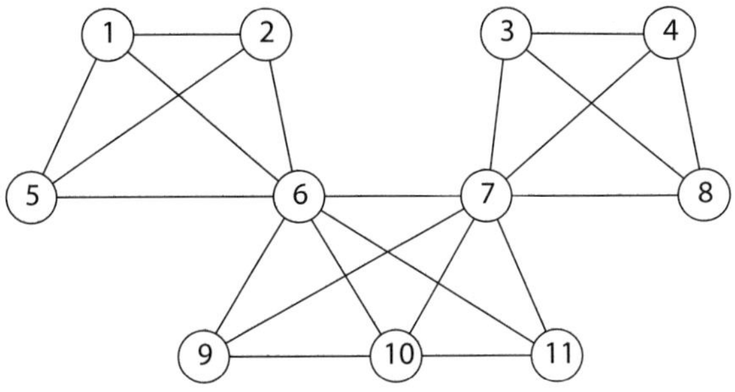

Figure 1. Demographic network.

product of the first generation and the average number of children per family in the second generation while the size of the third generation is the result of the subtraction of the population size minus the size of the first generation and minus the size of the second generation.

The 'marriage' in the first generation will be done by an algorithm which overwrites the zeros in the connections matrix by a one if two partners are found. In the next step another algorithm generates the offspring of the first generation and connects this second generation with the first generation. The marriage in the second generation is a little bit more complicated because it has to be excluded that there are no brothers and sisters marrying each other. Finally the grandchildren have to be included into the family structure, i.e. the third generation will be assigned to the second generation.

All these connections will be done in one adjacency matrix by overwriting the zeros by ones. In this way the demographic network will be set up, in parallel the agents will be initialised and assigned to the network.

Besides a family network each agent can also be part of a friend's network. These are random networks that are not determined like the family network, i.e. each agent randomly receives a variable number of links which connects him to agents that are no family members. The members of these social networks are again the members of the agent population, which already form the family (demographic) network. The connection to other agents can be chosen differently. It's possible to choose between different distributions of the links to the friends.

The uniform distribution for example sets up a connection matrix where the number of connections results from the product of the average connections and the population size. An algorithm then sets up the connections matrix according to the uniform distribution. Figure 2 shows a possible structure of this network.

Agents can be related in groups respectively clusters, where the nodes or agents are interconnected in a bidirectional way. Some agents have links to agents which are outside of the cluster but again in another cluster. In this way the friend's networks can be created.

Colleague's networks are technically similar, but they have a different weight in the information processing of the single agent, depending on the weight of the network. Both of these networks have a unique identifier so that the agents exactly know where their information comes from.

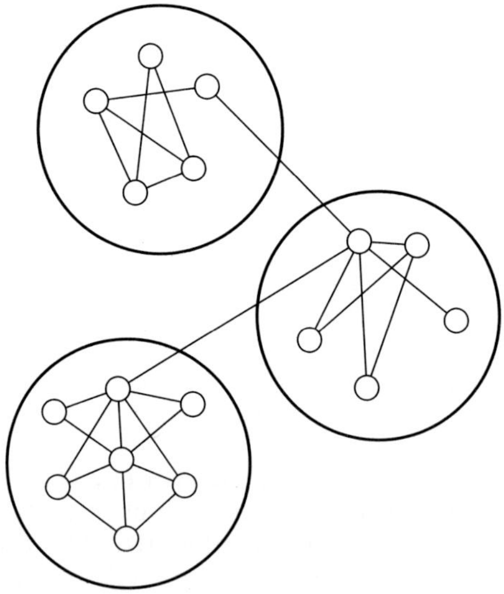

Figure 2. Social networks.

The centralised networks can be the shops, the media, the government etc. (see Figure 3).

We differentiate between global centralised and segment centralised networks. In the global centralised networks everybody receives a message which is released by these information sources. Whereas the segment centralised networks are just segment specific and segment exclusive. The social networks are fully decentralised, i.e. only the members of the single social networks will be informed. The intensity of the information received from the different networks is also different, depending on the information type, the information source (the weight of the network) and other

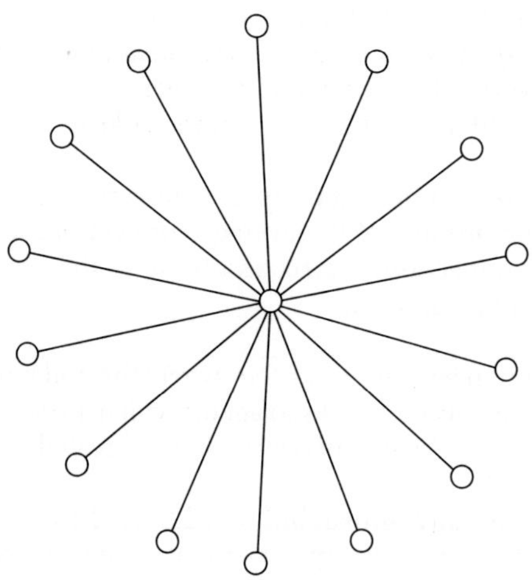

Figure 3. Centralised networks.

environmental influences. Media agents are newspapers, television, internet and radio. Television and Internet are global centralised networks whereas newspapers and radio can be either global or local or segmented depending on the definition respectively the declaration by the user.

The structure of the centralised network is different from the other networks. One reason is that it is responsible for the initial information release which will be diffused via the agent population and processed by each agent. Since we look at information which is related to food scares we have to consider negative and positive information. Bounded rationality is an aspect which has to be taken into account when looking at consumers which have to take decisions. The Prospect Theory of Kahneman and Tversky (1979) refers to that issue. Consumers evaluate negative information relative to a reference point higher than positive information, i.e. negative information has a higher weight than positive information. This point has to be taken into account with respect to information releases by centralised information sources.

Each centralised network also has a unique identifier. The information release distribution is a time related distribution - one iteration is one day. The user can chose between different media sources. Depending on the media source the data from the survey will be loaded into the text field. For these global media sources (they reach everybody) the data will be the same but of course different for the different media sources.

The segment centralised networks refer to specific segments of the population. The data analysis of the Trust-survey (Cavicchi *et al.*, 2005) identified socio-demographic and personality segments which determine differences among consumers' attitudes towards specific food risks and with respect to the trust they place upon various information sources. Four different consumer segments (4-segment case) were identified: 'the conservatives', 'the young intellectuals', 'the well-off' and 'the stop-workers'. For these four segments a cross country comparison was made, i.e. the five EU-countries (United Kingdom, Italy, Netherlands, France and Germany) have different sizes of the segments. Another segmentation analysis with special emphasis on trust leads to three different consumer segments. The simulation can be load either with the 4-segment case or the 3-segment case or can be freely equipped with up to six segments.

The population of consumer agents will be assigned to the segments according to the respective percentage distribution. The simulation starts to assign the agents of the population to the first segment then to the second segment on so on. If for example the percentage number for the first segment is 50 and the percentage number for the second segment is 51 then there are just two segments, i.e. the first segment with 50% of the population and the second segment with the remaining 50% of the population, the rest will be cut. And even if the other text fields are filled with number they won't be applied, i.e. if the hundred percent is assigned then the population is also fully assigned to the segments.

2.2. The agents

'An agent is a system that tries to fulfil a set of goals in a complex, dynamic environment. An agent is situated in the environment: it can sense the environment through its sensors and act upon the environment using its actuators.' (Maes, 1994: 2).

Starting point is an agent who maximises his expected utility (Boecker and Hanf, 2000). It is assumed that his purchase decision depends on four parameters: the utility from a safe unit of a certain product (UX+), the subjective probability to purchase a hazardous unit of that product

(PG)[1], the subjectively presumed disutility from consuming a hazardous unit of that product (UX-), and the expected utility from consuming a substitute which is perceived to be safe (UY). He only buys and consumes the product if its expected utility under consideration of the possible disutility is higher than the expected utility of the substitute (UY < UX+).

Further, the artificial consumer agent responds to information about risk and safety of the product. Every piece of positive or negative information changes the subjective probability that he relates to purchasing a hazardous unit of this product. For this updating of the prior subjective probability, the Bayesian updating is employed. Updating (trust) mechanism (revised trust in supplier J):

- Negative information:

$$P_{PJ} = \frac{P_J P(G|A)}{P_J P(G|A) + (1 - P_J) P(G|B)} \tag{1}$$

- Positive information:

$$P_{PJ} = \frac{P_J (1 - P(G|A))}{P_J (1 - P(G|A)) + (1 - P_J)(1 - P(G|B)} \tag{2}$$

Each agent is part in at least one network, the demographic network, but can also be part in other networks. The agent is registered in the networks where it belongs to. It can go through the networks and ask for information. Additional to the basic model there is the possibility to 'see' how other agents behave in risky situations, i.e. the trust value can be communicated which is a proxy for the demand. The third step is optional.

The agents have internal updating algorithms. These algorithms aim at the information on the one hand side and at the decisions taken by the related agents in the networks on the other hand side. In each step of the simulation run the basic agents' internal step method will be invoked. This method goes through the list of related agents, looks up what the trust value is, evaluates it and it also goes through the list of information sources in order to get information and to update its own information state.

One iteration/day:

- Step 1 - Information collection and processing: The agents collect in each time step information from their neighbours, i.e. from the decentralised information sources and also the agents collect information from the centralised information sources (media, government, shops, etc.) (see Figure 4).
- Step 2 - Bayesian updating: After the information collection the agents update their P_J according to Bayesian updating (see above). The old P_J enters into the equation and revises P_J which then again is the next value which enters in the following updating and so on.
- Step 3 - Trust communication (P_J): The third step is that the agents collect the P_J s from their related agents and aggregate the value according to an aggregation rule (mean, max, min) (see Figure 5). The model offers the possibility to assign different trust values to the four identified consumer segments. Together with the standard deviations of the segmented trust value the system assigns to the single agents normal distributed trust values around the mean value of the respective segment.

[1] With $P_G = P_J P(G|A) + (1\text{-}P_J) P(G|B)$ and $P(G|A) < P(G|B)$ *and* $P(G|A) + P(G|B) < 1$.

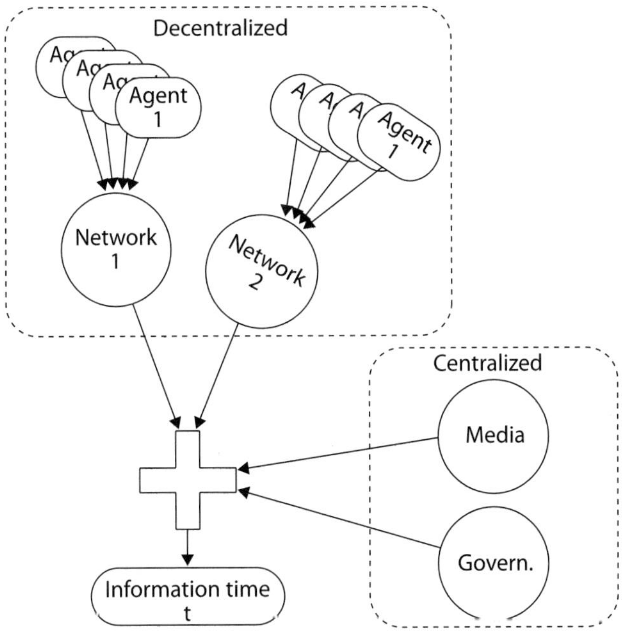

Figure 4. Information collection.

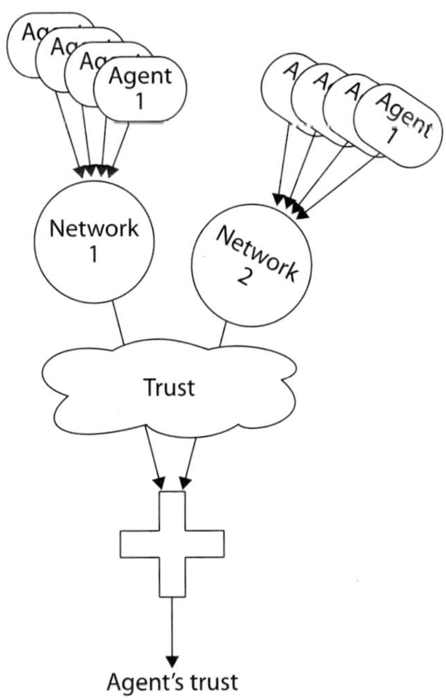

Figure 5. Trust collection.

3. Micro data of the simulation model

The information sources are different. One has to differentiate between the type of the network, the media source (credibility and influence) and the intensity of the information release and the range of the information release. Information sources in centralised networks like media have far more influence regarding these aspects than a single information source in a single social network.

3.1. Aggregation rule for getting weights

The weight of the network can be interpreted as a placeholder for the importance of this information source for the agents, i.e. the higher the weight is, the higher is the influence of this information source regarding the trust building of the agent. Radio has for example can have a higher weight than a local newspaper. The centralised information source can also be a shopkeeper; in this case the empirical data from the survey regarding trust of different information sources plays a crucial role, i.e. if the butcher is more trustful than the newspaper, then this weight value should be higher for the shopkeeper agent (e.g. Media 2).

The aggregation rule for getting weights is related to the collection of information values from the information sources. Each agent updates its information status randomly in the time horizon of the simulation run. The updating of the information status must result in the aggregation of only one information value. Based on this information value, i.e. positive or negative information value, the trust value *Pj* will be updated according to Bayesian updating - more trust or less trust compared to the previous trust value. For this reason, i.e. forming a unique opinion (*newInfoValue*) based on the information values which came from the information sources, the agent has to aggregate and weight the information values of the information sources: Each information source *i* (family, friends, media etc.) has its own information status and a network weight and are registered in the connection list of the respective agent. Both values will be asked by the updating agent. The information values and the network weights will then be multiplied and summarised ($\Sigma_i(inFoValues_i * weight_i)$). Additionally the network weights will be summarised ($\Sigma_i(weight_i)$). Finally the new information value (*newInfoValue*) will be computed (see Equation 3):

$$newInfoValue = \frac{\Sigma_i(inFoValues_i * weight_i)}{\Sigma_i(weight_i)} \quad (3)$$

The trust value will then be revised based on *newInfoValue*, i.e. if *newInfoValue* < 0 then the trust value will be decreased and vice versa.

3.2. Intensity of information releases

The simulation software works with numerical values hence the information regarding the safety of food has to be a numerical value. The reason for this technical issue lies in the nature of the decision for the agent when to increase the trust value or when to decrease the trust value: there are two equations for the Bayesian updating; the decision criterion between these forms of the revision of trust is based on the updated and aggregated information value. The range for the information values released by the centralised media agents on the one hand side (exogenously) and the agents in the networks on the other hand side (endogenously) is between -1 for negative information and 1 for positive information. These are the highest values for both directions. The lesser the values for both directions or say the absolute value, the lesser is the intensity of the information release. This is a second kind of weighting besides the weighting of the information source.

The intensity of the information can also be set explicitly. If a centralised information source like the shopkeeper should release information then the intensity of the information is probably less than information release from the television, e.g. for positive information - shopkeeper 0.5 and television 0.8 or another example within the same centralised information source like the television: 0.6 for the news reportage at the prime time and 0.3 for the news reportage in the afternoon.

With this variety of information release possibilities there can be predefined empirical information distributions and there can also be produced information strategies which can be tested with respect to the aggregate demand.

4. Information strategies (information policies)

One question in the beginning was which impact do different information strategies have on the aggregate demand, i.e. is it possible to evaluate risk communication strategies? Since we investigate food scares and the corresponding changes in the demand, we focus on information strategies starting with negative information releases regarding the food item under investigation followed by positive information releases by stakeholders using the media to spread information.

The distribution of the information releases by the centralised media can be chosen by the user of the simulation. Consider an example where centralised media agents spread negative information over the network respectively the population in the sense of observing a product failure regarding the food item under investigation, so that each agent receives this negative information signal. When time goes on, here the iterations of the communications steps, the intensity of the information release decreases. It follows an exponential distribution. In a certain point in time, we call it breakpoint, the information release changes from negative to positive information, e.g. it was discovered that the food is safe. It also follows an exponential distribution, beginning with a high intensity and decreasing intensity when time continues (see Figure 6).

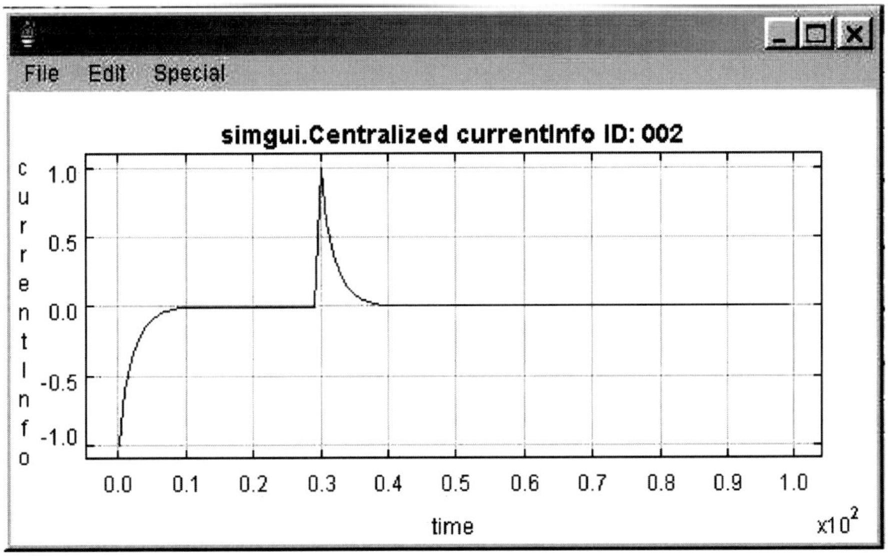

Figure 6. Information release strategy by one media agent.

The information release strategy was selected and parameterised by the user and is visualised by the simulation in an output graph.

This information release distribution starts with an intensity of -1, i.e. a very strong negative news reportage regarding a food scare. The time scale is the x-value times 102, e.g. 0.1*102=10 which means at 0.1 on the x-axis the information value (intensity) at day 10 can be derived from the y-axis. In this example at day 10 this information source respectively the centralised media agent does not longer report about this food safety issue whether negative nor positive. At day 30 (set by the user in the inversion text field) this media agent reports that the food under investigation is safe, i.e. positive information again with strong information intensity (may it be in the television at prime time).

The agents update their own trust value based on the new information and if selected also on the basis of the aggregated trust of its related agents.

The consumer agents update their trust value from time to time randomly; they are equipped with the starting parameters heterogeneously. In the example above there was an initial population of 100 agents which are represented each by a single coloured line.

On the aggregated level the development of the average trust value of the population emerge by taking the new information release into account on the micro level and communicating and updating accordingly (see Figure 7).

This double exponential distribution of information releases is just one possibility, there may be others. The user of the simulation can implement self created information distributions and can also test information strategies by directing specific consumer segments with information policies.

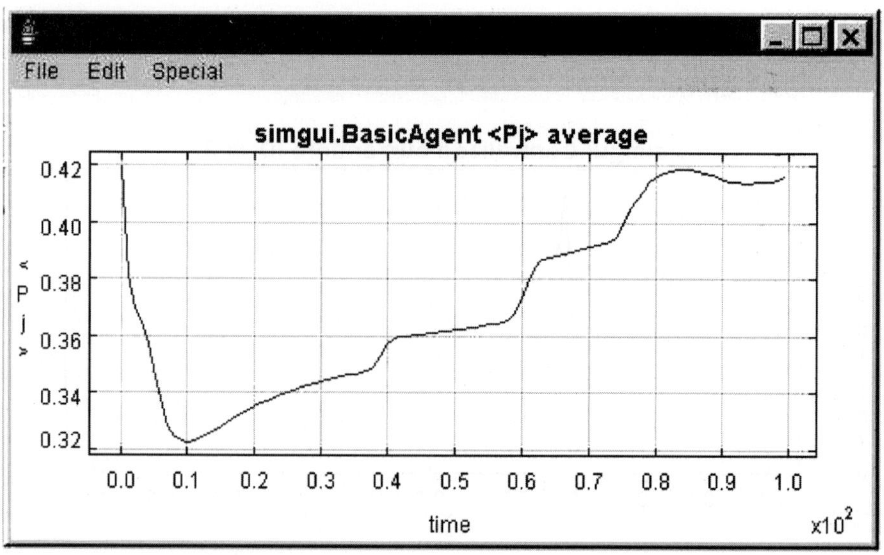

Figure 7. Aggregated (mean) trust.

4.1. Varying information strategies

Varying the breakpoint, the intensity and other parameters leads to different information strategies. These information strategies have different implications for the aggregate demand. The agents communicate about the new information and behave according to their internal processing mechanisms. The result emerges from the bottom up via the communication and the actions of the agents. It is planned to test different information strategies (risk communication strategies) with respect to the aggregate demand. We want to investigate how the different information strategies influence the dynamics of the system and the outcome.

Segment specific information policies can be applied. The segments receive global information by global media but can also be addressed by segment specific centralised information sources or campaigns.

4.2. Evaluating information strategies

Several simulation runs under controlled conditions have to be done in order to evaluate the different information strategies. We are now in the testing phase. The results will be discussed later on when we have tested the model in all areas. The idea is to evaluate risk communication strategies in cooperation with Partner 1. The model provides a tool to economically assess risk communication strategies with the corresponding consumption behavior and market outcome. A cost-effectiveness analysis of different risk communication procedures will be done within the Trust-project. A quantitative monetary measure of benefits will be provided by the outcome of the alternative communication strategies using the simulation model. The policy simulation activities are currently under construction.

5. Example scenario

The simulation consists of n consumer agents, m media agents and a number of auxiliary agents. The parameter space consists of x initial parameters and y structural parameters. The parameter exploration is directed to:
- initial parameters;
  - parameter distributions;
  - parameter distribution moments;
- structural parameters;
  - social networks;
    - link distribution;
    - link distribution moments;
  - network weights.

The analysis of the experiments:
- for every scenario running multiple simulations with different random number generator seed;
- for every scenario computing means and recovery rates;
- scenarios comparison: time series plots, distributions, etc.: comparing the effects of the parameter changes[2].

[2] This can just be done for a small number of scenarios.

The parameter space in its full size is not testable due to combinatorial reasons: more than 150 different parameters. For this reason just explicit scenarios can be tested by applying multiple simulation runs.

An example scenario was tested 13 times with 50 simulation runs each, i.e. 650 simulation runs in total with different seed. The mean recovery rate of all the runs together is 94.98% after 30 days of iterations. The data is shown in Table 1.

This part of the sensitivity analysis shows that the mean recovery rate is nearly always about 75% of the former level, in some cases the trust respectively the demand recovers completely. Due to the fact that the parameter space and the possible parameter constellations are very broad it is not possible to test all scenarios but if specific information policies and scenarios should be tested then multiple simulation runs under these conditions can help to discover the impact of these information policy scenarios for the aggregated demand.

6. Testing risk communication strategies

Risk communication strategies will be formulated in scenarios which should be tested. Depending on the form of the risk communication strategy, i.e. the duration of strategies, the segments addressed, the selected media etc., the recovery of the trust respectively the demand can be the measure for the effectiveness of the risk communication strategy.

Table 1. Recovery rates for an example scenario.

No. of runs	Seed	Mean recovery rate
50	1	110.9%
50	1	81.07%
50	24	115.66%
50	123	79.97%
50	37	108.4%
50	41	84.26%
50	55	55.83%
50	321	98.1%
50	152	110.44%
50	152	82.16%
50	37	112.27%
50	14	111.23%
50	8	84.51%
Σ=650	-	Mean recovery rate of all runs: 94.98%

Acknowledgements

This work belongs to a project of the European Union. 'Food Risk Communication and Consumers' Trust in the Food Supply Chain'. TRUST - QLK1-CT-2002-02343, http://www.trust.unifi.it

References

Boecker, A. and C.-H. Hanf, 2000. Confidence lost and - partially - regained: Consumer response to food scares. Journal of Economic Behaviour and Organisation 43, 471-485.

Cavicchi, A., A.E. Lobb, M. Mazzocchi, D. Romano, G. Stefani and W.B. Traill, 2005. Cross-country Segmentation, Deliverable no. 8, Firenze University Press.

Kahneman, D. and A. Tversky, 1979. Prospect Theory: An analysis of decision under risk. Econometrica 47, 263-291.

Maes, P., 1994. Modelling adaptive autonomous agents. Artificial Life Journal 1 & 2, No. 1-2.

Saggau, V. and P. Patelli, 2004. Information diffusion and decision making in artificial consumer societies. Second Conference of the European Social Simulation Association, ESSA'04, 16.-19. September in Valladolid (Spain). http://www.insisoc.org/ESSA04/.

The importance of emotional quality as a buying pattern argument: results of a consumer survey

Silvia Zenner[1], Bernd Wirthgen[1] and Marianne Altmann[2]
[1]University of Kassel, Steinstr. 19, 37213 Witzenhausen, Germany; zenner@wiz.uni-kassel.de
[2]Co Concept Marketingconsulting, Germany

Abstract

Because consumers´ face an increasing number of product alternatives to satisfy their needs, emotional wishes and psychological demands are the most important driving forces for consumers' behaviour. On the example of apparel produce it becomes evident, that especially aesthetic factors match consumers´ expectations for added values. The focus of this paper is to answer the question, whether emotional dimensions do play an important role in a food buying process, exemplified on food purchased directly from producers. The results of a consumer survey show the existence of an emotional quality factor as one attitude towards farm-gate produce. Nevertheless, with regard to its contribution to explain the real purchase intensity it becomes evident, that the emotional quality factor is not at all the driving force, whereas the price consciousness factor is the most important buying pattern argument when buying farm-food. In consequence, emotional buying pattern arguments are theoretically interesting for special produce, like e.g. apparel products, but its impact on a purchase decision in a food buying process seems to be relatively low.

Keywords: emotional quality, consumer survey, farm-gate produce, price consciousness

1. Theoretical background

Due to the change from sellers to buyers market, producers had to realise high food quality to survive in the strongly competitive food market, whereas consumers face an increasing number of differentiated product alternatives to satisfy their needs. However, not the physiological needs, but the psychological needs and especially emotional wishes and demands are the most important driving forces for consumers´ behaviour. Gaining deeper understanding of the dependencies between quality and consumers´ purchasing decision from a theoretical point of view, it is necessary to answer the question what quality is. Grunert *et al.* (1996 p. 76) point out, that there is no common definition of quality in literature. Therefore the authors define quality of a product as a multidimensional conceptual framework, which is influenced by the product itself, by its production process and finally by the consumers perception of the product. In addition Dilts (2004) shows two more dimensions of quality, the value-based quality and the so called transcendent-quality. Concerning consumer-based quality, Kroeber-Riel and Weinberg (2003) give deeper insight into consumer perception and consumers´ buying behaviour. One key-factor to understand purchasing decisions is the concept of emotions, motivations and attitudes, which belongs to the non-cognitive, but activating components of the Stimuli (S) - Organism (O) - Response - theory of consumer behaviour (Figure 1). In common use, emotions are a kind of psychological state of being, especially related to personal feelings, whereas motivations are not only feelings, but related to a kind of personal aim. Finally attitudes are motivations, which are related to a personal judgement over special objects (Kroeber-Riel and Weinberg, 2003 p. 53). Whereas Stimuli and Response are visible, emotions, motivations and attitudes belong to the so

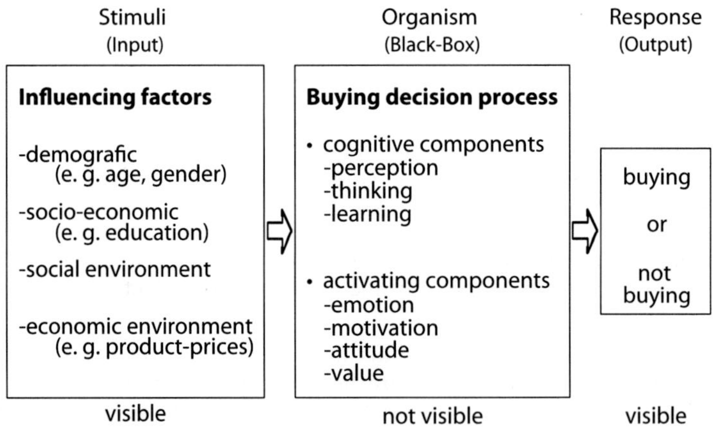

Figure 1. S-O-R-theory of consumer behaviour (according to Bänsch, 2002, p. 4).

called Black-Box of the organism. Within this Black-Box an invisible buying decision process took place and therefore the knowledge of attitudes, motivations and emotions is quite important to understand consumers′ buying decisions.

De Klerk and Lubbe (2004) present a theoretical overview about emotional dimensions in the purchasing process, exemplified on apparel products. It becomes evident, that functional quality of apparel seems to be no longer a buying argument, whereas the role of aesthetics in consumers′ quality assessment - and especially emotional factors as one key-dimension within the aesthetic framework - may play an important role in consumers′ purchasing decisions of apparel. More and more consumers ask the question 'What can the clothing do for me?' (Anonymous, 2004). But what about the importance of aesthetic factors in the food buying process? Do emotional dimensions play any role, when purchasing food?

Regarding mega-trends in food-consumption it is obvious that especially price consciousness is one of the main driving forces for German consumers when buying food (GfK, 2000). In consequence the question to be answered is, whether only low-price offerings match consumers′ expectations or not and whether emotional buying pattern arguments play no role at all in the food buying process. Consumer surveys underline the importance of knowledge of food origin as a buying argument for consumers. But consumers′ sympathy for food from well-known regions is based not only on the food-quality, but is influenced by consumers′ imaginations, expectations and finally emotional factors, related to the region of origin. Von Alvensleben (2002 p. 4) points out, that emotional quality based on an image transfer from producer to produce, this transfer changes the objective quality of this produce into emotional quality. To illustrate these phenomena, we use the example of farm-gate produce, because especially the personal contact between producers on the one hand and consumers on the other hand, might open the opportunity to match consumers′ expectations for special emotional values. If the consumer is in sympathy with 'his' farmer - the farmer as producer and seller of his farm-gate offering -, the farmer′ s produce must necessarily be good or better than food-offerings from unknown origin. So the individual and furthermore emotional imagination may influence the consumers′ perception of the farm-gate produce and transform its functional quality into emotional quality. Due to political and economical conditions, more and more farmers are forced to find new ways to ensure their farm-income. One income possibility is selling food and farm-produce from the own farm-gate. So a growing number of German farmers

- 30,000 (Recke and Wirthgen, 2004 p. 178) up to 60,000 (Produkt/Markt, cited in ZMP/CMA, 2002 p. 5) - use this marketing-possibility.

However, until now, no consumer survey has been carried out to identify the importance of emotional factors in the food buying process. In consequence, the aim of this paper is to give a deeper understanding of the importance of emotional quality as a buying pattern argument in the food buying process. We use the example of farm-gate produce, because farm-gate-offering may create a special emotional atmosphere to the consumers. Unfortunately there is no common definition of emotional quality in literature. Therefore we define emotional quality of food as the hedonistic imagination expected by the consumers of the food product.

2. Empirical context

To identify the importance of emotional quality in a food buying process, a consumer survey was carried out (standardised face-to-face interviews). The sample of 1,488 individuals integrates only purchasers of farm-gate produce; the survey-quota was based on the parameters of age and gender of the interviewees (70% females and 30% males). Special attention was given to the questioning regions within Germany (see Figure 2) to find out regional differences.

The survey took place in the year 2002 and was financed by the Federal Ministry of Consumer Protection, Food and Agriculture. The main objective of the survey was to get more information about consumer behaviour, when buying food directly from the farmer. Questions to be answered were for example buying-frequency, most important products, money spend for farm-food, attitudes towards farm-gate produce and finally preferences for different production modes and origins (Zenner, 2004).

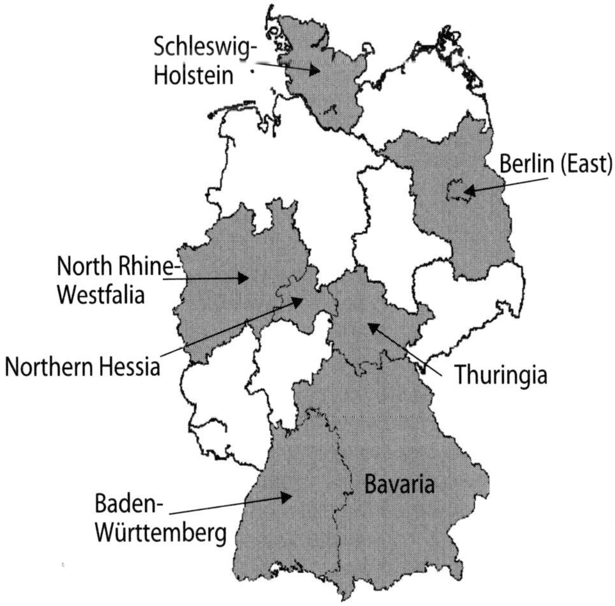

Figure 2. The questioning regions of the consumer survey within Germany (Zenner, 2004).

At first, a measurement of attitudes towards farm-gate produce took place (factor-analysis, main components). To identify differences within the questioning regions, cross-tables and factor-values had been used. To show dependencies between the different attitudes (e.g. the factor emotional quality) and food-purchase (measured by a purchasing index) a regression-analysis was carried out. To compare the impact of emotional quality in different groups of consumers, a cluster-analysis was executed.

3. Results

3.1. Attitudes towards food

To measure attitudes towards food, a matrix including 24 items concerning farm-gate produce, buying behaviour and nutrition in general, etc. were read to the interviewees. Special items were constructed, which integrated emotional buying pattern arguments. These items are: (1) farm-food is produced naturally and without chemical input; (2) farm-food is produced with consideration to animal welfare aspects; (3) in general, farm-gate produce is healthier than other food; (4) I am confident in farm-gate produce; (5) farm-gate produce is characterised by special freshness; (6) buying farm-food, I do protect our environment. Reflecting the presented items, it becomes clear, that farm-food is described not at all objective, but full of (emotional) ideas and expectations (e. g. naturally produced, healthier). The items were ranked by the interviewees on a scale from 1 (I disagree strongly) to 7 (I agree strongly).

The factor-analysis identified 6 different attitudes (Table 1). The main factor I (share of variance: 30.1%) integrates the emotional oriented items, so factor I is called `confidence in emotional quality´. Factor II integrates items like for example 'For me, farm-gate produce is too expensive' or 'When buying food, I am always looking for the cheapest offerings'. Therefore factor II is called 'price consciousness'.

Factor III 'social-political motivation' expresses the special experience and atmosphere when buying on a farm and the political aspects of the purchase of farm-gate produce. Factor IV `confidence in conventional produce´ integrates items expressing the confidence in the common food offerings and in the conventionally oriented farm-practise. Factor V `convenience oriented´ describes the attitude towards nutrition and consumption: fast, prepared and without much work to do on. And final factor VI integrates 'health consciousness' items, reflecting health aspects in the context of nutrition and purchasing.

Table 1. Attitudes towards food (n = 1,484)(according to Zenner, 2004, p. 107).

Attitudes towards food	
I	confidence in emotional quality
II	price consciousness
III	social-political motivation
IV	confidence in conventional produce
V	convenience oriented
VI	health consciousness

KMO: 0.911; Variance share: 62.5%.

3.2. Regional differences

As mentioned before, the survey was carried out in seven different regions of Germany. The background for that was the intention to identify regional differences - if they do exist. The following Table 2 shows the factor values within the questioning regions.

Especially two regions show surprising results; the sample from the northern questioning region Schleswig-Holstein and from the eastern questioning region Thuringia. The factor value -0.53 in Schleswig-Holstein shows that emotional dimensions seemed to play no important role for the northern interviewees. Different in Thuringia: the factor-value 0.35 shows that especially the eastern interviewees do agree on a relatively high level with emotional buying pattern arguments. This means, that especially the eastern interviewees are in confidence with products of their farmers. That does fit with results form Krippen (1992) and IM (1997), which point out that inhabitants of the eastern part of Germany prefer food out of their own regions.

3.3. The impact of emotional quality in different consumer segments

In further analysis, different groups of farm-gate buyers were identified. The cluster-analysis used the before mentioned attitudes as variables for segmentation. Four consumer segments with generally similar attitudes within one segment were found. Table 3 shows the results of the findings. Special interest is given to the factor I `confidence in emotional quality´: the factor-value differs in all 4 clusters. So cluster C1 shows the highest, whereas cluster C2 and C3 show the lowest factor-value concerning the factor `confidence in emotional quality´. In other words: the emotional quality aspect plays quite an important role in cluster C1 and - somewhat less so - in cluster C4. In clusters C2 and C3 the emotional criteria seems to be not at all important.

Furthermore, the purchase variables help to identify the different consumer segments. Without going into details, it becomes evident, that cluster C1 represents consumers, who are in general interested in organic food (low and medium users) and in farm-gate produce.

To analyse the dependencies between the attitudes and the real purchase behaviour of farm-gate produce, a purchase index was created. The purchase index is based on the following variables:

Table 2. Factor values within the questioning regions of the survey (n = 1,484)(according to Zenner, 2004, p. 110).

Questioning regions	n	Factor values of factor I: confidence in emotional quality***
Baden-Württemberg	272	0.13
Bavaria	221	-0.01
Berlin (Est)	233	-0.19
Northern Hessia	192	0.16
North Rhine-Westfalia	174	0.02
Schleswig-Holstein	185	-0.53
Thuringia	207	0.35

***p=0.000.

Table 3. Characteristics of the 4 clusterfactor-values and purchase intensity (n = 1,484)(Zenner, 2004, p. 127).

Factors		Cluster			
		C1 (n = 390)	C2 (n = 307)	C3 (n = 329)	C4 (n = 458)
I	confidence in emotional quality	0.91	-0.72	-0.79	0.28
II	price consciousness	0.17	0.62	-0.62	-0.11
III	social-political motivation	-0.27	-0.98	0.39	0.61
IV	confidence in conventional produce	-0.62	0.45	-0.82	0.82
V	convenience oriented	0.54	0.31	-0.09	-0.25
VI	health consciousness	-0.31	-0.07	0.56	0.27
Purchase variables					
Share of farm-gate buyers (%)*		70.3	30.0	76.6	78.6
Share of organic-buyers (%)*		59.0	28.3	79.3	48.3
Purchase index*		0.08	-0.69	0.22	0.24

*p=0.000

frequency of the farm-gate-purchase, number of shopping places, number of bought products, purchase intensity of the directly bought products, spent money for farm-gate-purchase. Especially in Cluster C1, the importance of the emotional quality factor is relatively high, but without being a driving force when purchasing farm-gate produce. The purchase index of 0.08 is surprisingly low. In addition, the factor value of factor V (convenience oriented) is extremely high and might therefore indicate a relevant purchase-barrier with regard to farm-gate produce. The purchase index of cluster C2 (-0.69) clearly identifies non-farm-gate-buyers. Especially for this consumer segment, the emotional quality factor is not at all a buying pattern argument, whereas factor I (price consciousness) plays the most important role. The remaining clusters C3 and C4 were identified as buyers of farm-gate produce (purchase index from 0.22 and 0.24). One main difference of C3 and C4 is the affinity to organic food, expressed by the share of organic buyers in these clusters (79.3% organic buyers in C3 and 48.3% in C4). The orientation towards organic food highly influences the importance of the emotional quality aspect: for organic oriented farm-gate buyers - which represent a group of regular users of organic food - emotional aspects are not as important as they are for conventionally oriented consumers. A reason for that might be the relatively high education standard of organic buyers, which helps to reflect emotional buying pattern arguments form a critical point of view.

Because of its relatively high level concerning the emotional quality factor I, especially Cluster C1 is analysed. The factors were examined with respect to their contribution to explain the purchase intensity of farm-gate-produce. How important is the emotional quality factor to explain the purchase intensity of farm-gate-produce in the highly interested Cluster C1? Table 4 presents the surprising results (beta-values) of the relevant regression-analysis (r^2= 0.31): with a beta-value of 0.16 the emotional quality factor is not at all one of the most important driving forces regarding the real purchase of farm-gate produce. Whereas factor II `price-consciousness´ is because of its beta-value from -0.48 the dominant argument in this food buying process.

Table 4. Dependencies between attitudes and behaviour in C1 (n = 388; r^2 = 0.31)(own survey, 2002).

Factors		Beta	Significance
II	price consciousness	-0.48	0.000
V	convenience oriented	-0.25	0.000
VI	health consciousness	0.18	0.000
I	confidence in emotional quality	0.16	0.000
III	social-political motivation	0.14	0.003
IV	confidence in conventional produce	-0.06	not significant

Analysing the factors′ contribution to explain the purchase intensity in the whole sample (n = 1,488), it becomes clear that the emotional quality factor reach a beta-value of 0.23 (Zenner, 2004 p. 123), whereas the factor price-consciousness is the most relevant dimension (beta-value: -0.42; r^2 = 0.32) in this food buying process. In consequence, emotional quality is quite an important, but not at all a sufficient buying pattern argument. From a critical point of view the outstanding importance of the price-factor might be a result of fears and debates during the Euro-implementation and might additionally express the economical stagnation of German industries and the growing 'discount-mentality' of German consumers in the food buying process.

4. Summary and conclusions

Because consumers face an increasing number of product alternatives to satisfy their needs, emotional wishes and psychological demands are the most important driving forces for consumers' behaviour. The example of apparel produce shows, that especially aesthetic factors match consumers′ expectations for added values, so consumers do know and estimate, what the apparel does for them.

The focus of this paper is to answer the question, whether emotional dimensions do play an important role in a food buying process, exemplified on food purchased directly from producers. We use the example of farm-gate produce, because especially the personal contact between producers on the one hand and consumers on the other hand, might open the opportunity to match consumers′ expectations for special emotional values.

The results of a consumer survey (n = 1,488) show the existence of an emotional quality factor as one attitude towards farm-gate produce. The importance of this factor does vary in the different German regions. Especially the interviewees in the eastern questioning region of Thuringia do agree on a relatively high level with emotional buying pattern arguments, whereas for the interviewees from the sample of Schleswig-Holstein, emotional dimensions seemed to play no important role. A cluster-specific analysis shows the impact of the emotional aspect in different consumer segments. Especially for the groups of price oriented non-farm-gate-buyers and highly educated intensive users of organic food, emotional buying pattern arguments do not play a role. However, they do play a role in that specific consumer segment, which is in general interested in questions of food origin and food production. But even if emotional arguments are relevant in this group, buying arguments like price consciousness and convenient shopping do influence the food purchase much more than emotional arguments do. In consequence, emotional buying pattern arguments are theoretically interesting for special groups of consumers. However, their impact on real purchase decision of food seems to be relatively low. So when buying food, consumers do not ask: what does this food do for me?

References

Anonymous, 2004. Matching consumers expectations with industry standards: The addad value of wrinkle resistence. Textile Consumer 34, Fall-Winter.

Backhaus, K., B. Erichson, W. Plinkre and R. Weiber, 2000. Multivariate Analyseverfahren. Eine anwendungsorientierte Einführung. Springer.

Bänsch, A., 2002. Käuferverhalten, Oldenbourg.

De Klerk, H.M. and St.J. Lubbe, 2004. The role of aesthetics in consumers′ evaluation of apparel quality: A conceptual framework. Journal of Family Ecology and Consumer Science 32, 1-7.

Dilts, D.M., 2004. http://www.kfmaas.de/qualidef.html.

GfK, 2000. Ein Volk von Schnäppchenjägern. http://www.gfk.com/index.php?lang= de&contentpath=http%3A//www.gfk.com/produkte/statisch/studien/produkt_1_2_4_059.php.

Grunert, K.G., H.H. Larsen, T.K. Madsen and A. Baadsgaard, 1996. Market orientation in food and agriculture. Boston.

Institut für Marktforschung (IM), 1997. Original Thüringer Qualität. Studie im Auftrag des Thüringer Ministeriums für Landwirtschaft. Leipzig.

Kribben, M.,1992. Entwicklung des Konsumentenverhaltens in den neuen Bundesländern. Ein entscheidungsorientierter Ansatz im Lichte des Bernoulli-Kriteriums. Lang.

Kroeber-Riel, W. and P. Weinberg, 2003. Konsumentenverhalten. Vahlen.

Recke, G. and B. Wirthgen, 2004. Betriebs- und Hofkundenanalyse in der Direktvermarktung. In: Recke, G., S. Zenner and B. Wirthgen, Situation und Perspektiven der Direktvermarktung in der Bundesrepublik Deutschland. Münster-Hiltrup.

ZMP/CMA, 2002. Direktvermarktung - Fakten zum Verbraucherverhalten. Materialien zur Marktberichterstattung 42. Bonn.

Von Alvensleben, R., 2002. Der Kommunikator ist oft wichtiger als die Botschaft. Ernährungsdienst. 06.04.2002, 4.

Zenner, S., 2004. Konsumentenverhalten bei der Nachfrage nach Lebensmitteln aus der Direktvermarktung. Eine verhaltenswissenschaftliche Analyse unter Berücksichtigung regionaler Besonderheiten. Dr. Kovac.

Benchmarking and harmonisation methods and their effects on the effectiveness of quality assurance schemes

Conception of integrated models for quality management in production chains in the agri-food sector

Stephanie Krieger and Gerhard Schiefer
University of Bonn, Chair for Business Management, Organisation and Information Management, Meckenheimer Allee 174, 53115 Bonn, Germany; st.krieger@uni-bonn.de

Abstract

Thinking about quality management started in the 1950s. Until now many different quality management standards have been developed. Deficits in food safety and the globalisation were reasons for the development of quality standards in the agribusiness and food industry. Sometimes several quality systems are relevant for firms. The consequence is that it is more efficient for firms to analyse requirements of different quality management standards to eliminate double requirements. The result would be an integrated quality management system. The aim of this paper is to describe the development of an instrument that simplifies the management of different quality systems in firms.

Keywords: quality management systems, cost/benefit, integrated description model

1. Introduction

Quality management is of paramount importance in all stages of the agri-food production and process chain. The approach of quality management has been changed in the past years due to effects of globalisation, numerous deficits in food safety and legislation such as the new European regulation 178/2002 concerning food safeties. A challenge was and is the integration of these changes into the sector and the design and development of appropriate information and communication systems. A trend is the development of several quality systems and norms in response to this challenge. Therefore, programmes will be developed and improved in the agri-food-industry further on.

There are general quality systems, which are applied in different countries and sectors, and country and product specific systems and programmes, which were developed by retail initiatives. The formulation of production processes and documentation is a central dimension of these different quality systems. Normally, most of their requirements are not harmonised and an acceptation often does not exist between different quality systems. However, in some cases the requirements of different quality systems are nearly the same. The result of this development is that farmers or firms who have implemented different quality systems have to fulfil a lot of requirements for their certification. And without an integrated accomplishment, much double work would be done.

Due to the variety of different quality systems, firms select quality management systems by comparing the requirements against the benefits. Hence, the question arises for firms how important quality management standards are and which advantages/disadvantages do they get with or without the certification.

The aim of the 'QUALINT' project is the development of an integrated description model to simplify the management of different quality systems. The formulation of this model is supported by a databank, which automatically generates operational system descriptions. A function of this model is also the calculation of costs, which would arise concerning the implementation of a quality system.

First this paper considers the initial situation of the changes in the food legislation, the development of quality standards and costs and benefits of quality systems (section 2). Section 3 includes the explanation of the developed instrument 'Qualint Sys' and analysis methods of different quality management systems to acquire the effectiveness of an integrated quality management system. These methods are used as a basis for a critical analysis of the efficiency of an integrated quality system (section 4).

2. Initial situation

2.1. Changes in the food safety legislation

Food legislation has changed during the past years. For example, in the year 1990 the product liability law was published. A key element of this law is that producers have to fulfil the due diligence of the product, what simply means that firms must take all relevant steps to assure the safety of the products. In the year 2000 the law redefined the legal meaning of due diligence (Krieger, 2002).

Another development was the publishing of the white book in the year 2000 by the EU. The new EU regulation 178/2002 with paragraph 18 concerning traceability of food is one result of the 84 actions of the white book (EU Commission, 2002). This regulation started in the year 2005 as the new subsidy payments did. Till last year the farmers got their payments according to the agricultural land size, since January 2005 farm subsidy payments have been linked to compliance with basic standards relating to environmental management, animal health and welfare and so on (Verbraucherministerium, 2004).

The Regulation 852/2004 aims to harmonise food hygienic legislation across Europe. This regulation lays down general requirements relating to food hygiene, clarifying the existing responsibilities of food businesses and is effective from January 2006. In a true ′farm to fork′ approach, primary producers are now subject to the hygiene requirements, but they must not fulfil the HACCP requirements (EU Commission, 2004).

These changes in legislations are also reasons for the development of quality standards.

2.2. History of quality management systems

At the beginning of the 1960s, the FAO and WHO developed the Codex Alimentarius regulation because of the expansion of the food trade. Moreover, until now this regulation has influences on quality and safety in the global food supply chain and is a basis for a ′fair′ international trade.

In the 1980s the development of systems with regard to process management (′Good practice′) started. Good practice (especially the good agricultural (GAP), good hygienic (GHP), good manufacturing (GMP) and good trade practice (GTP)) is basic for quality management. GAP is a guideline for the reduction of chemical, physical and biological hazards. GHP is obligatory for preventive hygienic arrangements in the firm and GMP is a basis for ensuring that products are consistently produced and controlled according to quality standards. GTP is a guideline for adequate transport of animals, raw materials and food (Krieger, 2002).

Since the 1990s, the international norm ISO 9000 has become popular. ISO (International Organisation for Standardisation) norms are international standards in order to achieve uniformity and to prevent technical barriers to trade throughout the world. The reason for the development of ISO 9000 was the publication of a consistent norm, which formulates the framework for quality management. The DIN EN ISO 9000:2000 norm includes basics and definitions of quality management (ISO 9000), makes demands on quality management (ISO 9001) and formulates a guideline for the improvement of quality system (ISO 9004). It is a sector independent standard (Pfeifer, 2001).

Since the middle of the 1990s, more and more systems with reference to the HACCP system are implemented in the agri-food sector. The main point of the HACCP concept is the identification of health hazards during the production. It includes the seven HACCP principles: conduct hazard analysis and identify control measures, identify critical control points (CCP), establish critical limits, monitor each CCP, establish corrective action to be taken when a critical limit deviation occurs, establish verification procedures and establish a record-keeping system (Luning *et al.*, 2002).

On account of increasing different national certification standards for HACCP and selective standards (e.g. the DS 3027 in Denmark, an HACCP standard in the Netherlands, an Australian norm) the international norm DIN EN ISO 22000 is directed for the standardisation of these different systems. The system's main point is the control of hazards with specific measures (SSM). The definition of the SSM is: 'Supportive safety measures: specified activities, other than critical control points, which affect food safety by preventing, eliminating or reducing the probability of hazards occurance' (Saferpak, 2004).

Furthermore, quality systems have been developed with specific demands for the agri-food industry and with a view on supply chains. Examples for systems of the agri-food industry are the Quality and Safety System (QS), the Quality Management Milk (QM) in Germany and the integrated chain control system (IKB) of the Netherlands; these are vertically oriented quality systems. Horizontal quality systems are for example the International Food Standard, the British Retailer Consortium, the European Food Safety Inspection Service and EurepGAP, which requirements were developed by retailer initiative (Krieger, 2004a).

2.3. Costs and benefits of quality management systems

To measure the effectiveness of quality management systems, costs and benefits are important aspects.In April 2003, questionnaires were sent out to the 300 biggest companies of the German food industry and 85 responses were received. The goal was to determine and analyse costs of quality management systems. Figure 1 shows costs, which are interesting for the internal work of a firm. Documentation and the high cost of entry checking and process analysis got the most criticism by the firms. Fault analysis causes 14%, quality checking 11% and training 10% of quality costs in firms (Beyer and Krieger, 2004).

Benefits of quality management have very different dimensions. Results from expert interviews and literature reviews present that some benefit aspects are more important and topical than others. Therefore, the following benefit aspects have been selected to evaluate the quality concepts:

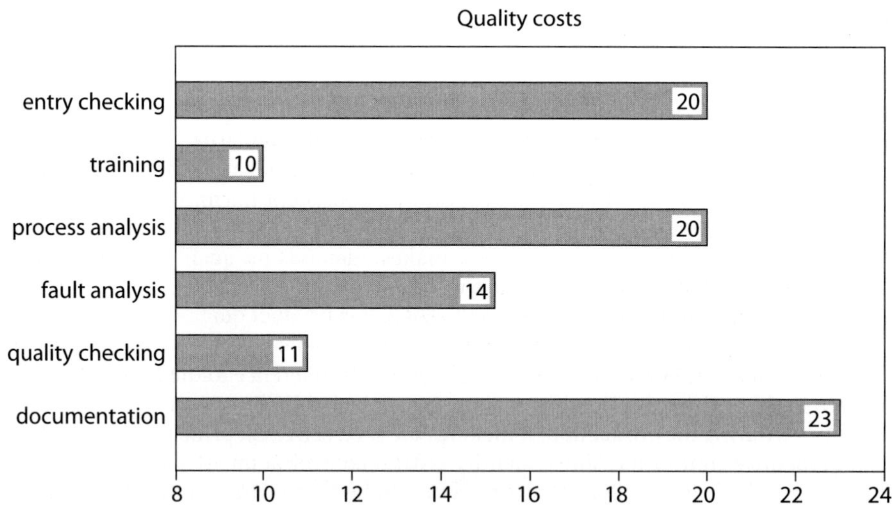

Figure 1. Costs of quality management in firms: result of a survey (Beyer and Krieger, 2004).

- *Market entry*: In some cases, a quality system certification is an entry to markets. The reason is that without a certification it is not possible to sell on specific markets. Standards can also be a barrier to trade for poorer developing countries because the cost of meeting them is assumed prohibitively high.
- *Product liability*: Since the year 2000, product liability has been a catchword not only in the agri-food industry. A key example is the legal standard to meet the due diligence requirements of the product liability law. The requirements that firms practice due diligence simply means that a firm must have taken all necessary steps to assure the safety of the products.
- *Cross compliance*: Cross compliance has been relevant for farmers since 2005. The subsidy payments will now be paid according to the fulfilment of 19 EU Regulations. In addition, in some cases the demands of those regulations have intersections with demands of quality management systems.
- *Process quality*: Process quality is the organisation of the internal process and transactions between firms. An optimal organisation of a process means lower costs. Moreover, the requirements of different quality systems have a special focus on the optimal organisation of the processes in firms.
- *Product quality*: Product quality concerns on the one hand physical product attributes (taste, shelf life, etc.) and on the other hand safety of a product with regard to health aspects.
- *Traceability*: The EU regulation 178/2002 contains general provisions for traceability (applicable since 1 January 2005), which cover all food and feed business operators, without prejudice to existing legislation on specific sectors such as beef, fish, GMOs, etc. (EU Commission, 2002). Importers are similarly affected, as they will be required to identify from whom the product was exported in the country of origin. Traceability has to be done one step down and one step up.

The evaluation of quality concepts is based in this study on the six above-mentioned benefit dimensions. The focus of different quality systems categories is not the same like the Figures 2 to 5 due to benefits of quality management present. The four figures also show the historical development of quality systems with the focus on their main benefits for the agri-food industry on the European market.

In the 1980s the HACCP-Concept became popular in the USA and later on in Europe. HACCP is widely recognised in the food industry as an effective approach to establish good production, sanitation, and manufacturing practices that produce safe foods (Pierson and Corlett, 1992). It establishes process control through identifying points in the production process that are most critical to monitor and control. HACCP´s preventive focus is seen as more cost-effective than testing a product and then destroying or reworking it. The system can be applied to control any stage in the food system and is designed to provide enough feedback to direct corrective activities. Like Figure 2 shows the main focus of HACCP is product quality of food. Food borne illness has been decreased since the integration of HACCP, which studies show (Unnevehr and Jensen, 1996). However, product quality can only be guaranteed if the process organisation is in a good order.

Adoption of HACCP as a regulatory standard has been motivated first by food safety concerns, and only second by a desire to facilitate trade (Caswell and Hooker, 1996). But the process of facilitating trade required mutual recognition of HACCP regulations across national boundaries, which shows that HACCP is internationally necessary for the market entry (Unnevehr and Jensen, 1999). In addition to greater food industry concentration, HACCP regulations may also create incentives for greater vertical coordination to control food safety throughout the production process. There is no necessary control if the product deliver and it is less expensive to contract or control production processes upstream (Mazzocco, 1996). A better vertical coordination can be guaranteed. Another benefit is seen in product liability. HACCP formulates no special requirements, which does fulfil Cross Compliance requests (Figure 2).

In the 1990s the ISO 9000 norm was developed. ISO 9000 is a set of international, voluntary quality management standards that ensure a consistent production process. The result is an improving in the efficiency (Böcker *et al.*, 2004). The ISO 9000 is a framework for a quality management system and the integration in a firm is very flexible. Therefore, the accomplishment of the ISO 9000 is not a guarantee of good product quality.

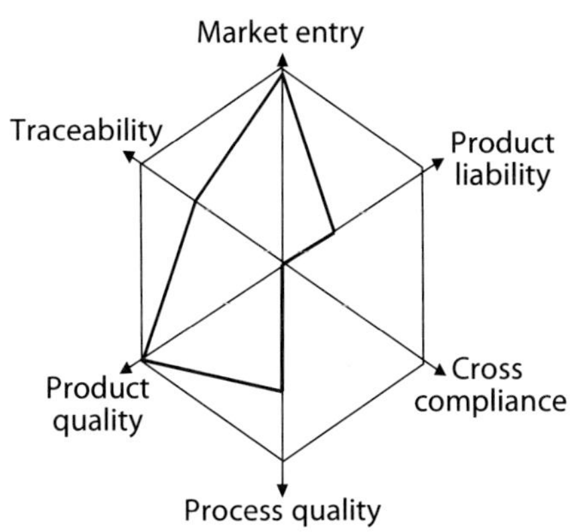

Figure 2. Benefits of HACCP.

This standard has an international acceptance and it was relevant for the market entry. A survey presents that international marketing aspects of the ISO 9000 certification and access to other markets have been regarded as one of the most important reasons to seek certification (Capmany *et al.*, 2000). A vertical traceability is not the main focus of the ISO 9000 like the product liability. The ISO 9000 has also no special requirements, which are important to fulfil Cross Compliance demands. However, it is possible to create the ISO 9000 on the farm with focus on their fulfilment (Figure 3).

Retailer initiatives developed quality systems for the food industry in the past years. On the one side there are quality systems like the International Food Standard (IFS), the British Retailer Consortium (BRC) and the European Food Safety Inspection Service (EFSIS) for the supplier to the retail and on the other hand, there is EurepGAP, a system for the farmers (Krieger, 2004b). These quality systems are often important for the market entry (see Table 1). Retailers ask for these systems also in due to product liability. Interventions exist between the requirements of the EurepGAP system and Cross Compliance.

The requirements of horizontal quality systems are mainly recording process quality (Krieger, 2002). Product quality is also in focus of these quality systems. An interaction between different quality systems over the stages of the agri-food supply chain would raise traceability. For example

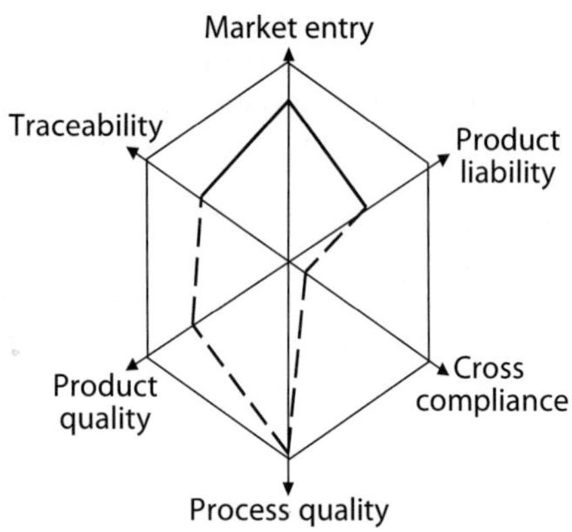

Figure 3. Benefits of ISO.

Table 1. Required quality assurance systems in the fruit, vegetable and potato sector (October 2004) (Sutor, 2004).

Retailer	Supplier	Farmer
ALDI	IFS	GMP
Metro	IFS	EurepGAP
REWE	QS	QS, discussion about the acceptance of EurepGAP
Globus	IFS or EFKO in Austria	EurepGAP
Netto	IFS	Legal requirements
EDEKA	IFS	EurepGAP

a combination of EurepGAP for farmers and of IFS for the supplier could increase a higher tracking and tracing between the stages of the agri-food industry (Figure 4).

The main focus of vertically oriented quality systems is traceability. The intensity of the cooperation between the different stages of the supply chain can be different. The supply chain can have an open character (e.g. QS), a semi closed character (e.g. IKB) and a closed character (e.g. regional quality systems). The result is that cooperation and traceability have a different intensity. Vertically oriented quality systems have problems to be accepted by the German retail (see Table 1), because the retail stage has designed its 'own' quality systems. The result was that only a few retailers ask for vertically oriented quality systems because they also have to fulfil special demands. Product liability plays also a rule like Cross Compliance. The directives included in the Cross Compliance demands are often addressed by farm assurance schemes.

Process quality is characterised by management routines that support the organisation and control of processes to assure desired process output (Schiefer, 2004). Points with this focus are also implemented in vertically oriented quality systems, but more important is the prevention

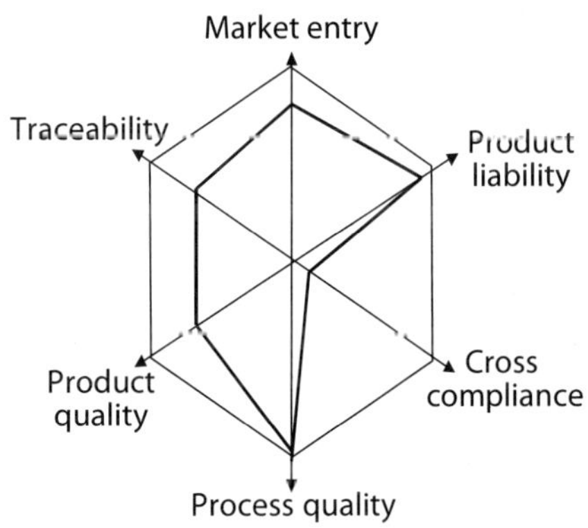

Figure 4. Benefits of horizontally oriented oriented quality systems.

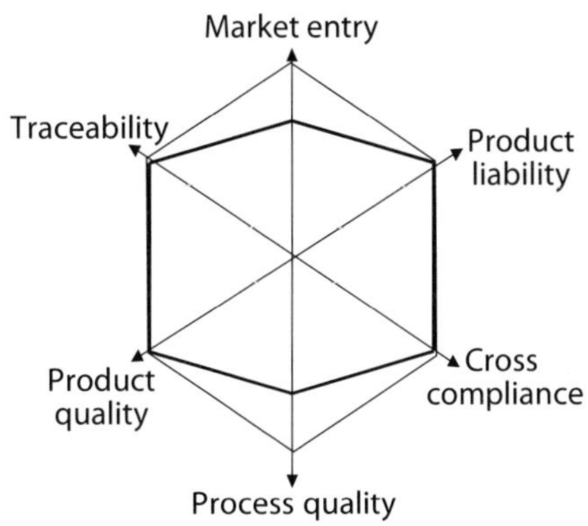

Figure 5. Benefits of vertically quality systems.

of product quality and safety (Figure 5). Another special benefit for each quality management in firms can be shown if firms do an integrated accomplishment of quality systems. Advantages of an integrated quality system are the use of synergies, reduction of time and cost in the application of quality systems and an easier integration of new quality systems (Schlüter and Petridis, 2000).

3. Methodologies

To answer the question about the optimal combination and implementation of quality management standards in a firm, an instrument was developed. This instrument demonstrates the requirements of quality management systems. The model utilises a databank, which automatically generates operational system descriptions. An important cost-value-analysis about special quality management scenarios in companies is the next step of this project. These scenarios will be analysed after a selection of marginal cost and benefit theories.

Marginal cost is the additional cost from increasing an activity. In production, marginal cost is the additional cost of producing one more unit of output (Varian, 1995). The firms' optimisation calculus can be represented as follows: the firm's marginal costs (MC) arise from the marginal costs of the fulfilment of the requirements of the new quality system (MC_N) minus the marginal costs of requirements, which are in the intersection of the existence and new quality systems (MC_{EN}). Another relevant parameter for the integration of a quality management system is the certification costs (MC_C).

$$MC = MC_N - MC_{EN} + MC_C \qquad (1)$$

Where:
MC = marginal costs;
MC_N = marginal costs of the requirements of new quality systems;
MC_{EN} = marginal costs of the requirements in the intersection of the existing and new quality system;
MC_C = marginal costs of certification.

The marginal benefit (MB) of a new quality system is the marginal sum of the advantages, which arise from an implementation for a firm like a new market entry (MB_M), more product liability (MB_P), fulfilment of demands for Cross Compliance (MB_C), improving in the process quality (MB_C), better product quality (MB_{PR}), enhance changes in the traceability (MB_T) and other special benefits for a firm (MB_F).

$$MB = MB_M + MB_P + MB_C + MB_{PQ} + MB_{PR} + MB_T + MB_F \qquad (2)$$

Where:
MB = marginal benefit;
MB_M = benefits for market entry;
MB_P = benefits for product liability;
MB_C = benefits for Cross Compliance;
MB_{PQ} = benefits for process quality;
MB_{PR} = benefits for product quality;
MB_T = benefits for traceability;
MB_F = benefits for firms.

From this initial situation a best combination of quality systems can be developed individually.

4. Integrated models for the application of quality standards

In this case, a computer based description model is the basic for the analysis of the marginal costs of a quality standard. This instrument is called Qualint Sys (Quality integration system), which is an instrument to measure the effectiveness of the implementation of a new quality system in the agri-food sector. Qualint Sys contains the requirements of different quality standards. An example is given in Table 2.

A mill has to fulfil four different quality standards: the QS, GMP 13, IFS and BRC standard. The requirements of these systems have a lot of intersections. HACCP is relevant for every quality system, an ISO 9000 or GMP certificate is also important for the QS system. A documentation of the HACCP concept is a requirement of every quality system, and training and pest control is important for the GMP 13, IFS and BRC. These are only some examples of requirements of these different quality systems. The following table shows that an integrated quality management system could reduce work.

But how does Qualint Sys exactly works? User could enter existing and new quality management systems into the description model. The additive requirements for the implementation of the new quality system are the result of the procedure. This database could be on the one hand a tool for advisory firms to have an integrated audit checklist and on the other hand firms can check, which requirements they have to fulfil to get the certification. The actors can also choose in which structure (integrated into the framework of the ISO 9000 chapter, hinge on department unstructured or a combination of ISO 9000 and department structure) the representation of the requirements will be illustrated.

5. Cost and benefit of quality systems: an example

A comparison of the British Retailer Standard (BRC) and the European Food Safety Inspection Service (EFSIS), two important standards on the Great Britain market, presents highly overlapping requirements. A BRC higher level certified firm, which would like to integrate the EFSIS higher level requirements in its quality management system has to fulfil only some new requirements like Table 3 shows. The firm has only to fulfil fifteen more requirements on a basic level and three

Table 2. Comparison of different quality systems.

Elements of quality systems	Quality Systems			
	QS	GMP 13	IFS	BRC
HACCP	required	required	required	required
ISO 9000	requirement (certification of ISO 9000 or GMP)	no requirement	no requirement	no requirement
documentation	- HACCP concept - ISO 9000 or GMP+	- HACCP concept - training - pest control	- HACCP concept - training - pest control	- HACCP concept - training - pest control - documentation of a QM system

Table 3. Integration of EFSIS in a BRC-certified firm.

Level of requirements	Additional requirements
basic level	15
higher level	3
recommendations	18

new requirements on the higher level. The marginal costs are the implementation of the eighteen new requirements plus the certification costs.

References

Beyer, J. and S. Krieger, 2004. Kunden und Mitarbeiter profitieren - Kosten-Nutzen-Analyse von Qualitätsmanagementsystemen in der Ernährungswirtschaft. Fleischwirtschaft 6, 59-60.

Böcker, A., M.E. Bredahl and J. Northen, 2004. ISO certification in British agribusiness: Motivations and performance impacts. In: Quality assurance, risk management and environmental control in agriculture and food supply networks. ILB, pp. 51-60.

Capmany, C., N.H. Hooker, T. Ozuna and A. van Tilburg, 2000. ISO 9000 - A marketing tool for U.S. agribusiness. International Food and Agribusiness Management Review 3, 41-53.

Caswell, J.A. and N.H. Hooker, 1996. HACCP as an international trade standard. American Journal of Agricultural Economics 78, 775-779.

EU Commission, 2002. Regulation (EC) No 178/2002 of the European Parliament and of the Council of 28 January 2002 laying down the general principles and requirements of food law, establishing the European Food Safety Authority and laying down procedures in matters of food safety ABI. L31/2002, 1-24.

EU Commission, 2004. Regulation (EC) No 178/2002 of the European Parliament and of the Council of 29 April 2004 of the hygiene of foodstuffs.

Krieger, S., 2002. Qualitätssysteme in der Agrar- und Ernährungsindustrie. B-02/4 ILB.

Krieger, S., 2004a. Quality management schemes in Europe and beyond. In: Zazueta, F., S. Ninomiya and R. Chitradon (eds.), Proceedings of the AFITA/WCCA2004 Joint Congress on IT in agriculture. Bangkok, Thailand, August 9-12, pp. 258-263.

Krieger, S., 2004b. Qualitätssysteme des Getreidesektors: Ein Ueberblick. B-04/2 ILB.

Luning, P.A., W.J. Marcelis and W.M.F. Jongen, 2002. Food quality management: A techno-managerial approach. Wageningen Academic Publishers.

Mazzocco, M., 1996. HACCP as a business management tool. American Journal of Agricultural Economics 78, 770-774.

Pfeifer, T., 2001. Qualitätsmanagement. Hanser.

Pierson, M.D. and D.A. Corlett, 1992. HACCP: Principles and applications. Van Nostrand Reinhold.

Saferpak, 2004. http://www.saferpak.com/iso22000.htm.

Schiefer, G., 2004. From enterprise activity 'Quality Management' to sector initiative 'Quality Assurance': Development, situation and perspectives. In: Schiefer, G. and U. Rickert (eds.), Quality assurance, risk management and environmental control in agriculture and food supply networks, Proceedings of the 82nd European Seminar of the European Association of Agriculture Economists (EAAE), May 14-16, 2003, Bonn, pp. 3-20.

Schlüter, S. and K.D. Petridis, 2000. HACCP umsetzen. Hanser.

Sutor, P., 2004. Qualitätssicherung bei Obst und Gemüse - international ein Vorreiter? In Bayerische Landesanstalt für Landwirtschaft (LfL), Qualitätssicherung in der Land- und Ernährungswirtschaft - Von der Vielfalt zum System, LfL.

Unnevehr, L.J. and H.H. Jensen, 1996. HACCP as a regulatory innovation to improve food safety in the meat industry. American Journal of Agricultural Economics 78, 764-769.

Unnevehr, L.J. and H.H. Jensen, 1999. The economic implications of using HACCP as a food safety regulatory standard. Food Policy 24, 625-635.

Varian, H.R., 1995. Grundzüge der Mikroökonomik. 3rd ed., Oldenbourg.

Verbraucherministerium, 2004. http://www.verbraucherministerium.de/index-0002A3407CD31022B9146521C0A816.html#03.

The organisation of private auditing systems and their limitations: a comparative analysis of the Eurep systems

Armelle Mazé[1], Myriam Aït-Aïssa[2], Nathalie Verjux[3], Georges Carrotte[4], Caroline Hédouin[5] and Aurélien Vaucelle[5]
[1]University of Paris I, Institut National de la recherche agronomique - INRA SAD-APT, 16 rue Claude Bernard, 75231 Paris, France; maze@inapg.inra.fr
[2]ACTA, 149 rue de Bercy 75595 Paris cedex 12, France;
[3]Arvalis Institut du Végétal, 3, rue Joseph et Marie Hackin - 75116 Paris, France
[4]Institut de l'élevage, 149 rue de Bercy, 75595 Paris, France
[5]Institut National Agronomique Paris Grignon, 16 Rue Claude Bernard 75231 Paris Cedex 05, France

Abstract

The increase of requirements in good agricultural practices (GAP) guidelines, quality certifications, environmental or food safety requirements at the farm level has created the need for more standardisation. Most of them are not harmonised and are sometimes based on inconsistent rules or elements. This article, based on the comparative approach suggested by transaction costs economics, analyses the benefits and limitations attached to several alternative standardisation strategies and benchmarking procedures developed in the European context regarding the implementation of quality assurance or environmental schemes. Our results suggest that, rather than a strict harmonisation of the technical content of these GAP guidelines, what is at stake is the coordination of private third party auditing systems at the farm level and its regulation by public authorities.

Keywords: quality audit, regulation, certification, standards, good agricultural practices, integrated agriculture, transaction costs

1. Introduction

Today, European farmers are faced with the increasing concerns of consumers and citizens with regard to the quality of food products and the impact of agricultural practices on the environment. More and more, large retailers and agro-food firms are requiring the farmers comply with the Good Agricultural Practices (GAP) guidelines, quality or environmental certifications and develop quality assurance systems for a specific production or at the farm level (Manhoult *et al.*, 2002; Mazé *et al.*, 2002; Verjux, 2003). These requirements are not always completely harmonised or standardised and may lead to inconsistent rules or interpretations when considered at the farm level (Vaucelle and Le Bail, 2004). Recent standard setting strategies implemented by agro-food firms or private benchmarking procedures by large retailers, like EurepGAP sytem, are intended to partially solve these impediments. However, these harmonisation strategies still remain difficult to implement for many firms or producer's organisations, especially when other quality or environmental schemes are already implemented. As well, they involve costly auditing processes (Garcia-Martinez *et al.*, 2004).

The purpose of this article is to analyse, based on a comparison of alternative private quality certifications and environmental schemes developed at the farm level in France, the actual benefits and limitations of alternative standard setting strategies or benchmarking procedures already implemented in agriculture to achieve a common definition of Good Agricultural Practices (GAP) and their codification. Our analytical approach is consistent with the assumptions and methodology of transaction cost theory by focusing on the efficiency of coordination mechanisms (Williamson, 1996; Arrunada, 1999). This research was conducted within an interdisciplinary context involving agronomists and economists.

The empirical data is based on three representative case studies including major quality or environmental schemes applied at the farm level in France, each focusing on alternative benchmarking strategies developed in agricultural sectors: the French label on 'integrated agriculture', the EurepGAP system and a collective quality charter to be applied for cereals. Our results shows the existence of alternatives procedures and the importance of the organisational design of private auditing systems and of their efficiency, i.e. their ability to keep auditing costs at a reasonable level while maintaining the quality of audit process.

2. Context of the study

With the development of private quality or environmental certifications, Good Agricultural Practices (GAP) guidelines and environmental or food safety requirements, or as the result of their voluntary involvement in private quality assurance schemes, almost all European farmers are subject to more controls or evaluations by their customers. This increase in requirements is especially problematic for farmers who do not specialise in one specific type of production, or who may trade with several customers, each having different quality or environmental requirements. This situation prevails in France where most individual farms are small to medium sized, geographically scattered, with diversified mixed crop-animal production farming systems[1]. As well, there is a well established tradition in France regarding official quality certifications dedicated to the agro-food sector.

Historically, several quality certifications, like 'Label Rouge', product certifications (CCP) as well as other regional labels and organic productions were developed since the 60's and have became very successful and acknowledged by consumers. Moreover, since the BSE crisis in 1996, a large number of the French cattle breeders (app. 110,000, 60% of dairy producers) are involved in an interprofessional quality charter, or a quality guideline used as a 'common base' for other quality certifications on beef products. For crop production, especially wheat, barley and maize, a special interprofessional quality charter Irtac-Arvalis is applied by more than 70 producer's groups or cooperatives and their 20,000 producers. As a matter of fact, most French agro-food firms are today developing also their own quality or environmental schemes with their suppliers.

As well, most French dairy farms are implementing HACCP and traceability systems under the supervision of agro-food firms, especially those involved in the production of raw milk cheese. Since the 90's, French cooperatives have developed a specific program based on an adaptation of ISO 9000 and ISO 14000, called *Agriconfiance*®, involving today about 30,000 farmers. Even small groups of farmers are now beginning to implement ISO 14000 systems on their own farms with independent third party certification. More recently, a new approach for the design of GAP guidelines appears based on a global approach to the farm level, like the French label '*agriculture*

[1] Exceptions are mostly related to fruit production and specialized vineyards in the south of France.

raisonnée'[2], including worker safety and information management, instead of focusing on one specific production. As a matter of fact, these global approaches of GAP guidelines at the farm level cannot also include all detailed and specific quality requirements (technological criteria or food safety). Hence, a better coordination of all these initiatives and their implementation at the farm level is still needed.

3. Analytical framework

In the literature, many studies still place priority on the agronomic evaluation and harmonisation of the technical contents of GAP guidelines (Mahouldt *et al.*, 2002; Bockstaller and Girardin, 2003). However, today, it is difficult to imagine a complete standardisation of all these requirements. As a consequence, alternative institutional arrangements emerge in order to facilitate their harmonisation and comparison in the European context.

3.1 Standardisation and diversity of benchmarking procedures

In economic literature, the study of standardisation strategies focused primarily on a distinction between *de facto* versus *de jure* standard setting strategies (David and Greenstein, 1990). In the first situation, market competition established the adoption of a specific technical standard. Another option is the use of a *de jure* standardisation process, either by regulation or the rise of formal institutions like national of international standard setting organisations (Foray, 1995)[3]. Between these two polar situations in the agro-food sector, we observe other hybrid arrangements which are trying to harmonise rather than fully standardise private quality or environmental requirements to be applied by farmers. Several harmonisation methods are already implemented by economic actors:

1. A simple '*equivalence principle*' based on a mutual recognition. A guideline is considered equivalent by the other party even if there are some (small) differences. There are no adjustments needed between the two GAP guidelines. The result is that the common items are not evaluated and controlled twice. The focus of stakeholders is more on the global guarantees, than an individual assessment of each individual item.
2. The definition of harmonised '*common base*' (like interprofessional cattle quality charter, etc.) used as a reference by other private technical guidelines. This is the option adopted, for example after the BSE crisis in 1996 for the certification of beef products.
3. A one-way recognition procedure (like the '*benchmarking procedures*' of the EurepGAP system), that include a formal evaluation and comparison of the technical content of the GAP guidelines and their adjustment and up-grading towards the required level.
4. A bilateral negotiation where parties try to improve the '*articulation*' or coordination of auditing activities, including external audits, internal audits, planning, statutes of auditors, formalisation of the audit report and its use as a reference for other audits, etc.

While the three first benchmarking procedures can be implemented for relatively narrow quality or environmental guidelines for the same agricultural production, the latter is required when there are few items in common or when attempts to harmonised failed. In another hand, these strategies still remain difficult to implement for many firms. There is no general methodology

[2] The name of this label can be translated in English by "Integrated Agriculture". However, it has a different a broader scope than the international definition used for fruit and vegetable.

[3] Recent examples in France of official standardization process realized within the French Association for standardization (AFNOR) concern a standard for potato farming, or an adaptation of ISO 9000 systems to the relationship between cooperatives and their farmers (Agriconfiance®).

available in order to improve harmonisation processes of quality or environmental requirements to be applied by farmers, at reasonable costs.

3.2. Actual limitations to benchmarking procedures

One of the main economic issues regarding the harmonisation of these quality or environmental requirements relies then on improved coordination of auditing systems. In order to evaluate the effect of this increase in quality requirements and controls, we identify, in the following diagram (Figure 1), the organisation of the three alternative auditing systems that are likely to coexist on the same farm in the French context:

1. On the upper left side, an official product certification, like the French Label Rouge, requiring an EN 45011 accreditation for the independent third party certifying body and an intermediate organisation (a dedicated marketing association, called 'quality group').
2. On the upper right, a normal private quality scheme implemented by an agro-food firm or a collective interprofessional organisation (like the collective French wheat quality charter IRTAC-Arvalis, etc.), a specific official accreditation of the independent third party control body is not required[4], and the audit system including a second party control.
3. On the lower middle section, a direct farm certification systems applied at the farm level, demanding an EN 45011 accreditation for the independent certifying body, with no intermediate organisation involved in the auditing system for a second-party audit or ('internal audit'). This type of design of the auditing system prevails for organic production and, in France since 2002, for the farm certification on '*Agriculture Raisonnée*'.

The situation described above (Figure 1) means that potentially one specific farm can be controlled by 5 different organisations for second or third party controls. Another source of difference can be created by the development of ISO 9000 or 14000 certifications by either the intermediate organisations (IO), or individual farmers (Wall *et al.*, 2001; Mazé *et al.*, 2002)[5]. This situation requires another accreditation for the independent third party certifying body (EN 45012, instead of EN 45011) and the objectives of the audit are different. Auditing for product certifications requires the verification of the auditee's compliance to detailed technical and quality requirements, the certification of ISO systems focusing primarily on the reliability of the organisation and the management of quality procedures applied.

Since the organisation of auditing activities is the sole responsibility of accredited certifying bodies and regulatory authorities, many studies still neglect this dimension. The consequence of all these characteristics is that the optimisation of actual benchmarking and harmonisation procedures of quality or environmental schemes applied at the farm level may also follow first-order efficiency considerations through economising on transaction costs (Williamson, 1996), i.e. a reduction of coordination and auditing costs.

[4] In this case the certifying body acts as an independent Third Party control Organisation (TPO).

[5] The independent certifying body is controlling (more in a logic of ISO system certification and EN 45012), i.e the goal of the audit is to evaluate the running of the overall system, instead of checking precisely at individual product level, like for usual product certification systems (EN 45011). This involves differences regarding the degree of legal responsibility of certifying body depending of what is audited and expected guarantees attached.

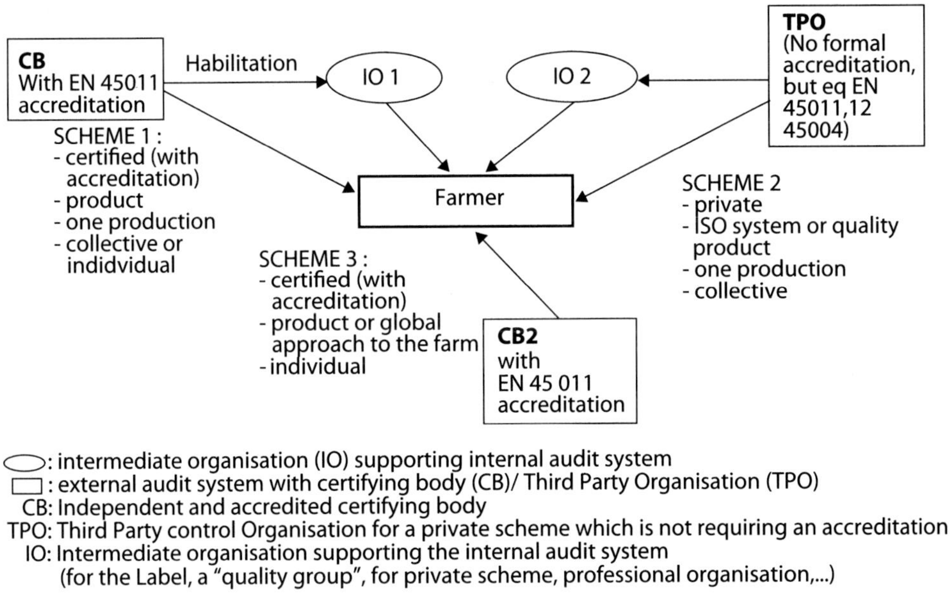

Figure 1. The coordination of auditing systems at the farm level and its limits.

3.3. Adjustments in the organisation of auditing activities

At the level of individual farmers, this increase of requirements also results in concerns and protests regarding the possibility of a drastic rise in the costs of controls, as well as their potential redundancies and the lack of real added value for farmers or for the whole supply chain. The acceptability by the farmers of these private audits, and as well of controls by public authorities, is a key issue, as they require a minimum of cooperation to be relevant for the quality of the audit itself. When considering a reduction of auditing costs, customers need to ensure that the level of the quality of the service delivered will not decline (Arrunada, 1999). Let us suppose that each firm, for each period of time, sells a single unit of product or service of certain quality, q, generating a cost which increases with the level of quality, given a function $c(q)$ such that $c'(q) > 0$ and $c''(q) > 0$ and providing the purchaser with utility that increases with the quality of the service. Under these conditions, the supply curve as a function of quality is shown as in Figure 2.

Improvements in auditing techniques are possible and require behavioural and organisational adjustments both for agro-food firms and farmers. There are productive economies of scope and then the cost function is shifted downwards from $c_1(q)$ to $c_2(q)$. This means that it is less costly to produce any level of quality. This reduction of auditing costs could be obtained through the reduction of potential redundancies among controls realised, especially a reduction of the number of audits on one specific farm and the duration of these audits. Quality safeguards on auditing activities are nevertheless required.

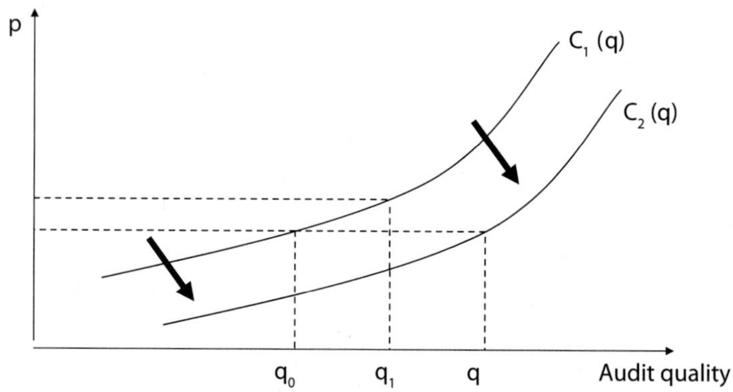

Figure 2. Effect of a production cost reduction.

4. Methodology and empirical data

In order to evaluate possible improvements and limitations to existing benchmarking procedures, our investigations were organised in two steps.

4.1. Identification of key attributes for GAP/quality/environmental guidelines

The first step of this research focuses on a diagnosis of the main critical points for the implementation of benchmarking procedures through the identification of key attributes on a sample of 14 representative quality and environmental schemes developed in France. These key attributes (Table 1) include a characterisation of their technical contents and objectives, an analysis of the internal and external auditing system and may or may not influence the feasibility of benchmarking procedures.

Solutions and problems encountered for the implementation of benchmarking procedures may vary according to the agricultural production and criteria like the nature and the seniority of collective or individual differentiation strategies developed by agro-food firms or producer's groups, and as well the nature of the production systems and degree of specialisation of the farms involved in a specific production.

4.2 Design and choice of representative case studies

The second step of this research uses the methodology of detailed cases studies based on a detailed characterisation of each quality scheme (Yin, 1984). The design of these case studies was chosen in order to grasp representative quality or environmental scheme developed at the farm level. This includes two crop production guidelines (interprofessionnal quality charter IRTAC-Arvalis and the Eurep fruit and vegetable GAP system) and one GAP guidelines at the farm level (official French Label 'Agriculture Raisonnée'). Their key attributes are summarised in Table 2.

The analysis of the organisation of auditing systems is important for the implementation of benchmarking procedures. Regarding the categorisations proposed in Figure 1, the French label 'Agriculture Raisonnée' is related to the described scheme 3, the interprofesional quality

Table 1. Analytical characterisation of quality or environmental schemes in agriculture.

Objectives of the scheme	B2B, consumer information, quality, environment, food safety
Nature of the technical content	content, number of items, type of items
Scheme holder	for example agro-food firm, producer's groups, professional organisations. This organisation is in charge of the whole coordination (negotiating contracts with third party certifying body, individual members)
Level of certification	official third party certification with a specific accreditation, simple third party control, second party control
Type of scheme	ISO system or product certification (involving EN 45011 or EN 45012 accreditation)
Perimeter of control	only one specific production (poultry, wheat, etc.) or the entire farm
Collective or individual membership of farmers	effects on control planning
Organisation of the auditing system (including independent third party and second party controls)	planning of controls, frequency, % of farmers controlled, sampling

Table 2. Characterisation of case studies.

Name	Key attributes
French Label 'Agriculture Raisonnée' (in English, 'Integrated agriculture')	official qualification/certification with independent CB
	at the farm level
	individual membership of farmers
Interprofessional quality charter IRTAC-Arvalis (cereal production: wheat and sweet corn)	voluntary private quality scheme with independent TPO/CB
	for one specific crop production
	collective certificates (delivered by third party)
Eurep'GAP system (fruit & vegetable production, that is relevant for sweet corn)	voluntary private quality scheme, with an independent CB
	for one specific production + general items at the farm level
	both individual or collective (depending of situation)

charter IRTAC-Arvalis, to the scheme 2, and the EurepGAP system to the scheme 1. We use for the analysis the more recent technical documentation available in 2004. Usual benchmarking procedures are based on a comparison of two (or more) different quality or environmental guidelines. It includes:

a. An evaluation of technical contents based on a comparison of: (1) the nature of each individual written items two by two; (2) the level of requirements for these items (minor, major, etc.) and possible differences in grading systems; (3) the nature of proofs to be controlled by the auditor or to be provided by farmers to demonstrate that they are operating according to the technical requirements.

b. An analysis of auditing systems and their architecture, i.e. second party and third party control, frequencies, % of farmers each year, sampling, objectives, levels of official accreditations, etc. The coordination of auditing systems is a more complex task than just considering the technical contents of quality or environmental guidelines. This includes especially differences between individual certificates for farmers (like in the case of 'Agriculture Raisonnée', organic production, etc.) and collective certificates (like the quality charter on cereal IRTAC-Arvalis) where only a small set of farmers is controlled each year by an independent third party certifying body, but involved an internal audit system (or second party control) that allow an annual follow up and evaluation of each farmers. They make their contribution to the reliability of the overall auditing system.

5. Results

Among the alternative harmonisation strategies analysed in this study, the attention is attracted by the solutions proposed by some large retailers. In some situations, there are other options that may be less costly and more relevant regarding the specific objectives of each individual quality or environmental guidelines. In this section, we analyse their respective advantages and limitations.

5.1 The importance of 'mutual recognition' and 'common base' mechanisms

Among the four methods of benchmarking and harmonisation between different quality or environmental schemes, the two first with the implementation of 'equivalence principle' and the definition of a 'common base' that can be used as a minimum standard for the design of others private quality or environmental schemes are the easiest to implement. One of the main advantage is to economise on the costs of design attached to individualised GAP guidelines, and as well on auditing costs through the elimination of redundancies. Many agro-food firms or retailers are already using these mechanisms. Even if there are slight technical differences in the technical content, the objectives, the organisation of the auditing system or the level of accreditation required, they are recognising it as equivalent:

- This 'equivalence principle' is used by some large agro-food firms as a support for the design of incentives mechanisms and the access of farmers to specific price premium for quality (for example, sugar beet producers in northern France). Another example is associated with the procurement policy for fruit and vegetable of the French retailer *Monoprix* considering as equivalent the implementation of EurepGAP or the French label '*Agriculture Raisonnée*'. As well, for some French potato buyers, their suppliers can implement either the official standard AFNOR NF V25-111, or the EurepGAP Guidelines. This equivalence among quality or environmental scheme allows more flexibility and is not eliminating arbitrarily producers.
- The definition of a '*common base*' for the design of quality and environmental scheme is also another spreading method. It was used for example, in the beef sector after the BSE crisis in 1996 in order to facilitate the access of a large number of breeders to higher standards regarding traceability systems and quality certifications (Label Rouge, CCP). It is the same for the GAP guideline Malster Malteurop used by individual agro-food firms for malting barley which is based on the French cereal guideline IRTAC-Arvalis.

The underlying idea with these two mechanisms is that farmers have to be involved in a quality assurance or environmental schemes small differences in content are not important. This implementation is promoting self-regulation mechanisms and an increased reflexivity of farmers regarding their own agricultural practices. In this perspective, internal audit system (or second party control) performed by habilitated organisations and their advisers (technicians of cooperatives, agro-food firms, extension services) may facilitate an increased commitment and a change in farmer's practices and behaviours[6].

5.2. Specificities of EurepGAP benchmarking procedure

Another example of benchmarking procedure is the 'one-way recognition'. This is applied, for example, by McDonald's France or the EurepGAP system. As suggested at the beginning, this system was developed under the initiative of Food Plus® and a large retailer's association located in Germany (Köln) is one benchmarking procedure among many others. First developed for fruit and vegetable productions imported from developing countries, the EurepGAP system[7] today covers a larger range of agricultural products (Garcia and Martinez 2004). Regarding another type of benchmarking observed in the other case studies, one advantage is the relative transparency about procedures to be followed by applicants. The usual steps for this procedure are the following:

- building of two tables for the comparison of auditing systems and technical content;
- a NZ-Australian accreditation organisation (Jaz-Anz) evaluates documentation;
- field audits are performed on a small set of farms;
- final decision of acceptance and contract signature including fees payment.

The technical items in EurepGAP guidelines are organised in 14 chapters and covering three domains: farm organisation, production organisation, and miscellaneous (waste treatment, worker safety and welfare, environmental and customer complaints management).

In order to evaluate the feasibility of this type of procedure and possible difficulties to be met by applicants during the process, we realise a specific field study in the fruit and vegetable sector with another quality guidelines already implemented by several French producer groups in the south of France (French sweet corn IRTAC-Arvalis). Two documents were fulfilled in order to proceed to the benchmarking, including general regulation and the description of technical items to be compared. The EurepGAP guidelines includes: Major must (44), Minor must (95), recommendation (61), for a total of 200 items for the combinable control points and compliance criteria.

The main difficulty met during the benchmarking procedure is related to the translation of technical items into combinable control points that simply double their number, and requires the identification of equivalent methods or indicators for the control. This is a long and drudgery job requiring a sentence by sentence comparison over more than 200 pages (EXCELL, see annex 2 for an example). The comparison of these check-lists is leading to the loss of the global coherence of control system attached to the GAP guidelines. Moreover, this approach of benchmarking is in contradiction with usual method of risk analysis and the need for a prioritisation of correctives actions (Aubry *et al.*, 2004).

The last point of our analysis is related to the accreditation system and the obtaining of a certification stricto sensu. Most of GAP guidelines developed in the European context by agro-food firms

[6] Audit regulation usually requires a strict separation of auditing and advising activities. According to this distinction, internal audit systems in the latter situation may not be taken into account for an official certification, even if they are contributing to the reliability of the quality system and the real commitment of farmers to it.

[7] In 2004, around the world, about 20,000 farmers were certified through this procedure.

or professional organisations are not requiring a specific accreditation for control or certifying bodies. The strategy developed by the EurepGAP system to require specific accreditation for the control of their GAP guideline is then introducing market restrictions on less costly solutions identified in this study[8].

5.3. Auditing global GAP guidelines at the farm level

The last category of benchmarking procedures identified in this study is related to the implementation of a better articulation and coordination of auditing systems. It is especially relevant when some items are in commons between two GAP guidelines, but not all, especially with the passage from specialised GAP guidelines or quality scheme for a specific production (wheat, barley, milk, etc.) involving the entire farm (like the French 'Agriculture Raisonnée' guideline, EurepGAP). Recent evolutions about the design of GAP guidelines in agriculture suggest that the latter may become the European standard for agriculture.

The analysis of our case studies (crop quality charter IRTAC-Arvalis and integrated agriculture) reveals several critical points, especially for crop productions that involve interdependencies between crops in the cropping systems (Papy, 2001). These interdependencies are less important for animal production. As a consequence, it is possible to audit animal workshops separately without redundancies. This is more problematic for crop productions. One of the central issues is that these new GAP guidelines, like the French 'Agriculture Raisonnée', cover the entire farm. Some items to be audited would not then be included in more specialised quality or environmental schemes. In another hand, these specialised quality schemes require more in-depth review of some technical items. These small differences are a first impediment and require a better articulation of auditing systems.

A solution could be based on the development of a joint/combined audit, i.e auditing several quality or environmental schemes at the same time (Mazé *et al.*, 2005). Instead of realising, for example, several audits at the same farm separately, on different dates, the auditor could perform all the audits of the farm on the same day. Another way is to take into account and acknowledge previous audit results and documents. The reduction of time spent during the audit process is also mainly related to the quality of the preparation by the farmer, and as well the existence of appropriate previous audit documentation and reports, both regarding third party and second party controls. In some of our case studies, we observe that sometimes individual farmers were not completely informed, especially breeders, that they were submitted to a second party audit required by some GAP guidelines. This situation is partly the result of the difficulty for adviser or technicians of cooperatives to adopt the attitude of a 'controller' during the interaction with the farmer (Maxime and Mazé, 2004).

6. Discussion and conclusion

Our study sustains the need for better coordination of auditing activities still leaved out of many theoretical and empirical studies. In the economic literature, most of analyses are focusing on mandatory environmental schemes and the optimal design of control mechanisms (Friesen, 2003; Spaeter and Verchere, 2004). The role of voluntary quality or environmental schemes

[8] Accreditations costs for certifying bodies are evaluated at 838 € for initial fixed accreditation costs (with an additional 305 € for each extra extension) and annual fees of 829 € (in addition to 2000 € for auditing costs).

implemented by farmers is also central[9]. Actual audit regulations in agro-food sectors are setting restrictions on alternatives to actual auditing practices that may facilitate the implementation of these benchmarking procedures. The methodology developed in this article identifies some of the organisational and regulatory factors that are actually restricting possible improvements of benchmarking procedures and the reduction of auditing costs. More theoretical and empirical studies are needed.

References

Aubry, C., M.B. Galan and A. Mazé, 2004. Garanties de qualité dans les exploitations agricoles. Cahiers Agricultures, in press.

Arrunada, B, 1999. The Economics of audit quality. Private incentives and the regulation of audit and non-audit services. Kluwer Academic Press.

Bockstaller, C. and P. Girardin, 2003. How to validate environmental indicators. Agricultural Systems 76, No. 2, 639-653

David, P. and S. Greenstein, 1990. The economics of compatibility standards. Economics of innovation and new technology 1, 3-41.

Foray, D., 1995. Coalitions and committees: How users get involved in information technology standardization. In: Hawkins, E., R. Mansell and J. Skea (eds.), Standards, innovation and competitiveness. The politics and economics of standards in natural and technical environment. Edwar Elgar.

Friesen, L., 2003. Targeting enforcement to improve compliance with environmental regulation. Journal of Environmental management and Economics 46, 72-85.

Garcia Martinez, M. and N. Poole, 2004. The development of private fresh produce safety standards: Implications for developing Mediterannean exporting countries. Food Policy 29, 229-255.

Klein, B., 1996. Why hold-ups occur: The self-enforcing range of contractual relationships. Economic Inquiry 34, 444-463.

Manhoult, A.G.E., G.W.J. Van de Ven, H. Udo A. de Haes and G.R. De Snoo, 2002. Environmental labelling in the Netherlands: A framework for integrated farming. Journal of Environmental Management 65, 269-283

Maxime, F. and A. Mazé, 2004. Designing and organizing auditing activities and competences to develop quality assurance systems in agriculture. A French case study, Working paper, INRA SAD, Paris, France.

Mazé, A., M.B. Galan and F. Papy, 2002. The governance of quality and environmental management systems in agriculture: Research issues and new challenges. In: Hagedorn, K. (ed.), Environmental co-operation and institutional change. Edwar Elgar, Cheltenham.

Mazé, A., N. Verjux, M. Ait-Aissa, A. Vaucelle and F. Maxime, 2005. Optimizing auditing techniques in agriculture: A controlled field experiment. Working Paper ACTA/Arvalis/INRA SAD, Paris.

Papy, F., 2001. Interdépendance des systèmes de culture dans l'exploitation. In: Malézieux, E., G. Trébuil and M. Jaeger (eds.). Modélisation des agro-écosystèmes et aide à la décision. Paris: Cirad, INRA.

Spaeter, S. and A. Verchere, 2004. Aléa moral et politiques d'audit optimales dans le cadre de pollution d'origine. Cahiers d'Economie et de Sociologie Rurales 71, 5-35.

Vaucelle, A. and M. Le Bail, 2004. A method to assess the effects of production guidelines on cropping systems. VIII ESA Congress European Agriculture in a Global Context, KVL Copenhagen, Denmark 11-15 July 2004.

Verjux, N., 2003. Qualitätsicherungs- / Umweltschutzmassnahmen bei den Produktzweigen des Ackerbaus in Frankreich. Internationale Ackerbautagung Qualitäts-Sicherungsstrategien im Ackerbau, 11/11/03, Hannover.

[9] Another point that is also under debate, regarding the design of GAP guidelines, is the inclusion in private and voluntary quality or environmental scheme of "*regulatory items*": (1) should private certifying bodies *de facto* control regulation and should they report non compliance to public authorities? (2) the existence of complementarities between law and private bilateral agreement (Klein, 1996). For example, in France, some items related to worker safety and welfare are not generally in GAP guidelines, as they are part of usual labour.

Verjux, N. and N. Bousquet, 2004. Brochure Démarches qualité en grandes cultures, des initiatives à découvrir. ARVALIS-Institut du végétal, octobre, Boigneville, France.
Wall, E., A. Weersink and C. Swanton, 2001. ISO 14000 and agriculture. Food policy 26, 35-48.
Williamson, O., 1996 The mechanisms of governance. Oxford University Press.
Yin, R., 1984. Case study research, design and methods. Sage.

Appendix 1. Detailed characterisation of case studies

	French label on 'integrated agriculture'	EurepGAP System (fruit and vegetables)	Cereals Quality Charter IRTAC/Arvalis
General characteristics			
Objectives	consumer information	B2B	B2B
Main orientation	environmental and farm management (traceability, security, etc.)	product quality and security	product quality and good environmental practices
Type of certificates	individual certificate/ farmer	both	collective certificate
Period of validity	5 years	1 year	1 year
Technical content			
Type of GAP	farm level	specific product and general items	only cereals
General structure	14 chapters on 9 themes and regional specifications	14 chapters	7 chapters on parcels characteristics, soil preparation, fertilisation, phyto, irrigation, harvest, storage, interculture
No of items	98 items, including: 18 commitments, 47 regulatory items and 37 others	max 143 (major and minor) and 65 encouraged	max 100 (if irrigation and storage)
Requirements levels	compulsory, commitment	minor, major, encouraged	A (major), B and C
Perimeter of application	all parcels and workshops on the farm	one production (all parcels)	one production (one charter/cereal), parcels involved
Auditing system			
CB accreditation	EN 45011	EN 45011	no specific program, but EN45011 required
External control	initial audit and an intermediate control in the 5 years period	each year if individual	each year 10% of farmers/group; 100% co-op or firms
Internal control (technicians)	none	depending	each year 100% farmers
Specific documentation	farms records	plots records; self diagnosis	plot records; self diagnosis

Appendix 2. Example of table to be fulfilled with the EurepGAP benchmarking procedure

EurepGAP		Applicant scheme
Section	General regulations point	Clause
6.3	The certificate is the document that a Farmer holds to show he/she has been certified, and the licence is a contractual relationship that EurepGAP* and the Farmer or Farmer Group enter into by means of a Sub-licence Agreement signed between the farmer and EurepGAP approved CB	Agreement Third party protocol (VI.1)
6.5	EurepGAP* is a registered trademark and the use is regulated by the EurepGAP* secretariat.	-
6.6	The Normative documents that conform the EurepGAP* scheme are the following:	-
6.6.1	EurepGAP* General Regulations: provides instructions as to how the certificate can be applied for, obtained and the rights and responsibilities involved.	Agreement (cf. preamble)
6.6.2	EurepGAP* Control Points and Compliance Criteria for Fruit and Vegetables: Contains all the control points and compliance criteria that must be followed by the applicant Farmer/Farmer Group and which are audited to verify compliance.	Sweetcorn follow-up guide Protocol of third party verification
6.6.3	EurepGAP* Checklist for Fruit and Vegetables: Contains the control points and is a tool for inspecting and evaluating compliance.	Follow-up guide and self-inspections

		Compliance		
General regulations point	Original version	Yes	No	Justification
An agreement is signed between the group, ARVALIS and IRTAC. A letter of commitment or a contract is signed by the producer and the group. The award of certification is sent to the group by ARVALIS. A contract is set up between the group and the certifying body taking part in the third party inspection.	Une convention est signée entre le groupement, ARVALIS et l'IRTAC. Une lettre d'engagement ou un contrat est signé entre le producteur et son groupement. Le certificat de référencement est délivré au groupement par ARVALIS. Un contrat est établi entre le groupement et l'organisme certificateur intervenant dans le contrôle tierce partie.	X		-
-	-		X	No direct licence agreement to date.
-	-			-
'The aim of the current contract is to establish how the Institutes and the group wishing to take part in the procedure for sweetcorn production should work together.'	'Le but du présent contrat est de fixer les conditions de la collaboration entre les Instituts et le Groupement qui souhaite entrer dans la démarche pour la production de maïs doux.'	X		
These two documents state the points to be checked.	Ces 2 documents mentionnent la liste des points à contrôler.	X		-
The follow-up guide provides the rules for internal inspection by the group. Self-inspections allow the producers to evaluate themselves against the Standard and its annex.	Le guide de suivi donnent les règles de contrôle interne pour le groupement Les autodiagnostics permettent aux producteurs de s'évaluer sur la charte et son annexe.	X		-

Appendix 3. List of technical documentations used for this study

EurepGAP

EurepGAP, 2004. Control points for fruit and Vegetables. Version 2.0, January 2004.
EurepGAP, 2004. General regulations, Fruit and Vegetables, Version 2.1, january 2004.
EurepGAP, 2003. Eurepgap Benchmarking Procedure, may 2003.
EurepGAP, 2004. Benchmarking Cross reference checklist (excel document), Version 2.0, january 2004.

French cereal Charter IRTAC-ARVALIS

IRTAC, ITCF[10], 2000. Chartes de production du blé tendre (mise à jour en 2004).
ARVALIS-Institut du végétal, 2003. Guide de suivi et de contrôle, Version 3, juin 2003.
ITCF, 2002. Chartes de production des céréales à paille, protocole de vérification tierce partie, Version 1, juin 2002.
AGPM-technique*, IRTAC, 2001. Charte de production du maïs doux.
ARVALIS-Institt du végétal, 2004. Chartes de production Maïs, Protocole de véréification tierce partie, Version 1, janvier 2004.

Agriculture raisonnée

Arrêté du 30 avril 2002 relatif au référentiel de l'agriculture raisonnée, J.O. Numéro 104 du 4 Mai 2002.
CNAR, 2003. Recommandations générales en matière de plan de contrôle, Rév 06 02/07/03.

[10] AGPM-technique and ITCF merged on 18 December 2002 to form ARVALIS-institut du vegetal.

Analysis of the compatibility of selected quality systems in agriculture and food production

Bettina Rother
University of Munich, Bavarian State Institute for Agriculture, Institute for Food Economy and Agri-Markets, Menzinger Str. 54, 80638 München, Germany; bettinarother@lfl.bayern.de

Abstract

In recent years certification systems have been widely introduced into the European agrofood sector. This has resulted in considerable overlap between quality systems and requires benchmarking of these systems. The purpose of this study is to compare selected certification systems on the basis of their checklists in order to show where their requirements differ and where they are identical. The results will help farmers to avoid duplications of effort and better manage similar requirements.

Keywords: quality systems, compatibility analysis, COP software, farmers' checklists

1. Introduction and problem

In recent years all those involved in the food production chain have become increasingly aware of the demand for more information about the origin and nature of products and the production process. The buyers of products have realised the necessity for a systematic and methodical approach to risk reduction. And as a result, quality systems have been simultaneously developed in the various production areas.

In response to the increasing demands of the consumer, the food law has been altered to create uniform standards at European level. These form the basis of many national and international quality systems and customer-specific requirements for agriculture and food production.

In Germany the crucial law is the product liability law valid since 1 December 2000, while at the European level this area is governed by EU Regulation No. 178/2002. In order to comply with the law, food and feed business operators are in fact required to keep records so that food and feed can be traced through all stages of production, processing and distribution. They are thus forced to document the various processes and the flow of goods to, from and within the business so that they can provide clear information in the case of any claims.

In addition, the cross compliance provisions introduced as part of the European agricultural reform, with 19 comprehensive regulations from the areas of the environment, food and feed safety, animal and plant health and animal welfare, have established new standards of quality assurance in agriculture.

Many different mandatory and voluntary quality systems in agriculture and the food industry are now in existence throughout Europe, which is not only confusing for producers and pro-cessors but also expensive and time-consuming.

2. Aim

The individual quality systems have a wide range of criteria and requirements, structures and checking regulations. It can however be assumed that they also have many parallels, since they frequently have a legal foundation and many of the requirements are similar.

The purpose of this study is to compare selected major quality systems on the basis of their farmers' check lists, and show where their requirements differ and where they are identical.

The check lists are used because they help the farmers to document their flow of goods and processes and help them prepare for an audit. In addition, these check lists enable farmers to minimise the risk of claims against them by drawing their attention to the legal requirements and helping them to systematically observe them. The purpose of this compatibility analysis is to show where the systems are similar and recommend ways of coordinating them.

3. Method used for the compatibility analysis

The compatibility analysis was conducted with the help of COP software (Coordinated Program Management System) which was developed by mais GmbH (Mitteldeutsche Agentur für Informationsservice GmbH). This software displays the differences and similarities of quality systems (Karge *et al.*, 2003; Sillat *et al.*, 2004; Branscheid and Miosga, 2004).

3.1. Entering the criteria

First the requirements of the selected quality systems, in this case the criteria from the farmers' checklists, are recorded in a database. Where possible, the original farmers' checklists are used and the criteria entered exactly as they are in the original document. The data are entered by ifta AG (Institute for Animal Health and Agricultural Ecology) in Berlin.

3.2. Assignment to hierarchy levels

The criteria of the individual quality systems are depicted in a hierarchical structure (area - complex - sub-complex - category - criterion). This method makes it possible to classify the individual requirements according to subject matter and compare the criteria of a particular subject group.

In the case of quality systems which differ considerably with respect to the number of criteria, the hierarchy of the more comprehensive system is taken over and the system with the smaller number of criteria is classified according to this structure.

With quality systems that have a more or less equal number of criteria, a single hierarchy is developed for comparison purposes and the systems are classified according to this new hierarchical structure.

3.3. Criteria derivation

If the requirements of the quality systems to be compared are similar, or in other words if they are pursuing the same goal, then a link can be established between these requirements. What this link indicates is that although the criteria that the farmer is expected to meet may differ in detail, comparison of the systems shows that the requirements may be considered to be identical. This link is termed 'criteria derivation'.

For example, in one particular subject area of a quality system (Programme 2) a large number of criteria have to be met, whereas in a comparable quality system (Programme 1) the same area is more abstract, in other words contains fewer criteria. If it is assumed that some of the criteria of Programme 2 are contained in the requirement of Programme 1, with the COP software the similarities and differences of the two systems can be illustrated by means of criteria derivation. This is however a subjective portrayal and thus maybe subject to error.

3.4. Form of comparison

The criteria entered from the individual checklists form the basis of a number of analytical procedures. The COP software performs a compatibility analysis, identifying the smallest common denominators and showing the plus factors of the individual systems. It also produces individual checklists for each of the quality systems involved and combined checklists for several quality systems.

4. The problems of compatibility analysis

4.1. Comparability

Some quality systems are product-oriented and others process-oriented, which makes comparison difficult. The authorities of the systems are also in competition with one another and hence deliberately avoid comparison.

4.2. Coherence

Ideally, quality systems have an explanatory manual or guidelines, checklists for the farmers and checklists for third party inspection. Not all quality systems are however structured in this way. In addition, some farmers' checklists do not fully correspond to the requirements of the system as set out in the guidelines. In such cases the guidelines are used to supplement the farmers' checklists. For systems which have no checklist for the farmers, the guidelines or the checklists for third party inspection are used for the compatibility analysis. These aspects must be taken into account when interpreting the results.

5. System comparison

An overview of system comparisons used in this study is given in Table 1. The quality systems GQ, IKB, EurepGAP, BQM and KKM were compared on the basis of the farmers' checklist. In all other cases the systems' guidelines were also used for the comparison. Quality systems are constantly updated. The results of this study are based on the farmers' checklists in use at the time of comparison.

As an example of the above this article illustrates the system comparison in the product area of milk. Comparisons of systems for the other product groups will be published shortly in relevant journals.

Table 1. Overview of the system comparisons.

Product groups	Quality systems						
	GQ	QS	IKB	EurepGAP IFA	BQM	QM Milch	KKM
Beef	x						
		x		x			
Pork							
		x	x				
Milk						x	x
Fruit and vegetables		x		x			
Grains				x	x		

6. Milk

In this product area the quality systems KKM and QM Milch are compared.

6.1. Nature of the quality systems

KKM ('Keten Kwaliteit Melk') is a Dutch quality system which was developed in 1995 as a project of the Dutch Agricultural Association (LTO) and the Dairy Association (NZO). The aim is to achieve the participation of all Dutch milk producers in the quality assurance system and ensure that milk products are only made from milk from approved KKM farms. Today 99% of Dutch milk producers meet the KKM criteria. Milk from the remaining producers which does not conform to the KKM criteria is dealt with separately and used for the animal feed industry (see reference Internet 01).

QM Milch (Qualitätsmanagement Milch) was initiated by the German Farmers' Association, the German Farmers' Cooperative Association and the Milk Industry Association, which formed a working group to produce the 'QM Milk - Uniform National Guidelines for Milk Production' of 4 December 2002 and recommend their application throughout the country. These guidelines specify the requirements for milk production and provide checklists so that the milk producers can monitor their own practices. QM Milch is incorporated into the milk supply regulations or milk purchasing contracts and will be mandatory for all producers (see references Internet 02 and 03).

Both quality systems are continually adapted by their authorities to the changing conditions of the market. While the German QM Milch is process oriented and also to some extent product-oriented - or at least this is the intention of the uniform national guidelines - the KKM system is clearly only process oriented. Product quality in Holland is guaranteed by observance of the technical standards for milk supplies.

6.2. Method

To enable the systems to be compared, the KKM programme was translated into German. A hierarchical structure that corresponded to both quality systems was then produced with the COP software and the criteria of KKM and QM Milch were classified according to this hierarchy. Since in QM Milch not all the requirements of the guidelines are included as criteria in the farmers' checklist, additions were made to this list. When the KKM standards were entered, several very comprehensively formulated requirements had to be divided up into a number of individual criteria. This made it easier to compare the quality systems with one another.

It was beyond the scope of this project to check whether the Viehverkehsverordnung (VVVO) (used in the QM Milch programme) conforms with the requirements of the relevant EU regulations such as (EU) Regulation No.1760/2000 (used in the KKM programme), but conformity can be assumed.

6.3. System comparison

The quality system KKM has a higher number of criteria than QM Milch (Table 2). The two systems have 41 criteria in common; 26 QM Milch criteria and 49 KKM criteria are not comparable.

The KKM system has more detailed requirements primarily with respect to the milking pro-cess and storage of the milk, with particular emphasis on cleaning/disinfection and dairy/milk storage facilities. In the milking and storage complex, the QM Milch system, on the other hand, focuses more on milk hygiene (Table 3).

The systems overlap considerably in the *animal medication* complex, with all the QM Milch criteria also being included in KKM. In addition (6 criteria) KKM requires, for example, that every veterinary surgeon who supplies prescription medication must produce a GVP certificate (Good Veterinary Practice; a code which guarantees conscientious and ethically responsible work on the part of the veterinary surgeon.) For cattle only medication specifically intended for cattle may be used which is either registered in the EU, registered for the Netherlands, or requires no registration. This also applies to homeopathic medication.

In the area of *animal health and welfare* both systems attach importance to the observance of the legal regulations for identifying and registering cattle and the existence on each farm of a system for registering animal diseases. Where udder disease is suspected, the animals concerned must be examined and treated or designated as chronically affected with udder disease or resistant to treatment. Only milk from healthy animals may be supplied to the buyer. Both systems also require ration calculation and feed analyses (nutriment analyses) to ensure the feed is suitable for the animals and environmentally compatible.

Table 2. Overview of the number of criteria used in QM Milch and KKM.

Quality system	Total number of criteria	Number of criteria in common	Number of criteria in one system only
QM Milch	67	41	26
KKM	90		49

Table 3. Overview of the number of criteria used in QM Milch and KKM for the individual complexes.

	Number of criteria		Number of criteria in common
	QM Milch	KKM	
Area			
Milk production	67	90	41
Complex			
Animal medication	13	19	13
Animal health and welfare	19	18	9
Feed	8	6	4
Milking and storage	25	45	14
Sub-complex			
Milking technique	1	9	1
Cleaning and disinfection	7	16	4
Milk hygiene	4	0	0
Dairy / milk storage facilities	11	17	7
Milk transportation and sale	2	3	2
Environment	2	2	1

KKM also stipulates under animal health and welfare that the farm must be free of leptospira hardjoe. Until 1 January 2005 livestock owners simply have to implement measures introduced by the animal health service against this disease. In addition, every animal that looses calves between the 100th and 260th day of gestation must be examined by means of a blood test for brucellosis. And since 1 January 2004, every farm is required to implement the preventive measures of the Dutch paratuberculosis programme. KKM prohibits the use of substances to stimulate milk production and increase growth. The animals must also be regularly examined by a veterinary surgeon (four times a year.)

In the same complex QM Milch stipulates that the cow stalls must have sufficient room for the animals to lie down and be adequately lit and ventilated. In the case of an epidemic, the farm must have its own protective clothing for visitors.

In the area of *feed*, both systems require that all cattle-feed be stored in a clean dry place on the farm and stipulate that it must not be mixed with other feeds. When feed is purchased, the last three invoices or a quarterly overview from the feed supplier must be produced. In addition, QM Milch requires feed producers and suppliers to comply with a feed agreement. KKM requires GMP certification.

In the *environment area* both systems require nutriment comparisons. KKM also requires the dairy farms to produce a valid certificate of approval with respect to their stock and environmentally compatible practices. QM Milch stipulates that there must be no contamination of the ground or surface water by liquid manure.

KKM has considerably more criteria in the *milking and storage complex* than the QM Milch system (Figure 1), with the exception of the milk hygiene area.

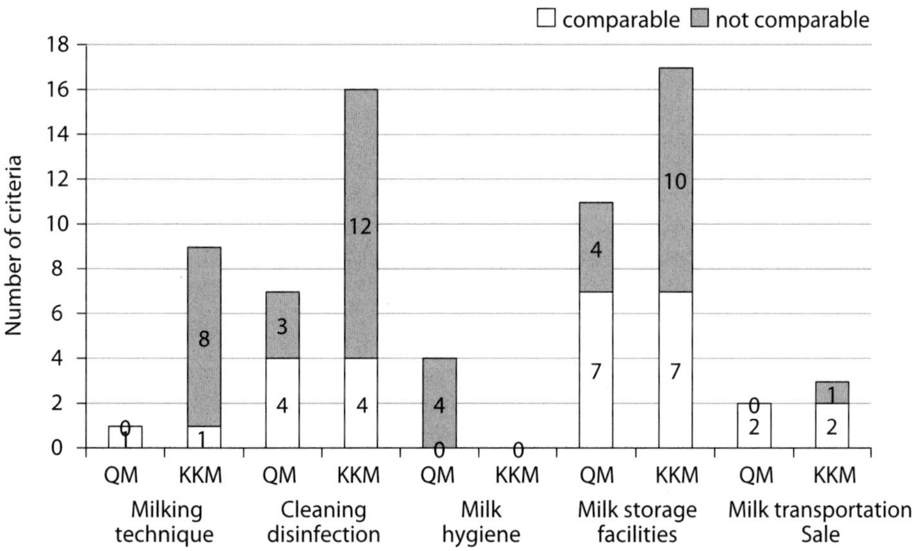

Figure 1. Comparison of QM Milch and KKM; milking and storage complex.

In this complex the two systems differ considerably (8 additional criteria in the KKM system) with respect to *milking technique*. KKM in particular has strict requirements with respect to the maintenance of automatic milking systems.

Under *cleaning and disinfection*, both systems for example require that the milking area and milking system can be properly cleaned. Feed, milk and traces of excrement must be removed after every milking session and the milking area must be sufficiently lit and ventilated.

In this category KKM also focuses particularly on water quality. Drinking water quality is required and must be certified as such on the basis of laboratory analyses. Surface water may not be used. The fact that QM Milch has no criteria for this may be explained by the fact that in Germany this area is covered by compliance with the drinking water regulations and is hence not included in the checklist of this system.

There is a noticeable absence of criteria for *milk hygiene* in KKM. While QM Milch lists personal hygiene, care during milking (cleanness of the udder, pre-milking check) and the handling of the milk (stipulations for cooling according to whether the milk is sent out daily or less frequently), KKM has no comparable criterion.

Under *dairy and milk storage facilities* both systems require the storage area to be closed and able to be properly cleaned and aired. The drains which are connected with the liquid manure channel or waste water drains must be sealed off against smell. Cleaning and disinfectant must be clearly marked and only stored in the quantities needed. Other substances may not be kept in the milk tank storage area.

In addition KKM, has detailed requirements in this complex with respect to the standard and monitoring of the technical equipment. The milk cooling tanks, for example, must be maintained at least once a year by a recognised service provider and corresponding records must be kept.

6.4. Summary

The KKM system and QM Milch have a relatively high degree of congruence (41 criteria). The differences ultimately lie in the detail with which farmers are asked about the various areas of milk production. While KKM also covers technical details, QM Milch has open, comprehensive questions with less emphasis on the production technology.

Independently of this, QM Milch should include all the criteria of its guidelines in the farmers' checklist. The following criteria from the guidelines should be included in the checklist:

- Where udder disease is suspected, the animals concerned must be examined and treated or designated as chronically affected with udder disease or resistant to treatment (Guidelines item 1).
- Where feed is suspected to be contaminated the farm's feed must also be checked for residues (Guidelines item 3).
- The livestock owner must document all medication administered to the animals in the stock book (on paper or in electronic form) (Guidelines item 5.1).
- The medication received from the veterinary surgeon or pharmacy must be stored in accordance with the instructions printed on it (Guidelines item 5.2).

To make the systems compatible, some of the criteria specific to KKM would have to be added to QM Milch. These include GVP certification by the veterinary surgeon, participation in a preventive paratuberculosis programme and the maintenance of the automatic milking systems or milk storage facilities. The KKM system would need corresponding criteria for ensuring milk hygiene. The system authorities would also both have to agree to accept the feed agreement (QM Milch) and GMP certification (KKM).

By including milk hygiene, QM Milch is both process- and product-oriented, but in practice the quality of the milk and its suitability for supply are primarily based on analyses in accordance with the Milk Quality Regulation and evaluation according to the Milk Regulation. Since the analysis of the milk to determine germ count, cell count and inhibitors is not carried out on the spot, these results are not currently included in a QM Milch audit. A milk-producing farm can thus be successfully certified according to the criteria of QM Milch even though it does not fulfil the requirements of § 17 of the Milk Regulation. It is thus very important to make sure that the evaluation according to the Milk Regulation is taken into account in the QM Milch certification process, which would not be very difficult from the organisational point of view.

To sum up, in spite of the existing deviations in individual areas, a benchmarking of the QM Milch and the KKM system could be possible.

These conclusions coincide with results from the University of Bonn (Krieger and Schiefer 2004). As part of the research project 'QUALINT', the quality standards QM Milch, IKM, KKM, NDFAS and EurepGAP were analysed and compared from the point of view of system structure, checking criteria and evaluation standards. These standards were shown to correspond to a high degree with respect to their main priorities (health and welfare of the animals, milking and the environment). According to the 'QUALINT' study, KKM and QM Milch differed with respect to the detail in which their requirements were formulated.

7. Overall conclusion and recommendations

The following conclusions may be drawn from the compatibility analysis of selected quality systems in agriculture and the food industry:

- *Comparability of the quality systems:* It is perfectly possible to compare quality systems. However, their comparability is complicated by the differing structure of the systems (systems are either based on one area of production or are constructed in a modular fashion and hence combinable), which makes it difficult to show how the contents compare with one another (because, for example, comparable criteria are assigned to different areas).
- *Availability and completeness of farmers' checklists.* The document structure used in the various systems is not identical. Not all systems provide checklists for the farmers. In other cases the checklist provided does not cover all the requirements of the system as set out in the guidelines.
- *Incorporation of national and European law.* Internationally oriented quality systems usually also include the legal requirements in the farmers' checklists. This is not normally the case with the national systems. Here it is assumed that the legal regulations are observed, so they are not included as separate criteria.
- *Formulation of the criteria.* The systems sometimes differ considerably with respect to the formulation of the criteria (e.g. summary of several requirements in one criterion or listing of individual criteria; general or specific formulation of the criteria). Nationally oriented quality systems tend to formulate the criteria more abstractly and more like statements. Internationally oriented systems, on the other hand, frequently ask in considerable detail about the fulfilment of specific criteria.

On the basis of these conclusions the following recommendations are made:

- *Standardise the structure of quality systems.* The structure of the quality systems should be standardised to make it easier to compare the individual production areas of the various systems. A modular structure is almost certainly the best solution because it enables product areas to be combined.
- *Provide complete checklists for the farmers.* Every quality system should provide the farmers with their own checklist to help them prepare for audits. This must include all the requirements of the programme. The ideal quality system would consist of guidelines, a farmers' checklist and a checklist for third party inspection.
- *Include legal regulations as criteria and identify them accordingly.* It cannot be assumed that system users know all the legal regulations. If a certificate of conformity to the law is issued, the user could be liable if any regulations are not in fact observed. It would thus be useful if quality systems also included the legal regulations as criteria and identified them accordingly. In addition, it would then be easier to see where the laws of the various countries differ and to improve the comparability of international quality systems.
- *Evaluation of criteria.* The severity of the requirements represented by a quality system is not only determined by the number of criteria. These say more about the detail in which the user is asked about particular topics and requirements. Nevertheless it can generally be assumed that detailed requirements help farmers carry out their own checks better than abstractly formulated ones. Concrete criteria reduce the extent to which a requirement is subjectively interpreted.

It would be interesting to analyse the quality systems involved to discover which of the variously formulated criteria have exactly the same content and which are fundamentally different. It would also be useful to investigate which criteria should be formulated in detail and which should be more concise. The ultimate object would be to establish a basis of material common to all systems from which a modular quality system could be created, including modular guidelines and checklists for the farmers and for third party inspection.

A second approach would be to compare quality systems according to the principle of the HACCP concept and investigate agricultural primary production on the basis of quality and risk factors. In this way it would be possible to determine the critical control points which established quality in each area of production and make these obligatory for every quality system. These core criteria would then be supplemented by additional requirements from the various quality systems. The additional criteria could be audited separately and serve to distinguish a system authority from the competition. Consumer protection should be the main purpose of such an analysis.

References

Internet 01. http://www.ketenkwaliteitmelk.nl/

Internet 02. http://www.mpr-bayern.de/3.1/mpr.de/index.php?StoryID=2175

Internet 03. http://www.milchwirtschaft.de/Download/QM-Milch%20Leitfaden%20.PDF

Karge, J., K.-H. Miosga and M. Sillat, 2003. Im Spannungsfeld der Systeme. Neue Landwirtschaft 12, 67-69.

Sillat, M., K.-H. Miosga and T. Hille, 2004. Ein Europa - viele Qualitätsstandards. Lebensmittelzeitung 36, 74-75.

Branscheid, W. and K.-H. Miosga, 2004. Im Spannungsfeld der Systeme. Fleischwirtschaft 9, 93-95.

Krieger, B. and G. Schiefer, 2004. Europäische Prüfzeichen des Milchsektors - Modell zur integrierten Durchführung. Deutsche Molkerei-Zeitung 23, 32-35.

Farmers' perceptions and motivation on food safety and quality

Enhancing the acceptance of quality systems by German farmers: the case of quality management and quality assurance

Ulrich Enneking[1], Thomas Obersojer[1], Richard Balling[1], Marion Kratzmair[2] and Anton Kreitmeir[2]
[1]Technische Universität München, Marktlehre der Agrar- und Ernährungswirtschaft, Alte Akademie 14, 85350 Freising-Weihenstephan, Germany; u.enneking@fh-osnabrück.de
[2]Bayerische Landesanstalt für Landwirtschaft, Institut für Ernährungswirtschaft und Markt, Menzinger Str. 54, 80638 München, Germany

Abstract

In order to ensure and to improve quality in the agro-food sector quality systems have to undergo a permanent change: simple end-of-the-pipe monitoring is therefore being replaced with management oriented concepts. Farmers are also increasingly requested to implement quality systems. This paper gives an overview of quality regimes in the agro-food sector and compares three systems in the domains of pork production, potato growing and the cultivation of hop. 315 Bavarian farmers were interviewed in order to identify factors which may influence the acceptance of quality systems. Differences between the three quality regimes were moderate with respect to the overall acceptance. Hop growers judged their system to improve production efficiency while the others did not to this extent. Apart from age, socio-demographic factors do not have any significant impact on farmers' satisfaction with the quality systems. Key factors are gains in image, in sales and in production efficiency.

Keywords: quality assurance, quality management, acceptance of quality systems, DIN EN ISO 9000

1. Introduction

In order to ensure and improve quality in the agro-food sector, quality systems are having to undergo constant change: simple end-of-the-pipe monitoring is being replaced by management-oriented regimes. Increasingly, farmers, too, are required to implement quality systems. A variety of quality-assurance systems for use at the primary-production level have therefore been developed over the last few years. They differ widely, however, with respect to their requirement levels, regional scope of application, degree of integration into the supply chain or their sponsorship. The supposition that certified farmers' reasons for participating in the various systems and their assessment of the latter are not necessarily identical was the reason for this study. Apart from measuring differences in assessments of individual quality systems, the aim of the study was also to ascertain non-system-specific factors influencing the acceptance of quality systems. The following article provides an overview of selected results from a project with the aim of evaluating three quality systems implemented in Bavaria at the agro-food production level. The article is structured as follows: after a brief description of the general path of development of quality systems, the second section discusses the quality regimes investigated in the study. Section 3 then goes on to describe the sample populations, survey design and evaluation method. Section 4, which presents the results, first compiles partial aspects with respect to assessment of the individual systems by the interviewees before proceeding to derive globally valid success factors for system acceptance. Section 5 summarises the most important results of the study.

2. Quality systems

2.1. General path of development

The development of different quality strategies aimed at achieving a desired quality level can best be described in terms of their temporal evolution (Lehnert *et al.*, 1995: 2). Although the general course of development of the various quality systems is more or less identical in all economic sectors, differences in timing can nevertheless be ascertained. The literature shows that the development of quality systems in the agricultural sector is characterised by a certain time lag as compared with the food industry (Menke, 1999: 12; Weindlmaier, 2002: 8). Figure 1 gives an overview of how quality regimes have evolved in the agro-food sector. A similar evolution several years down the line can also be assumed for the agricultural sector.

Quality control focuses solely on product-oriented detection of inferior quality or defects in the form of final inspection after completion of the production process, which means that it is no longer possible to influence product or defect generation (Schneider, 1995: 10). In the case of quality assurance, such inspection of end products is extended to cover the entire production process (Amelung *et al.*, 2002: 50). This approach focuses mainly on process and flow control to achieve defect prevention and thus quality assurance of the final product (Amelung *et al.*, 2002: 50). Continuous control of work flows in conjunction with suitable documentation for traceability purposes accompanied by process improvements enhance product quality, make it more transparent and prevent defects (Menke, 1999: 10f.). The next higher quality-regime level is called quality management. Quality is hereby seen as a management task. All parties involved must take all quality-assurance measures for each process in order to control all aspects that may influence product quality (Erling, 1999: 25). Defects in final products thus occur less frequently, as they are avoided from the start. In Europe and North America, Total Quality Management currently occupies the highest position in the quality-system hierarchy. The definition of Total Quality Management, as per DIN EN ISO 8402, is as follows: 'A management approach of an

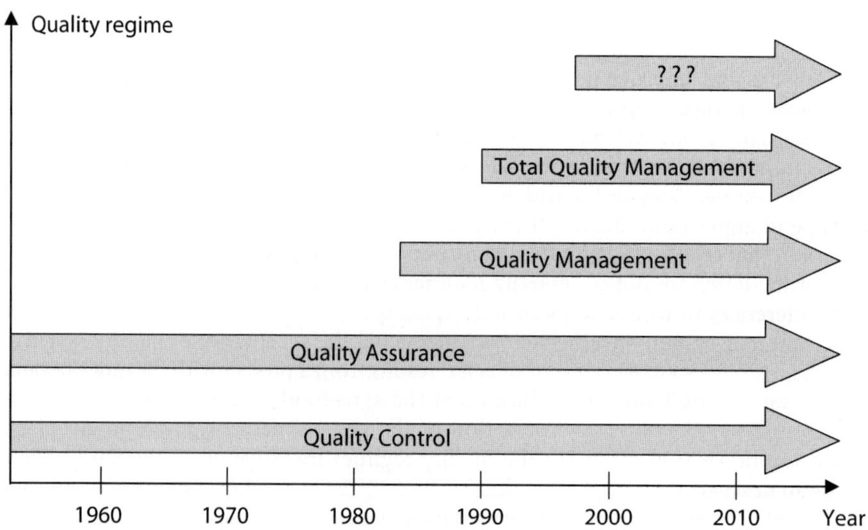

Figure 1. Development of quality systems in the agro-food sector (Weindlmaier, 1999: 97).

organisation, centred on quality, based on the participation of all its members and aiming at long-term success through customer satisfaction, and benefits to all members of the organisation and to society.' (Geiger, 1994: 204). The three basic pillars are thus strict customer, employee and process orientation. At present, most of the quality regimes established at the farm level do not go beyond quality assurance (Jahn *et al.*, 2003: 3), and even at that level, acceptance of farmers is low, resulting either in a slow diffusion of systems or in relatively low quality standards.

2.2. Quality systems evaluated

This study compares three different quality systems in the domain of pork production (QS-System)[1], potato growing (KVA)[2] and hops cultivation on the basis of surveys involving 315 Bavarian farmers. Producers' unions of potato growers and customers are contractually integrated into the quality system, 'Kontrollierter Vertragsanbau' (KVA). The aim is reciprocal guarantees regarding quantities and prices, taking quality-assurance aspects into account. 'Qualität und Sicherheit' (Q&S) is a vertically adjusted quality-assurance system in the pork sector covering all production, processing and distribution levels. The purpose of this system is thorough quality and information assurance along the supply chain. The hops growers' system is a quality management system (EN ISO 9001:2000) in which participating farmers have to undergo matrix certification organised by the hop growers' association, 'Hopfenring Hallertau'. The aim is to improve international competitiveness, exploit managerial advantages and increase the reputation of hops production. While the hops system is based on ISO certification, the quality assurance systems are much less demanding on farmers in financial and in organisational terms. Further differences and similarities between the three systems are displayed in Table 1.

The choice of systems to be evaluated within this study reflected different requirement levels; we assumed that such differences would influence acceptance of the systems by farmers. An empirical comparison of quality systems within a single product group was impossible, due to lack of adequate systems in Bavaria.

In the first step, this paper provides an overview of the core acceptance of each system; in the second step, it identifies general factors crucial for quality-system acceptance by farmers. The question of whether acceptance of a quality management system differs from that of quality assurance systems was also investigated.

3. Study method

Telephone interviews with agricultural managers based on standardised questionnaires, mainly with closed questions, were conducted in the period from January to March 2004. In the case of the DIN EN ISO 9001:2000 system in hops cultivation, a complete survey of all associated farmers was carried out (74) and in the case of the KVA-potato and QS-pork systems, random samples of 98 and 143 valid interviews respectively were taken. The total population in each case derived from address lists made available by the relevant cooperation partners[3]. The interviews were conducted with the help of the Ci3 Computer programme. The data obtained was then transferred to the

[1] www.q-s.info.

[2] Controlled cultivation bound by contract.

[3] For KVA: Landeskuratorium für pflanzliche Erzeugung; for DIN ISO: Hopfenring Hallertau; for QS: Fleischprüfring Bayern.

Table 1. Comparison of designs of the selected systems.

	KVA	QS	DIN EN ISO hop
Quality regime/level	quality assurance	quality assurance	quality management
	Documentation for traceability and product-liability purposes[1]		
Scope of application	Bavaria	Germany/Europe	Bavaria
Integration into the supply chain	only producer level	entire chain integrated	entire chain integrated
Initiative	LKP[2], LfL[3], industry	industry and politicians	Hopfenring, political support

[1]Legally valid and acceptable documentation is the basis of future marketability. This must be guaranteed by controls.
[2]Landeskuratorium für pflanzliche Erzeugung in Bayern e.V.
[3]Bayerische Landesanstalt für Landwirtschaft.

SPSS 11.5 and STATA statistical software packages for evaluation. The questionnaire consisted of the following question sets:

- assessment of the individual systems according to the following criteria: reasons for participating and system strengths and weaknesses;
- assessment of quality systems on the basis of predefined criteria on expectations and consequences of system participation (e.g. safeguarding of sales, image gains, etc.);
- sociodemographic questions.

The first two sets of questions will be discussed in detail below.

Differences in perception and acceptance of the three quality systems investigated are probably connected with their design and importance in the relevant marketing system. With respect to the latter, participants' main reason for participating may be an important indicator. Open questions regarding the strengths and weaknesses of individual systems were intended to highlight inter-system differences. The second and third question block were intended to identify factors influencing acceptance of quality regimes in agriculture across all systems. It can be assumed, for example, that less implementation-related effort or improved sales potential in the value-added chain will result in increased acceptance. However, it is also conceivable that differences in assessment might arise because participants vary from system to system in their personal characteristics and farm structure. As far as the design of the questionnaire is concerned, this meant that farm-related and personal questions had to be included as well as questions on the quality system[4].

As so-called 'fact questions'[5] on farm-related and personal data are easy to collect, we will not discuss them in greater detail here. In contrast, when it comes to the design of the set of questions relating to system assessment, which is mainly based on opinion questions, two central tasks have to be solved: (1) identification of key criteria for assessing quality systems, and (2) measurement of these criteria in behavioural science terms.

[4] Within the framework of the farm and personal data, size of farm, age and education of the farm manager, main branch of farming and experience with other quality systems were ascertained.

[5] See also Converse and Presser (1986) on the basic difference between fact and opinion questions.

1. In order to identify assessment criteria, the few papers dealing with farmers' acceptance of quality systems were consulted (Ellner *et al.*, 2004; Fachhochschule Weihenstephan, 2004; Jahn *et al.*, 2003). In addition, discussions were held with experts from the agro-administration, political and scientific sectors concerning the key criteria, leading to the following seven assessment criteria:
   - safeguarding of sales potential;
   - internal management advantages;
   - implementation effort;
   - higher sales revenues;
   - high running costs;
   - high control quality;
   - image gains with consumers and colleagues.

   One difficulty in measuring these seven assessment criteria is that two of the three systems to be assessed are relatively young and the farmer-interviewees have not all had the same amount of experience with them. It is thus difficult for farmers to judge whether the safeguarding of sales potential is a key element of the system, for instance, if they have not already experienced concrete success in this respect themselves. In order to make allowances for this, the criteria 'safeguarding of sales potential', 'increased profits' and 'internal management advantages' were measured twice. On the one hand, the farmer-interviewees were asked to assess them on the basis of their own concrete experiences (empirical dimension) and on the other, to assess the quality systems from the point of view of these criteria (expectation dimension). The other criteria were only measured once. Implementation effort, running costs and control quality are familiar to farmers and it is unlikely that expectations will differ from previous experiences. The opposite applies to the criterion 'image gains with consumers and colleagues'. This is hard for individual farmers to judge, so this criterion was measured solely on the basis of expectations. Altogether, there were thus 10 assessment criteria.

2. A rating scale enabled interviewees to show their degree of approval with respect to the given statements (cf. Table 2). Full-number answer options of -3 (full rejection) to +3 (full approval) were available (Mummendey, 1995: 55)[6]. The zero corresponds to neutral. The analyses were carried out with the help of the statistic programmes SPSS 11.5 and STATA. In total, 74 interviews with ISO hops farmers, 98 interviews with 'KVA Potato' participants and 143 interviews concerning the 'Qualität und Sicherheit' pork system were evaluated. In order to increase measurement precision, two to three statements were formulated for each assessment criterion, with the exception of 'increased profits', and subsequently combined to create an index. With the help of item consistency analysis (reliability analysis), it is possible to verify statistically whether combining the relevant items to create an index is statistically justified. Cronbach's Alpha is a measure of the internal consistency of a scale and should attain a value of 0.7 (Brosius, 2002: 764). The criteria and pertinent statements and Cronbach's Alpha can be found in Table 2.

[6] For telephone interviews, the partially verbalised rating scale has proved its worth, due to its survey efficiency. Only the end points of the scale - here minus three and plus three - and possibly the zero, too, were explained to interviewees.

Table 2. Allocation of items (own computations as per Angermeyer, 2004; Obersojer, 2004; Zapilko, 2004).

Criteria of expectation	Cronb. Alpha	Pertinent statements
Safeguarding of sales potential	0.61	The systems is safeguarding my future sales potential
		Global competitiveness demands quality assurance systems
Improvements in internal management	0.75	It makes my workflow more efficient
		This system helps me to reduce operating costs
Image gains with consumers and colleagues	0.63	With quality assurance I want to be an outrider compared to my colleagues
		The system is important because of the product's image
Higher sales revenues		I expect higher sales revenues/more profit
Criteria of experience		Statements of impacts
Safeguarding of sales potential	0.69	Sales ability/sales potential has increased
		Contact with trading partners is closer now
Advantages in internal management	0.81	Work-flow is more efficient now
		Weak points were discovered
		Operating costs have decreased
Implementation efforts	0.62	There were high investment costs to fulfill the system's efforts
		Modifications in internal workflow have been required
Higher sales revenues		I have an additional profit/higher sales revenues
Running expenditure	0.70	Requirements are overdone
		Documentation is immoderate
		The multi-level control-system is overstated
Quality of control (audit)	0.79	Auditor was competent
		Auditor was objective
		Auditor was consultative

4. Results

4.1 Evaluation of the three systems

4.1.1. Reasons for participating

At the beginning of the interview, interviewees were asked a closed question requiring them to select one out of four motives for participating in the respective system. Figure 2 shows that for KVA participants, the safeguarding of sales potential (41%) and consumer image gains (34%) are important reasons for participation. As regards the QS system, consumer image gains (35%) and higher sales revenues (31%) are the main motivating forces. For the ISO hops growers, more

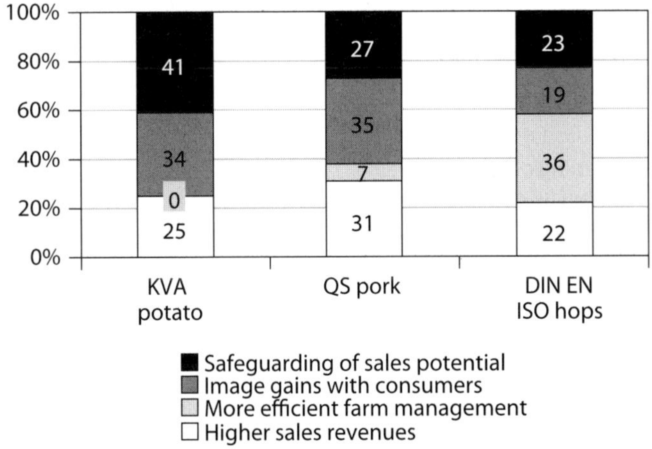

Figure 2. Main reasons for participating by system (own computations as per Angermeyer, 2004; Obersojer, 2004; Zapilko, 2004).

efficient farm management weighs strongest (36%), followed by safeguarding of sales potential (32%). These inter-group differences are significant (χ^2-test).

4.1.2. Strengths and weaknesses of individual systems

An open question regarding the strengths and weaknesses of the individual systems offered interviewees an opportunity to mention and include aspects of their own not investigated by the closed questions. At the same time, the answers provided an initial assessment of the individual systems.

To some extent, the strengths ascertained in this way varied quite considerably from system to system. Table 3 shows the five most frequent responses for each system. In the case of QS pork and KVA potato, the most prominent criteria are largely matching, viz. the safeguarding of sales potential, greater transparency, increased consumer confidence and higher sales revenues. These aspects are further supplemented by positive assessment of documentation and controls. In contrast, the hops growers underscored systematic documentation as a key factor. Observance of all valid statutory requirements, more efficient and transparent farm management and the system-administration services provided by the Hopfenring (hops growers' association) were also seen as positive aspects. Although structured documentation was seen as a positive factor in all systems, market-related benefits are the main criteria with respect to the QS pork and KVA potato systems, whereas with DIN EN ISO hops, management advantages tend to play a more important role.

With respect to system weaknesses, participants in the QS pork and DIN EN ISO hops systems, largely agreed that the additional time and documentation effort without commensurate financial reward (i.e. higher prices) was a negative effect of system participation. The main weakness according to the KVA potato farmers was the lack of financial reward. QS pork and DIN EN ISO hops were frequently seen as too theoretical and impracticable. The QS pork system participants also criticised the lack of effective consumer advertising in particular. The other weaknesses referred to tended to be one-off responses and can be seen in Table 4.

Table 3: Strengths of individual systems (own computations as per Angermeyer, 2004; Obersojer, 2004; Zapilko, 2004).

Ranking	Strength	No. of responses
KVA potato		
1	safeguarding of sales potential	19
2	transparency/ consumer confidence / image gains	17
3	safeguarding of prices/ higher sales revenues	14
4	external control/ quality control	13
5	documentation / traceability	11
QS pork		
1	competitive advantage/ boosting and/or safeguarding of sales	27
2	transparency / consumer image gains and protection	27
3	higher sales revenues	11
4	traceability of previous decisions/ documentation	11
5	cross-chain quality assurance	10
DIN EN ISO hops		
1	systematic, complete documentation	20
2	observance of all valid statutory standards	10
3	more efficient farm management	9
4	improved farm management transparency	7
5	services provided by Hopfenring	7

Table 4. Weaknesses of individual systems (own computations as per Angermeyer, 2004; Obersojer, 2004; Zapilko, 2004).

Ranking	Weakness	No. of responses
KVA potato		
1	no additional revenues/inadequate financial reward	16
2	cut-off date control too imprecise	4
3	lack of advice at audits/controls	3
4	binding contract with purchaser	3
5	obligation to purchase plants	3
QS pork		
1	no additional revenues/inadequate financial reward	42
2	documentation effort / duplications	31
3	lack of consumer awareness / insufficient advertising	22
4	time effort	17
5	impracticable / inefficient	14
DIN EN ISO hops		
1	documentation effort / bureaucracy	22
2	additional time effort	14
3	no additional revenues / no acknowledgement	11
4	too theoretical / too complex	10
5	additional costs	3

Whereas, with respect to system strengths, the benefits that can be derived from the quality regimes in Section 2.1 largely predominate, the additional revenues that generally fail to eventuate are participants' greatest problem. With QS pork and DIN EN ISO hops, the documentation effort was a further source of considerable dissatisfaction.

As already described in Section 3, investigation of 4 of the 10 assessment criteria concerned participants' expectations. It was established that the expectations of the participants in the DIN ISO hops system, the only quality management system investigated, were much higher, except with respect to potentially higher prices, than those of participants in the other two systems. Above all, the latter did not expect any management benefits through system participation. Otherwise, expectations are in the vicinity of 1 (Figure 3).

It is interesting to note that there were no great differences between the three quality regimes with respect to overall acceptance (Figure 4). However, hop growers saw their system as having improved production efficiency, whereas the others did not do so to the same extent. On the other hand, hop growers did experience lower additional sales revenues compared to farmers taking part in the two quality assurance schemes. Finally, potato growers saw their system as protecting sales, which the farmers participating in the other systems did not. It can thus be hypothesised that overall acceptance of quality management systems is not negatively affected by whether they are more demanding on farmers. This result might be explained by internal management improvements experienced by the participants.

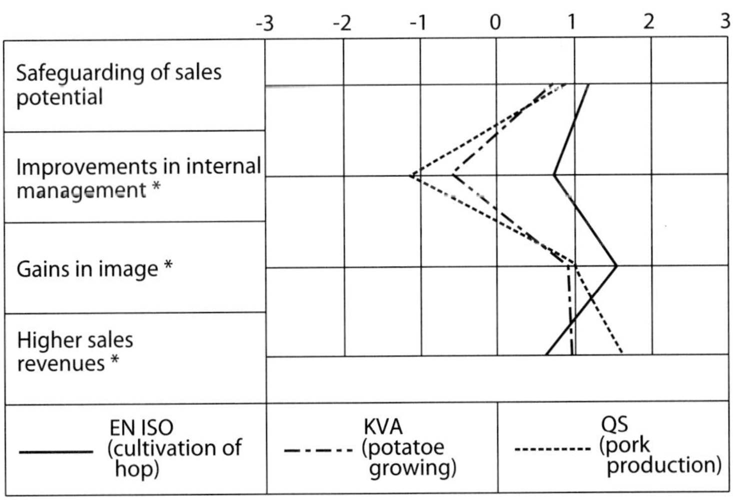

*Figure 3. Evaluation of three different quality systems on the basis of farmer's expectations (own computations as per Angermeyer, 2004; Obersojer, 2004; Zapilko, 2004). All factors were measured on a 7-point rating scale. * means that the differences between the three quality regimes were significant using the Duncan test.*

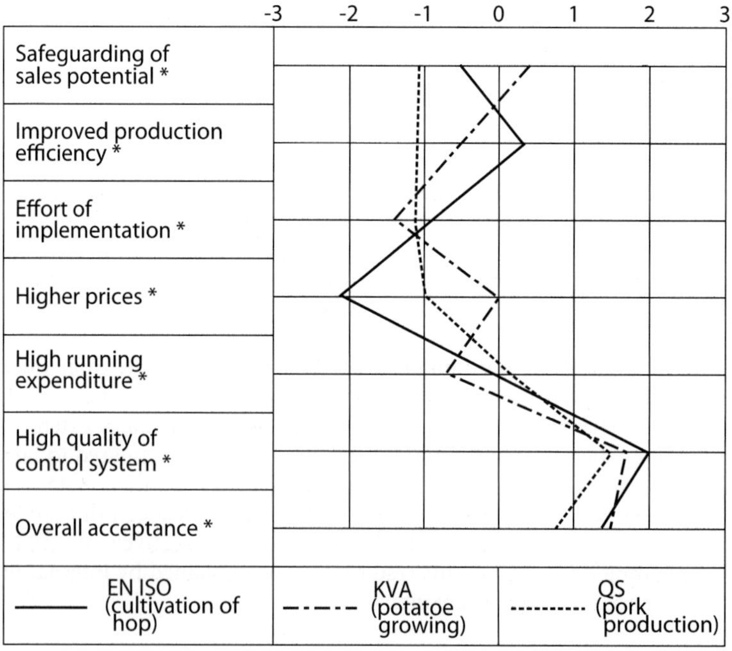

*Figure 4. Evaluation on the basis of farmer's experiences. All factors were measured on a 7-point rating scale (own computations as per Angermeyer, 2004; Obersojer, 2004; Zapilko, 2004). All factors were measured on a 7-point rating scale. * means that the differences between the three quality regimes were significant using the Duncan test.*

4.2. General success factors

The three systems investigated are, of course, rather heterogeneous and focus on very different agricultural products. Thus, the system differences discussed above should be interpreted with caution. In order to obtain more reliable information on factors determining the acceptance of quality systems we made use of the heterogeneity within the three sub-samples. In an ordinal logit model, overall acceptance measured on a 7-point rating scale was regressed on factors potentially influencing acceptance (Table 5). Apart from the above described expectations and experiences, the differences between the study dimensions sales, efficiency and revenues with respect to expectations and experiences were ascertained and integrated in the solution of the model as additional variables. In addition, personal and farm-related variables, such as farm manager's age, were included in the model.

Apart from age, socio-demographic factors do not have any significant impact on farmers' acceptance of quality systems. Key factors are gains in image, sales and production efficiency. Acceptance of quality systems is higher among farmers who believe that the pertinent quality system will improve the producer's image than among those who do not believe so. It is interesting to note that 'expected image gains' is the most influential factor followed by 'experienced sales' and 'experienced efficiency'. It should also be noted that the difference between experienced and expected improvements in sales has a significant impact on overall acceptance of quality systems. Farmers whose expectations have not yet been fulfilled but who expect increasing sales

Table 5: Results of ordinal logit model significant (own computations as per Angermeyer, 2004; Obersojer, 2004; Zapilko, 2004).

Variable	Coefficient	Standard error	t-value[1]
Age	0.019	0.011	1.72*
Experienced effort	-0.233	0.088	-2.65***
Expected image	0.620	0.102	6.06***
Experienced sales	0.512	0.121	4.23***
Experienced efficiency	0.377	0.098	3.84***
DIFF sales[2]	0.363	0.097	3.75***

Obs=315; LR chi^2 = 179.06; Prob > chi^2 = 0.0000; Log likelihood = -411.30679; Pseudo R^2 = 0.1788
[1]Significance * = at the 0,1 level, ** = at the 0,05 level, *** = at the 0,01 level
[2]Difference between effectively experienced sales and expected sales in the future.

in the future obviously show greater acceptance of quality systems than farmers who have already experienced higher sales but assess future sales more pessimistically.

5. Conclusions

The strengths of the systems are clearly system-specific ones. With the KVA potato and QS pork quality assurance systems, external and market-oriented parameters such as safeguarding of sales, transparency and consumer confidence are acknowledged system strengths, whereas with DIN ISO hops, internal, farm-related systematic documentation is of importance. Of course, the systems also show weaknesses. The QS pork and ISO hops participants criticised the additional time needed for documentation without commensurate financial reward. Besides, both systems were partly classified as too theoretical.

The results of this study also show that quality management systems are not generally rejected by farmers because of higher costs or organisational effort. Efficiency improvements, gains in image and sales may compensate for costs associated with the introduction of quality systems. As such, higher demands may even motivate farmers to continue participating in quality systems. The possibility of enhancing acceptance of the QS and KVA systems by an add-on management module should therefore be investigated. Finally, farmers must be convinced of the benefits of any quality system.

References

Angermeyer, M., 2004. Die Qualitätssicherungssysteme KVA, EUREPGAP und IFS bei Speisekartoffeln - Eine Bewertung aus Sicht der Primärerzeuger und der abnehmenden Hand. Master thesis, Freising.

Amelung, C., S. Kiefer, T. Scherb and J. Schwerdtle, 2002. Qualitätssicherung bei Schweine- und Geflügelfleisch - Konzepte und praktische Umsetzung. In: Landwirtschaftliche Rentenbank (Ed.): Lebensmittelsicherheit und Produkthaftung - Neue Entwicklungen in der integrierten Produktion und Vermarktung tierischer Erzeugnisse. Landwirtschaftliche Rentenbank, 43-92.

Brosius, F. 2002. SPSS 11. VS Verlag für Sozialwissenschaften.

Ellner, R. and T. Huke, 2004. Abschlussbericht zum Pilotvorhaben 'Umsetzung von QM Milch in Bayern'. München.

Erling, P., 1999. Qualitätsmanagement in landwirtschaftlichen Erzeugergemeinschaften - Analyse und Konzeption am Beispiel der Produktion und Vermarktung von Brotgetreide. Agrimedia-Verlag.

Fachhochschule Weihenstephan, 2004. Akzeptanz des Programms Geprüfte Qualität - Bayern, Befragung von Landwirten und weiteren Marktbeteiligten Gruppen. Bayrische Landesanstalt für Landwirtschaft.

Geiger, W., 1994. Qualitätslehre - Einführung, Systematik, Terminologie. In: Deutsche Gesellschaft für Qualität (Ed.), DGQ-Schrift 11-20. Vieweg Verlag 1986.

Jahn, G., M. Peupert and A. Spiller, 2003. Einstellungen deutscher Landwirte zum QS-System: Ergebnisse einer ersten Sondierungsstudie, Working Paper 0302. Institut für Agrarökonomie, Göttingen.

Lehnert, S., B. Petersen, B. Kolb and P. Jürgens, 1995. Klein- und Mittelständische Betriebe auf dem Weg zur Zertifizierung. In: Forschungsgemeinschaft Controlling in der Landwirtschaft (Ed.), FCL-Schriftenreihe Band 3. Forschungsgemeinschaft Controlling in der Landwirtschaft.

Menke, P., 1999. Qualitätsproduktion aus Sicht der Landwirtschaft. In Kuratorium für Technik und Bauwesen in der Landwirtschaft e.V. (ed.), Integrierte Qualitätsproduktion in der Landwirtschaft - Verfahren, Techniken und Bewertungskriterien - KTBL/DLG-Vortragstagung am 21 April 1999, Veitshöchheim.

Mummendey, H.D., 1995. Die Fragebogenmethode, 2nd Edition. Hogrefe-Verlag.

Obersojer, T., 2004. DIN EN ISO 9001 Zertifizierung von Hopfenerzeugern - Systemkonzeption und dessen Bewertung aus Sicht der Teilnehmer. 92nd SeminarQuality Management and Quality Assurance in Food Chains 2-4 March 2005, Göttingen.

Schneider, B., 1995. Wettbewerbsfaktor 'Qualität' - Eine Herausforderung für die deutsche Landwirtschaft, Nutzen, Ziele und Hintergründe der ISO-Normen. Frankfurt (Main).

Weindlmaier, H., 1999. Ökonomische Bewertung und Perspektiven des Qualitätsmanagements. In: Kuratorium für Technik und Bauwesen in der Landwirtschaft e.V. (Ed.), Integrierte Qualitätsproduktion in der Landwirtschaft - Verfahren, Techniken und Bewertungskriterien - KTBL/DLG-Vortragstagung am 21. April 1999 in Veitshöchheim.

Weindlmaier, H., 2002. Transparency landwirtschaftlicher Produktionsprozesse erhöhen. VDL Journal Magazin für Agrar, Ernährung, Umwelt 52, No. 10, 8-9.

Zapilko, F., 2004. Das Qualitätssicherungsprogramm Qualität und Sicherheit am Beispiel Schweinemast - Eine Bewertung aus Sicht der Primärerzeuger. Master thesis, Freising.

Controversial positions about the QS System in agriculture: an empirical study

Gabriele Jahn and Achim Spiller
University of Goettingen, Department of Agricultural Economics and Rural Development, Platz der Goettinger Sieben 5, 37073 Goettingen, Germany; a.spiller@agr.uni-goettingen.de

Abstract

A drop in consumer confidence stimulated private initiatives to enhance quality assurance throughout the entire German agri-food chain. Hence, representatives of the meat sector came together to establish the so-called QS system for meat products. The positions about the adoption have been controversial since then. A better understanding of the farmers' attitudes is necessary to increase acceptance and to guarantee the longer-term success of the QS system. Against this background we designed a theoretical framework based on the popular behavioural research approach of Fishbein and Ajzen. A survey was conducted in which farmers were asked about their attitudes towards the QS system. Results highlight that intention is significantly affected by the attitudinal factors, whereas behaviour is not. The high influence of the perceived coercion for adoption on both intention and behaviour reveals that the current high adoption rate is linked to external incentives and not to farmers' beliefs.

Keywords: certification, adoption, attitudes, Germany

1. Introduction

Certification systems, such as ISO 9000, are widespread and well established in almost all industrial sectors. In German agriculture, however, most quality assurance and certification schemes are in an initial stage of implementation. The main reason for this time lag has been the broadly accepted public responsibility for food quality and safety. Due to consumer protection laws, governmental controls have played a major role in the agri-food sector compared to other industrial sectors. Food scandals, however, have highlighted the fact that systems based on public monitoring and end-product control neither sufficiently ensure food safety nor allow the production of quality products. Hence, initiatives driven by the private sector have come to the fore.

As a result of such an initiative, representatives of the German meat sector established the QS system ('System Qualität und Sicherheit'). The main objective of the QS is to regain consumers' trust in food products by ensuring transparency and traceability throughout the entire meat supply chain. The first QS certified products were launched in 2002. Despite an increasing number of participants, the QS system has been discussed controversially ever since its introduction.

So far, only a few studies have dealt with farmers' attitudes towards quality assurance systems in the German food sector. Referring to experiences in other countries, two main problems have been revealed: (1) the cost-benefit ratio is often negatively evaluated, and (2) communication, which is necessary for successful implementation, is neglected (Fitzgerald *et al.*, 1999; Böcker *et al.*, 2003). As a consequence, a 'gap of acceptance' is emerging among the adopters.

The research objectives of this study have been formulated against this background. Firstly, it was aimed at gaining a more detailed insight into the adoption process regarding the implementation of the QS system. This is particularly important for longer-term success of the QS system in the German meat sector. Secondly, a theoretical framework was designed based on two very popular behavioural research models - the Theory of Reasoned Action and the Theory of Planned Behaviour - exploring the relation between attitudes, intention and behaviour (cf. Fishbein and Ajzen, 1975; Ajzen, 1991). The study transferred these research approaches to farmer's perception during the adoption process focusing particularly on the intention-behaviour relation. The theoretical model was then tested empirically by conducting a survey.

2. The QS certification approach

2.1. Certification as a quality tool within the food chain

The origin of the current quality assurance systems can be traced back to the statistical product controls, which were already widespread in the middle of the 20th century (Sinha and Willborn, 1985). Increasing globalisation, e-procurement and just-in-time production, however, have set a stimulus towards more process-oriented concepts. With the establishment of the ISO 9000, a first certification system was designed. This was the beginning of a broad differentiation process of more specific certification systems. Compared to the earlier audit approaches (first and second party audits), a certification approach (third party audit) is characterised by inspections carried out by neutral institutions on an accredited basis (Luning *et al.*, 2002).

In comparison to many other industrial sectors, quality assurance within the food sector is quite complex. Most food products are characterised by quality attributes that cannot be tested by the customers (credence quality). A lack of corresponding quality signals can result in market failure (Akerlof, 1970; Nelson, 1970). 'Reliable' certification systems, however, can bypass these information asymmetries and increase transparency in markets (Auriol and Schilizzi, 2002; Hobbs, 2004). The quality level can be communicated to the customers by awarding the respective certificates and labels (Jahn *et al.*, 2005).

Certification schemes are suitable particularly in spot markets since specific investments are not required. A necessary assumption is the acceptance of the certificate by all the customers and that it is treated as a baseline standard for market access. Consequently, a high number of adopters increase market transparency and thus the benefits of all the participants (network effects).

Finally, it has become obvious that in spite of the diffusion throughout all the industrial sectors, the effectiveness of many certification approaches is contested. While many managers associate the ISO 9000 with a purely paper-driven and bureaucratic process, to others the implementation appears as a suitable starting point in achieving a comprehensive quality management system in the firm (Curkovic and Pagell, 1999). Although on the one hand, increasing transparency and positive effects on business performance and quality cost have been reported (Jatib, 2003), so that as a consequence, particularly bigger firms might realise cost savings. It has also been found that smaller firms expect a competitive advantage as quality pioneers (Caswell *et al.*, 1998). It is known that on the other hand, certification systems are often less flexible than first or second party audits, so that competitive advantages might vanish with increasing adoption. After all, empirical studies have often highlighted that adoption is not caused by cost-benefit considerations, but by the pressure of the customer (Chow-Chua *et al.*, 2003; Martinez-Lorente and Martinez-Costa, 2004; Sun *et al.*, 2004).

2.2. Quality assurance in the meat sector

The preconditions in the German meat sector for establishing quality assurance systems is quite unfavourable compared to the situation in other countries. A low degree of vertical integration characterises the market. Short-term contracts are common leading to spot markets. The lack of trust and commitment between the slaughterhouses and the farmers might be the main reason why slaughterhouses return only a low fraction of the recorded data. Only a few quality assurance schemes that included vertical integrated contracts and internal audits were established before the introduction of the QS system. However, the over-capacities of the slaughterhouses make it impossible to build up enough pressure towards their suppliers to implement quality management on their farms.

Among the different types of meat production in Germany, the poultry sector forms an exception. One major processor (PHW group) covers 48% of German poultry production (Lüth and Spiller, 2005). This structure simplifies the enforcement of quality assurance within the poultry supply chain. Thus, a quite comprehensive system of quality controls to ensure traceability has already been implemented independently of other segments of the meat sector.

In succession of the recent food scandals, public attention and political players have exerted pressure on the meat sector. Opponents had taken up the above-mentioned arguments against other certification systems, such as the ISO 9000. They particularly feared an increase in bureaucracy on farms. Nevertheless, representatives of the entire meat supply chain agreed to found the QS Company as the core institution of a new quality assurance system for food products. Even though a consensus might sometimes be difficult to achieve within the company's heterogeneous board consisting of representatives from the entire food chain, this structure guarantees traceability throughout the whole supply chain.

The QS system is a private certification approach and thus suited for spot markets. In comparison to concepts in other countries (e.g. the Netherlands or the United Kingdom), the German QS system includes the entire food sector (retailers as well). Since its establishment, the number of participants has been continuously increasing and in the meantime 56,000 locations have been certified, including 46,300 farms. Forty certification bodies (464 auditors) have a license to carry out the inspections (QS, 2005). In spite of the increasing diffusion, a significant uncertainty about the actual willingness of the farmers to participate in the scheme has remained. As this might increasingly influence the motivation to enhance product and process qualities, it was decided to investigate the factors which determine attitudes and intention towards the adoption of the QS system.

2.3. Conceptual framework of the study

The following section deals with the conceptual framework of the study, which is focussed on the acceptance of the QS certification at farm level. Behavioural research offers a broad variety of approaches for analysing the relationship between attitudes and behaviour, which can be applied in such a study. These models are used to predict and understand behaviour. Their core hypothesis is the causal effect of attitude towards behaviour (Raden, 1985). Seminal in this field of research are Fishbein and Ajzen's Theory of Reasoned Action (TRA) (Fishbein and Ajzen, 1975) and Ajzen's Theory of Planned Behaviour (TPB) (Ajzen, 1988, 1991). Both theories are quite easy to implement and can be applied to a broad range of behavioural domains (Leone *et al.*, 1999). The direct predictor of behaviour is intention as the cognitive dimension of the attitude construct. Thus, 'the intention is interpreted as a transition between the cognitive and evaluative components of attitude and behaviour and intentions are direct functions of both individual and social related variables [...]. Moreover, intentions mediate the influence of the attitudinal variables on the behaviour totally so that no direct path from attitude to behaviour is hypothesised' (Leone *et al.*, 1999: 162).

In applying TRA and TPB to the implementation of quality assurance systems, the research model had to be modified. The main reason for this is that in the case of the German meat sector the adoption is not voluntary, but nearly necessary for market access. The voluntariness is often an implicit assumption of TPB and TRA. The focus must therefore be moved towards the attitude-intention relation. Actual behaviour does not present a suitable indicator for the degree of acceptance. Nevertheless, a view on the relation between attitudes and behaviour can still be justified: Pioneers who are highly motivated and who have a positive attitude might be explicitly represented during the current initial stage of the adoption (Rogers, 1983).

Similar findings due to the mandatory use of innovations have been revealed by Brown *et al.* (2002) in applying an acceptance model on the use of information systems. The authors also suggest focusing on attitudes and intentional behaviour. Figure 1 depicts the TRA and the TPB in comparison to the present approach.

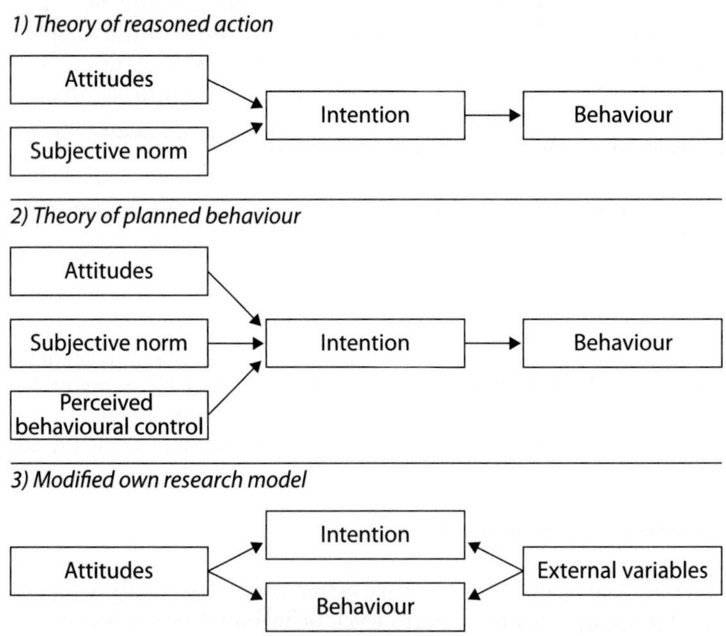

Figure 1. Research models on attitude-intention-behaviour relationships.

3. Methodology

3.1. Procedures and hypotheses

The methodology of the analysis consists of different parts whereof the first part described the pattern of the farmers' attitudes by clustering the sample; with this the farmers' positions could be revealed. The preliminary stages of the analysis already indicated the suitability of the data set for the application of a cluster analysis (high standard deviations among others). Before doing this, a factor analysis was applied in order to reduce the number of variables and so, determine the main attitudinal components. These factors were the base for the further steps of the analysis (cf. Figure 2).

A second part dealt with the relationship between intention and attitude. In practice, this consideration is necessary to guarantee a successful implementation on a long-term perspective. Intention is defined as the dependent variable. The independent variables are the attitudinal factors and control variables. The core hypothesis (*H1*) describes the impact of a positive attitude towards a positive intentional behaviour according to TRA and TPB:

H1: A positive attitude yields a positive intention towards the introduction of the QS system on the farm.

In the third part, the analysis considered the relationship between attitudes and the actual implementation of the QS system. A logistic regression model underlined this analysis. Analogue to H1, a guiding hypothesis (H2) could be formulated:

H2: A positive attitude is conducive with the implementation of the QS system

Finally, the analysis examined the relation between behavioural intention and behaviour and the separated considerations as suggested in our conceptual framework. The following hypothesis is formulated:

H3: The intention does not contribute to explain and predict behaviour.

In order to test this hypothesis a chi square difference test was conducted analysing the differences between the attitude-behaviour regression model and an extended model, which comprises the same independent variables and in addition behavioural intention. Figure 2 summarises the procedures and the applied statistical methods.

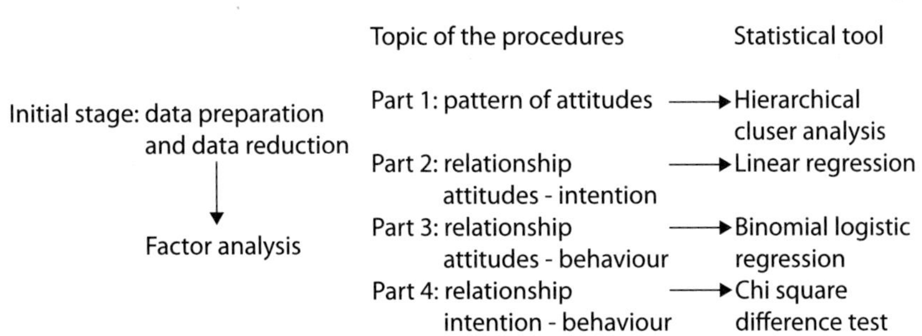

Figure 2. Procedures involved in the analysis.

3.2. Data sample

A pilot study on the acceptance of the QS system was conducted in 2003, in which a group of 94 farmers with animal husbandry were interviewed, mostly in Lower Saxony. The sample of farmers interviewed (n=94) is a 'convenience sample' and does not fulfil all the criteria of representativeness. It includes more 'future-oriented' and bigger farms than the average in Germany. However, these farms might be decisive for future developments as it is the larger farms that gain more importance due to the structural changes in German agriculture. Altogether 31.9% (n=30) of the interviewed farmers were already certified, 38.3% (n=36) were preparing for the certification process and 29.8% (n=28) were not willing to participate. Eight of the farms were organic and rest were conventionally farmed.

3.3. Measures

For measurement, attitudes are divided into an affective and cognitive component: 'the affective-cognitive dichotomy reflects a duality between heart and mind that has long been assumed in Western culture' (Giner-Sorolla, 2004: 762). Affective attitudes are defined 'as the summed valence of specific emotions and feelings associated with the object', and cognitive attitudes as 'the summed valence of the object's perceived non-emotional properties' (Giner-Sorolla, 2004: 762). For further operationalisation, theoretically derived cognitive and affective variables have been utilised. In addition to the cost-benefit considerations mentioned by Chow-Chua *et al.* (2003), suitability and usefulness have been added (Davis, 1989; Szajna, 1996; Mathieson *et al.*, 2001). Further variables refer to the reliability of the QS system and the perceived fairness of the introduction (Frey and Oberholzer-Gee, 1997; Frey and Jegen, 2001). All the variables were measured by means of a 5-point Likert scale.

4. Results

4.1. Initial stage: data reduction

As a first step, 14 theoretically derived statements were reduced by means of a factor analysis. Five factors were revealed as determining attitudes of the interviewed farmers: (1) perceived necessity and effectiveness of the QS system, (2) expected benefits, (3) perceived external pressure, (4) expected structural impact, and (5) perceived importance (cf. Table 1).

The attitude dimension 'perceived necessity and effectiveness' (15.78% of the total variance explained) covers all those statements, which the farmers used to evaluate the necessity and suitability of the system at the farm level and the standard itself.

The second factor 'expected benefit' (13.01% of the total variance explained) refers to the potential benefits, which the farmers expected, i.e. a higher profit, better collaboration with the slaughterhouses and a positive effect on the farm's image.

The third component, the 'perceived external pressure' (12.18% of the total variance explained), includes those statements that primarily refer to the affective dimension: the perceived fairness associated with the introduction of the QS system, the perceived coercion of the decision related to the reliability of the label.

Table 1. Factor analysis.

Components	Variables	Factor loadings
Perceived necessity and effectiveness (PNE)	Necessity of quality assurance	0.782
	Effectiveness of the criteria	0.719
	Profitability of quality assurance	0.639
	Assessment of the cost-benefit ratio	0.597
Expected benefit (EB)	Improvement of farm's image	0.816
	More trust towards customer	0.728
	Expected gains	0.620
Perceived external pressure (PE)	Label's reliability	0.826
	Fairness of the introduction	0.652
	Participation compelled	0.625
Expected structural impact (SE)	Acceleration of structural changes	0.759
	Initiation of structural change	0.755
Perceived importance (PI)	Necessity of an international standard	0.795
	Importance of quality problems	0.750

Sample size n=94, method of factor extraction: Principal Component Analysis, Rotation: Varimax Total variance explained: 62.28, KMO = 0.63, Cronbach's alpha: $\alpha = 0.6$.

The fourth component refers to the expected structural impacts (10.68% of the total variance explained), i.e. an increased exclusion of low-competitive farms and an acceleration of the structural changes in total.

Finally, the fifth factor 'perceived importance' comprises an evaluation of the quality problems in general and the international relevance of a QS system (10.37% of the total variance explained).

Summarising the results of the factor analysis, it becomes apparent that the first two factors and Factor 5 refer to the benefits related to the QS system. Together they explain about 40% of the total variance and determine primarily attitudes towards the QS system.

These five attitudinal factors represent the base of the following analysis. The reliability of the factor analysis (KMO=0.63, α=0.6) is considered as being adequate given the explorative character of this study (Hair *et al.*, 1998).

4.2. Pattern of attitudes

Before analysing the assumed hypotheses and causal effects of the attitudinal variables, the first module allows a detailed insight into the different attitudes of the farmers. By means of a hierarchical cluster analysis (Ward method), the sample was divided into four segments. In addition, the ANOVA (cf. Table 2) and a discriminant analysis (Wilk's Lambda: $p < 0.001$), confirmed the selected clusters. Outliers were detected by the Single Linkage method.

Only the 'Rejecters' (n=10) would not recommend the introduction of the QS system. The three other clusters would advise its adoption. Nevertheless, the necessity and effectiveness of the QS system were clearly contested. The group of 'Indifferents' (n=33) discerns neither a necessity nor

Table 2. Results of the cluster analysis.

	Indifferents (n = 33)	Intrinsically motivated supporters (n = 33)	Extrinsically motivated supporters (n = 17)	Rejecters (n = 10)	F-Statistics (ANOVA)
Behavioural intention	0.29 (slightly positive)	1.17 (clearly positive)	0.41 (positive)	-0.13 (slightly negative)	5.486 (p=.002)
Clustering variables					
Perceived necessity and effectiveness (PNE)	-0.83 (negative)	0.69 (positive)	0.49 (positive)	-0.38 (rather negative)	26.685 (p=.000)
Expected benefit (EB)	-0.07 (indifferent)	0.66 (positive)	-1.39 (clearly negative)	0.30 (slightly positive)	32.733 (p=.001)
Perceived external pressure (PE)	0.04 (indifferent)	0.07 (indifferent)	0.03 (indifferent)	-0.49 (negative)	0.872 (p=.459)
Expected structural impact (SE)	0.48 (positive)	-0.48 (negative)	0.01 (indifferent)	-0.23 (slightly negative)	6.320 (p=.000)
Perceived importance (PI)	0.47 (positive)	0.23 (slightly positive)	-0.09 (indifferent)	-1.85 (clearly negative)	33.344 (p=.000)
Control variables					Chi^2 test
QS certified farms	6[1] (10.3)[2]	14 (10.3)	9 (5.3)	0 (3.1)	12.824 (p=.005)
Farms prepared to participate	24 (22.7)	26 (22.7)	13 (11.7)	1 (6.9)	19.668 (p=.000)
Organic farming	0	3	0	5	26.514 (p=.000)

Sample size n=94; hierarchical cluster analysis, Ward method; Outliers detection: Single Linkage method.

[1]Observed number of cases within the sample.

[2]Expected number (cf. results of cross tabulation).

any benefits associated with QS. The 'Intrinsically motivated supporters' (n=33) are convinced of the benefits and the effectiveness of an introduction of the QS, whereas the 'Extrinsically motivated supporters' (n=17) do not recognise any benefits, but accept the necessity of the QS system. The number of QS certified farms is higher in the last two clusters than in the others, while the majority of the organic farmers belong to the 'Rejecters'.

4.3. Attitude-intention relationship

To continue with the analysis as illustrated in Figure 2, a linear regression was applied by using the method of least squares (OLS) as estimation procedure. Besides the attitudinal factors and the perceived external pressure, two control variables (membership in a producer association and organic farming) were tested. Table 3 summarises the results of the attitude-intention model.

As the results show, the basic hypothesis (*H1*) is supported, i.e. attitudes and behavioural intention are positively correlated. Furthermore, the regression model reveals the following three relationships:

- The higher the perceived necessity and effectiveness, the higher the intention to adopt the QS system.
- The higher the expected benefit, the higher the intention to adopt the QS system.
- The higher the perceived external pressure, the higher the intention to adopt the QS system. This factor is the one with the most important effect on the intention.

An influence of the two attitudinal factors referring to the expected structural impact and the perceived importance had to be rejected. The two control variables, organic farming and producer association, had no significant effect either.

4.4. Attitude-behaviour relationship

The third past of the analysis dealt with a binomial logistic model testing the relationship between the attitudinal factors and behaviour. Therefore the farmer were divided into two groups (Table 4): All the farmers who were participating in the QS system or who were preparing for their first audit were compared with those farmers who were not trying to participate up to the time of data collection. We integrated the same control variables as in the linear model from above.

As the first hypothesis about the attitude-intention relation (*H1*) the second hypothesis about the attitude-behaviour relation (*H2*) can be confirmed, but not as clear as the first one. Only the perceived external pressure affects significantly behaviour. The other factors indicate no or only weak correlation. Instead, the structural variables have a high impact: While the membership in a producer association influences behaviour positively, organic certification has a negative impact on the introduction of the QS system. Results highlight thus that in the current stage of adoption external factors play the decisive role and not the attitudes.

Table 3. Attitude-intention model. Linear regression: dependent variable - behavioural intention.

Independent variables	Coefficients (beta)	Standard error	t-values
Perceived necessity and effectiveness (PNE)	0.369	0.088	4.207 (p=0.000)
Expected benefit (EB)	0.370	0.086	4.312 (p=0.000)
Perceived external pressure (PE)	0.532	0.087	6.124 (p=0.000)
Expected structural impact (SE)	0.071	0.091	0.785 (p=0.435)
Perceived importance (PI)	0.123	0.090	1.367 (p=0.176)
(Intercept)	0.558	0.089	6.239 (p=0.000)

Sample size n=94, R = 0.702, adj. $R^2 = 0.460$; F = 14.956, p = 0.000.

Table 4. Attitude-behaviour model. Binomial logistic model: dependent variable - behaviour (Behaviour coded into participating/preparing to participate (yes: n = 66 / no: n = 28)).

Independent variables	Coefficient (beta)	Standard error	Wald
Perceived necessity and effectiveness (PN)	0.654	0.340	3.703 (p=0.054)
Expected benefit (EB)	0.468	0.357	1.557 (p=0.212)
Perceived external pressure (PEP)	1.315	0.383	11.780 (p=0.001)
Expected structural impact (SE)	-0.302	0.289	1.096 (p=0.295)
Perceived importance (PI)	0.303	0.310	0.953 (p=0.329)
Control variables			
Organic certification	-3.370	1.307	6.650 (p=0.010)
Member of a producer association	2.264	0.750	9.112 (p=0.003)
(Intercept)	0.723	0.727	0.359 (p=0.549)

Sample size n=94, McFadden's R^2 = 0.355, Chi^2 (df = 7) = 40.6; p = 0.000.

4.5. Intention-behaviour relationship

The conceptual framework finally suggests that behaviour should be treated separately from intention due to the mandatory adoption of the QS system (*H3*). To examine the relation between intention and behaviour we included intention into the estimated binomial logistic model from above. The chi square differences test between the extended model and the model as summarised in Table 4 reveals a Chi^2 of 0.928 (df=1) resulting in a non-significant p value higher than 0.3. Intention therefore does not contribute significantly to explain or predict behaviour. Other variables such as external pressure or membership in a producer association are more important.

5. Implications

Although the adoption of the QS system was in an initial stage as the survey was conducted, our results highlight that farmer's convictions rarely play a role. These findings are contradictory to experiences in adoption theory. Particularly during the initial stage the adopters are normally pioneers who are characterised by high motivation and beliefs. This underlines what we suggested in our conceptual framework: Key determinants and interdependencies, which are normally discussed in regard to voluntary adoption processes, have to be analysed carefully and not all of them are suitable for analyses about the acceptance of systems with mandatory use.

Nevertheless, the constructs and measures from the above-mentioned behavioural research models present starting points for further research on the mandatory use of quality assurance systems. In our study we concentrated on attitudes, intention and behaviour. Further studies should consider more constructs. The integration of additional external variables such as image or output control (Chau, 1996; Venkatesh and Davis, 2000; Mathieson *et al.*, 2001) might be useful. They are often applied to extend the technology acceptance model for information systems developed by Davis (1989). TPB and TRA also provide possible constructs such as subjective norms or perceived behavioural control, which can influence attitudes and intentional behaviour decisively.

Finally, the set of measures on attitudes and intentional behaviour towards the QS system should be extended. By means of confirmative approaches the best indicators can be selected. Such an instrument could be the base for regular evaluation and the development of the guidelines associated with the QS system. Only suitable and accepted quality assurance systems permit improvements in the effectiveness of data exchange between farmers and their customers and enhances cooperative behaviour and the relationship on longer-term.

6. Conclusions

Our pilot study about the newly introduced QS system in the German meat sector provides already a detailed view of farmers' acceptance. Although only a few farmers are convinced of its benefits and effectiveness, the majority would recommend the implementation of this system.

By investigating the pattern of attitudes it became apparent that, in practice, there are at least two starting points for enhancing the system and increase the acceptance:

1. Effectiveness. Farmers claimed that they could neither reduce their costs nor optimise the processes on their farms by implementing the QS guidelines. Instead of that, they had higher ongoing costs associated with documentation. A first option would therefore be a reduction and/or revision of the QS criteria; not all might be necessary or suitable.
2. Communication. So far, the aims and benefits of the QS system have not been properly communicated - neither to the farmers, nor to the consumers. External pressure is still the main reason for implementing the QS system. In the long run, imposing a system without any convincing arguments is a threat to the system as this will provoke opportunistic behaviour. Improved communication could clearly increase acceptance and enhance the relationship between supplier, customer and system owner.

The application of the conceptual framework in this study highlights the fact that the relation between intention and behaviour must be considered carefully in acceptance studies on quality assurance systems. In our case a separated analysis is important due to the very early stage of implementation. Contrarily to more 'classical' behavioural research approaches, behaviour is not or only little influenced by behavioural intention: In our case intention is closely related to almost all attitudinal factors, while behaviour is not. An exception is the high importance of the perceived external pressure in both regression models. The consideration of such a construct appears to be suitable for studies about the adoption of quality assurance systems as a necessary requirement for market access.

In conclusion, this research shows that although the majority of the farmers questioned accepted the introduction of the QS system, they were not convinced of its cost-benefit relationship. The current high adoption rate is more associated with the requirements of customers and external pressure. An enhanced effectiveness and communication is necessary in terms of increasing the intentional behaviour of the participants in such a scheme and to ensure diligence in implementing the standards.

References

Akerlof, G.A., 1970. The Market for 'Lemons': Quality uncertainty and the market mechanism. Quarterly Journal of Economics 84, 488-500.

Ajzen, I., 1988. Attitudes, personality and behaviour. Open University Press.

Ajzen, I., 1991. The theory of planned behavior. Organisational Behavior and Human Decision Processes 50, 179-211.

Auriol, E. and S.G.M. Schilizzi, 2002. Quality signaling through Certification. Theory and an application to agricultural seed markets, Research paper, University of Toulouse, Toulouse, 1-33.

Böcker, A., M.E. Bredahl and J. Northen, 2003. ISO 9000 certification in British agribusiness: motivations and performance impacts. In: Schiefer, G. and U. Rickert (eds.), Quality Assurance, Risk Management and Environmental Control in Agriculture and Food Supply Networks. Proceedings of the 82nd Seminar of the European Association of Agricultural Economists (EAAE), University of Bonn, Bonn, Germany, pp. 51-60.

Caswell, J.A., M.E. Bredahl and N.M. Hooker, 1998. How quality management systems are affecting the food industry. Review of Agricultural Economics 20, 547-557.

Chau, P.Y.K., 1996. An empirical assessment of a modified technology acceptance model. Journal of Management Information Systems 13, 185-204.

Chow-Chua, C.M.Goh and B.W. Wan, 2003. Does ISO 9000 certification improve business performance?. International Journal of Quality & Reliability Management 20, 936-953.

Curkovic, S. and M. Pagell, 1999. A critical examination of the ability of ISO 9000 certification to lead to a competitive advantage. Journal of Quality Management 4 (1), 51-67.

Davis, F.D., 1989. Perceived usefulness, perceived ease of use, and user acceptance of information technology. MIS Quarterly 13, 319-340.

Fishbein, M. and I. Ajzen, 1975. Belief, attitude, intention and behaviour: An introduction to theory and research. Addision-Wesley.

Fitzgerald, A.I., C.E. Storer and M.J.M. Bent, 1999. Impediments to adoption of on-farm quality assurance. Proceeding of the 12th International Farm Management Congress (IFMA 99), July 18-24, Durban, South Africa.

Frey, B.S. and R. Jegen, 2001. Motivation crowding theory. Journal of Economic Surveys 15, 589-611.

Frey, B.S. and F. Oberholzer-Gee, 1997. The costs of price incentives: An empirical analysis of motivation crowding-out. The American Economic Review 87, 746-755.

Giner-Sorolla, R., 2004. Is affective material in attitudes more accessible than cognitive material? The moderating role of attitude basis. European Journal of Social Psychology 34, 761-780.

Hair, F.J., E.R. Anderson, L.T. Ronald and C.W. Black, 1998. Multivariate data analysis. Prentice-Hall.

Jahn, G., M. Schramm and A. Spiller, 2005. The reliability of certification: Quality labels as a consumer policy tool. Journal of Consumer Policy 28, 53-73.

Jatib, I., 2003. Food safety and quality assurance key drivers of competitiveness. International Food and Agribusiness Management Review 6, 1-19.

Leone, L., M. Perugini and A.P. Ercolani, 1999. A comparison of three models of attitude-behaviour relationships in the studying behaviour domain. European Journal of Social Psychology 29, 161-189.

Luning, P.A., W.J. Marcelis and W.M.F. Jongen, 2002. Food quality management: A techno-managerial approach. Wageningen Academic Publishers.

LZ, Lebensmittelzeitung, 2004. Industrial profiles in industry. Firmenprofile Industrie: PHW Gruppe, http://www.lz-net.de/; access date: 2004/03/01.

Martinez-Lorente, A.R. and M. Martinez-Costa, 2004. ISO 9000 and TQM: substitutes or complementaries? An empirical study in industrial companies. International Journal of Quality & Reliability Management 21, 260-276.

Mathieson, K., E. Peacock and W.W. Chin, 2001. Extending the technology acceptance model: The influence of perceived user resources. The DATA BASE for Advances in Information Systems 32, 86-112.

Nelson, P., 1970. Information and consumer behaviour. Journal of Political Economy 78, 311-329.

QS, Qualität und Sicherheit, 2005. Rund 1.100 Systempartner im QS-System. QS Infobriefe - Aktuell. http://www.q-s.info/de/news, access date: 2005/18/02.

Raden, D., 1985. Strength-related attitude dimensions. Social Psychology Quarterly 48, 312-330.

Sinha, M.N. and W.W.O. Willborn, 1985. The management of quality assurance. John Willey.

Sun, H., S. Li, K. Ho, F. Gertsen, P. Hansen and J. Frick, 2004. The trajectory of implementing ISO 9000 standards versus quality management in Western Europe. International Journal of Quality & Reliability Management 21, 131-153.

Szajna, B., 1996. Empirical evaluation of the revised technology acceptance model. Management Science 42, 85-92.

Venkatesh, V. and F.D. Davis, 2000. A theoretical extension of the technology acceptance model: Four longitudinal field studies. Management Science 46, 186-204.

Growers' perceptions about EurepGAP in developing countries: results of a survey carried out in Peru

Angelo Lazo, Gabriele Jahn and Achim Spiller
University of Goettingen, Department of Agricultural Economics and Rural Development, Platz der Goettinger Sieben 5, 37073 Goettingen, Germany; a.spiller@agr.uni-goettingen.de

Abstract

Quality assurance systems have become of increasing concern for suppliers in both developed and developing countries. Most of these systems have been designed to satisfy consumers' quality demands, but little evidence could be found about considering suppliers' perspectives. The implementation of quality assurance systems involve benefits and disadvantages that turn into adoption dilemmas, which should be taken into account in diffusion strategies. The assessments of the impact of quality assurance systems on suppliers can be argued from a cognitive perspective. The following study provides a conceptual framework and empirical data which analyse the perceptions of growers in relation to the implementation of such a quality assurance system, focusing particularly on the adoption of EurepGAP by export-growers in Peru. The results demonstrate that there is a high perceived importance of EurepGAP among Peruvian growers. This is explained not only by the expected benefits, but also because of the principal business motivation and customer orientation of the growers.

Keywords: EurepGAP, Peru, adoption, market orientation

1. The importance of EurepGAP in the Peruvian export market

The increasing demand for new food safety and quality standards - driven by the public or private sector - is playing an important role for the marketing and trading of products in many export-oriented countries. Contracts or special agreements with customers require the implementation and compliance of food safety quality management and assurance systems. If suppliers do not fulfil with these requirements, their chances for market access and product exports will be diminished in the future.

Although such standards are aimed at establishing control systems that contribute to evoke customers' trust and satisfy consumers' quality demands, their implementation has often caused problems during the adoption process. On the one hand, the adoption of quality management systems is commonly related to the generation of process optimisation, cost reduction and competitive advantages. In addition, delivering guaranteed safe products has also been recognised as a way of maintaining or even assuring access to markets. On the other hand, these benefits have been controversially discussed by managers: bureaucratic nonsense ('Give me papers') is associated with quality management systems rather than the optimisation of business processes. Additionally, quality and assurance systems have been foreseen as technical barriers to trade (Curkovic and Pagell, 1999; Henson and Loader, 2001; Luning *et al.*, 2002).

The evaluation of these dilemmas is important in order to understand the suppliers' adoption perspective and to prove the feasibility of a quality assurance system's introduction within the different production sectors among the differing conditions found in various countries. A cognitive assessment of perceptions, motivations and implications of adoption could be a useful tool of initially investigating such adoption dilemmas.

The following research, thus, explores the adoption process of the recently introduced global quality assurance standard EurepGAP. While previous studies have already focussed on the cost-benefit ratio of quality standards in the food sector, our analysis explores the grower's perception and acceptance of the adoption process and the impact of business characteristics and market orientation. This study provides a conceptual framework based on behavioural research and Adoption Theory. Its core is a survey carried out in the fresh fruit and vegetable agro-export sector in Peru. Two linear regression models were estimated for data analysis. At the end, the results are discussed and conclusions are drawn in order to achieve a better understanding of the adoption process and growers' perceptions.

In the European Union, consumer's demands for preventing food-borne scandals generated a proliferation of stricter national and private schemes. Complying with these safety and quality standards is becoming a major concern in developing countries. Quality assurance systems have been designed to guarantee that agreed-upon characteristics and production procedures are being complied with. This helps to avoid information asymmetries of non-visible product attributes such as food safety, organic production, denomination of origin, etc. (Holleran *et al.*, 1999).

The implementation of quality assurance systems has not even been an easy process in developed countries. New demands imply that organisational changes and motivations for adopting these systems might differ among food chain operators. Although some studies have mentioned that suppliers could find having to deal with the paper work demanded by the certification systems discouraging, consequent internal administration changes can contribute to quick procedure responses and competitive improvements (Caswell *et al.*, 1998; Jatib, 2003). Furthermore, not only the expected benefits explained the motivations for implementing quality assurance systems, but also external pressures, such as national regulations and commercial requirements, have an important role in the adoption of quality assurance standards (Henson and Holt, 2000).

Recently, new quality assurance systems have appeared at the pre-farm gate section which might be eliciting controversial reactions among growers. EurepGAP, for example, has become of increasing importance for the food industry. It is a global programme that assures good agricultural practices in order to achieve food safety, social welfare, environmental protection and animal welfare within primary production. It was developed by retailers, suppliers, and producers (all members of the Euro Retailer Produce Working Group - Eurep) as a response to the negative impact of multiple certifications, which had affected all stages of the food chain. EurepGAP is a business-to-business product certification based on international accreditation to EN 45011 and/or ISO Guide 65.

The programme started covering fruit and vegetable production, and today it has been extended to other areas such as livestock, flowers, coffee and aquaculture. EurepGAP is supported by important retailers such as Edeka, Metro, Tesco, Migros, Sainsbury, Eroski, etc.; and recently it has become a basic requirement for products entering the European Market. The certification of EurepGAP is currently carried out by 80 accredited certification bodies (CBs) in more than 60 countries worldwide, most of which are in developing countries (EurepGAP, 2005). The introduction of new certification schemes, such as EurepGAP, forces the local suppliers in developing countries to make fast operational changes in the short term in order to maintain and/or improve market access. Moreover, the problems of quality assurance implementation might be stronger in this situation since suppliers can face considerable disadvantages due to local conditions such as a lack of adequate food safety regulations, immature certification systems and a fragmented market structure (Farina *et al.*, 2005; Garcia and Poole, 2004); as well as more specifically speaking, problems of farm certification, the atomisation of production areas, local barriers of information and inappropriate infrastructure. All or any of these factors could place suppliers behind in facing any new challenges associated with EurepGAP.

The following study focuses on growers' perceptions of the adoption process in Peru. Peru has around 5 million hectares of cultivated land, which represent 16% of the total country's surface. Small-scale farming is the predominant characteristic of this sector and most of the country's production is traded to local markets. Although there are no clear figures to help identify how many small growers participate in the export sector, there are some indications that those who are involved in international trade grow mainly traditional crops such as sugar cane, coffee and cotton. The non-traditional export sector is dominated by medium to large growers who are few in number (around 500 growers), and who recently have became the main protagonists of agricultural exports in Peru contributing 85% of the total exports. The change to non-traditional products began in the 1990's, with asparagus as the main crop. Other crops which have been gaining importance in recent years are paprika, mangoes, avocados and citrus fruit (Alza and Vasquez, 1998; PROMPEX, 2004; Portal Agrario, 2004).

Typically of developing countries, Peru has not developed a local certification infrastructure apart from HACCP, which is part of the national food safety law and is controlled by the official organisation, SENASA (Servicio Nacional de Sanidad Agraria - National Service for Agriculture Safety). Otherwise there is only the private implementation of the ISO 9000 series, Organic Farming, BRC, EurepGAP, etc. This occurs only within the agro-export sector as the local markets have no particular demand with respect to quality. Most of the latter are district markets supplied from one wholesale market. The few supermarket chains in existence have their own particular set of quality demands. Thus, it is obvious that any implementation of food safety standards is only driven by the exporters (Portal Agrario, 2004; SENASA, 2004).

EurepGAP certification was first implemented in the year 2003 and by the time of the study there were more than 200 individually certified growers. No clear strategy had been put in place at national level for the implementation of this standard. Diffusion had gradually been effected by the efforts of local individual producer associations, like the Peruvian Asparagus and Vegetables Institute (IPEH - Instituto Peruano del Espárrago y Hortalizas) that certified all their members in 2004 with the support of the Inter-American Development Bank. There are eight certification bodies (CBs) operating in the country, though few of them are owned locally. These CBs supply services for certifying the growers and only four of them come from EurepGAP accredited offices. The main constraints found within the certification of this standard were initial investment, waste disposal (agrochemical containers) and worker training (Lazo, 2004).

2. Conceptual framework

Different theoretical approaches could have been used as a starting point for analysing the adoption process of EurepGAP. A very popular approach is the Adoption Theory, which was mainly developed by Rogers (1983). Following this theory, an 'innovation-decision process can lead to either adoption, a decision to make full use of an innovation as the best course of action available, or to rejection, a decision not to adopt an innovation' (Rogers, 1983: 21). The key element of empirical studies based on Adoption Theory is the period in which the adoption process is going on. In most cases, the process follows a S-shaped curve regarding the rate of adoption; however, the slope of the 'S' might differ significantly among the adoption of different innovations (Rogers, 1983).

Adopters are classified according to the stage when they introduce an innovation. The five categories include (1) innovators, (2) early adopters, (3) early majority, (4) late majority and (5) laggards; all of which have different motivations for adoption the innovation. Considering the implementation of EurepGAP, we might obtain an atypical beginning of the adoption process as EurepGAP is a requirement demanded by some European customers and therefore compulsory for their suppliers. This could result in a curve with a very high slope at the beginning due to a rapid adoption process. In this situation, the first category of adoption will therefore be more typical of the 'early adopters' than the innovators. The typical starting situation of an adoption process is characterised by a high degree of uncertainty because of the lack of experience in implementing standards. The adopters of EurepGAP, however, would be forced to 'innovate' later on in order to maintain competitiveness and market access. In the current stage of adoption, it remains uncertain as to how many Peruvian farms producing non-traditional crops will introduce EurepGAP in the future. This would depend on the expansion of the local export sector and its future supply orientation towards the European market.

In addition to the Adoption Theory, behavioural approaches can also be used for analysing the EurepGAP adoption process. Whereas Rogers primarily deals with the technical side of an adoption process, behavioural research approaches are focussed on the 'soft' determinants of adoption. The most popular approach in this field is the Theory of Planned Behaviour (TPB) developed by Fishbein and Ajzen (Ajzen, 1988; Fishbein and Ajzen, 1975). This theory considers the impact of attitudes, subjective norms, and perceived behavioural control on the intentional and observable behaviour towards the adoption of an innovation. The attitude construct is obtained from the aggregation of positive and/or negative behavioural beliefs regarding the observable behaviour. The subjective norms are obtained from the aggregation of normative beliefs describing the external pressure generated by actors, which might influence the performance of behaviour. The perceived behavioural control is an aggregate of control belief variables describing the presence or absence of resources and opportunities, which facilitate/impede the performance of the behaviour.

There are, however, several difficulties in applying TPB in explaining the adoption of a quality assurance system. The most important difference is the fact that TPB focuses mainly on voluntary behaviour rather than on quasi-voluntary or mandatory adoption as described by Caswell *et al.* (1998). As a consequence, observable behaviour is no longer a suitable variable for measuring acceptance. Given the fact that Peruvian growers intend to provide the European market, we have to assume that this implies a quasi-voluntary behaviour. Thus, two constructs appear to be interesting: (1) the attitudes toward EurepGAP as a quality standard itself, which may indicate the motivation of the growers in implementing the criteria on their farm, and (2), the perceived necessity and importance of the adoption revealing behavioural intention and willingness to adopt EurepGAP (Figure 1).

Lazo (2004) first tested the TPB model for analysing the adoption of EurepGAP, and although he could not explain empirically the influence of the three TPB constructs on the intentions for adopting this scheme, there were some indications that attitudes related to business expectations (e.g. costumer trust, market access, etc.) could be influencing the adoption behaviour. This was our starting point for further conceptual work in order to analyse the effect of business characteristics and orientation on the acceptance of EurepGAP. The following criteria were utilised to measure the determinants:

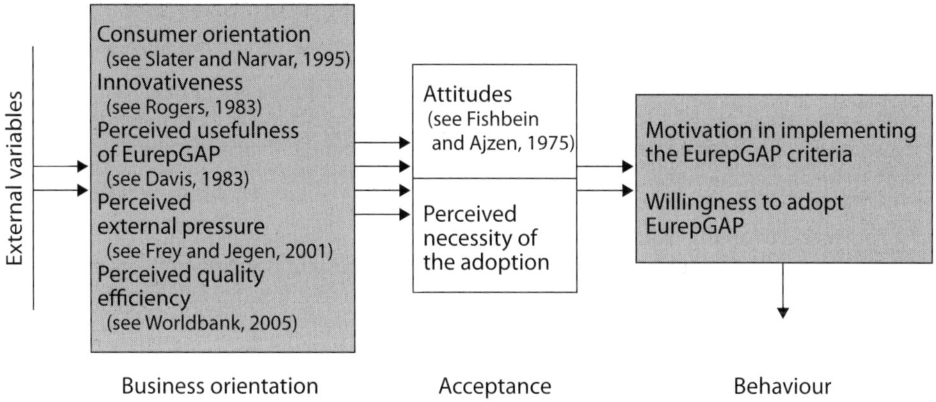

Figure 1. Research model: hypothetical determinants of the perceived importance.

1. Customer orientation. Grunert *et al.* (2004) differentiated business basic orientation into several categories. The most popular are production orientation and its counterpart, market orientation. While the first category is rather technically and operationally motivated, the second one is customer-based and strategy-oriented. As revealed in previous empirical studies, a well-developed market orientation significantly influences business performance (Slater and Narver, 1995). Our construct measures the importance of the relationship with customers and the relevance of getting access to new markets. We hypothesised that a high customer orientation of the growers will lead to an increasing acceptance of EurepGAP.
2. Perceived external pressure. Another key construct is the perceived pressure driven by the customers. The effect assumed is double-sided: while external pressure may lead to a negative attitude, the perceived necessity of the adoption might be positively influenced. The hypothesised effects can be traced back to the crowding-out theory developed by Frey and Jegen (2001). They presumed a trade-off characterising the relationship between external incentives and intrinsic motivation. In transferring this theory to our conceptual framework, we implicitly assumed that attitudes are rather intrinsically driven and the perceived necessity is more motivated by external factors.
3. Innovativeness. This construct can be ascribed to Rogers (1983) and focuses on its importance for the adoption process. He defined innovativeness as 'the degree to which an individual or other unit of adoption is relatively earlier in adopting new ideas than the other members of a system' (Rogers, 1983: 22). In our case, this construct was measured by asking growers about their actual willingness to innovate and about the importance of production efficiency and administration in their management procedures.
4. Perceived usefulness of EurepGAP. Perceived usefulness is a key determinant applied in acceptance studies (Davis, 1989). The respective evaluation of the improvements in competitiveness, productivity and a farm's administration by implementing EurepGAP may not only influence acceptance, but may also be an expression of the farm's main business orientation.
5. Quality efficiency of the farm. A last determinant is related to the perceived quality efficiency of the farm, which can differ significantly among farms. Particularly smaller farms may have less infrastructure and knowledge than bigger farms. Therefore, the different cost of compliance related to the starting organisational structure and financial situation is often perceived as a decisive barrier for implementing safety and quality standards (Worldbank, 2005). Figure 1 illustrates the conceptual framework.

3. Methodology

The main part of the analysis was the testing of our conceptual framework as depicted in Figure 1. Two linear regression models were applied, the first one using attitudes as a dependent variable and the second one perceived necessity of adoption. As independent variables the four constructs referring to business characteristics and perceptions were utilised.

All the constructs evolving from the theoretical study were measured by means of a five-point-Likert scale (-2 to +2). Descriptive analysis and an explorative factor analysis were used to explore the sample. After minor modifications, a principal component analysis was applied for data reduction and to build up factors according to the hypothetical constructs. These constructs were tested with Cronbach's alpha (α) (see Tables 1 and 2). All the constructs revealed reliability, i.e. an α higher than 0.6. Additionally, an item-to-item correlation analysis was calculated for discarding any problem of multicolinearity between the factors.

The factor scores of these constructs were the base of the two linear regression models. Innovativeness (I), customer orientation (C), perceived quality efficiency of the farm (Q), perceived usefulness (U), and perceived external pressure (P) served as independent variables in both models. In addition, the market of destiny to Europe was added as a dummy variable. The dependent variables included the attitudes towards EurepGAP (A) and the perceived necessity of the adoption (N). Linear regression was applied by using the method of least squares (OLS) as the estimation procedure.

4. Results

4.1. Description of the sample

A purposive sampling method was implemented for sampling the agro-export growers' population, relying on member lists provided by ProCitrus (Peruvian Citrus Producer Association) and ProHass (Peruvian Hass Avocado Producer Association), as well as through direct personal contacts with producer members of the IPEH. Face-to-face interviews with a total of 62 field growers took place on farms located in eight different production regions along a 500-km stretch of the Peruvian coast: Viru, Lima, Chancay, Huaral, Huacho, Chincha, Pisco and Villacuri. The sample consisted of medium and large growers producing for the international trade. The size of their farms was between 15 to 1200 hectares. The main crops produced were asparagus, avocado and citrus (see Figure 2). While most citrus within the sample was exported to the European market, most of the asparagus was for the USA.

The results revealed that only a small number of growers had received EurepGAP certification (11%), though most of them were preparing for their first audit (71%). The other 18% had not taken the decision to implement EurepGAP. The certified growers belonged therefore to a pioneer group, since the introduction of the standard at the moment of research was in its initial phase. For this reason, the sample cannot be considered as being representative of the country as a whole, but the results could be considered as an indication of the initial situation when implementing a quality assurance standard at the farm level in the export sector.

4.2. Attitudes and perceived necessity of the EurepGAP adoption

The grower's attitudes towards EurepGAP as the most important quality assurance system in Peru are surprisingly positive (see Table 1). The standard deviations are comparatively low, i.e.

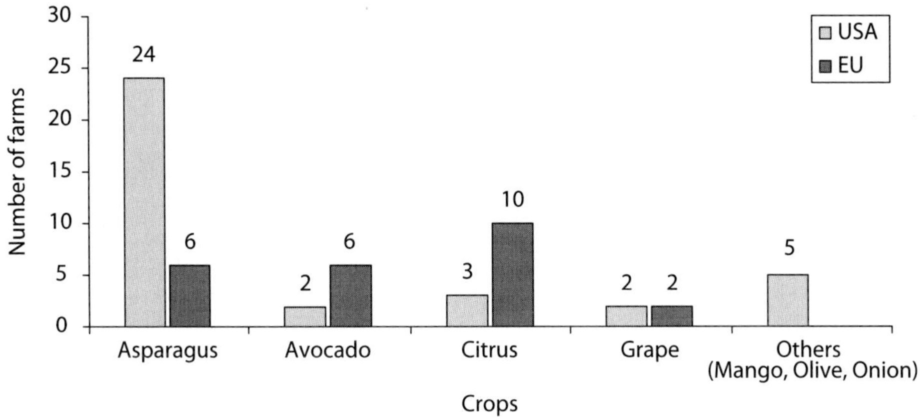

Figure 2. Main crops cultivated and their market destiny.

Table 1. Attitudes and perceived necessity of adoption.

Construct	Mean	Standard deviation	Cronbach's alpha
Attitudes (A)			0.712
A1: Implementing quality assurance standards is nothing beneficial - extremely beneficial	1.54	0.657	
A2: Implementing quality assurance standards is bad - good	1.67	0.476	
A3: Implementing quality assurance standards in the process is not valuable - valuable	1.60	0.634	
Perceived necessity of implementation (N)			0.642
N1: For you the implementation of EurepGAP is essential - not essential for the continuity of your farm in the agro export sector?	1.61	0.754	
N2: According to the trading position of your farm, what degree of relevance has the implementation of EurepGAP completely irrelevant - completely relevant	1.53	0.863	

the growers interviewed in Peru agreed on this. This result stands in contrast to the experiences in other countries where its introduction has been controversially discussed and has often been perceived as a bureaucratic burden (Fitzgerald and Storer, 1999; Jahn and Spiller, 2005a).

A similar result was obtained regarding the perceived necessity of the implementation. However, here the standard deviations were a little higher. This difference may be due to the crops produced. A cross tabulation (see Appendix 1) highlights the fact that particularly growers of citrus and asparagus perceived the adoption as extremely important, whereas growers of other fruits did not perceive the same high degree of necessity. The market of destiny also influenced the necessity of adoption; this can naturally be traced back to EurepGAP being a requirement mainly for the European market.

4.3. Business characteristics

In the following, the different growers' perceptions with respect to their farms and their management are considered (see Table 2). A comparison of the means identifies that most growers perceived their farms as being sufficiently prepared to implement EurepGAP. The standard deviations, however, revealed that there were slight differences with regard to the starting conditions. While the knowledge appears to be sufficient in most cases, the financial situation and particularly the infrastructure were more critically evaluated. This result demonstrates two issues: (1) none of the growers interviewed belonged to the very small growers, who had no access to knowledge about EurepGAP, and (2) the basic quality efficiency varies among the farms in the sample.

The cross tabulation in Appendix 1 illustrates that quality efficiency is correlated to the crop cultivated. Particularly the citrus growers were well-prepared, whereas several asparagus growers stated that they were not prepared. These differences could be due, for instance, to sectoral differences or the availability of information.

The majority of the growers perceived the need for adopting new techniques, increasing production efficiency, and the improvement of administrative processes. Such a high innovativeness and the importance of managerial issues is a first indication that the basic business orientation among the Peruvian growers is different compared to German farmers. In Germany, innovations in administrative processes are often neglected, less appreciated and conducted unwillingly (Jahn and Spiller, 2005a). This Peruvian willingness to innovate and realise the importance of managerial issues is underlined given the high importance these farmers place on their relationship with their customers. The low standard deviations demonstrate that there was agreement among the growers with respect to this question of customer relations. All of the growers were primarily export and customer driven. Most of them agreed that the pressure to implement EurepGAP was extremely high and was mainly driven by the customers and consumers (i.e. market access would be lost, if they did not implement the system). This result corresponds to the situation worldwide: customers and their requirements are the main drivers of any implementation process.

4.4. Regression models: determinants of acceptance

The following part aims to analyse the impact of different business characteristics. Two regression models were calculated; whereby the first deals with attitudes and the second, with perceived necessity and the importance of adoption (Table 3). Both models are significant (F-test). Whereas the variance of the attitudes (Model A) is explained with 29% (adj. R^2), the explained percentage of the perceived necessity is higher, being 40%.

The results of Model A highlighted the fact that neither the efficiency level of the firm nor the growers' innovativeness had a significant effect on attitudes. Instead, the most important factor was customer orientation. This result underlines the high importance of export and the requirements of the respective markets. In contrast, the majority of German farmers perceive the introduction of quality assurance systems as a bureaucratic burden and they rarely discern any improvements associated with documentation and administration (Jahn and Spiller, 2005a,b). Requirements and innovations driven by customers tend to be more criticised and negated by German farmers.

Table 2. Constructs related to business characteristics and perceptions (n=62).

Construct	Mean	Standard deviation	Cronbach's alpha
Perceived quality efficiency of the farm (Q)			0.673
Q1: I consider myself to have sufficient experience in agriculture production in order to implement new quality assurance standards on my farm totally disagree - totally agree	1.42	1.10	
Q2: I consider that the financial situation of my farm is sufficient in order to implement new quality assurance standards totally disagree - totally agree	0.75	1.43	
Q3: I consider that there is sufficient infrastructure to implement new quality assurance standards on my farm totally disagree - totally agree	0.33	1.42	
Innovativeness (I)			0.696
I1: Nowadays, it is extremely unnecessary - extremely necessary for my farm to adopt other techniques for managing and conducting production	1.12	0.77	
I2: Nowadays, it is extremely unnecessary - extremely necessary for my farm to improve production efficiency	1.44	0.65	
I3: Nowadays, it is extremely unnecessary - extremely necessary for my farm to improve its administrative and production management	1.15	0.93	
Perceived usefulness of EurepGAP (U)			0.707
U1: Implementing EurepGAP on the farm would improve productivity efficiency totally disagree - totally agree	1.18	1.21	
U2: Implementing EurepGAP on the farm would improve administrative and production management totally disagree - totally agree	1.47	1.00	
U3: Implementing EurepGAP on the farm would improve competitiveness totally disagree - totally agree	1.32	1.13	
External pressure (P)			0.860
P1: International retailers wish that you shouldn't - should implement EurepGAP on your farm	1.53	0.97	
P2: International consumers wish that you shouldn't - should implement EurepGAP on your farm.	1.22	1.14	
P3: Your clients wish that you shouldn't - should implement EurepGAP on your farm	1.44	1.13	
Customer orientation (C)			0.712
C1: Nowadays, it is extremely unnecessary - extremely necessary for my farm to improve my clients' trust	1.42	0.766	
C2: Nowadays, it is extremely unnecessary - extremely necessary for my farm to access new markets	1.45	0.753	
C3: Nowadays, it is extremely unnecessary - extremely necessary for my farm to comply with my main clients' requirements	1.63	0.581	

Table 3. Linear regression models.

Independent variables	Model A: dependent variable - attitudes (A)				Model B: dependent variable - perceived necessity (N)			
	Standard β-Coeff.	Standard error	t- values	Significance	Standard β-Coeff.	Standard error	t-values	Significance
Perceived quality efficiency (Q)	-	-	-	-	0.358	0.095	3.035	0.005**
Innovativeness (I)	-	-	-	-	-	-	-	-
Perceived usefulness (U)	0.292	0.122	2.307	0.026*	0.372	0.095	3.203	0.003**
External pressure (P)	-	-	-	-	0.298	0.114	2.446	0.020*
Customer orientation (C)	0.417	0.130	3.313	0.002**	0.283	0.094	2.343	0.026*
Market of destiny: EU	0.189	0.257	1.503	0.140*	0.464	0.203	3.953	0.000***
(Constant)		0.167	-0.973	0.336*		0.127	-2.181	0.037*

*** = $p < 0.001$; ** = $p < 0.01$; * = $p<0.5$.

Model A: R=0.58 (Adj. R^2 =0.29), F = 7.40 (p = 0.000).

Model B: R=0.70 (Adj. R^2 =0.40), F = 5.74 (p = 0.001).

Obviously compared to German farmers, market orientation is on quite a high level among the Peruvian farmers interviewed. Hence, it is not surprising that Peruvian farmers perceive the adoption of EurepGAP as useful with regard to improvements in competitiveness, administration and production efficiency. The usefulness has a positive effect on the Peruvian farmers' attitudes. In Germany, the low evaluation of these issues leads to a poor acceptance of the adoption of quality assurance standards in this country (Jahn and Spiller, 2005a,b).

Analogous to attitudes, customer orientation is a determinant of the perceived necessity. The more export-oriented growers are, the higher they rank the need for introduction. While attitudes are not affected by perceived external pressure, perceived necessity is weakly and positively correlated: necessity increases when the perception of external pressure is high.

Furthermore, the market of destiny (namely, the EU market) has a high impact. The main reason for this could be that EurepGAP is perceived as a requirement solely for the European market and not for other important markets of destiny for Peruvian growers, such as Asia and the USA. However, there are differences regarding the requirements of the USA and the EU markets in terms of food quality standards. While in Europe private responsibility and certification is already widespread, in the USA food quality is still driven by the public sector. Given the increasing demand for food quality, a discussion about privatisation has started in the USA. Public-private partnerships are being increasingly established in order to regulate quality assurance more efficiently and comprehensively (Henson and Reardon, 2005).

In contrast to attitudes, the quality efficiency level of the farm determines perceived necessity of adoption. The better the growers evaluated their own 'starting position', the higher they ranked the need for adoption. This result is congruent to what Adoption Theory says: if producers have lower costs of compliance, they will adopt the standard more rapidly (Rogers, 1983).

4.5. Limits of the analysis

Before concluding the study, it should be said that the conducted survey is an explorative one and has several limitations. The most 'critical point' is the sample size. It was not possible to choose more growers due to the low adoption rate at the time of the survey. In order to reveal changes over time, the adoption process should be evaluated continuously. Such information will allow the standard owner and certification bodies to react accordingly with respect to communication and the development of the standard.

On the whole, the conceptual framework was applicable and well suited for the analysis. The results emphasise the necessity for separate considerations of the perceived necessity of adoption and attitude. TPB and TAM cannot be transferred to a mandatory adoption process without considering the differences between perceived necessity and attitude. Although growers perceived a high necessity for the adoption (market access), attitude may have a different impact towards the implementation of EurepGAP. More indicators for each construct would have been necessary to test these relationships and clarify these differences in detail.

5. Conclusion

The adoption process of EurepGAP was only slightly contested among the Peruvian growers interviewed and its importance was perceived as high. The attitude towards the system was much better than that found in other comparable studies on the acceptance of recently introduced quality assurance systems. The farmers neither perceived the introduction of EurepGAP as a bureaucratic burden nor only implemented it because of external pressure. The basic attitude may be a suitable indicator for the business orientation of the growers interviewed. The growers were characterised by their market orientation considering the importance of the relationship towards the customer and the relevance of entering new markets. Table 4 summarises the key elements of the two basic orientations. The focus of the production-oriented farmers is normally more operational and short-term based. Only the data related to the quality of the internal production process is possibly the main interest of such production-oriented farmers. Corresponding information and communication strategies towards the farmers should have an important role for the acceptance during the adoption process. Market-oriented farmers are more broadly focused, i.e. towards the customer and new markets. Data obtained from the certification process could be useful to enhance the relationship towards customers, increase transparency and ensure the access in export markets.

In principle, EurepGAP's goals are much more market-oriented and customer-driven than production-oriented. Therefore, a higher effectiveness could be obtained by distributing information for benchmarking and increasing the competitiveness of their farms. Local stakeholders could be the main drivers of such initiatives.

Finally, our study should not be closed without pointing out the possible problem areas which are related to the adoption of EurepGAP. Our sample is not representative for the whole agricultural sector in Peru as all the interviewed producers belonged to the export sector. These farmers had direct access to information through their international clients and belonged to a preferred group compared to other growers in Peru. For them, EurepGAP may have increased competitiveness and enhanced their customer-supplier relationships. However, the 'standards-as-catalyst' perspective stands *vis-à-vis* to the 'standards-as-barriers' perspective (Worldbank, 2005). Particularly less-informed farmers owning smaller farms may not be able to implement EurepGAP without the support of local agencies (e.g. government organisations, NGOs, private sector, etc.) as they do not have the necessary communication network and technical capacity (Kleinwechter, 2005). Nevertheless, in the current stage of EurepGAP adoption the discussion about its effects are still hypothetical, and its actual impact will also be influenced by the market demand and the international retailers.

Table 4. Production orientation versus market orientation.

	Production orientation	Market orientation
Strategic objective	cost leadership	expansion and/or differentiation
Entrepreneurial culture	technical-rational, short-term planning horizon	less technical, often power or success oriented, varied planning horizon
Farmers' interests in implementing EurepGAP	focus on improvements in product quality and process management	focus on improvement of the relationship towards customer and better access to export markets
Farmers' interest in receiving quality data	internal key data concerning process quality to increase business performance	key data for benchmarking to increase competitiveness and market access

References

Ajzen, I., 1988. Attitudes, personality and behaviour. Taylor & Francis Group.

Alza, M. and V. Vasquez, 1998. La agroexportación en el Perú. 3rd ed., Lima, Peru, pp. 15-36.

Caswell, J.A., M.E. Bredahl and N.H. Hooker, 1998. How quality management metasystems are affecting the food industry. Review of Agricultural Economics 20, 547-557.

Curkovic, S. and M. Pagell, 1999. A critical examination of the ability of ISO 9000 certification to lead to a competitive advantage. Journal of Quality Management 4, 51-67.

Davis, F.D., 1989. Perceived usefulness, perceived ease of use, and user acceptance of information technology. MIS Quarterly 13, 319-340.

Farina, E.M.M.Q., G.E. Gutman, P.J. Lavarello, R. Nunes, and T. Reardon, 2005. Private and public milk standards in Argentina and Brazil. Food Policy 30, 302-315.

Fishbein, M. and I. Ajzen, 1975. Belief, attitude, intention and behavior: An Introduction to Theory and Research. Addison-Wesley.

Fitzgerald, A.I. and C. Storer, 1999. An exploration of impediments to the adoption of on-farm quality assurance: Implications for cattlecare and flockcare in Western Australia. Cattlecare Steering Committees, South Perth.

Frey, B.S. and R. Jegen, 2001. Motivation crowding theory. Journal of Economic Surveys 15, 589-611.

Garcia, M. and N. Poole, 2004. The development of private fresh produce safety standards: implications for developing Mediterranean exporting countries. Food Policy 29, 229-255.

Grunert, K.G., L.F. Jeppesen, K.R. Jespersen, A.-M. Sonne, K. Hansen, T. Trondsen, und J.A. Young, 2004. Four cases on market orientation of value chains in agribusiness and Fisheries. Working paper no. 83-2004, February 2004, Aarhus, School of Business, Denmark.

Henson S.J. and G. Holt, 2000. Exploring incentives for the adoption of food safety controls: HACCP implementation in the U.K. Dairy Sector. Review of Agricultural Economics 22, 407-420.

Henson S.J. and R.J. Loader, 2001. Barriers to agricultural exports from developing countries: The role of sanitary and phytosanitary requirements. World Development 29, 85-102.

Henson, S. and T. Reardon, 2005. Private agri-food standards: Implications for food policy and the agri-food system. Food Policy 30, 241-253.

Holleran, E., M.E. Bredahl and L. Zaibet, 1999. Private incentives for adopting food safety and quality assurance. Food Policy 24, 669-683.

Jahn, G. and A. Spiller, 2005a. The adoption of the QS system in agriculture: Exploring the attitudes and the intention-behaviour relation. Conference proceeding presented at IAMA's 15th World Food and Agribusiness Forum, June 26-28, Chicago, United States.

Jahn, G. and A. Spiller, 2005b. Acceptance of a processor-driven quality management system by dairy farmers: A structural equation model, conference proceeding presented at 92nd EAAE Seminar on 'Quality Management and Quality Assurance in Food Chains', 2-4 March 2005, Goettingen, Germany.

Jatib, I., 2003. Food safety and quality assurance key drivers of competitiveness. International Food and Agribusiness Management Review 6, 1-19.

Kleinwechter, U., 2005. The significance and adoption of quality and safety standards in the international agricultural trade. A developing country case study of the Mango export sector in Piura, Peru. M.Sc. Thesis. Georg-August University, Goettingen, Germany.

Lazo, A., 2004. Implementation of EurepGAP standard in the agro-export sector of Peru: A case study. M.Sc. Thesis. Georg-August University, Goettingen, Germany.

Luning P.A., W.J. Marcelis and W.M.F. Jongen, 2002. Food quality management: A techno-managerial approach. Wageningen Academic Publishers.

Portal Agrario (Ministry of Agriculture, Government of Peru), 2004. Vision General del Agro Peruano. http://www.portalagrario.gob.pe/agricola.shtml, access date: 2005/04/24.

PROMPEX, 2004. Situación de las Exportaciones Peruanas. http://www.prompex. gob.pe/prompex/, access date: 2005/04/24.

Rogers, E.M., 1983. Diffusion of innovations. New York.

Schramm, M., A. Spiller and T. Staack, 2004. Brand Orientation in der Ernährungsindustrie. Erfolgsdeterminanten der Markenführung am Beispiel genossenschaftlicher Hersteller. Deutscher Universitätsverlag.

SENASA (Servicio Nacional de Sanidad), 2004. Plan Estratégico: Proceso de Planeamiento Estratégico. http://www.senasa.gob.pe/organisacion/principal.htm; access date: 2005/03/12.

Slater, F.S. and J.C. Narver, 1995. Market orientation and the learning organisation. The Learning Organisation 59, July, pp. 63-74.

Worldbank, 2005. Food safety and agricultural health standards: Challenges and opportunities for developing countries. Washington D.C.

Appendix 1. Cross tabulations

Cross tabulation 1. Crops and N1 (construct perceived necessity, N).

For you the implementation of EurepGAP is ____ for the continuity of your farm in the agro-export sector	Which is the main crop that you produce for exportation?					
	Citrus	Avocado	Asparagus	Grapes	Others	Total
not essential at all	0	0	0	0	1	1
not essential	0	0	0	0	1	1
neither/nor	0	0	1	0	0	1
essential	3	0	9	2	1	15
extremely essential	11	8	21	2	2	44
Total	14	8	31	4	1	62

Cross tabulation 2. Crops and A1 (construct attitudes, A).

Implementing quality assurance standards is	Which is the main crop that you produce for exportation?					
	Citrus	Avocado	Asparagus	Grapes	Others	Total
not beneficial	0	0	0	0	1	1
neither/nor	0	0	1	1	0	2
beneficial	6	2	8	2	1	19
extremely beneficial	8	5	19	1	2	35
Total	14	7	28	4	4	57

Cross tabulation 3. Crops and perceived quality efficiency (infrastructure, Q3).

Actually, I consider that I have sufficient infrastructure in order to implement new quality assurance standards on my farm	Which is the main crop that you produce for exportation?					
	Citrus	Avocado	Asparagus	Grapes	Others	Total
totally agree	5	2	8	1	0	16
agree	7	0	9	0	0	16
neither/nor	0	5	3	0	1	9
disagree	2	1	3	2	2	10
totally disagree	0	0	6	1	2	9
Total	14	8	29	4	5	60

Cross tabulation 4. Crops and external pressure (P3).

Actually, your clients wish that you ____implement EurepGAP on your farm	Which is the main crop that you produce for exportation?					
	Citrus	Avocado	Asparagus	Grapes	Others	Total
definitely should	12	6	23	3	0	45
should	1	1	4	1	0	8
neither/nor	0	0	1	0	0	1
shouldn't	0	0	2	0	1	4
definitely shouldn't	1	1	0	0	0	3
Total	14	8	30	4	1	61

Dairy farmer's acceptance of a processor driven quality management system: a structural equation model

Gabriele Jahn and Achim Spiller
University of Goettingen, Department of Agricultural Economics and Rural Development, Platz der Goettinger Sieben 5, 37073 Goettingen, Germany; a.spiller@agr.uni-goettingen.de

Abstract

Representatives of the German dairy sector gathered in 2002 to establish national guidelines for a new quality management system on dairy farms (QM). Subsequently, QM was introduced and the majority of German dairy farmers have been compelled by their dairy to establish new quality management procedures, otherwise they would have lost market access. This situation was the initiating cause for conducting the following study about the factors influencing the acceptance of QM. The relationships and effects with respect to acceptance were analysed by applying a structural equation model. The empirical basis of the study was a survey conducted with farmers (n=209) in Northern Germany. The results revealed that the most important determinant is the impact of QM on product quality yielding, which led to the conclusions that QM has to be more effective and the guidelines more elaborated to increase acceptance. Even though the current situation of low acceptance is not decisive for QM's diffusion, it certainly endangers the farmer's diligence in its implementation and the long-term success of QM.

Keywords: quality management, technology acceptance model, theory of bureaucracy

1. Introduction

In Germany milk has for a long time been one of the products with the highest density of product controls. Public inspection combined with the monitoring of dairy companies has guaranteed milk quality and only a few scandals have occurred. Despite this good record, in 2002, a process-oriented quality management system was introduced to increase transparency in the processes on German dairy farms ('Qualitätsmanagement Milch', QM). Since then, German farmers have been compelled to adopt the system because QM must be integrated as a necessary requirement into milk delivery contracts.

So far everywhere in Germany, an apparently inevitable controversial discussion has been associated with the adoption of QM. Farmers do not perceive the necessity of its introduction. This situation of low acceptance surely endangers QM's long-term success and its diligent implementation on dairy farms.

This paper begins to address the shortfalls in understanding farmers' attitudes towards the adoption of QM; only little is known about the actual factors provoking it. In order to overcome this lack of understanding, structural equation modelling techniques were applied to uncover the relationships between those factors pertinent in the farmers' opinions. Answers are given to the question as to how dairies could intervene more effectively in the implementation of QM and enhance its comprehension by conducting suitable communication strategies. Generally, the paper presents new insights into the ongoing debate on private food safety standards and their impact on supply chain structures (Henson and Reardon, 2005).

2. Conceptual design of QM

Before presenting the theoretical foundations of the study, the following chapter provides some insights into the conceptual design of QM in the German dairy sector. Quality management systems are already well-established in almost all other industry sectors. In the agri-food system, due to the long history of public responsibility for food quality and safety in Germany, private players did not see the necessity for adopting on-farm quality management until the end of the 1990's. However, food scandals did occur and consumers increasingly demanded more safety and traceability. As the existing public controls were not able to adequately ensure these requirements, the private sector established its own quality initiatives (Farina *et al.*, 2005; Henson and Reardon, 2005).

In contrast to wide-spread certification systems such as EurepGAP, QM in Germany represents a second party audit approach carried out by control associations (administered by the state, but ordered by the dairies) or directly by the dairies. QM was developed by the German Dairy Industry Association (MIV; Milchindustrieverband), the German Association of Farmers (DBV; Deutscher Bauernverband) and representatives of the German animal feed industry (Raiffeisenverband). The guidelines are common principles on a national level to which all should refer to, but the concrete implementation and communication is left to the dairies and the individual German states themselves. Whereas in Southern Germany it is the states that primarily ensure adoption, in Northern Germany it is the private sector, namely the different dairy companies.

The recently introduced second party audits associated with QM constitute the third and last pillar of the existing quality assurance system in the German dairy sector (Figure 1). The first well established pillar is the direct control of product quality (fat, protein, cells, etc.) and the second traditional pillar is concerned with the monitoring of milk and animal feed for Aflatoxins, PCBs, pesticides, etc.; both procedures are carried out on behalf of the dairy companies. The main purpose of the new control and documentation procedures in the third pillar is to ensure process quality during production. The criteria that have been developed include animal registration, medicine, fertilisers, management, animal feed, etc. All the criteria fulfil the legal requirements in Germany, and as the level of these criteria is comparatively low in sense of the legal basis, all German dairy farmers should have already fulfilled the majority of them.

The adoption of QM in Northern Germany has been mainly driven by the dairy companies. The German dairy sector is characterised by stable relationships between the farmers and the dairies. The contracts are executed for a period of two years and dairy farmers have only a few (and sometimes no) options of changing their dairy company. All the dairies - with the exception

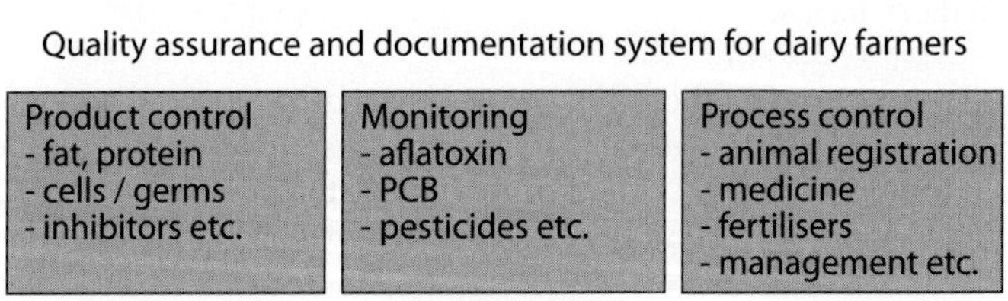

Figure 1. The three pillars of quality assurance in the German dairy sector.

of Humana - support the adoption of QM. Humana, the third largest dairy in Germany, even argues against implementation: 'official regulations would ensure milk safety sufficiently. QM, in contrast, only increases bureaucracy and does not ensure either a better product quality or safety in the production processes' (Anonymus, 2003).

The dairy farmers, however, were not asked about their willingness to introduce QM, but were forced to do so due to their contracts with their dairy. As a consequence, the dairies have so far not seen the necessity to promote QM (for instance, communicate its usefulness to the farmers). The upshot of this lack in promotion is that acceptance and motivation with respect to QM have been low among the farmers involved.

3. Focal constructs

Our conceptual framework is based on behavioural research. So far, few studies have dealt with the acceptance of quality management systems; this is particularly true for the food sector with respect to recent developments. Our hypothetical model is primarily based on the Technology Acceptance Model (TAM) developed by Davis (1989). It is aimed at explaining and predicting the acceptance and usage of information systems. TAM is a well-known and important modification of the Fishbein and Ajzen Theory of Planned Behaviour and the Theory of Reasoned Action (Fishbein and Ajzen, 1975; Ajzen, 1991).

TAM's main focus is on measuring both the perceived usefulness as 'the degree to which a person believes that using a particular system would enhance his or her job performance' and the perceived ease of use, i.e. 'the degree to which a person believes that using a particular system would be free of effort' (Davis, 1989: 320). Both factors determine the attitudes, the intentional behaviour and finally, the observable use of the information system (see Figure 2). In the last few years, TAM has been applied to various studies and has become a powerful model for predicting user acceptance (Chau, 1996; Szajna, 1996; Venkatesh and Davis, 2000; Mathieson *et al.*, 2001).

One theoretical foundation of the TAM is the Cost-Benefit Paradigm that explains a company's choice among various decision-making strategies. Such analyses focus on the (cognitive) trade-off, evoked by 'the effort required to employ the strategy' on the one hand and 'the quality (accuracy) of the resulting decision' on the other (Davis, 1989: 321). Cost-benefit studies are normally rather formal and rely on objective figures. Although these may be the theoretical basis for decision-making, the perceived and much more subjective cost-benefit considerations are finally decisive for acceptance and actual adoption (Davis, 1989).

While the basic TAM represents a suitable starting point for the development of a research model for QM, some characteristics specific to the current situation associated with the

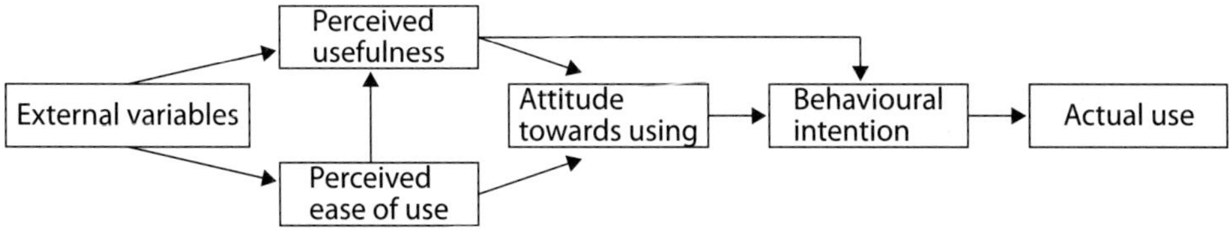

Figure 2. Technology Acceptance Model (Davis, 1989).

implementation of quality assurance systems must be considered. The key difference is the fact that the adoption of QM in most cases is not voluntary, but a necessary requirement for market access. TAM models with regard to mandated use have been discussed by Brown *et al.* (2002). Their results showed that ease of use was the primary determinant of behavioural intention in systems with mandated use, whereas usefulness is only of secondary importance. This is in contrast to research on volitional settings where perceived usefulness is normally the primary antecedent to behavioural intentions.

Furthermore, the latter study revealed that behaviour in mandated settings is complex; particularly the attitude-behaviour relation is often not significant and has to be regarded carefully (Brown *et al.*, 2002). This is what was also shown in an exploratory study about the mandatory implementation of a certification system in the German meat sector (Jahn and Spiller, 2005). To compensate for the complexities of the mandatory use scenario, Brown *et al.* (2002) suggested taking out the attitude construct and focussing on behavioural intention.

According to TAM, the ease of use construct and the usefulness are regarded as determinants of acceptance. While an extension of the usefulness construct has been discussed often (Chau, 1996), a modification of the ease of use has not been undertaken. The main reason for this is that TAM has been applied to information systems where the analysis of perceived costs can be reduced to an analysis of usability and user-friendliness. In contrast to information systems, the adoption of quality management systems may cause significant costs. Hence, not only does the usefulness of the quality management system have to be extended and discussed, but it also has to be determined whether the construct ease of use is comprehensive enough. The starting point of such a modification is the fact that the adoption of a quality management system is associated with the institutionalisation of quality controls and documentation procedures. The increasing formalisation of the management process results not only in costs, but also benefits such as better internal communication structures or specified task responsibilities.

This consideration leads to another theoretical approach which deals primarily with formalisation procedures in enterprises: the Theory of Bureaucracy. Max Weber (1864-1920) was one of the first prominent representatives of this theoretical direction ('Economy and Society'(1968)). He developed his theory against the background of Taylorism and industrialisation. Since then, the effects of bureaucratic procedures have been discussed controversially. During the last decades, it has become popular to criticise the 'bureaucratic burden': inefficiencies linked to very formal and rigid organisational structures (Curkovic and Pagell, 1999; Gotzamani and Tsiotras, 2002; Chow-Chua *et al.*, 2003). However, Adler (1993) recommends not only focusing on the negative impact of bureaucracy, but also on the positive one. Positive assessments highlight that 'well-designed procedures would facilitate task performance and thus augment employee's pride of workmanship' (Adler and Borys, 1996: 64). Other studies also mention bureaucracy's positive influence on job satisfaction and innovativeness (Adler, 1999; Beck and Walgenbach, 2002; Theuvsen, 2004).

Based on these considerations, we introduced three determinants of QM's acceptance. Two of them mainly refer to the direct impact of increasing bureaucracy procedures at farm level, while the third one is more generally related to improvements on performance. These modifications are based not only on the Theory of Bureaucracy, but also on the TAM modification mentioned by Chau (1996) and Venkatesh *et al.* (2003). These authors proposed splitting the original usefulness into two constructs: with reference to Chau (1996), we named them *usefulness* and *effectiveness*. In comparison, Venkatesh *et al.* (2003) introduced the second construct as *performance expectancy*.

Our effectiveness construct is comprised of those improvements in performance a farmer perceives through implementing QM. Numerous studies on quality management systems offer broad insights into their positive impact on a firm's success and competitive power (Antony *et al.*, 2002; Lin *et al.*, 2004; McAdam and Henderson, 2004). Brah *et al.* (2002), for instance, classified success factors into supplier performance, employee service quality, product quality, employee satisfaction, customer satisfaction, and manufacturing/service process quality. We, therefore, defined effectiveness as 'the degree to which a person believes that using a particular system would enhance corporate quality and process management'.

Against this background, we hypothesised a basic acceptance model which considers not only the effects on the acceptance, but also the effects among the constructs themselves (see Figure 3). All these effects reflect paths assumed in the TAM (Figure 2). Beginning with the bureaucratic cost defined as 'the effort, which a farmer perceives by employing QM's formal requirements' the following hypothesis was formulated.

H1a: The higher the perceived bureaucratic costs of documentation, the lower the acceptance.

This path is analogous to the path between 'attitude' and 'ease of use' in Figure 2. The following two hypotheses deal with the effects of the usefulness of QM and so, with the benefits of formalisation procedures at farm level. Hypothesis H2a again corresponds with the path between 'ease of use' and 'attitude':

H2a: The higher the perceived usefulness, the higher the behavioural intention towards the adoption of QM.

The second effect is equivalent to the TAM path lying between 'ease of use' and 'usefulness'.

H3: The higher the perceived bureaucratic costs, the lower the perceived usefulness.

Both usefulness and effectiveness are related to the positive effects of QM and so they can be interchanged leading to the following two hypotheses, where perceived effectiveness has been exchanged for perceived usefulness:

H4: The higher the perceived effectiveness of QM, the higher the behavioural intention towards the adoption of QM.

H1b: The higher the perceived bureaucratic costs, the lower the perceived effectiveness of QM.

Whereas the usefulness refers to the firm or lower level, the effectiveness measures the performance of QM on a more general and higher level. Therefore a path between these two constructs was considered to be necessary, starting from the lower level, usefulness, and ending at the higher level, effectiveness. This, therefore, resulted in the following hypothesis:

H2b: The higher the perceived usefulness, the higher the perceived effectiveness of QM. Figure 3 sums up these six hypotheses in a basic acceptance model.

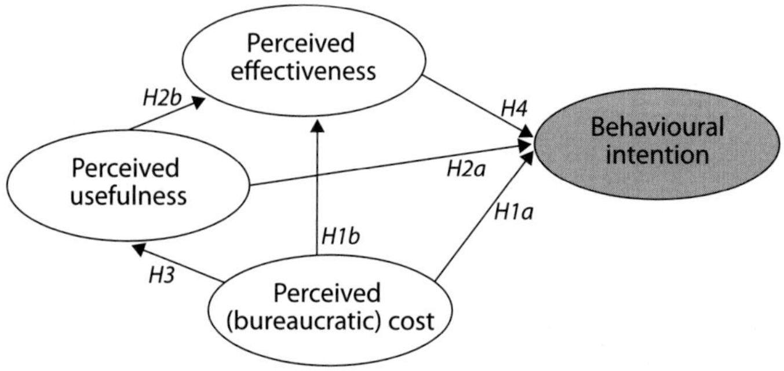

Figure 3. Basic acceptance model.

4. Methodology

4.1. Data collection

The analysis reported in this paper was conducted on data obtained from a sample of 209 dairy farmers in Northern Germany. In summer 2004, the farmers were questioned in face-to-face interviews. The average interview took about 45 minutes. All in all, the sample includes larger sized farms (73 cows per farm) than the average in Germany (35 cows per farm; Eurostat, 2004). The average milk quota per farm was comparatively high (600,000 kg p. a.) and the quoted milk quality is much higher than the required official minimum standards. The data was collected at an early stage in QM adoption: 38.9% of the dairy farms had already been audited; 39.9% had not been audited at the time of data collection, but planned to do so in the future; and 21.2% were not planning to participate.

The respondents were on average 42 years old and well-educated (only 3.4% had no agricultural education). The majority of these farmers (60.6%) were active members in agricultural associations. In conclusion, the sample includes presumably more opinion leaders than the average of the total population. 84.1%, of the respondents intended to continue milk production in the future. The others were not sure about the future, but only two of them had already decided to give up their farm. All in all, the sample represents bigger and future-oriented farms.

The farmers interviewed delivered their milk to 22 different dairies: 'Nordmilch e.G.' (Nordmilch) was the customer of 35.4%, 'Humana Milchunion e.G' (Humana) of 22.5%, and 'Campina GmbH Deutschland' (Campina) 14.8%; the remaining farmers (27.3%) delivered to smaller companies. Humana is the biggest dairy in Germany (sales volume: 2,444 Mio. €; milk production 2,175 Mio t p.a.) followed by Nordmilch. Campina is on fourth place (MM, 2004).

4.2. Measures

To capture the latent variables, different measurement scales that had been partly tested in a previous survey about the attitudes towards the QS system in the German meat sector in 2002 were used (Jahn and Spiller, 2005). Thus, acceptance was measured by using indicators rating the willingness to adopt QM and the perceived necessity for it.

Due to QM's characteristic of being a baseline standard close to the minimum standards of legal requirement, there can rarely be an actual investment in implementing QM. The bureaucratic costs, therefore, include two aspects: the first one considers the perceived complicatedness of QM, the second one the perceived burden and effort in working time. As a result, the measurement scales used are a reversion of Davis's measures of ease of use. Perceived usefulness implies three statements also closely related to the measures developed by Davis (1989): (1) improvements in process management by increased transparency, (2) more efficiency and (3) better process management. The effectiveness constructs also consist of three indicators: (1) improvements in product quality, (2) improvements in process quality and (3) an enhancement of the image of dairy products. These variables and their importance for improvements in product quality have been described by Dunk (2002), as well as by Sebastianelli and Tamimi (2002).

The indicators used in the survey are either Likert-scaled or semantic differential items. All of them were examined using a factor analysis. After minor modifications for double loading and nonloading items, the measures demonstrated acceptable levels of fit and reliability (see Appendix 1).

5. Results

Generally, the farmers were rather sceptical about QM. Only 32.6% would have introduced it, if they had had free choice; 62.7% would not have adopted it. However, the high standard deviations already indicated that there were clear differences in the degrees of acceptance (see Appendix 1).

To test the six hypotheses concerning their causal effect on acceptance a structural equation model was applied including the behavioural constructs as suggested in Figure 4. The overall fit of the model was excellent with a χ^2 (38) = 36.713, p = 0.529, GFI = 0.967, AGFI of 0.942 and a RMR of 0.076 (Hair *et al.*, 1998; Kline, 2005). Two of the hypothesised paths were not significant, all the others indicated t values higher than ±1.96. We further analysed the discriminant validity. Only one standardised residual covariance was lower than -1.

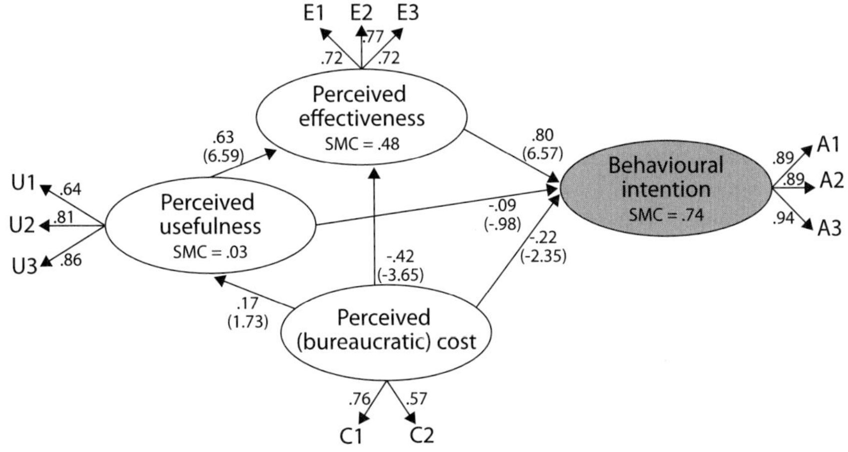

Figure 4. Estimated model of QM acceptance.

All the suggested factors had a causal effect on acceptance. The most important effect, that of perceived effectiveness on acceptance, is 0.80 (*H1*). Evidently, the perceived usefulness affects the acceptance more than the perceived bureaucratic costs, which equals -0.22. There was only an indirect effect of the perceived usefulness towards the acceptance, but this effect (via effectiveness) was quite strong ($0.63 \times 0.80 = 0.50$). *H2a* must therefore be rejected, whereas *H2b* is confirmed. The perceived (bureaucratic) costs have a direct and an indirect causal effect on the acceptance, therefore hypotheses *H3a* and *H3b* suggested above are confirmed. The total effect is -0.56 [$-0.22 + (-0.42 \times 0.80)$].

We further estimated a constrained model of the hypothesised model, where the non-significant path of the estimated model was set to zero. In the estimated model from above, this would be the path between the bureaucratic usefulness and the acceptance. By conducting a chi-square difference test for nested models, we compared the alternatives. The constrained model did not demonstrate a significantly better fit than the hypothesised model data (nested model test: χ^2 [1 df] = 1.764; $p > 0.1$). The hypothesised model fitted the data better than the constrained one. A model combining the 'perceived usefulness' and 'perceived effectiveness' towards one construct demonstrated a poor fit and had therefore to be rejected. These results suggest that the hypothesised model and paths are preferable compared to other models.

6. Discussion

Interpreting the results of the hypothesised model, the farmers' conviction and their acceptance of QM is higher, if they perceive an increase in the quality of the product through the control procedures. This result may reflect a problematic understanding of the self-conception of the farmer himself. Gasson (1973) highlighted in a study about the goals and values of British farmers that a 'good' farmer is the one who produces the best crops or livestock. This criterion was ranked much higher than other criteria such as preserving the beauty of the countryside or making the most money (Gasson, 1973). A similar result was pointed out in a more recent survey analysing the entrepreneurial behaviour of Dutch farmers (Bergevoet *et al.*, 2004): producing a good and safe product was ranked highly, whereas realising an income as high as possible was ranked much lower.

All in all, the results support the argument that there is only a low level of market orientation among the farmers whereas product and production orientation are important drivers of the farm management (Grunert *et al.*, 2004). Market requirements are perceived as necessary constraints, which do not increase intrinsic motivation for quality management on farms.

Farmers' goals and values play obviously a decisive role in the decision-making process. Cost-benefit considerations, which may lead to higher efficiency in production processes, are therefore less important for the evaluation of QM than the actual improvement and assurance of product quality. To reach this aim, the audit quality of QM has to be ensured. Only a system, which is credible, will be able to convince the farmers of its worth. This result may also indicate that farmers are in principal motivated to produce high quality products, despite the fact that there has been a long trend towards higher quantities in production. In regard to future strategies for quality management systems in agriculture, the relation of product and process management should be linked more closely in order to prevent pure 'give-me-paper' procedures. For that purpose, laboratory analyses of milk quality and management metasystems must be combined.

Another negative influence on acceptance is associated with the bureaucratic burden involved in the documentation and formalisation procedures of QM. This effect is quite typical at the beginning of a diffusion process and indicates resistance of the 'mass' against the pioneers and promoters (Rogers, 1983). Two factors could reduce this: time and communication. The efficacy of the latter can be clearly influenced by the differences between the dairies. As revealed in our study, Humana's rejection of QM, for instance, has led to a strong negative evaluation of QM by its suppliers. This result reflects the high importance of dairies' involvement in the adoption process. If they support QM and provide suitable information, farmers' acceptance can increase considerably. However, one problem may have complicated communication: the current pricing pressure driven by retailers, and also by the dairies. This situation could have distorted the perception of the introduction of QM. Farmers may have been more concentrated on their 'survival' and they may have negativing new investments in general, particularly if they are suggested by their customers. Given a more relaxed situation the farmers may have demonstrated a higher acceptance of QM.

7. Conclusions

Our contribution highlights a variety of starting points for the further development of behavioural research models about the acceptance of quality management systems in food supply chains. A first acceptance model was designed revealing the effects among different behavioural constructs and indicating which theoretical approaches are valuable. For this purpose, the basic TAM was modified by splitting the original construct of usefulness into usefulness and effectiveness. Instead of implementing the 'ease-of-use' construct, we utilised the inverse 'bureaucratic cost' in accordance with the Cost-Benefit paradigm and the Theory of Bureaucracy.

The results demonstrated clearly that the transfer of Davis' TAM to the acceptance of a quality management system is possible and applicable. However, as Adams *et al.* (1992) highlighted in their empirical studies, the relationships among the constructs in reality may be much more complex than suggested in the basic TAM; therefore, further studies should not only examine these relationships again, but also focus on extensions of the basic acceptance model of quality management systems in mandated settings.

For this purpose more external variables could be included. Venkatesh *et al.* (2003) discussed the integration of constructs such as gender, age or perceived behavioural control, derived from the Theory of Planned Behaviour. Motivational aspects or management orientation could be included. A special focus of future research should be on the attitude-intention relation as discussed by Brown *et al.* (2002). The second focus for extension could be on the measures themselves. Effectiveness, for instance, could include improvements in employee service quality or employee satisfaction as well (Brah *et al.*, 2002). These variables were not considered in our study due to the small-structured business environment - mainly family farms.

Besides the theoretical considerations, the study allows conclusions with respect to communication strategies and the development of QM. Processors have a decisive impact, which enforces the implementation of QM and provides information. Although the guidelines are the same, the perception of QM differs significantly between the dairies. This fact indicates not only the future challenges, but also the potentials for communication strategies utilising the dairies.

Interestingly, the farmers particularly considered the impact of QM on their product in evaluating QM's value in general. This result may indicate that one of the farmers' main values is still the production of high quality products. They identify themselves as specialists in production and not as market oriented supplier. These values must be considered when planning communication to achieve higher acceptance. In contrast, improvements at the farm level do not affect

QM's acceptance directly and may not be suitable as the main base for communication. The perceived bureaucratic burden of QM, however, decreases acceptance. The understanding of the documentation procedures is low. A better communication of the positive effects of QM may increase comprehension.

It became apparent that the majority of the farmers would not implement QM unsolicited. However, successful implementation is only possible if QM is accepted in the long-term. A higher conviction and motivation is necessary in the farmers to ensure their diligence in the implementation of the guidelines. These results correspond to research emphasising that farmers in developed countries like Western Europe are in some parts less market-driven and quality-oriented than the export sector in developing countries in Latin America or Asia (Lazo *et al.*, 2005; Farina *et al.*, 2005). The low acceptance of new quality management metasystems is therefore not only a question of market access for developing countries and their small farm business, but also a signal for the low level of market orientation in some European farm sectors with a high level of subsidies.

References

Adams, D.A., R.R. Nelson and P.A. Todd, 1992. Perceived usefulness, ease of use and usage of information technology: A Replication. MIS Quarterly 16, 227-247.

Adler, P.S., 1993. The 'learning bureaucracy': New united motor manufacturing, Inc. research in organisational behaviour, JAI Press, pp. 111-194

Adler, P.S., 1999. Building better bureaucracies. Academy of Management Executive 13, 36-47.

Adler, P.S. and B. Borys, 1996. Two types of bureaucracy: Enabling and coercive. Administrative Science Quarterly 41, 61-89.

Anonymus, 2003. Brauchen Milcherzeuger QM? DLG-Mitteilungen 8/2003, 24.

Ajzen, I., 1991. The theory of planned behavior. Organisational Behavior and Human Decision Processes 50, 179-211.

Antony, J., K. Leung and G. Knowles, 2002. Critical success factors of TQM implementation in Hong Kong industries. International Journal of Quality & Reliability Management 19, 551-566.

Beck, N. and P. Walgenbach, 2002. ISO 9000 and formalization - How organisational contingencies affect organisation responses to institutional forces. Schmalenbach Business Review 55, 293-320.

Bergevoet, R., C.J.M. Ondersteijn, H.W. Saatkamp, C.M.J. van Woerkum and R.B.M. Huirne, 2004. Entrepreneurial behaviour of Dutch dairy farmers under a milk quota system: goals, objectives and attitudes. Agricultural Systems 80, 1-21.

Brah, A.S., S.S.L. Tee and B.M. Rao, 2002. Relationship between TQM and performance of Singapore companies. International Journal of Quality & Reliability Management 19, 356-379.

Brown, S.A., A.P. Massey, M.M. Montoya-Weiss and J.R. Burkman, 2002. Do I really have to? User acceptance of mandated technology. European Journal of Information Systems 11, 283-295.

Chau, P.Y.K., 1996. An empirical assessment of a modified technology acceptance model. Journal of Management Information Systems 13, 185-204.

Chow-Chua, C., M. Goh and B.W. Wan, 2003. Does ISO 9000 certification improve business performance? International Journal of Quality & Reliability Management 20, 936-953.

Curkovic, S. and M. Pagell, 1999. A critical examination of the ability of ISO 9000 certification to lead to a competitive advantage. Journal of Quality Management 4, 51-67.

Davis, F.D., 1989. Perceived usefulness, perceived ease of use, and user acceptance of information technology. MIS Quarterly 13, 319-340.

Dunk, A.S., 2002. Product quality, environmental accounting and quality performance. Accounting, Auditing & Accountability Journal 15, 719-732.

Eurostat, 2004. Agrarstatistik - Vierteljährliches Bulletin, 1-2004. http://epp.eurostat.cec.eu.int/cache/ITY_OFFPUB/KS-NT-04-001-3A/FR/KS-NT-04-001-3A-FR.PDF; [10.09.2004]

Farina, E.M.M.Q., G.E. Gutman, P.J. Lavarello, R. Nunes and T. Reardon, 2005. Private and public milk standards in Argentina and Brazil. Food Policy 30, 3, 302-315.

Fishbein, M. and I. Ajzen, 1975. Belief, attitude, intention and behaviour: An introduction to theory and research. Addision-Wesley.

Gasson, R., 1973. Goals and values of farmers. Journal of Agricultural Economics 24, 521-533.

Gotzamani, K.D. and G.D. Tsiotras, 2002. The true motives behind ISO 9000 certification Their effect on the overall certification benefits and long term contribution towards TQM. International Journal of Quality & Reliability Management 19, 151-169.

Grunert, K.G., L.F. Jeppesen, K.R. Jespersen, A.-M. Sonne, K. Hansen, T. Trondsen and J.A. Young, 2004. Four cases on market orientation of value chains in agribusiness and fisheries. Working paper no. 83-2004, February 2004, Aarhus, School of Business, Denmark.

Hair, F.J., E.R. Anderson, L.T. Ronald and C.W. Black, 1998. Multivariate data analysis. Prentice-Hall.

Henson, S. and T. Reardon, 2005. Private agri-food standards: Implications for food policy and the agri-food system. Food Policy 30, 241-253.

Jahn, G. and A. Spiller, 2005. The adoption of the QS system in German agriculture: Exploring attitudes and the behaviour-intention relation. Paper presented at the 92nd Seminar 'Quality Management and Quality Assurance in Food Chains', 2-4 March 2005, Goettingen, Germany.

Kline, R.B., 2005. Principles and practice of structural equation modeling. Guilford Press, New York [and others].

Lazo, A., G. Jahn and A. Spiller, 2005. Grower's perception about EurepGAP in developing countries: The results of a peruvian study. Paper presented at the 92nd Seminar 'Quality Management and Quality Assurance in Food Chains', 2-4 March 2005, Goettingen, Germany.

Lin, C., C.N. Madu, C.-H. Kuei and M.H. Lu, 2004. The relative efficiency of quality management practices: A comparison study on American-, Japanese-, and Taiwaneses-owned firms in Taiwan. International Journal of Quality & Reliability Management 21, 564 577.

Mathieson, K., E. Peacock and W.W. Chin, 2001. Extending the technology acceptance model: The influence of perceived user resources. The DATA BASE for Advances in Information Systems 32, 86-112.

McAdam, R. and J. Henderson, 2004. Influencing the future of TQM: internal and external driving factors. International Journal of Quality & Reliability Management 21, 51-71.

MM, Milch & Markt, 2004. Die groessten deutschen Molkereien 2004. http://www.milch-markt.de/de/teaser/die_groessten_deutschen_molkereien_2004/; [10.11.04].

Rogers, E.M., 1983. Diffusion of innovations. Free Press, New York.

Sebastianelli, R. and N. Temimi, 2002. How product quality dimensions relate to defining quality. International Journal of Quality & Reliability Management 19, 442-453.

Szajna, B., 1996. Empirical evaluation of the revised technology acceptance model. Management Science 42, 85-92.

Theuvsen, L., 2004. On good and bad bureaucracies: Designing effective quality management systems in the agrofood sector. 84th EAAE Seminar Food Safety in a Dynamic World.

Venkatesh, V. and F.D. Davis, 2000. A theoretical extension of the technology acceptance model: Four longitudinal field studies. Management Science 46, 186-204.

Venkatesh, V., M.G. Morris, G.B. Davis and F.D. Davis, 2003. User Acceptance of information technology: Toward a unified view. MIS Quarterly 27, 425-478.

Weber, M., 1968. Economy and Society. New York.

Appendix 1. Measurement scales, reliabilities, means and standard deviations

Construct	Cronbach's alpha	Mean	Standard deviation
Acceptance[1]	0.934		
A1: If you had the free choice, how would you decide?[2]	-1.07	1.741	
A2: Which attitude do you have towards QM? not necessary - a must[3]		-0.39	1.719
A3: Which attitude do you have towards QM? nonsense - makes sense[3]		-0.95	1.586
Perceived (bureaucratic) costs	0.608		
C1: Which attitude do you have towards QM: simple - complex[3]	-1.40	1.474	
C2: QM means a lot of work for myself	-0.47	1.659	
Perceived usefulness	0.800		
B1: Business processes will be more transparent.	-0.40	1.668	
B2. Procedures on the farm will be more efficient.	-0.86	1.522	
B3: Procedures on the farm will be more thought through.	-1.53	1.311	
Perceived effectiveness	0.782		
U1: QM is important because it improves milk quality.	-1.07	1.545	
U2: QM is important because it improves the quality of production (such as animal welfare, environment).		-0.30	1.581
U3: QM is important because it improves/enhances the image of dairy products.	-0.23	1.669	

[1]The items were measured on a 7-point Likert scale, where

Strongly disagree	-3	-2	-1	0	1	2	3	Strongly agree
[2]Certainly no introduction	-3	-2	-1	0	1	2	3	Introduction for sure
[3]Strongly agree	-3	-2	-1	0	1	2	3	Strongly agree (semantic differential)

Quality management: a supply chain perspective

Comparison among quality assurance and management schemes regarding the ability to enhance suppliers-retailers relationships in Emilia-Romagna (Italy)

Maurizio Canavari, Roberta Spadoni and Erika Pignatti
Dip. di Economia e Ingegneria agrarie, Alma Mater Studiorum-Università di Bologna, Viale Giuseppe Fanin, 50 - 40127 Bologna, Italy; maurizio.canavari@unibo.it

Abstract

This study is aimed at evaluating the ability of voluntary standards to satisfy the needs of large retailers towards their suppliers. Large retail companies operating in the Emilia-Romagna region were asked to participate in the survey. An analysis of the main requirements set by large retail chains to their suppliers, of the underlying needs these retailers want to satisfy, and of the role of standards and certification in the selection of suppliers process was performed. A perceptual map of different quality assurance schemes was also drawn. The results of the analysis suggest that, while the ability to apply the quality assurance methods is crucial, certification is generally not seen as a discriminant attribute for the supplier.

Keywords: quality assurance systems, certification, retailer

1. Research background

The agri-food sector is increasingly interested by a process of creation of specific models of quality management/certification systems, involving the production process, the product, as well as the management system as a whole, both on a firm and supply-chain basis. This is an opportunity for many firms, to try to select the best fit solution according to their needs, but also create a problem of overlapping and sometimes contrasts between the certification protocols.

Some of the more recent standards are strongly promoted by large retail chains and retailers associations, that are increasingly interested in making the quality assurance level higher and the burden of quality control more easily manageable on their side. The market power they are able to exert in many markets let them put on their suppliers additional requirements on the adopted standards or to adopt chain-specific standards, more suited to satisfy retailer chains needs.

The regulations background for the analysis is given by the sector specific set of standards, which have been developed in the last decades, regarding quality management systems, food safety, traceability, social responsibility, etc.

The variety of choices given to the food firms in terms of adoption of those kinds of standards raises the need to compare advantages and problems given by the adoption of each of them. A conceptual model aimed at measuring the effectiveness of quality management systems in the food industry was developed by Van der Spiegel *et al.* (2003). This conceptual model may represent a starting point for a specific comparison of different food industry-based or general quality management systems.

Additionally, the tools adopted in marketing research (Molteni and Troilo, 2002) may be suitable to help comparing the several certification schemes available in the food industry.

2. Objectives

This study is aimed at evaluating the ability of voluntary standards, as well as contractual production rules to answer to the needs expressed by large retailers towards their suppliers. This objective may be specified through the following steps:

- Analysis of the main requirements set by large retail chains to their suppliers.
- Analysis of the underlying needs these retailers want to satisfy (customers demands, actions to comply with the law, company goals, etc.).
- Analysis of the role of standards application and certification in the selection of suppliers process.

Then, the research tries to examine the existing relationship between retailers and suppliers, in terms of requirements and eventually offered pay-offs and the influence exerted by the certification schemes in this regards.

3. Materials and methods

A preliminary step of the analysis is to define a surveying basis. It was decided to work on a selection of retail chains, chosen between those operating in the Emilia-Romagna region, according to Fanfani and Pieri (2004), as shown in Table 1.

The retail chains have been contacted and their web-sites have been explored, in order to refine the knowledge of the industry structure. It was ascertained that not all the above mentioned retailers actually managed outlets in the Emilia-Romagna region, but they were considered in any case. A large majority of retailers claimed they were available to participate in the survey, since at present time not all of them actually responded to the questionnaire.

The data collection has been performed following a two-step procedure. First, a qualitative analysis of requisites and criteria adopted by large retailers to select and evaluate the suppliers has been performed. Second, starting from this basis and from the knowledge of quality standards, a quantitative analysis aimed at understanding the reciprocal proximities (similarities) of standards and at obtaining their relative position according to requisites, factors, and performance, as well as the importance of third-party certification for suppliers selection.

Selection criteria for suppliers have been hypothesised using previous literature and researchers' a priori knowledge, a tentative check-list was built and it was validated during a small number of de-structured in-depth interviews with experts, i.e. the quality assurance managers or CEOs of six retail chains, selected in order to assure the presence of different types of retailers (consumer co-ops, retailer co-ops, privately owned firms, etc.). The in-depth interviews started from a single specific and direct question on the 'main criteria used to select suppliers', and were conducted going progressively in-depth, according to the topics and issues mentioned by the interviewee. The check-list for relevant topics was prepared exclusively to let the interviewer note the issues mentioned by the respondent; it was not shown in order to be sure the interviewee were not influenced. In particular, the requirements in terms of quality assurance, as highlighted and demanded by the retailers, were compared with the certification standards characteristics, as perceived by the interviewee.

Table 1. Emilia-Romagna retail chains and their selling areas in 2003 (modified from Fanfani and Pieri, 2004).

Chain	Retail brand	Surface area	
		m²	%
COOP		263,545	32.7%
CONAD		151,773	18.8%
SIRIO	SIGMA	84,605	10.5%
	PICK-UP	1,195	0.1%
ESD ITALIA	SELEX	59,887	7.4%
	ESSELUNGA	22,360	2.8%
	AGORA' (Poli, Iperal, SoGeGross, Tigros, Sermark)	2,310	0.3%
MECADES	SINTESI/DESPAR	27,857	3.5%
	INTERDIS	15,243	1.9%
	SISA	11,345	1.4%
	CRAI	18,942	2.3%
INTERMEDIA/ RINASCENTE	PAM	25,115	3.1%
	LOMBARDINI	3,730	0.5%
	BENNET	13,600	1.7%
	AUCHAN	18,483	2.3%
CARREFOUR	CARREFOUR	16,130	2.0%
	FINIPER	16,770	2.1%
	IL GIGANTE	3,700	0.5%
LIDL		20,014	2.5%
STANDA-REWE	Penny Market, Billa Italia/Standa	16,809	2.1%
CORALIS/MDO		13,485	1.7%
Total		806,898	

A list of 15 factors/requisites used to evaluate the potential of a supplier to be selected, as well as a list of 11 characteristics of the different quality standards were the result of this first phase.

Afterwards, a structured questionnaire was submitted to several executives within the retail chains organisation, e.g. quality assurance managers (for a general overview on the inquired aspects), fruit and vegetables buyers, dairy products buyers, and retailer own brand managers (with the aim to highlight specific aspects). The strategy to address different subjects within the same organisation was chosen in order to assure a wider view on several aspects of supplier selection, which may be influenced by industry-specific factors and by subjective perceptions.

The questionnaire was developed in order to get information concerning suppliers characteristics and the importance assigned by retailers to specific requirements, and is structured in 8 sections to let us get information on the following aspects:

1. Role of the respondent within the organisation.
2. No. of suppliers managed by or under the responsibility of the respondent.
3. Types of quality certification schemes adopted by the suppliers.
4. Self-evaluation of the knowledge of the 9 quality assurance schemes (QAS).

5. Evaluation matrix of the proximities (similarities) between the QASs.
6. Evaluation matrix of the performances of each QAS with respect to 11 factors and of an overall performance rating.
7. Evaluation vector of the importance of 15 criteria for the selection of suppliers.
8. General information on the retailer.

Both sections 5 and 6 are suitable to let the researcher draw a perceptual map of the QASs from the retail professionals. Actually, considering the small number of responses obtained to date, section 5 was privileged for the following data analysis.

The chosen methodology was the decompositional approach to multi-dimensional scaling (Hair *et al.*, 1998), which requires only the collection of similarity data among the evaluated objects (a set of 9 certification schemes) by the respondent. About the data collection method used for the similarity evaluations in the MDS positioning the paired comparisons one was chosen. It represents, in relationship with the present survey condition, the more reasonable compromise between completion time, fatigue and boredom of subjects, information for individual subjects, which makes it possible to derive individual-level perceptual maps and levels of error variance in the MDS solution (Bijmolt and Wedel, 1995). We have, though, limited control towards the formation of the similarity judgments, then possible biases in perceptual maps caused by various factors are not taken into consideration in this study and requires caution in interpreting the maps.

However, since also a preference evaluation was considered of great importance, data were collected in section 6 on ratings assigned by the respondent to several attributes and the performance of the certification scheme.

In section 7 interviewees (responsible of quality management and chief buyers) are asked to highlight the requirements their suppliers have to comply with. The identified requirements are graded according to the importance in relationship with the strategic goals of the retailer (e.g. supplier organisation control; chain-wide traceability; international acknowledgement; supplies regularity and reliability).

The data collected in section 5 have been processed using a multidimensional scaling approach (INDSCAL package) in order to get 2 and 3-dimensional perceptual maps. In order to compare the maps and reinforce the interpretation of these results an attribute-based positioning using discriminant analysis of the data collected in question 6 was also performed. Finally, in order to study the relationship between the overall performance rating of the QASs considered with the chosen attributes, a linear regression analysis (OLS) was also carried out on the results of a previous factor analysis (principal component method). This procedure was chosen in order to summarise the original factors into a reduced set of variables, to be used in the following step. In a further stage of the study the factor scores, together with retailers variables may be used as input for a cluster analysis, aimed at identifying homogeneous groups of retailer strategies towards quality management characteristics of the suppliers.

4. Results

The survey has interested 27 retailing companies; 22 of them agreed to participate, but actually only 15 of them sent at least one filled questionnaires back, for a total amount of 24 questionnaires. One respondent operated with a double function in the organisation, so the available data are, in some cases, 25. The data form was in many case incompletely filled, sometimes with the motivation of a specific privacy policy on certain data, sometimes because the respondent claimed he was not aware of the data or he considered himself not qualified enough.

The data collection results were not encouraging. Anyway, some interesting information may be extracted by the data collection process and the analysis of the data. First, the decisional process for supplier selection is to be considered extremely variable across the companies: in some cases the responsibility is concentrated, while in other cases the responsibility for any single sector is split between the quality and commercial divisions. In the latter case, knowledge and competence on quality standard may be unequally distributed among the company executives. Presently, 10 QA managers, 1 fruit and 5 dairy products buyers and 5 retailer own brand manager returned the questionnaire, together with 3 executives of different type. The number of managed suppliers (14 answers) vary between 36 and 4,000 with a median value of 120.

The survey was aimed at giving a picture of the situation of the selected retail chains about their suppliers' qualification: answers given about the actual suppliers' situation will give a hint on the penetration of different certification standards (ISO 9000; ISO 14000; EMAS; BRC; IFS; EurepGAP; etc.), and the quota of third-party certified suppliers.

About the certification schemes applied by the suppliers, 17 respondents completed section 3 of the questionnaire; some results are summarised in Table 2. It must be noted that the most popular certification scheme is ISO 9000, followed by geographical indication and organic. Anyway, the only certification adopted by a relevant quota of suppliers is ISO 9000, all the others being quite marginal.

The survey allowed us to give a picture of the degree of knowledge, stated by the retail industry operators, of set of norms in many cases specific for the agri-food sector, and the degree of awareness of certification standards and of their implications at a managerial level.

A self-evaluation was asked to the interviewee, using an increasing 1-10 scale with a central pivoting point, considering the ratings between 1 and 5 as insufficient to express a judgement on the standard, and the ratings over 5 as sufficient knowledge. In Table 3 the results of this question are showed.

Table 2. Presence and relative quota of certified suppliers, by scheme (survey data).

	Certified suppliers			Quota of certified suppliers				
	Answer	Yes	Percent	Answer	Mean	Median	Minimum	Maximum
ISO 9000	17	17	100.00	8	56.25	62.50	10.00	90.00
ISO 14000	17	13	76.47	7	7.43	10.00	0.00	15.00
EMAS	17	6	35.29	6	2.33	2.00	0.00	5.00
PDO or PGI	17	15	88.24	5	11.00	5.00	0.00	30.00
Organic	17	15	88.24	6	9.50	5.00	0.00	30.00
IFS	17	10	58.82	6	9.67	3.50	0.00	30.00
BRC	17	14	82.35	7	10.71	5.00	2.00	30.00
EurepGAP	17	8	47.06	5	6.20	5.00	2.00	17.00
SA 8000	17	5	29.41	4	1.75	1.00	0.00	5.00

Table 3. Descriptive statistics of the knowledge on the certification standards self-evaluation by the respondents (1-5=insufficient knowledge, 6-10 sufficient knowledge) (survey data).

	Count		Minimum	Maximum	Median	Mean	Std. deviation
	Valid	Missing					
ISO 9000	21	4	3.00	10.00	8.00	7.90	2.1887
Organic	21	4	1.00	10.00	8.00	7.71	2.1712
PDO/PGI	21	4	1.00	10.00	8.00	7.67	2.1525
BRC	21	4	1.00	9.00	6.00	5.86	2.4756
ISO 14000	21	4	1.00	10.00	6.00	5.57	2.5213
IFS	21	4	1.00	9.00	6.00	5.38	2.5588
EurepGAP	21	4	1.00	9.00	5.00	4.86	2.4142
SA 8000	21	4	1.00	10.00	5.00	4.52	2.4621
EMAS	20	5	1.00	8.00	4.00	3.90	2.3819

The median values denote that the respondents generally evaluate themselves at a high competence level for few certification schemes: ISO 9000, Organic, and PDO/PGI, the first being a transversal competence, since it is probably taken as reference for the internal procedures of the company, the others being highly specific for market niches and speciality products. The median knowledge of food chain specific standards (BRC, IFS, EurepGAP) is considered at the border line for a sufficient rating, and a great variability must be highlighted. This is a signal of a still scarce penetration of these schemes at the retail level.

Other standards (ISO 14000, EMAS and SA 8000), non specific for the food chain, obtain quite low knowledge ratings, as expected. The ratings assigned obviously vary across the different type of executives interviewed. The QA managers are generally more competent than the other operators.

A certain level of knowledge of the certification schemes is required for the MDS techniques, so this situation had an impact on the quality of data collected in the following parts of the questionnaire. For example, in many cases the comparisons and evaluation regarding the SA 8000 scheme were left blank. Sometimes the same happened for other schemes.

Since the quality of data is crucial in the MDS technique, we decided to drop many of the incomplete evaluations, using only a subset of 14 paired comparisons proximity matrices for the analysis. The INDSCAL model (Hair *et al.*, 1998) contained in the ALSCAL procedure of the SPSS software package was used, in order to draw a map on a common multidimensional space for all the respondents, but also to have the chance to register the differences among the respondents. In this paper only the overall results of the procedure are discussed, leaving the comment of individual aspects to a further analysis.

The paired comparisons of the 9 certification schemes plus a hypothetical (and subjective) ideal point made by the respondents allowed us to draw both a 2-dimensional and a 3-dimensional perceptual map. The appropriate dimensionality was chosen looking at the average measures of fit and at the improvement of Stress (Kruskal's stress formula 1, Kruskal and Wish, 1978) and R squared parameter as the number of dimensions increases (Table 4).

Table 4. Assessment of the overall fit and determination of the appropriate dimensionality (survey data).

Number of dimensions in the solutions	Stress	% change	R Squared (RSQ), quota of expl. similarity	% change
6	0.12239		0.63240	
5	0.15000	18.41%	0.57978	9.08%
4	0.18326	18.15%	0.51671	12.21%
3	0.21661	15.40%	0.52570	-1.71%
2	0.30943	30.00%	0.34970	50.33%

The 2-dimensions solution (Figure 1) is easier to describe, and the results are quite consistent with the expectations. The image of certification schemes differs sensibly across the two selected dimensions. The group of schemes on the right side of the map are characterised by a 'system' approach, they are not focused on the food chain and are not intended to influence the product in itself, even if they may have an impact on some product (credence) attributes.

The schemes on the left, on the contrary, are all focused on the food chain and are intended to influence the product characteristics and to give information and assurance on food quality and safety issues. The intermediate position of the Organic scheme appears consistent with this interpretation.

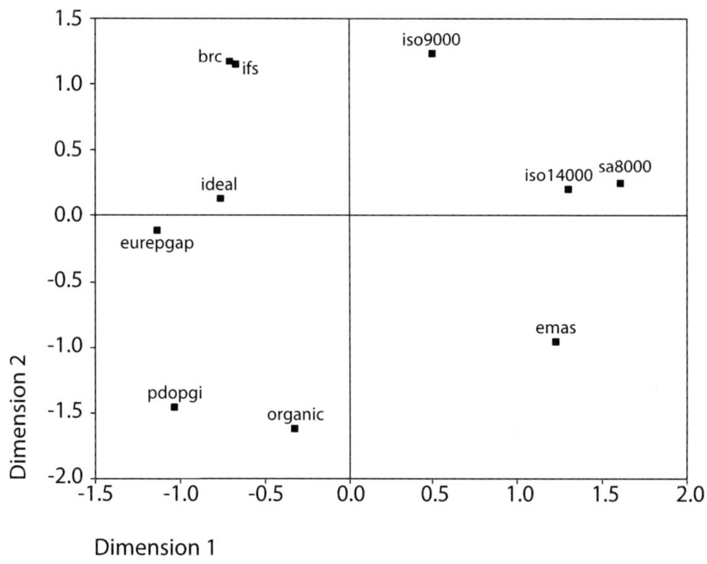

Figure 1. Perceptual map of the certification schemes on 2 dimensions. Derived Stimulus Configuration, individual differences (weighted) Euclidean distance model. The map is based on the similarity judgements given by 14 retail chains executives.

Looking at the vertical dimension, it may be noted that on the top there are more 'global' schemes, aimed at increasing the effectiveness and efficiency of organisations and supply chains through standardisation, while on the bottom they may be considered more 'local and traditional', aimed at enhancing the value through differentiation. The Organic certification is actually not strictly local, but it could be perceived as a means to enhance the value of local traditional productions. Contrasting the EMAS and ISO 14000 certification we know that the former is strongly linked to a site, while the latter is related to the organisation.

The relative positions of the certifications schemes are also interesting. First, we may note that IFS and BRC are almost overlapping, thus giving us the information of an almost complete 'substitution' relationship.

The position of the ideal point deserves a short explanation. The ideal point showed in the figure is to be considered the more preferred position of an average respondent. Assuming a point representation of the ideal point, the closer the scheme the more is it preferred. In our map EurepGAP, IFS, and BRC (ISO 9000 is a little farther) seems to be the ones that represent the solution closer to the ideal.

However, in case differences among respondents are relevant, it would be useful to look at the individual ideal points, in order to understand the attitudes and preferences of single subjects. This is not the aim of the present study, then the issue will be explored in a following paper.

Furthermore, significant differences among the subjects may also be found looking at the weights derived by the subject's responses (Table 5). Subject weights measure the importance of each dimension to each subject. RSQ is the sum of squared weights. The subject weights are used to express the 'weirdness' of the subject in comparison to the average solution. A subject with weights proportional to the average weights has a weirdness of zero, the minimum value; a subject with one large weight and many low weights has a weirdness near one; a subject with exactly one positive weight has a weirdness of one, the maximum value for nonnegative weights. In this analysis, we may note that the rate of weirdness is not very high, so the judgements have been rather homogeneous.

The discussion of the 3-dimensions solution (Figure 2) is a little bit more difficult, since a the interpretation of the graph is less straightforward. Moreover, several alternate solutions were found through rotation of the axes. The map showed in Figure 2 is the one in which the positions of the QASs on dimensions 1 and 2 are about the same as in the 2-dimensions solution. It is possible to highlight the fact that on the third dimension we find the ideal point on a central position, close to PDO/PGI, Organic and ISO 9000, while we find SA 8000 on the bottom and all the other schemes on the upper side. As a common trait for SA 8000, PDO/PGI and Organic we may perhaps identify the 'ethical' values, but this may on some extent be shared also with ISO 14000 and EMAS. The latter two are probably far less known by the consumers and probably are more addressed to public administrations and local citizens, so we are considering the third dimension as a consequence of the communication power owned by the certification scheme, business-to-consumer downwards versus the business-to-business or business-to-government upwards.

The last two questions were aimed at going more in-depth in the understanding of the criteria and attributes, both for certification schemes and supplier requisites.

Table 5. Single matrices measures of fit and subject weights by dimension (survey data). The map is based on the similarity judgements given by 14 retail chains executives. Overall importance of each dimension: 1 = 0.1838; 2 = 0.1659.

Subject	Measures of fit		Dimension Weights		
Number	Stress	RSQ	1	2	Weirdness
1	0.182	0.793	0.777	0.435	0.314
2	0.337	0.260	0.437	0.263	0.272
3	0.368	0.078	0.215	0.179	0.073
4	0.289	0.434	0.522	0.402	0.123
5	0.215	0.676	0.493	0.658	0.220
6	0.358	0.141	0.279	0.252	0.022
7	0.292	0.411	0.402	0.499	0.177
8	0.211	0.691	0.489	0.672	0.238
9	0.329	0.254	0.353	0.360	0.054
10	0.329	0.269	0.410	0.317	0.121
11	0.331	0.249	0.362	0.345	0.011
12	0.376	0.039	0.146	0.134	0.015
13	0.327	0.270	0.385	0.348	0,023
14	0.311	0.330	0.382	0.430	0.117

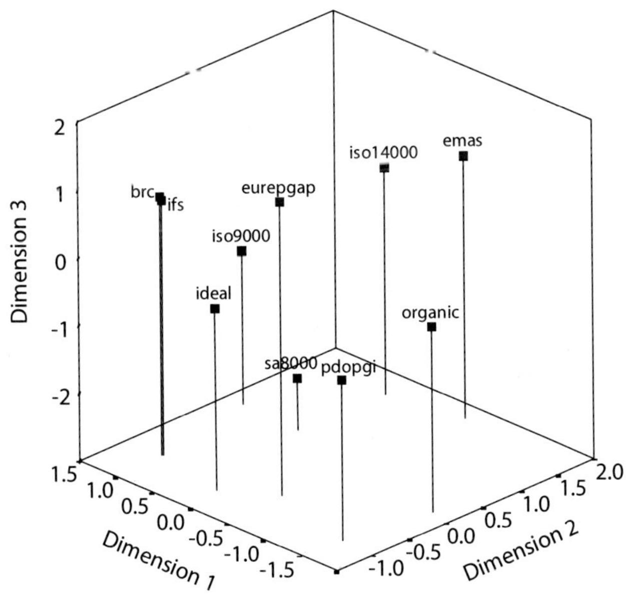

Figure 2. Perceptual map of the certification schemes on 3 dimensions. Derived Stimulus Configuration, individual differences (weighted) Euclidean distance model. The map is based on the similarity judgements given by 14 retail chains executives. Overall importance of each dimension: 1 = 0.2135; 2 = 0.1614; 3 = 0.1507.

The identification and importance evaluation of retailers' strategic requirements as well as the comparison between these requirements and their implementation into the different standards, should let us single out many elements representing a starting point to analyse the degree of 'completeness' of several standards/norms towards the specific aspects taken into consideration.

In question 6, the respondents were asked to assign ratings to a list of factors, selected during the qualitative phase of the survey, that should represent the main elements for comparing the performance of a certification scheme (Table 6).

Looking at the descriptive statistics of the ratings assigned by the respondents, we may note that the average scores seems quite consistent with a priori expectations. Another aspect to be pointed out is the relatively high level of variability within the ratings assigned for the same factor to the same certification scheme: the high level of standard deviation testify the presence of a strong heterogeneity in the evaluations.

Table 6. Mean and standard deviation of the ratings assigned to each quality standard on the performance factors (1-10 increasing scale) (survey data).

		ISO9000	ISO14000	EMAS	PDOPGI	Organic	IFS	BRC	EurepGAP	SA8000
Connection with the territory	Mean	1.06	2.19	2.43	9.75	5.59	2.25	2.19	3.00	2.00
	Std.Dev.	0.250	1.905	1.785	.447	2.347	1.770	1.760	2.309	1.881
Hygiene control	Mean	6.07	3.79	3.15	3.86	4.20	8.80	8.80	5.60	1.21
	Std.Dev.	3.081	3.512	3.313	2.797	2.513	1.146	1.146	2.414	0.802
Supplier control of raw material	Mean	8.13	3.00	2.62	7.80	8.25	8.73	8.73	6.29	1.00
	Std.Dev.	1.598	2.660	2.063	2.884	2.176	1.280	1.280	2.946	0
Supplier control of final product	Mean	8.27	3.00	1.92	7.87	7.47	8.67	8.73	6.57	1.21
	Std.Dev.	2.219	2.602	1.321	2.588	3.165	1.234	1.280	2.564	0.579
Production process management	Mean	8.75	5.86	4.92	6.60	6.63	8.07	8.13	6.43	1.77
	Std.Dev.	1.483	3.348	3.068	2.667	3.096	2.577	2.615	2.980	2.204
System reliability	Mean	8.44	6.14	5.15	6.14	6.13	8.60	8.47	7.29	2.62
	Std.Dev.	1.712	3.060	3.051	2.568	2.532	0.986	1.060	1.638	2.725
Documentation	Mean	9.07	8.50	7.23	6.64	7.00	8.00	7.93	6.93	4.00
	Std.Dev.	1.163	1.286	2.048	2.205	2.070	1.604	1.710	1.981	3.162
Identification and traceability	Mean	7.56	3.33	2.69	7.13	7.50	7.93	7.80	6.86	1.77
	Std.Dev.	2.309	2.870	2.529	2.066	2.033	1.335	1.320	1.610	1.536
Ethical behaviour	Mean	1.40	4.53	4.85	2.36	5.13	2.07	2.00	3.86	9.79
	Std.Dev.	0.737	3.502	3.313	2.240	3.815	1.580	1.558	2.984	0.426
Impact on price	Mean	5.17	4.73	4.45	5.82	7.33	4.00	4.00	5.40	5.22
	Std.Dev.	2.553	2.970	3.078	2.483	2.640	2.749	2.749	2.757	2.539
Communication to consumer	Mean	5.88	5.88	4.29	8.25	8.24	3.00	3.00	4.25	7.79
	Std.Dev.	2.713	2.335	2.302	2.463	2.223	2.280	2.280	2.840	2.547
Overall performance rating	Mean	7.31	7.07	6.64	7.40	7.44	8.13	8.00	7.69	6.79
	Std.Dev.	1.815	1.534	1.906	1.724	1.094	1.784	1.713	1.353	1.968

Finally, while the single factor ratings are strongly differentiated across the certification schemes, the overall performance rating is much more homogeneous. This suggests that a compensation logic is perhaps adopted by the respondents, thus they consider that each standard being not able to cover all the issues, all of them still maintain their importance and usefulness.

The ratings assigned by the respondents to the certification schemes have been used to perform a discriminant analysis, in order to compare the positioning solution obtained by a decompositional approach to that obtained by a compositional approach.

The perceptual map derived by the performance variables seems to be very close to the ones obtained from the distance matrices. It is a very good element for the validation of the MDS solution, and also a confirmation of the ability of the 11 variables chosen to describe the performances of the certification schemes.

In this paper, for the sake of brevity we do not report the result of the alternate MDS solution, but it is available under request by the authors. The performance ratings have also been used to try to understand the relationship between the rated variables and the overall performance rating.

A previous factor (principal component) analysis was necessary in order to reduce the dimensionality of the problem and to solve the problem of strong colinearity among the variables. The results of factor analysis allowed us to create three performance factors, which may be summarised as follows: the first one, highly correlated with system and product management performance variables (PERF02 through PERF08) is labelled as 'system effectiveness and efficiency'; the second, mainly linked to communication to consumer and price impact (PERF10-11) and at a lower extent to traceability (PERF08) is labelled as 'differentiation'; the third one, positively linked to the connection with the territory (PERF01) and negatively linked to documentation (PERF07) is a 'trust' factor, that may be labelled as 'reputation versus evidence'.

The three factors have been used to explore the relationships with the overall performance ratings obtained by the certification schemes. An exploratory stepwise regression analysis including the overall rating as dependent variable and the three factor scores as independent variables was performed. It succeeded finding a significant relationship with the first two factors, giving account of about 27% (adj. $R^2 = 0.271$) of the total variance in overall ratings.

This result suggests that while the first two factors are positively linked to the 'satisfaction' for the performance given by a certification scheme, the third one may work in both directions, then a linear model may not be the more suitable to explore its relationship with the scheme appreciation.

Finally, the questionnaire coped with the issue of the relative importance of several factors, used as criteria to consider a producer as potential supplier, identified during the exploratory qualitative phase of the study. A simple descriptive statistics table is already enough to point out that the third-party certification is considered as a 'second level' choice criterion for the retail chains (Table 7). Instead, many elements that characterise the diverse quality assurance schemes (supplier controls, system reliability, documentation, etc.) are seen as extremely important, and it is apparent that the retailing organisations actually do prefer to set their own standards and check for compliance to standards by themselves.

The most important requisites according to the claims of the respondents are focused on the product, considering first the respondence to customer needs, then the hygiene, raw materials,

Table 7. Descriptive statistics of the importance assigned to factors used for suppliers selection (ordinal increasing semantic scale 0-5)(survey data).

	Count		Min.	Max.	Median	Mean	Std. Dev.
	Valid	Missing					
Connection with the territory	24	1	0	5	4.00	3.75	1.422
Hygiene control	25	0	0	5	5.00	4.56	1.121
Supplier control of raw material	25	0	0	5	5.00	4.44	1.158
Supplier control of final product	25	0	0	5	5.00	4.32	1.108
Production process management	25	0	0	5	5.00	4.44	1.158
System reliability	25	0	2	5	4.00	4.20	0.866
Documentation	25	0	1	5	3.00	3.40	1.225
Identification and traceability	25	0	1	5	5.00	4.36	1.036
Ethical behaviour (envir., labour, solidarity)	25	0	1	5	3.00	3.16	1.546
Third-party certification	25	0	0	5	3.00	2.76	1.300
Brand awareness	25	0	1	5	3.00	3.04	1.457
Product suitable to customer needs	25	0	3	5	5.00	4.72	0.542
Price at supplier level	25	0	0	5	5.00	4.24	1.128
Sales expectations	25	0	0	5	5.00	4.04	1.428
Range of offer completion/widening	24	1	0	5	3.50	3.67	1.167

and final product quality control. Obviously, the commercial issues (sales expectations, together with customer needs and brand awareness, completion of the range of offered products, price) play a relevant role, thus giving a confirmation of the crucial element represented by the marketing strategies applied by the producers to maintain the relationships with the final customer.

The distribution of answers regarding the importance of third-party certification is the last interesting information that is possible to extract from the survey data (Figure 3). It confirms that the comments made above are valid for a majority of the respondents, whose ratings for the specific factor are mainly concentrated around the 'medium importance' point in the semantic scale no/extreme importance.

5. Final remarks

Despite the fairly low level of participation, the analysis performed gave us a better insight on the large retailers' attitudes towards the certification schemes. This may give to the producers some suggestions on the strategies of adoption of one or more standards/norms.

As found in many literature studies, retailers could demand such standards to their suppliers in order to optimise the selection processes and to satisfy their needs more carefully.

However, it seems clear that the retailers are renitent to rely on certification bodies to perform the controls and to assure quality. They still prefer to set their own standards and to perform second-party controls made by their personnel or in outsourcing by controlling companies.

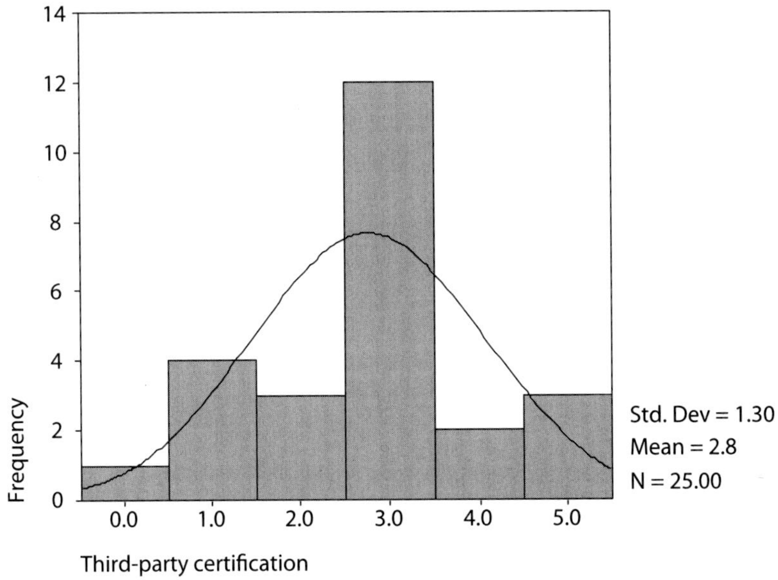

Figure 3. Distribution of the importance level assigned to third-party certification in the selection process of suppliers (ordinal increasing semantic scale 0-5).

Then, the quality assurance certificate given by a third-party certification body on the base of an external set of standard rules seems not to represent a real competitive advantage. On the contrary, the ability to comply with norms is a necessary and highly appreciated requisite. Definitely, it is more important for a company to be aware and actually able to apply the quality assurance methods and principles, rather than to be certified. According to these results, it seems wiser for a producer to look for personnel with knowledge and skills to be applied in the improvement of the firm organisation, rather than looking for 'certificates'.

The study seems to confirm that the retailers-borne standards are closer to the ideal situation, but the penetration level of these standards is still low. Moreover, the Italian retail executives asked for an opinion did not actually perceived a real difference between the two standards, implicitly suggesting that it should be merged in a single standard. The talks taking place within the Italian Association of Food Distributors (FAID) may be a step forward in this direction, or at least towards the choice of one of the two standards as a reference basis for the setting of retailers' production rules and the organisation of the control procedures, as well as for the simplification and improvement of the supplier selection process.

The information provided by this study suggests that the roadmap to the unification of controls and quality assurance strategies is still long and difficult. This information could benefit agri-food operators (Certification bodies, Consultants, Producers, Suppliers) in terms of definition of strategies and optimisation of resources devoted to the qualification processes.

Acknowledgements

We wish to thank all the retail chain executives who gently agreed to participate into our study, namely the retail companies CONAD S.c. a r.l., Gruppo PAM S.p.a., ESSELUNGA S.p.a., SIGMA S.c. a r.l., SERMARK S.p.a., BENNET S.p.a., CRAI S.c. a r.l., AGORA' Network, AUCHAN, COOP S.c. a r.l., IL GIGANTE S.p.a., FINIPER S.p.a., Gruppo Commerciale SELEX S.r.l., IPERAL S.p.a., Standa. The analysis was designed and conducted jointly by the authors. However, Roberta Spadoni wrote sections 1 and 2, Maurizio Canavari performed the data processing and wrote section 4, Erika Pignatti, performed the data collection and wrote sections 3 and 5.

References

Bijmolt, T.H.A. and M. Wedel, 1995. The effects of alternative methods of collecting similarity data for multidimensional scaling. International Journal of Research in Marketing 12, 363-371.

Fanfani R. and R. Pieri, 2004. Il sistema agroalimentare dell'Emilia-Romagna - Rapporto 2003. Franco Angeli, Milano.

Hair, J.F., R.E. Anderson, R.L. Tatham and W.C. Black, 1998. Multivariate data analysis. Prentice Hall.

Kruskal, J.B. and M. Wish, 1978. Multidimensional scaling. Sage University Paper Series on Quantitative Applications in Social Sciences, 07-011, Sage.

Molteni, L. and G. Troilo, 2003. Ricerche di marketing. McGraw-Hill, Milano.

Van der Spiegel, M., P.A. Luning, G.W. Ziggers and W.M.F. Jongen, 2003. Towards a conceptual model to measure effectiveness of food quality systems. Trends in Food Science and Technology 14, 424-431.

Prevention of food scandals by means of contracts

Alexander M. Hinrichs
University of Kiel, Institute for Agricultural Economics, Olshausenstr. 40, 24098 Kiel, Germany; ahinrichs@agric-econ.uni-kiel.de

Abstract

Generally food scandals do not only have negative effects on the consumer-health but might also trigger massive financial losses regarding concerned business entity (Hinrichs, 2004). Regarding food-grievances, particularly the recent food scandals like BSE or the Nitrofen scandal, for example, sensitised both the participating business entities and the consumers. The misconduct of individual participants might affect the up- or downstream stages of a product-value chain. This might result in a decline in sales and additional costs for the respective business entities. Hence, all participants of a food-value chain have an legitimate interest in the prevention of food scandals. Yet, recent incidents revealed that as well single quality management systems as quality assurance systems that involve the entire value chain might indeed contribute to the quality and security of aliments but cannot guarantee them. Often, the reason for grievances is an opportunistic behaviour of individual participants, which hardly can be prevented by quality management systems. In order to prevent food scandals which are caused by opportunistic behaviour, a revised principal-agent approach is regarded. Based on a model of Becker (1968) and Hanf (1993) a contract-demand-approach will be developed.

Keywords: food scandal, opportunistic behaviour, principal-agent theory, penalty fees

1. Introduction

The agricultural and food industry in the highly developed nations obtained a high quality level in the recent decades. The high prosperity level generated high expectations in quality and security of food. Unobjectionable products are taken for sure and as well the pressure groups of the food sector, the food industry as the traders want to suggest that the food is safe. So it is quite scandalous if the media gives an account of grievances in the food sector. In a way how the consumers feel threatened by the grievances, they react in a different way. Some ignore the whole situation and some boycott the affected products. But all great and small grievances have in common that they have economic consequences for the food sector. They can vary in the nature and extent of the grievance.

2. Causes of food scandals

In a closer analysis of past food scandals (Hinrichs, 2004), we can assert three different causes for the grievances: Human failure, technical failure and criminal behaviour. Human failure and even technical failure cannot be avoided but should be anticipated by well implemented quality assurance instruments. Hence human and technical failure will not be regarded in this paper.

Below criminal behaviour as the third cause for grievances will be analysed. Some participants of the food supply chain offend against existing agreements or standards for their own advantage and improve the asymmetric information of the other involved participants. This behaviour, in literature called moral hazard, can be divided into three aspects: Negligent acting, grossly negligent acting and finally consciously opportunistic (criminal) acting. Opportunistic (criminal) acting is the conscious acting of participants against existing contracts, laws or regulations, hoping to achieve an economic advantage or hoping not to be convicted or the impending sanctions are minor than the short-term advantages of the opportunistic acting. Williamson (1975) said that some people are opportunistic and want to exploit the no-opportunistic participants. Hirschauer *et al.* (2003) point out that the risk for opportunistic behaviour will be greater the more costs can be saved and the less the likelihood to be disproved. Unlike human or technical failure criminal acting cannot be controlled by quality assurance systems. The conscious and unconscious disregard of laws or contracts cannot be prevented by letters of intents or setting of standards for the whole supply chain. Only special contracts between the supply chain participants can prevent opportunistic behaviour.

3. Contracts as a chance for quality assurance

An effective quality assurance by stipulations has to comprise four basic assumptions:

1. One level of the supply chain must be in a position to compile product - quality- and safety-guidelines for the whole supply chain.
2. A well implemented quality assurance system must be implemented in every level of the supply chain.
3. The compliance with regulations must be controlled by the overbearing level.
4. On contravention of these regulations economic sanctions must be defined.

But this system can only prevent criminal human misconduct, if the subordinated partner in the supply chain receives a higher benefit (G_V) by compliance with the guidelines than he would receive at the free market or at a alternative purchaser (G_M) (Holmstrom, 1982). Thereby it has to take in consideration that the production costs in regarding the quality specifications (K_E) will be higher then the production costs without this regarding (K_N). E_V is the reward for the product within the supply chain - quality assurance - system and E_M is the reward for a product without regarding of the quality specifications:

$$G_V > G_M \qquad (1)$$

with

$$G_V = E_V - K_E \qquad (2)$$

and

$$G_M = E_M - K_N \qquad (3)$$

Below it is shown how a contract between one principal and one agent and one principal and several agents can be configurated.

3.1. One principal and one agent

Base of the following model is model of Becker (1968). This paper keeps with the paper of Hanf (1993). The following approach should be applicable along the whole supply chain. The principal provides the quality-guidelines. The agent only signs the contract if

$$G_V > G_M \quad (1)$$

and

$$G_E > G_N \quad (4)$$

with G_E = Agents benefit by compliance with the contract, and
G_N = Agents benefit by non-compliance with the contract.

For the contract conditions some restrictions must be made:

a. With compliance with contract the agent will have costs of $K_E > 0$. These costs comply with the higher production-expenditures for contractually high quality food. But will the agent not adhere to the contract conditions and produces non-contractual food, the expenditures are only K_N with $K_E > K_N > 0$.
b. The contract even comprises the agent's reward of $E > 0$. The expected penalty for the agent in case of proved breach of contract is $S > 0$ and consists of three aspects: The recall-costs of the defective products (S_R), the contractual fixed contractual penalty (S_K) and the costs for the supply-exclusion of the principal in the following periods (S_A). Thus is:

   $$S = S_R + S_K + S_A \quad (5)$$

   The extent of S_R depends on the expenses of the recall. S_K is the only contractual fixed parameter of S. Whereas S_A is variable and addicted to E:

   $$\partial S_A / \partial E > 0 \quad (6)$$

   On the other hand the extent of S_A depends on the homogeneity of the produced goods and the number of alternative potential principals. The extent of S_A finally is determined by the relation of GV > GM. In the long run S_A is the expected value of all capitalised future contract-benefits, calculated to the moment of contract-cancellation.
c. The contract-control by the principal occurs spot checked. The likelihood to detect a breach of a contract is $p < 1$. It is assumed that the subjective assumed likelihood of the agent agrees with p.
d. The model assumes that both agent and principal are risk neutral.

The expected benefit in case of contract-compliance of the agent is:

$$G_E = E - K_E \quad (7)$$

The expected benefit for agents concluding but not keeping the contract is:

$$G_N = (1 - p) E - K_N - pS \quad (8)$$

The contract will be closed, if $G_E > 0$ or $G_N > 0$. But the contract only will be abided if $G_E > G_N$. As you can see in Figure 1 there are in principle four possibilities for the contractual relationship in respect of the extent of E with given values for K_E, K_N, S and p. If the agents reward less than in an alternative contract ($E_0 < E_1$), the contract will be declined. If the agents reward between E_1 and E_2, the contract will be concluded, but not abided. At this the sector II can be divided into sector IIa, where only the breach of contract is profitable and into sector IIb, where the compliance with the contract is profitable, but the breach of contract is more profitable. Is the reward is higher

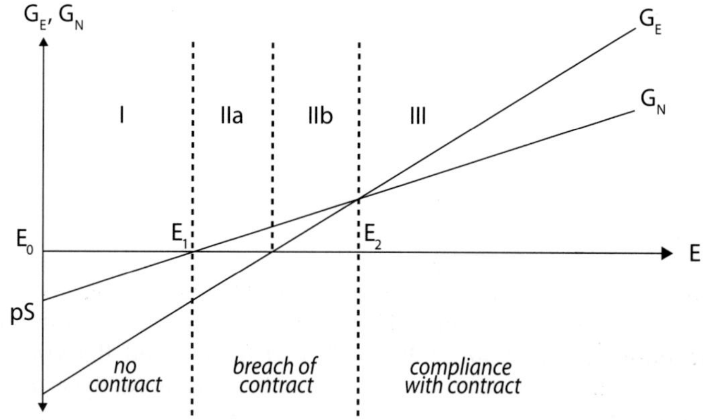

Figure 1. Agents bearing with different E.

then E_2 the contract will be adhered. E_2 is the intersection of the straight lines G_E and G_N. The equation of G_E and G_N arises:

$$E_2 = (K_E - K_N)\, p^{-1} - S \tag{9}$$

E_2 is the minimum reward the agent is about to agree to the contract with the principal. There are possible other runs of G_E and G_N with no intersection, but this would not accord to the existing problem.

Summary of the basic microeconomic model:

a. As it is shown in Equation (9) the extent of the necessary reward increases when K_E increases. Declines K_N, the extent of the necessary reward increases too and if K_N arises, the extent of the reward decreases.
b. The necessary reward decreases if S_K increases and if the controls and thus the likelihood p increase.
c. In case of different K_i of different potential agents an enhancement of S_K and/or an intensification of the controls respectively an enhancement of p, causes a decline of the number of contract-signings and an increase of compliance with contracts.
d. Finally the contract-parameters E, S and p can be set that the appeal for breach of contract decreases. At the same time the agent's reward by the contract must be higher than the alternative reward without this contract ($E > E_1$) and $E_2 > E_1$ ($E_V > E_M$). There is no incentive for a breach of contract, if:

$$S > (K_E - K_N)\, p^{-1} - E \tag{10}$$

resp.

$$E > (K_E - K_N)\, p^{-1} - S \tag{11}$$

resp.

$$p > (K_E - K_N)\, (E + S)^{-1} \tag{12}$$

According to Stigler (1970) agents willing to maintain the contract (A_E) can be discouraged by too high threats of punishment. The potential agents would assume a probability value w ($0 \leq w \leq 1$) that a breach of contract is assumed untruly. For this reason the expected benefit by compliance with the contract will change from:

$$G_E = E - K_E \quad (7)$$

to

$$G_E = (E - K_E)\ (1 - w)\ wS \quad (13)$$

Less loyal agents would sign the contract.

3.2. One principal and many agents

Normally the principal signs contracts with several agents. On the one hand the principal does not want to depend on only one agent and on the other hand only one agent often could not satisfy the supply of the principal. With this new assumption, the reactions of all involved agents must be aggregated in the previous approach. In this new approach it is assumed that the agents differ in their individual costs (K_i), their subjective control-likelihood (p_i) and even in their individual penalty S_{Ri} and S_{Ai}. For simplification it is assumed, that the likelihood of an error-exposure and the extent of the expected penalty is equal to all agents. They differ only in the individual costs K_i.

For further considerations subsequent assumptions are made:
1. In total there are M agents, being able to supply the principal.
2. K_i are uniformly distributed in a range between $K_U \geq 0$ and $K_O \geq K_U$.

The intention of the model is the design of contracts between principal and agents in the way, that all agents, signing the contract will maintain it ($A_N = 0$). In Figure 2 the agent's single optimal behaviours with different K_i and given p and S_K are shown.

The straight line K = E divides the agents if the contract with the principal is profitable. If E* is given and K_i is below this straight line, the agents would achieve a profit of $G_E > 0$ in maintaining the contract. The straight line K = pS + pE divides the agents if it is more profitable to maintain the contract or to breach it. All agents with a K_i below this straight line achieve a higher profit in maintaining the contract. Whether agents breach contracts or prefer not signing the contracts, depends on the extent of rewards in ratio to p and S.

Is $E \leq p\ (1 - p)^{-1}\ S$, no potential agent would sign the contract. Recapitulating following relations between the contract-parameters and individual K_i can be deflected:

a. Acceptation and abidance of the contract, if:

   $K_i < E$ and $K_i < p\ (E + S)$

b. Acceptation and non-abidance of the contract, if:

   $K_i > p\ (E + S)$ and $S < (1 - p)^{-1}\ K_i$

c. No contract acceptation in all the other cases.

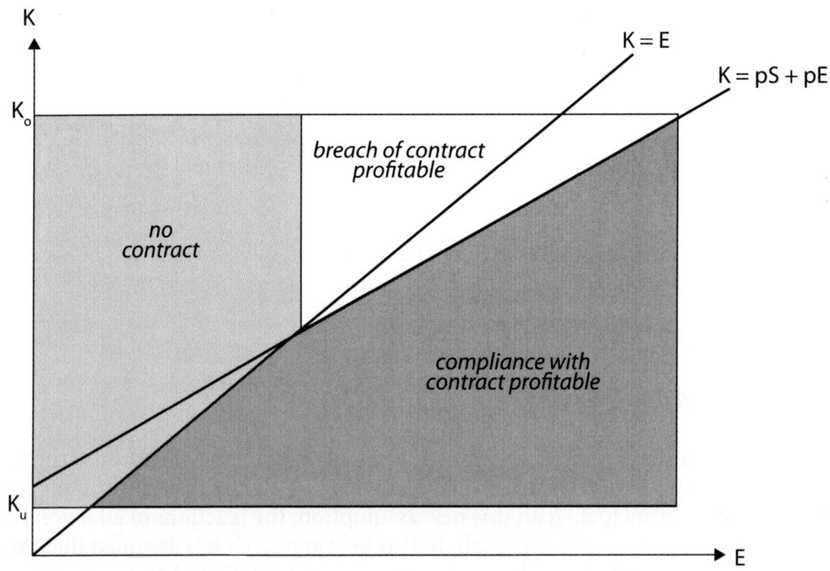

Figure 2. Agent's single optimal behaviour.

Figure 3 shows when no agent wants to sign a contract with the principal, when all signed contracts will be preserved, when some agents will breach the contract and some maintain the contract and finally, when all M agents will sign and maintain the contract.

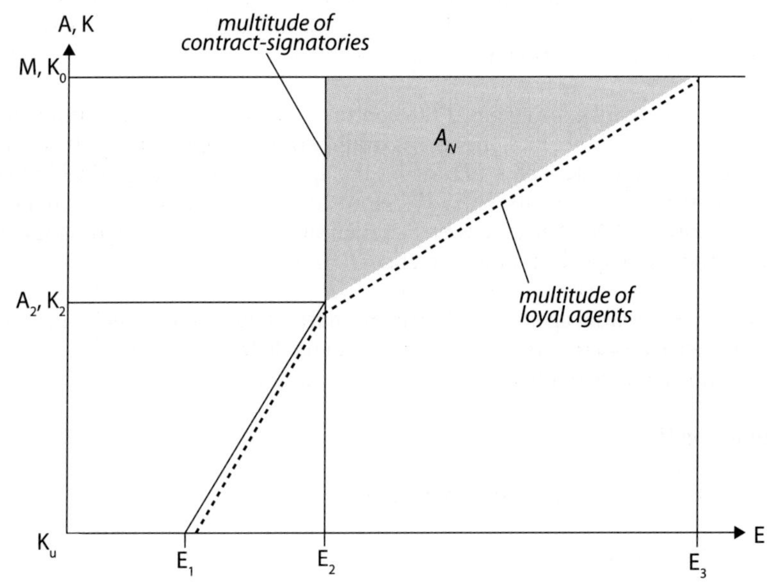

Figure 3. Multitude of loyal and disloyal agents.

K^* symbolises the randomised amount of the determined agent-reward. K^* is conform to E^*. With the assumption that $K_i \leq K^*$ the following can be noticed: Is the reward for $K^* \leq K_U$, no potential agent will sign a contract with the principal. Is the reward $K_U < K^* \leq K_2$, every signing agent will maintain it. The multitude of loyal agents would be A_2. K_2 complies with the reward E_2 arising from the intersection of K = E and K = pS + pE. For $K_2 < K^* < K_O$ some signing agents will be loyal (A_E), but some will not (A_N). Is $K^* \geq K_O$, all potential M agents will sign and maintain the contract. One constraint for no breach of contract at $E^* = K^* \geq K_O$ is $S > K_O$.

Among following constraints, all signing agents will maintain the contract:

$$A_E \mid A_{N=0} = \begin{cases} E_1 < K^* \leq E_2, \text{ for } A_E < M \\ K^* \geq E_3, \text{ for } A_E = M \end{cases} \tag{14}$$

Should the multitude of AE be larger or smaller, the intersection of K = E and K = pS + pE must be displaced by changing the parameters E, S and p. So the principal has three control parameters to arrange an optimal form of supply-contract for high-quality and secure products. Such an optimisation assumes, that:

a. an optimisation-purpose is defined;
b. the relation between the control-parameter E, S and p and the satisfaction level are determined; and
c. the demand-function for contract conclusions resp. the supply-function for contract-performance is quantified.

To prevent food scandals, all signing agents must maintain the contract ($A_N - 0$). As Figure 3 shows all signing agents maintain the contract if $E_1 < K^* \leq E_2$ or $K^* \geq E_3$. E_1 comes up to K_U and E_3 to K_O. A maximum of A_E can be achieved by the maximisation of E^* by the principal. But this would be contrary to the economic principles of a principal-company. The most important company-principle is the asset maximisation. So the principal's expenses for the agents must be minimised. The principal's objective function is:

$$\text{Min } B = Z + C_{\ddot{U}} + C_D + C_N + Q - D \tag{15}$$

with:
B = principals contract-expenses for the agent;
Z = contractual determined principal-payments to the agent;
$C_{\ddot{U}}$ = principals expenses for quality check and agent-control;
C_D = expected expenses for sanction-executions;
C_N = expected principals expenses for new contract-initiation with new agents;
Q = principals loss in case of an agents mistake (e.g. out-of-stock, cross-damage);
D = expected principals revenues from S_K of convicted agents A_N.

It must be noticed that:

$$Z = (A_E + A_N)\, E \tag{16}$$

and

$$p = p\,(C_{\ddot{U}}) \tag{17}$$

The principal's control-expenses affect the likelihood p. It is an inverse notation of $C_{Ü} = f(p)$ - the control-costs will be influenced by the likelihood p.

Now B should be minimised under the constraint that all agents should be loyal

$$A_E = A^* \quad (18)$$

and that the number of agents is determinated exogenous. The consideration of this constraint is important as otherwise A* could be reduced by the principal to only few agents, which can be controlled easily. But in reality the principal requires a lot of agents. Normally the agents adjust their output.

3.3. Function specification

Respectively to the impacts of the principals control expenses on p (17), it is assumed that:

$$\partial p / \partial C_{Ü} > 0 \quad (19)$$

Higher control expenses cause a higher control-likelihood:

$$\partial pp / \partial C_{Ü}^2 < 0 \quad (20)$$

Additional principal's controls cause only a sub-proportional likelihood for the agent to be controlled, if the selection of the controlled agents occurs accidentally. Becker (1968) pointed out that in case of additional controls the principal's administration expenses increase.

The principal's cost-minimisation in the context of the agent-contracts is also affected by the number of A_E. A_E in turn is affected by the parameters E, S and p. Subject to these parameters and the agents K_i, four different sub areas can be distinguished (see Figure 1):

a. No agent signs the contract ($E^* \leq E_1$).
b. Some agents sign and maintain the contract ($E_1 < E^* \leq E_2$).
c. Some agents sign and maintain the contract, the other agents sign but breach the contract ($E_2 < E^* < E_3$).
d. All agents sign and maintain the contract ($E^* \geq E_3$).

For the existing problem only the alternatives (b) and (d) are relevant. As it is explicated above, the following constraints for (b) and (d):

$$E > K_U \text{ and } E < p\,(S + E)$$

resp.

$$E > K_O \text{ and } K_O < p\,(S + E)$$

The principal expense-function (15) should be minimised with the constraint $A_E = A^*$. Therefore E^* must be chosen that $K_U < E^* < K_2$ or $E^* \geq K_O$. In consideration of Equation (14) A_E is:

$$A_E = J\,[K_2\,(E_2 - E_1) / 2] + (1 - J)\,M \quad (21)$$

with:

$J = 1$, if $E \le p\,(S + E)$ and $E_1 < K^* \le E_2$ and
$J = 0$, if $K_O \le p\,(S + E)$ and $K^* \ge E_3$

For the minimum-definition of B, the parameter-combination of E, S and p, causing minimal expenses for the respective segment, must be defined. Subsequent the absolute expense minimum in comparison to both values will be defined. In doing so, the penalty-limit (q.v. 3.1) must be taken into account. A penalty-limit of S_{Kmax} is defined; otherwise a high S_K could cause a p near zero.

3.4. Discussion of the optimisation

The minimum of B will be achieved, if S_K is S_{Kmax}. In case b) E^* must be equal to K_2, as in consequence of the extent of the expected penalty-payment, there should not be an incentive to breach the contract. K_2 is the cost-level, falling below by $A^* = A_E$. Finally p_{min} must be chosen, whereas p_{min} must fulfil the Equation (12) with given S_{max} and K^*. Thus, in case (b), there is a minimal-value for B of:

$$\text{Min } B_{\text{case b)}} = A_E K^* + C_Ü\,(p_{min}) + C_D + C_N + Q \qquad (22)$$

In case (d) the minimum-value is calculated with the constraint A_E = M agents. For such a solution the parameters must be defined that

$$E^* = K^* \ge K_O = E_3 \qquad (23)$$

The result from this restriction is:

$$\text{Min } B_{\text{case d)}} = M\,E_3 + C_Ü\,(p_{min}) + C_D + C_N + Q \qquad (24)$$

Due to the fact that $A_{E\text{ case b)}} < M$ and $E_2 < E_3$, the difference of Min $B_{\text{case d)}}$ - Min $B_{\text{case b)}}$ is always more than zero. The principal total expenditures in connection with the agent's contracts, with the constraint $A^* = A_E$, are always lower when $A^* < M$.

4. Conclusion

The model points up, that an appropriate form of contract and the creation of according control mechanisms can influence the probability of breach of contracts fundamentally. Initially a microeconomic model from the agent's point of view was designed. This resulted from a modification of neoclassic theory. That followed a simplified sectoral contract-demand model, which provided a basis for definition of the optimal contract design. Would the contracts be designed in the above pictured mode, the risk of hidden action could be decreased. This would be an enormous advance for the food safety.

This model does not consider the financing of the quality assurance along the supply chain. The production of high-quality food is aligned with higher costs. In this context it should be mentioned that the consumers are only up to pay for improved food quality if the improvements are obvious. Is, e.g. the trader (principal) able to communicate the better quality of a product, he may achieve a higher sales price, which he could pass down to the producer (agent). This would be a motivation for the agent to maintain the high-level production. Brand- or trademarks-producer in the food sector can use their trade name to communicate quality and security (e.g. 'Wiesenhof' or 'Gutfleisch'). If the brand automatically stands for high quality the consumer is possibly poised to pay higher prices. The holder of a brand (branded goods or trademark goods) has an economic interest to prevent food scandals of their products. A detected grievance with their products would cause slump in sales of their whole product range. Hence trade companies or producers with well-known brands have the possibility and the interest setting standards for residual supply chain.

References

Becker, G.S., 1968. Crime and punishment: An economic approach. Journal of Political Economy 76, 169-217.

Hanf, C.-H., 1993. Ökonomische Überlegungen zur Ausgestaltung von Verordnungen und Verträgen mit Produktionsauflagen zum Umwelt- und Naturschutz. Agrarwirtschaft 36, 138-147.

Hinrichs, A., 2004. Verhinderung von Lebensmittelskandalen mittels vertraglicher Bindung und vertikaler Integration. PhD Thesis, Kiel.

Hirschauer, N., M. Odening and C. Oertel, 2004. Moral Hazard in Wertschöpfungsketten des Agrar- und Ernährungsbereiches - Ableitung anwendungsorientierter mikroökonomischer Modelle. In: Schriften der Gesellschaft für Wirtschafts- und Sozialwissenschaften des Landbaues 39 (ed.), 169-178.

Holmstrom, B., 1982. Moral hazard in teams. Bell Journal of Economics 13, 324-340.

Stigler, J.G., 1970. The optimum enforcement of laws. Journal of Political Economy 67, 526-536.

Williamson, O.E., 1975. Market and hierachies: Analysis and antitrust implications. New York.

Globalisation, quality management and vertical coordination in food chains of transition countries

Liesbeth Dries and Johan F.M. Swinnen
Centre for Agricultural and Food Economics, K.U.Leuven, De Croylaan 42, B-3001 Leuven, Belgium; Liesbeth.Dries@agr.kuleuven.ac.be

Abstract

A major problem in the agricultural sector in countries in transition is the breakdown of relationships of farms with input suppliers and output markets. As a result, many farms face serious constraints in accessing inputs and in selling their products. In search for guaranteed and high quality raw materials new investors in food companies have introduced programs and contracting arrangements to enhance the quality of supplies and ensure their reliability. This paper presents unique evidence for Bulgaria, Poland and Slovakia of the introduction of high quality standards throughout the dairy chain and how this has resulted in increased vertical coordination.

Keywords: transition countries, dairy sector, vertical coordination

1. Introduction

There is a growing literature on quality standards implementation and enforcement for agricultural and food products. Studies have focused on the application of sanitary and phytosanitary standards by developed countries on exports from developing countries (e.g. on fresh fruits and vegetables or fish products), or for example, on the application of private standards by supermarkets (in a collective way: EurepGAP; or through individual chain standards: e.g. Tesco) on suppliers in developing countries (Henson *et al.*, 2000; Unnevehr, 2000; Dolan and Humphrey, 2000). Furthermore, there has been some research on the role of supermarket investments on setting and imposing quality standards for local fruit and vegetable producers in Latin America (e.g. Balsevich et al, 2003). The present study will look at how the introduction of quality management in transition countries' food companies is affecting the structure of the supply chain and economic activities of suppliers.

A major problem in the agricultural sector and in rural areas in countries in transition is the breakdown of the relationships of farms with input suppliers and output markets. Pre-transition systems were strongly vertically integrated. The central planner provided the information and enforced contracts involving exchanges between the various agents in the chain. The removal of the central planning and control system, in the absence of new institutions to enforce contracts and to distribute information and finance caused serious disruptions throughout the economy (Blanchard, 1999; Konings and Walsh, 1999). The result is that many farms and rural households face serious constraints in accessing essential inputs (feed, fertiliser, seeds, capital, etc.) and in selling their products.

The inflow of foreign investment in the food chains, combined with it the pressures of global competition and quality standards, is causing a dramatic restructuring of the food chains in these countries. Because of both pre-transition low quality standards, and transition-specific problems, agribusiness and food processing companies have difficulties to obtain sufficient high quality products. In search for guaranteed and high quality raw materials new investors in food companies have introduced a series of programs and contracting arrangements to enhance the quality of supplies and ensure their reliability (Gow and Swinnen, 2001). For example, food processors have negotiated contracts with banks and input suppliers to provide farms with inputs that enable them to deliver high quality products to their company. Similarly, input supply firms have been involved with assisting farms to find guaranteed outlets for their products in order to stimulate farms' demand for the company's products.

This paper presents evidence of how food companies have tried to introduce high quality standards throughout the food chain and how this has resulted in increased vertical coordination. We provide unique evidence of the impact on farm investments, productivity and quality standards, based on farm surveys and company interviews in several countries, including Bulgaria, Poland and Slovakia.

With its 12 million ton of milk in 2003, Poland is by far the most important dairy producing CEE country. Slovakia and Bulgaria both produced about one tenth of this amount: 1.1 and 1.2 million ton respectively (FAO, 2004). Both Poland and Slovakia have joined the EU in May 2004, meaning that the process of EU integration has already proceeded further in these countries than it has in Bulgaria (to join the EU in 2007).

Furthermore, the three countries present an interesting mix with respect to the structure of their dairy farms. Both in Poland and in Bulgaria, the dairy sector consists of mainly small-scale household production as more than 85% of all milk producers have a herd size of less than 5 cows. A high share of their production is used for self-consumption and the remainder is often sold to dairies through village collection points. The situation is completely different in Slovakia. A 1999 representative survey (ACE) shows that only 10% of family farms have dairy cows and more than half of the milk produced in these farms is used for self-consumption. On the other hand, 81% of farming enterprises have dairy cows and 100% of this milk production is sold (of which, 87% directly to a dairy processing company).

The processing sector is much more fragmented in Poland and Bulgaria than it is in Slovakia. For example, in Bulgaria, 95% of all dairy processing units are classified as so-called mandras (mini-dairies) with a capacity of less than 10 ton/day (FAO, 2000). This in contrast to Slovakia where the ten largest dairy companies hold about 60% of the market.

The data used in this paper are the result of interviews that were conducted in the three countries between July 2001 and March 2003. The next section presents evidence of changes that have occurred and policies that are implemented at the level of the dairy processing companies. The last section summarises the impact of these changes and policies at the farm level.

2. Data

We interviewed six Polish companies, which were selected in the north-eastern region of Warminsko-Mazurskie and provide an interesting mix. Four are medium size companies (50-70 million litres of milk) with one large (420 million litres) and one small (2.5 million litres). Three are cooperatives, two private, and one a joint venture of a cooperative and a private company. In terms of foreign investment, two are majority foreign owned, and two have important links to foreign companies. The six Slovakian dairy companies are located in West and Central Slovakia.

Three are medium size companies (25-42 million litres per year) and three are large companies (75-155 million litres per year). Four dairies are majority foreign owned, two are domestically owned companies. The eleven Bulgarian companies are spread over the whole country. All are small to medium size companies. Two dairies are majority foreign owned.

The Polish dairy companies purchase milk mainly from individual farmers, either through village collection points or directly at the farm in case the farmer has invested in on-farm cooling equipment. Figure 1 shows the share of milk that these companies buy through village collection points. We see that although collection point supplies are decreasing in importance, they nevertheless still make up an important share of total milk deliveries.

In contrast to the situation in Poland, we see that Slovakian dairy companies are buying milk exclusively from large scale milk producers. This is in line with the observation that only a small share of family farms have dairy cows and those households that have cows, mainly use the milk for self-consumption.

In Bulgaria, like in Poland, an important part of milk is delivered through collection stations (only Meggle is following a policy that excludes the smallest producers from their supply base). In contrast however, cooperatives have a higher share in milk deliveries in Bulgaria than in Poland. This difference is mainly due to the fact that collectivisation of the agricultural sector was much less successful in Poland than in most other Central and Eastern European countries so that collective farms were practically non-existent even before the start of the reforms. It should be noted however that the importance of cooperatives in the dairy sector in Bulgaria is also declining. FAO (2000) state that the ownership structure of cooperatives hinders sound management and decision-making mechanisms. As a consequence, most co-operatives are posting chronic losses and are also decapitalising.

3. Quality policy dairy companies

In the case of Poland, foreign investors in the dairy sector have played an important role by setting an example strategy for improving quality. When Land O' Lakes invested in ICC Paslek in 1994 milk quality of its supplying farms - as everywhere in the region - was poor. From the start, ICC

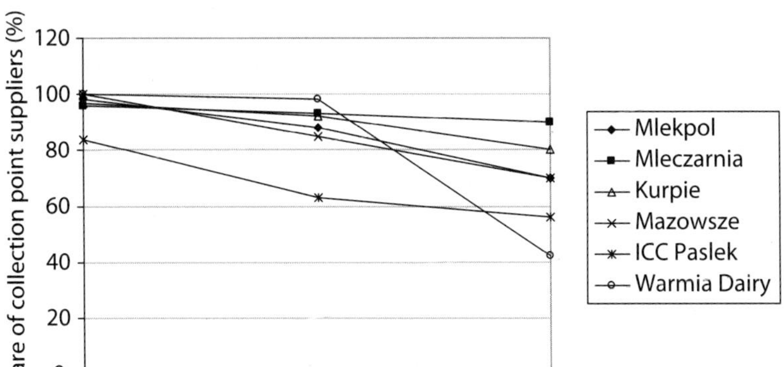

Figure 1. Share of milk purchases from collection points - Poland.

Paslek set out a clear strategy to increase the quality of delivered milk. One of their requirements was that the cooperative - from which they lease collection stations - should install cooling tanks in these collection points. Furthermore, they invested in agricultural extension to raise farmers' awareness of the importance of milk quality and to improve quality through basic hygienic rules for farmers handling the milk. From the beginning, ICC Paslek also required germ count and cell count tests (in accordance with EU standard tests for milk quality classification). Farmers were also allowed to have their milk tested for antibiotic residues free of charge in the dairy's laboratory. This was especially helpful for farmers who had had a cow disease in their farm and who needed to make sure that no antibiotics residue was left in the milk. Soon after Land O' Lakes set up its quality improvement programs, local dairies started to copy these practices and by doing so have created an important spill over effect as shown by the dramatic milk quality improvement throughout the region since 1995.

Milk quality improvements were primarily driven by export strategies of the dairy companies, rather than by domestic/foreign ownership. Furthermore, government regulations also had an impact. Since 1999, Poland has implemented the EU classification system of milk quality grading and as a consequence has obliged dairies to do specific tests (e.g. germ count, cell count, etc.) to classify their supplied milk in Extra (highest), First, Second, and Third (lowest) class milk. It was only at this moment that Warmia Dairy, Kurpie and Mleczarnia also started implementing this grading system. In January 2000, a Polish law was passed which specified that second class milk is the minimum milk quality that may be used by dairies for further processing. All dairies still accepted third class milk at the moment the law was passed which excluded third class milk from the processing sector. In 2003, dairy companies in Poland were no longer allowed to accept second class milk either.

Table 1 shows the evolution of the quality of milk deliveries to the selected Slovakian dairy companies. The quality of the milk delivered to the six dairy companies is for more than 90% up to EU standards, and for most of the companies, this was already the case in 1997. The six companies are not exceptional in this respect in Slovakia. Table 2 shows that in the whole of Slovakia 95% of all milk deliveries to dairy processors is of EU standardised quality. The high share of EU quality milk in Slovakia relatively early on in the transition period is somewhat different from the Polish case, where in 2001, still only 50% of milk purchases were up to EU standards. The reason for this may be that the large dairy farms in Slovakia that dominate the milk market, already had basic investments such as on-farm cooling equipment, a milk line (that ensured that milk would flow directly from the cow to the cooling tank minimising the contact with possibly contaminated sources), while the Polish milk producers were in general too small to make these investments that required scale economies. Furthermore, cooperative farms also had a trained veterinarian staff readily available that would allow a quick response in case of an infection or other veterinary problem.

However, milk quality remains an important issue for all dairies since all of them offer a quality premium for high quality milk. Hence, while the numbers do not clearly show this, it seems that important quality improvements have occurred after the start of the reforms. In this sense, the situation is similar to that in Poland although the starting point for quality improvements may have been less problematic in Slovakia because of the initial farm structure.

Table 3 presents the share of milk deliveries in different quality classes for the Bulgarian dairy companies in our sample. One should be careful when interpreting these results since there is

Table 1. Quality of milk deliveries to the six Slovakian dairy companies.

		Q-class	1st class	2nd class	3rd class	Unstd.
Liptovska	1995	-	-	-	-	-
	1997	75%	25%	-	-	-
	2000	80%	20%	-	-	-
	2003	80%	20%	-	-	-
Mliekospol	1995	-	-	-	-	-
	1997	50%	43%	-	-	7%
	2000	55%	48%	-	-	2%
	2003	55%	48%	-	-	2%
Rajo	1995	-	-	-	-	-
	1997	-	-	-	-	-
	2000	-	-	-	-	-
	2003	98%			2%	
Levicka	1995	96-98%			2-4%	
	1997	96-98%			2-4%	
	2000	96-98%			2-4%	
	2003	98%			2%	
Tatranska	1995	20.1	28.7	38.6	12.3	
	1997	-	-	-	-	-
	2000	59.7	28.2	6.4	5.7	
	2003	55	42			4
Nutricia Dairy	1995	-	-	-	-	-
	1997	-	-	-	-	-
	2000	70%	28%	-	-	2%
	2003	70%	28%	-	-	2%

Remarks: Tatranska used the 2nd / 3rd quality classification system until 2002; other dairies stopped reporting these quality classes since they obtained an EU export license. Milk of lower quality than Q or 1st class is classified as unstandardised milk. Q and 1st quality milk is of acceptable quality according to EU standards.

Table 2. Milk deliveries in Slovakia according to quality class, 2003 (VUEPP, 2003).

	million ton	%
Total	506	100
Class Q + I.	480	95
Class Q	271	53
Class I.	209	41
Unstandardised	27	5

Table 3. Quality of milk deliveries to the Bulgarian dairy companies.

		Extra class	1st class	Unstd.
Merone	1997	-	-	-
	2000	-	-	-
	2003	15%	60-65%	20-25%-
Fama	1997	50%	30%	20%
	2000	50%	30%	20%
	2003	50%	30%	20%
Mlekimex	1997	30%	70%	-
	2000	50%	50%	-
	2003	50%	50%	-
Danone	1997	-	-	-
	2000	3-4%	-	-
	2003	10-15%	65%	20%
Iotovi	1997	-	-	-
	2000	10%	80%	10%
	2003	40%	50%	10%
Milky World	1997	-	-	-
	2000	10%	80%	10%
	2003	10%	80%	10%
Markelli	1997	-	-	-
	2000	-	-	-
	2003	50%	50%	-
Mandra Obnova	1997	-	50%	50%
	2000	30%	70%	-
	2003	30%	70%	-
Meggle	1997	-	-	-
	2000	5-10%	85-92%	3-5%
	2003	35-37%	60%	3-5%
PRL	1997	-	-	-
	2000	-	-	-
	2003	75%	20%	5-6%
Serdika 90	1997	0%	70%	30%
	2000	10%	70%	20%
	2003	10%	70%	20%

no standard quality classification system for milk in Bulgaria[1]. The definition of the different quality classes may therefore be based on different indicators and limits depending on the company that was interviewed. As a result, cross-country and even inter-company comparisons may be misleading.

[1] Before the introduction of the EU standard classification system of milk quality in Poland and Slovakia milk quality tests in these countries were not based on germ and cell counts (the basis of the EU classification system) but on other indicators - like currently is the case in Bulgaria. However, since the introduction of the EU classification system in 1999/2000, all dairy companies in Poland and Slovakia had to implement the same tests and classify their milk according to the same system.

However, like in Poland and Slovakia, improving the quality of milk deliveries is one of the main objectives of dairy companies in Bulgaria. To achieve this goal, all dairy companies pay price premiums for high quality milk. Due to the absence of a standard classification system, these premiums are paid on the basis of different indicators (fat content, protein content, dry defatted residual, germ and cell counts) for suppliers to different companies.

Government regulation is playing a role as well. Recently, the government has regularly circulated new regulations with respect to milk quality to all dairy companies (according to some of the respondents these policy changes occur much too frequently). For example at the end of 2002 a communication was circulated indicating that from now on dry defatted residual should be used as a proxy for quality rather than fat content. Furthermore, farmers that are applying for milk price support have to present the results of germ and cell count tests and apparently the extent of support is based on these results (i.e. better test results give right to higher subsidy levels). Finally, the government is also involved in quality tests itself as all dairy companies send samples of milk deliveries to the State Veterinary Service at least twice a month. The State Veterinary Service laboratories are equipped with devices to perform germ and cell count tests (these tests are only rarely performed in the dairy company labs or at delivery but samples are normally taken and tested for other indicators by the dairy company on a daily basis).

The two foreign owned companies, Danone and Meggle, both have a quality policy that seems to go further than that of domestic companies. Danone uses its own quality classification system, based on germ counts, where the top-category (lowest germ count) is labelled 'Danone class'. Farmers delivering Danone class milk are rewarded with a price bonus. Furthermore, the company has introduced a system of 'traceability' for their suppliers, the Danone Quality Control System (imposed by the International Danone Group). Once per year all suppliers are visited and evaluated on the basis of 26 criteria related to quality and safety of milk production. Suppliers that are evaluated positively on all 26 criteria are labelled 'Traced and Controlled Danone' and are preferred suppliers for Danone. Suppliers that do not fulfil all 26 criteria are labelled 'Traced and Controlled', 'Traced and not-controlled', and 'Not-traced and not-controlled'. Collection stations are mainly located as 'Traced and not-controlled' (the manager of the collection station is required to have a detailed list of all cows that supply milk to this station). Meggle has a somewhat similar traceability requirement for its suppliers. All Meggle suppliers have to keep a diary reporting for each of the animals: identification, inseminations, veterinary services, feeding details, quality and quantity of milk.

4. Investment assistance to suppliers

We find that in all three countries dairy companies are offering assistance programs to their suppliers. Assistance is provided most often under the form of facilitating access to inputs and investment credit (mainly for dairy-specific investments) as well as the provision of extension services. Often these types of assistance are complemented with the provision of a bank loan guarantee for farmers that are applying for a bank loan.

Access to these assistance programs is in some cases limited. For example in Slovakia, several dairies indicated that they offer programs only to large-scale producers, delivering best quality milk and in the best financial position (in other words, the most 'valuable' suppliers for the dairy). In Bulgaria, a similar restrictive policy is applied by the foreign investors. Only in Poland all suppliers seemed to have a reasonably equal chance of getting access to the assistance programs offered by the dairy companies.

Finally, the reason for offering these programs was in most cases twofold. First, assistance programs are meant to help farmers to upgrade the quality of their milk production. Second, assistance is offered because it is felt to be necessary to keep suppliers from delivering to the competitor (who is offering these programs as well). Price premiums and the length of payment period are other tools that are used to compete with other dairies in the area for milk supplies.

Table 4 shows how the share of dairies that offer assistance programs in these countries has changed over time. Dairy companies in Poland seem to have been faster in implementing assistance programs for their suppliers. In Bulgaria, gradually more dairy companies started to offer assistance. In Slovakia however, the increase in assistance to local farmers has been stronger in the period 1998-2002 than in the previous four years. The inflow of foreign investments in the dairy sector in Slovakia also surged in the period 1998-2002. On the one hand, foreign investors may have played an example role by implementing these assistance programs and local diary companies copied these policies. On the other hand, foreign investors may also have increased competition in the market and assistance programs could be used as a tool to compete with other dairies.

While the share of dairy companies that are offering assistance to their suppliers has increased in all three countries, the content of the programs may have changed over time. For example, one of the Polish dairy companies explained that currently they were no longer supporting suppliers that want to buy a cooling tank (because most of their suppliers had bought a cooling tank with aid from the dairy in previous years), however, the investment credit program was now mainly used for buying new cows. Similarly, while extension services in earlier years may have been focused mainly on improving hygiene conditions during the milking process, these days, farmers are for example being informed about upgrading the genetic material of their herd.

5. Impact at the farm level

Evidence on the impact at the farm level is limited and is often based on case-study evidence only. In this section, we present results of a representative survey of Polish dairy farmers, preliminary evidence of a survey of Bulgarian dairy farmers.

Evidence from a 2001 survey of 290 Polish rural households shows that the assistance programs have a significant positive impact on on-farm investments. More than three quarters (76%) of

Table 4. Share of interviewed dairy companies having assistance programs, in %.

Year	Country	Credit	Inputs	Extension	Veterinary	Bank	Total
1994	Poland	50	67	50	0	50	43
	Slovakia	0	0	83	17	17	23
	Bulgaria	9	18	9	0	0	7
1998	Poland	83	100	83	17	83	73
	Slovakia	17	17	83	17	33	33
	Bulgaria	45	64	18	18	18	33
2002	Poland	83	100	83	17	83	73
	Slovakia	100	33	83	17	50	57
	Bulgaria	82	91	73	18	36	60

all households in the survey made investments in the past ten years (see Table 5). Of those who invested, 58% used loans. Further, the reason why loans come from dairies or from banks is determined by the type of investment rather than farm characteristics. Dairy loans are used almost uniquely for investments in enlarging and upgrading the livestock herd (30%) and cooling tanks (56%). Together these account for 86% of all dairy loans. In contrast, only 29% of all bank loans are used for these types of investments. Note that the loans from dairies are only a partial indicator of the financial assistance offered by dairies. As explained above, part of their assistance is under the form of loan guarantees with the banks. Hence, part of the loans given by the banks are indirectly due to these loan guarantee programs of dairies. They are important. Almost half (45%) of the households who could not obtain preferential bank loans identified lack of sufficient collateral as the main reason.

Furthermore, the programs which assist farms in accessing inputs (mainly feed) enhance investment indirectly by lowering input costs, or reducing transaction costs in accessing inputs, and consequently, through improved profitability.

Evidence suggests that foreign investment has played a more important role early on in transition as an initiator of change and institutional innovation. We found no significant difference in 2001 of assistance programs provided by foreign owned companies and domestic dairies, except for the loan guarantee programs, which were more extensively provided by the foreign dairies. The survey also shows that the share of farms delivering extra class milk (the highest quality by EU standards) was significantly larger among farmers delivering to foreign owned dairies (58% versus 38% among farmers delivering to domestic dairies) in 1995. However, by 2000 this gap had almost disappeared: 83% versus 79% of farms delivering to foreign versus domestic dairies supplied extra class milk (see Figure 2). This is in line with qualitative evidence that foreign companies have played a role in providing an example in quality improvement strategy.

A key issue is how opening of the dairy sector to foreign competition and increased quality requirements has affected the survival and growth of dairy farms. It is often argued that such forces can drive local companies out of business, in particular the smallest. The latter may result directly from their inability to compete in a liberalised market or because restructuring of the processing companies induces the restructured companies to drop small suppliers and to prefer fewer but larger suppliers to reduce transaction costs.

The Polish household survey provides findings that do not support these arguments, quite the contrary. 283 households in our sample delivered milk to dairy processing companies in 1995. Of

Table 5. Investments and loans of farm households in Poland (Dries and Swinnen, 2004).

Size (# of cows)	Invests (% of total) A	Uses loan to invest (% of A) B	Uses dairy loan (% of B) C	Uses bank loan (% of B) D	Uses dairy loan (% of A) E	Uses bank loan (% of A) F
1-5	52	54	41	50	21	26
6-10	78	51	43	70	22	36
>10	92	74	43	75	31	54
All	76	58	43	69	25	40

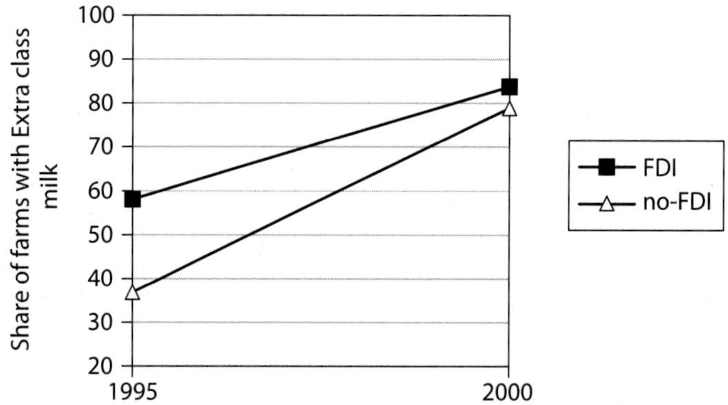

Figure 2. Change in share of highest quality milk (EU Standard) in the farm survey in Poland.

these, only 36 (13%) stopped delivering milk between 1995 and 2000. Ten of them (4%) stopped producing altogether while the rest kept some cows for home consumption. Hence, 87% continued delivering to dairies despite radical restructuring of the dairies and tightened quality demands. Moreover, some of those who stopped delivering might have stopped anyhow: the average age of those who stopped producing is 56 years, compared to 45 years for the entire sample.

The size distribution changed, but only gradually (see Figure 3). Three quarters of the households (211) had between 4 and 12 cows in 1995. The share of farms in the 4-12 cow category has reduced significantly with about the same amount upgrading to a larger size as falling back to smaller, presumably subsistence farms producing solely for home consumption. More specifically, of the 211 household farms, 135 (65%) had still between 4 and 12 cows in 2000; 35 (17%) had less than 4 cows in 2000, while 41 (19%) had more than 12 cows in 2000. Farmers with growing farms were significantly younger (42 years on average) than those whose farm size declined (51 years on average).

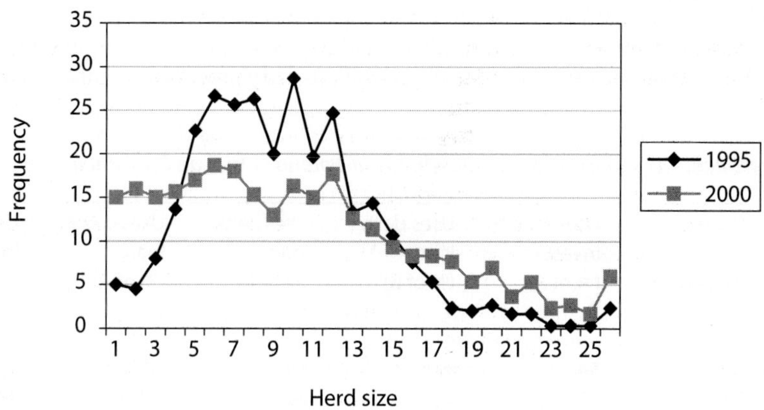

Figure 3. Size distribution of dairy farms in total survey sample in Poland.

Table 6 shows evidence from a 2003 survey of 240 Bulgarian dairy farms (that are producing milk since 1994 or earlier). We see that like in Poland, the majority of households (63%) have made on-farm investments in the past 10 years. Furthermore, we find no evidence that large-scale producers are more likely to invest than small-scale producers. However, looking at the source of finance that was used to make investments we see an important disparity with the results from the Polish survey. Access to supplier credit from the dairy company is very limited; only 2 households have ever used credit from the dairy to make investments. Furthermore, households do not have access to bank loans either. Under these circumstances, the high level of investments is quite surprising.

Table 6. Investments and loans of farm households in Bulgaria.

Size (# of cows)	Invests (% of total) A	Uses dairy loan (% of A) B	Uses bank loan (% of A) C
1	62	0	2
2	66	0	4
3	64	6	6
4	65	0	0
5	70	14	0
>5	56	0	0
All	63	1	3

References

Balsevich, F., J.A. Berdegue, L. Flores, D. Mainville, T. Reardon, L. Busch, and L. Unnevehr, 2003. Supermarkets and produce quality and safety standards in Latin America. American Journal of Agricultural Economics 85, 1147-1154.

Blanchard, O., 1999. The economics of post-communist transition. Clarendon.

Dolan, C. and J. Humphrey, 2000. Governance and trade in fresh vegetables: The impact of UK supermarkets on the african horticulture industry. Journal of Development Studies 37, 147-176.

Dries, L. and J.F.M. Swinnen, 2004. Foreign direct investment, vertical integration and local suppliers: Evidence from the Polish dairy sector. World Development 32, 1525-1544.

FAO, 2000. Bulgaria: Investing in private dairy farming. Sector Review / Identification Report, FAO Investment Centre Divisions, Rome.

FAO, 2004. FAOSTAT [Online available: http://www.fao.org].

Gow and J.F.M. Swinnen, 2001. Private enforcement capital and contract enforcement in transition countries. American Journal of Agricultural Economics 83, 686-690.

Henson, S.J., Loader, R.J. and A. Brouder, 2000. Food safety standards and exports of perishable products from developing countries: Fish exports from East Africa to the European Union. American Journal of Agricultural Economics 82, 1159-1169.

Konings, J. and P.P. Walsh, 1999. Disorganisation in the process of transition: Firm-level evidence from Ukrain. Economics of Transition 7, 29-46.

Unnevehr, L.J., 2000. Food safety Issues and fresh product exports from LDCs. Agricultural Economics 23, 231-240.

The consequences of heterogeneous agents and moral hazard on food safety and trade

Norbert Hirschauer and Oliver Mußhoff
Humboldt-Universität zu Berlin, Department of Agricultural Economics, Luisenstraße 56, 10099 Berlin, Germany; n.hirschauer@agrar.hu-berlin.de

Abstract

The probability that buyers are being deceived with regard to the quality or safety of products increases with the profits that sellers can earn through opportunistic behaviour. It decreases with the probability and level of losses that result from disclosure of malpractice. A systematic prevention of malpractice in food chains needs to eliminate misdirected incentives along the whole chain through a suitable contract design. The scope of this paper is more limited. It demonstrates how moral hazard models can be used for behavioural risk assessment and exemplarily investigates the case of grain producers who might be tempted to breach the minimum waiting period after fungicide application.

Keywords: contract design, fungicides, incentives, moral hazard, risk management

1. Background and objectives

Food quality and safety regulations as well as voluntarily contracted standards may have significant effect on trade. This is obvious for private contractual arrangements governing transactions in the food sector and defining the specific quality requirements of individual buyers. However, it also applies for regulatory measures implemented by public authorities. In international food trade, for instance, there have been many cases where countries have been denied market access due to their failure to meet standards and/or document their compliance with regulations which are in force in the relevant export markets.

While the WTO Sanitary and Phytosanitary (SPS) Agreement promotes harmonising national with international standards such as those set by the Codex Alimentarius Commission, countries are permitted to take measures which are more stringent than these international benchmarks. However, it requires scientific justification to do so (Wilson and Otsuki, 2003). This formulation is a source of trade dispute, especially if national standards significantly exceed the international benchmarks. Hence, new regulatory measures must be based on risk assessment which provides scientific evidence of their benefits. Such evidence can refer to public health and technological aspects (which levels of hazardous agents can be tolerated? which technological procedures are acceptable? what kind of process controls are needed?), behavioural aspects (which incentive and control systems are needed to induce compliance?) and informational aspects (what labelling or other information should accompany products?).

Food quality measures are linked to these aspects through product, process, control and information standards. They may, for instance, require that tolerance standards are met with regard to pesticide or drug residues, pathogens, toxins etc. (product standards). They may also define the conditions of food production such as sanitary conditions, pest management procedures etc.

(process standards). Although end-of-pipe controls are still important, the emphasis has shifted from product to process standards in recent years, enhancing the need for integrated systems and shared information. Instead of simply inspecting and controlling the physical or bio-chemical qualities of the end product, buyers require that compliance with prescribed production and control procedures, including the application of company-based safety assurance systems such as HACCP (Hazard Analysis and Critical Control Points system), is adequately documented and communicated to trading partners (control and information standards).

Three decisive reasons for this development can be distinguished: (1) consumers increasingly value credence qualities which do not represent physical attributes inherent to the product, but which are rooted in the very way of production (e.g. organically grown products, adherence to social standards, animal protection standards etc.). (2) End-of-pipe controls are incomplete and costly at the same time. It is not feasible, for instance, to test for all hazardous agents such as pesticide residues and the like out of the nearly unlimited number of harmful substances. However, it may be feasible to document a proper and professional use of pesticides in an integrated approach which shares procedural information between the different stages of production down to the retail market. (3) There is a growing scientific understanding that food-borne health problems cannot be excluded by simply guaranteeing tolerance standards with regard to hazardous agents. In most cases, even exposure to low levels of these substances is detrimental. Furthermore, these levels are always subject to stochastic influences. Hence, business operators must be made to choose the right procedures which 'move distributions of unwanted qualities (e.g. toxin levels) to the left', instead of simply 'truncating them at arbitrary tolerance levels' through end-of-pipe controls and a rejection of objectionable batches.

However, the specification of suitable technological procedures and quality management systems does not suffice to guarantee compliance because documents themselves may be subject to fraud by opportunistic economic agents. That is why it is essential to design incentive-compatible contracts which induce compliance with process standards as well as with quality assurance systems. Incentive compatible contracts work independent of the moral attitudes of economic agents because they eliminate economic temptations to infringe rules and thus replace the need for character trust by situational trust (cf. Noorderhaven, 1996).

Utilising a case study from agriculture, this paper focuses on the behavioural aspect of risk assessment by demonstrating how food risk stemming from opportunistic behaviour of heterogeneous agents can be identified through systematic moral hazard analyses. We exemplarily investigate the incentive situation of grain farmers with regard to their compliance with the minimum waiting period after pre-harvest fungicide use. Specifically, our moral hazard investigation aims to assess the behavioural risks (positive analysis) by answering the following questions: (1) do misdirected economic incentives persist in the actual contractual arrangements in that it is more profitable to infringe the waiting period than to comply with it? (2) Which consequences result from farmers being heterogeneous with regard to their individual perception of economic parameters, cost estimates and risk attitudes?

Besides risk assessment, moral hazard models have also the potential to support the management of behavioural risks in that they can be used for weighing and selecting appropriate prevention measures. In other words: if misdirected economic incentives are found to exist, the question which contractual arrangement should be designed by the downstream buyer could be answered through a model-based moral hazard analysis (normative analysis). However, this requires that the costs of measures that aim to change incentives (e.g. paying premiums, increasing control intensities, rising information standards and increasing traceability, imposing sanctions etc.) are estimated as well. The necessary cost estimates, however, exceed the scope and the aim of this paper whose objective is limited to the assessment of the existing behavioural risks. For further use, we will nonetheless present the complete moral hazard model which is suited for positive as well as for normative analysis.

While not trying to determine the optimal design of the contract between grain dealer (buyer, principal) and grain producer (seller, agent) in this paper, we use the grain producer example to discuss which kind of rule is needed in the first place to turn buyers on all chain levels into 'responsible principals'. Responsible principals would indeed act on behalf of the entire downstream chain and introduce behavioural risk management systems in order to design incentive-compatible contracts for their purchasing transactions. The 'making of responsible principals' requires that they are forced, in turn, to internalise societal costs resulting from downstream diseconomies and finally from consumers' exposure to increased residue levels. In this context, we comment on trade implications resulting from an introduction of behavioural risk management standards, especially in the light of international trading agreements.

2. The moral hazard model for behavioural food risks

Principal-agent models (also referred to as PA- or moral hazard models) tackle the problem of information asymmetries arising in transaction. They assume that a less well informed principal and a better informed agent have conflicting interests. While both maximise their respective objective functions, the principal has the power to design the contract and to take account of the expected actions of the agent within the limits of his informational constraints.

With a view to empirical application, we develop an adequate approach for the analysis of moral hazards in food chain transactions by using a general PA-model (as found, for instance, in Kreps, 1990: 577) as a starting point (cf. Hirschauer, 2004). The model assumes that an agent has the choice between discrete actions $a_1, a_2, \ldots, a_N$ and corresponding efforts $k_1 < k_2 < \ldots < k_N$. In a stochastic environment, these actions result - with given probabilities π_{nm} - in discrete outputs $y_1 < y_2 < \ldots < y_M$ and output-dependent remunerations $w_1 < w_2 < \ldots < w_M$. The agent's utility depends on his remuneration and effort (), where $u(w_m)$ represents a von Neumann-Morgenstern utility function. If the principal is presumed to be risk-neutral, his design problem can be stated as the following constraint optimisation problem:

Step 1: determine the minimum wage costs $w_{min}(a_n)$ for each possible action:

$$\underset{w}{Min} \sum_{m=1}^{M} \pi_{nm} w_m = w_{min}(a_n) \tag{1}$$

$$s.t. \sum_{m=1}^{M} \pi_{nm} u(w_m) - k_n \geq \mu \quad \text{(participation constraint)} \tag{2}$$

$$\sum_{m=1}^{M} \pi_{nm} u(w_m) - k_n \geq \sum_{m=1}^{M} \pi_{n'm} u(w_m) - k_{n'}, \; n' = 1, \ldots, N \quad \text{(incentive compatibility constraint)} \tag{3}$$

Step 2: determine the maximum payoff over all actions a_n:

$$\underset{a_n}{Max} \left(\sum_{m=1}^{M} \pi_{nm} y_m - w_{min}(a_n) \right) \tag{4}$$

The specific structure of PA-problems which is caused by information asymmetries and a stochastic environment is illustrated by the fairly general problem formulation used above. While the meaning of model parameters varies with investigated contexts, PA-models have the capacity to provide valuable insights into the structure of many real-life problems that involve transactions under information asymmetries, including behavioural food risks. In the food risk problem, the principal is to be considered the buyer of a product whose uncertain qualities depend on the actions of the seller (i.e. the agent) and a stochastic influence.

However, empirical estimations of parameters such as prices, costs of compliance, frequency of control, traceability, level of sanctions etc. are needed to provide intelligence for specific action situations (i.e. specific food chain activities and transactions). That is, if model calculations are to facilitate practical conclusions, they must be simple enough to be 'filled with empirical data' from the chain activity under investigation. Bearing the applicability of the model in mind, a number of modifications are made to model (1) to (4):

1. Instead of using a generally discrete formulation, we reduce the model to a binary perspective. That is, we consider only two possible actions (a_1 = non-compliance; a_2 = compliance), two effort levels ($k_1 < k_2$), two outcomes ($y_1 < y_2$), and two remunerations ($w_1 < w_2$). The binary perspective allows us to estimate and use simple binomial distributions for stochastic variables such as outcome and remuneration.
2. Instead of accounting for risk aversion endogenously, we assume risk neutral principals and agents in model calculations. Therefore, optimal risk sharing will not be our concern here. However, due to a costly and incomplete output observation (see 5.) we still have the non-trivial problem of how to design an optimal control and remuneration scheme.
3. Instead of accounting for a positive reservation utility μ, we assume a reservation utility of zero. This matches a situation where there are binding rules on how food processing activities have to be carried out: if the agent does not officially 'participate', he does not have the choice to produce a lower quality category and to sell it at a lower price, but has to refrain from production altogether.
4. Instead of accounting for a principal who maximises his utility by selecting the agent's optimal effort level, we assume that the principal knows a priori that his maximum utility results from the higher effort level (i.e. from compliance on the part of the agent). Therefore, he is determined to induce compliance and only strives to do so at minimum costs. Hence, the second step of the optimisation can be omitted and the principal's problem is reduced to cost minimisation for action a_2.
5. Instead of assuming that the output can be verified without costs, we take the characteristics of the food risk problem (credence qualities) into account and consider that observation is costly and that it can only take the form of random sampling inspections carried out with a control intensity $s \leq 100$ %. This results in incomplete output information.
6. The standard PA-model does not account for multiple agent settings. Incentive problems resulting from incomplete output information will be aggravated if identified properties cannot be retraced to a single upstream seller. We consider such situations which are frequently found in food chains by accounting for a limited traceability $z \leq 100$ %.

Instead of simply reformulating the model for these modifications, we turn to its explicit food risk interpretation and a handier notation for the binary incentive problem (see Table 1).

Abstracting at first from incomplete inspection and traceability and following the notation of Table 1 the principal's minimisation problem may be restated as follows:

$$Min\ w(a_2) = Min\left(-(1-q)\,S + qP\right) = Min\ (P - (1-q)(P+S)) \tag{1'}$$

Table 1. Notation for the binary food risk model.

w_1	=	$-S$	: sanction (loss) inflicted on the agent if the undesired/hazardous quality y_1 is detected
w_2	=	P	: price paid for a product of the desired quality y_2
$k_2\text{-}k_1\text{=}k_2$	=	K	: agent's cost of compliance with regulations[1]
π_{11}	=	r	: probability of undesired product quality y_1 in case of action a_1 (i.e. non-compliance)
π_{12}	=	$1\text{-}r$	: probability of desired product quality y_2 in case of action a_1 (i.e. non-compliance)
π_{22}	=	q	: probability of desired product quality y_2 in case of action a_2 (i.e. compliance): $q > 1\text{-}r$
π_{21}	=	$1\text{-}q$	: probability of undesired product quality y_1 in case of action a_2 (i.e. compliance)
		s	: intensity (frequency) of random controls ($0 < s \leq 100$ %)
		z	: probability that responsible sellers are traced ($0 < z \leq 100$ %)

[1]We replace k_2-k_1 by the costs K of compliance. It is unrealistic to assume that food business operators produce the unauthorised quality at cost $k_1 = 0$. For the sake of simplicity we normalise k_1 to zero and thereby avoid having to carry an extra variable through the analysis without impeding the general insights into the structure of the problem. A consideration of $k_1 \neq 0$ in applications will be easy. It is only necessary in normative analysis.

$$s.t.\ w(a_2) - k_2 = -(1 - q)\,S + qP - K = P - (1 - q)(P + S) - K \geq 0 \tag{2'}$$

$$w(a_2) - k_2 - w(a_1) = -(1 - q)\,S + qP - K + rS - (1 - r)P = (q + r - 1)(P + S) - K \geq 0 \tag{3'}$$

In the next step we need to account for incomplete inspection and traceability: prohibitively high costs of complete inspection (e.g. because the product is destroyed by testing) force the principal to resort to partial and random controls. Control intensities $s < 100$ % result in incomplete information about the relevant output (product quality). The limited traceability problem arises whenever an undesired product quality cannot be traced back to a single seller out of the many in a supply chain. This is regularly the case if the principal is dealing with multiple agents without having established an absolutely reliable traceability system.

Both a control intensity $s < 1$ and a traceability coefficient $z < 1$ change the expected remuneration for non-compliance $w(a_1)$ as well as for compliance $w(a_2)$. Independent of the agent's action or the product quality, the principal has to pay P whenever the quality is not ascertained or cannot be ascribed to a single agent. The agent can at best be made to pay a sanction S if the undesired quality y_1 is evident and clearly due to his making. Contrary to complete inspection and traceability where output probabilities coincide with remuneration probabilities, partial (sampling) inspection and limited traceability entail remuneration probabilities according to Table 2. If we additionally consider the control costs depending on the intensity $c(s)$, the costs for achieving different levels of traceability $c(z)$, and the costs for imposing different sanctions $c(S)$, the principal's incentive problem needs to be restated as follows:

$$Min\ w(a_2) = Min\left(P - sz\,(1 - q)(P + S) + c(s) + c(z) + c(S)\right) \tag{1''}$$

$$s.t.\ w(a_2) - k_2 = P - sz\,(1 - q)(P + S) - K \geq 0 \tag{2''}$$

Table 2. Output and remuneration probabilities.

	output probabilities for		remuneration probabilities for	
	y_1	y_2	$w_1 = -S$	$w_2 = P$
a_1 = non-compliance (low effort)	r	$1-r$	szr	$1-szr$
a_2 = compliance (high effort)	$1-q$	q	$sz(1-q)$	$1-sz(1-q)$

$$w(a_2) - k_2 - w(a_1) = sz\,(q + r - 1)(P + S) - K \geq 0 \qquad (3")$$

$$0 < sz \leq 1$$

3. Perspectives and scope of behavioural risk investigations

Any investigation into behavioural risks can be related to the definition of the risk analysis process according to regulation EC 178/2002 which defines that 'risk analysis means a process of three interconnected components: risk assessment, risk management and risk communication'. Furthermore, analogous to the business analysis and planning process, the time horizon determines which parameters are 'givens', and which are 'decision variables' that can be influenced by the decision-maker (i.e. the designing principal). Table 3 visualises how the time horizon and the three components of the risk analysis process are interconnected with the parameters of the moral hazard model. In the overview, the technological parameters of the production process (costs of compliance K; stochastic linkage between the agents' action and the outcome, represented by q and r) are assumed to be constant over all perspectives.

In behavioural risk assessment, the stochastic links between behaviour and technological outcomes as well as the economic parameters (compliance costs, control intensities, etc.) need to be assessed for each activity and for each group of actors in the present situation. Using the binary model, there are only few parameters to be considered. In behavioural risk management, the decision variables which are under the control of the principal vary with the perspective: (1) a short term given state of a control system can be seen, for instance, as being equivalent to

Table 3. Time horizon and the components of the risk analysis process.

	Short-term perspective	Medium-term perspective	Long-term perspective
Risk assessment	all parameters that are in force in the present situation are considered in the analysis of the existing incentive situation		
Risk management	givens: K, q, r, s, z, $[c(S)]$	givens: K, q, r, $c(s)$, z, $[c(S)]$	givens: K, q, r, $c(s)$, $c(z)$, $[c(S)]$
	decision variables: P, $[S]$	decision variables: P, s, $[S]$	decision variables: P, s, z, $[S]$
Risk communication	all risk assessment and risk management findings are interactively exchanged among stakeholders		

a given control intensity and a given traceability. (2) In a medium-term perspective, it may be possible to change the control intensity by stocking up on control personnel and equipment. Solving the principal's constraint minimisation problem consequently implies that control costs $c(s)$ are considered. (3) In a long-term perspective, the traceability that results from the structure of transactions along the chain may be changed by restructuring and documentation efforts. This implies that traceability costs $c(z)$ for achieving different traceability levels are taken into account. Furthermore, it is necessary to consider the costs for imposing sanctions $c(S)$ whenever the sanction is treated as a decision variable. The brackets in Table 3 indicate that the sanction is only partially under the control of the principal. Due to legal constraints, its range is usually limited by an upper level. In behavioural risk communication, a clear conceptualisation of the problem provides a 'language' for its perception, description and exchange of information, thus improving the interactive communication of behavioural food risk findings among all actors along the food chain.

While there are only few parameters to be considered in the model, their estimation still represents a formidable task. It is not trivial, for instance, to define different control alternatives and provide their cost estimates (let alone intensity-dependent control cost functions $c(s)$ for different control systems and technologies). It is even more daunting to provide reliable estimates for the costs of increasing traceability or for imposing increased sanctions. As has been mentioned before, we therefore limit our investigation to the assessment of behavioural risk in the present system. While using critical value analyses to find out which changes of contract would get incentives 'right', we do not try to optimise the system as a whole. Technically speaking, we estimate the parameters K, q, r, s, z, P, and S from within the food chain and then use Equation (3") to quantify the incentives in force.

4. The situational incentives of farmers after fungicide use

4.1. The situational background

Grain farmers regularly apply a last dose of fungicides approximately five to six weeks before harvesting. Applied products are labelled for control of fungal infections which could otherwise significantly reduce the quality and quantity of harvested grain. Under certain weather conditions, profit maximising farmers might be tempted to breach the minimum waiting period of 35 days. This is particularly tempting if, a few days before the end of the waiting period, the weather is ideal for harvesting, whereas a long period of rain is expected afterwards.

The individual farmer's incentives depend on the contract, i.e. the overall conditions of the transaction including control and tracing activities on the part of the buyer. The contract details reflect the quality requirements of the grain buyer as well as the degree of trust he puts on the farmer. Since contracts are never completely enforceable (e.g. due to costly controls), farmers are left with opportunities to hide rule-breaking behaviour.

Aiming to assess the rough dimension of the incentives in force, we use stylised facts from a case study instead of collecting data in an extensive survey. The farmers who participated in the case study sell their wheat to a local corn dealer who takes and stores samples from all individual trailer loads, tests them for their technological qualities (humidity, protein content etc.) and differentiates prices for different quality categories. However, before testing for pesticide residues, the corn dealer blends the 'loads' from different farmers into 'lots' according to the specific quality requirements of his downstream trading partners. Because residues are monitored at downstream control points only, farmers might be tempted to infringe the waiting period. Infringements are only detected if blended lots exceed the tolerance standards. This is only the case if a critical number of farmers break the rule. Otherwise, free-riding farmers stay undetected since residues resulting from their premature harvest are 'sufficiently' diluted. That is, actual tracing does not take place although complete ability to trace is assured through stored individual samples. In other words: the free-riding opportunity arises precisely because the confided group appears trustworthy on the whole, but is in fact (morally) heterogeneous. Despite a complete ability to trace, and even if 100 % of blended lots are monitored, there is only a small probability that the harmful behaviour of a minority of shirking farmers triggers the testing of stored individual samples. Technically speaking we might say that the free-riding problem 'moves the distribution of the unwanted quality to the right'.

4.2. Assessing farmers' situational decision parameters

The economic parameters determining the farmers' incentive situations were assessed in oral interviews with three farmers in a grain producing area in the federal state of Brandenburg, Germany (see Table 4). Additionally, the local corn dealer was asked to appraise the situation. Information is uncertain, and resulting data give evidence of the individual perception of the parameters. Since only discrete data can be gained in a survey, the interviewees were asked to assess the economic parameters for four discrete types of weather, implying, in turn, four different 'technologically optimal' harvest dates: 10 days, 6 days and 2 days prematurely (i.e. before the end of the waiting period) as well as an optimal harvest date after the expiration of the waiting period. The term 'technologically optimal' implies that, in the absence of a prescribed waiting period, a farmer would harvest because he expects economic losses due to a reduced quality, and/or quantity, and/or increased costs for any posterior date.

Table 4. Parameters determining the profitability of shirking as perceived by interviewed farmers.

	x-days	parameter	Farmer A	Farmer B	Farmer C
1. probability that the farmer exceeds	10	*r*	15%	95%	33%
the residue limit in his individual load	6	*r*	5%	50%	20%
if he harvests x-days before the end of	2	*r*	0%	0%	0%
the waiting period	0	*1-q*	0%	0%	0%
2. the farmer's probability of being detected if his individual load exceeds the residue limit		*s*	5%	50%	5%
3. losses in sales and additional costs	10	*K*	200	260	200
(€/ha) if the waiting period is met in spite of weather conditions making it optimal to harvest x-days prematurely	6	*K*	100	130	100
4. losses in sales (€/ha) if non-compliance is proven		*P*	984	984	984
5. 'sanctions' (€/ha) if non-compliance is		*S*	1,100	20,750	13,375
proven thereof: short-term sanctions			350	20,000	13,000
(fines, damages, etc.); capitalised long-term losses in the market			750	750	375
6. probability that the farmer can be traced		*z*	100%	100%	100%

Ad 1. Although none of the farmers had access to results of scientific tests concerning the decomposition of fungicides, they all agreed that there is no risk of exceeding the tolerance standard if they comply with the waiting period of 35 days. Trusting that a safety margin had been built into the prescribed waiting period, they also agreed that harvesting two days early would still involve a zero probability of exceeding the limit. According to this perception, a 2 days-infringement of the waiting period has the same results as compliance ($r = 1\text{-}q = 0\%$). The farmers' assessments of the decomposition process before that date differed widely, however. Farmer A, for instance, estimated that the probability of exceeding the limit would rise to 5% (15%) in the case of a 6 days- (10 days-) infringement. In wide contrast to that, farmer B, for instance, believed this probability to rise to 50% (95%).

Ad 2. In the considered situation, the parameter s represents the aggregate 'probability of being detected' if the individual load exceeds the residue limit. This probability results from the joint effect of two components: (1) the control intensity (i.e. percentage of blended lots which are controlled), and (2) the dilution effect caused by the fact that loads from different farmers are blended before being tested for residues. All farmers ignored the actual control intensity. They had no information as to whether all blended lots are inspected, or whether only random controls are being made. They likewise ignored the details responsible for the physical dilution effect[1]. However, their ad hoc estimations differed widely with regard to the overall effect of these two factors, i.e. the probability that an infringement would be detected if their individual load exceeded the limit.

Ad 3. Reduced sales resulting from suboptimal harvest dates are treated as opportunity costs forming the major part of the compliance costs K: expected losses resulting from technologically suboptimal harvest dates are mainly due to a threatening decline of quality which could force farmers to sell their wheat as animal feed grain at 80-90€ per metric ton, instead of food grain at 110-120€ per ton. If it is technologically optimal to harvest 10 days before the end of the waiting period, the three farmers expected an almost certain loss of sales of 175-210€ per hectare (or 25-30€ per ton) due to the degradation of grain to feed quality. If it is technologically optimal to harvest 6 days prematurely, they commonly expected the loss to occur with a 50% probability only. Besides these opportunity costs, farmers estimated machinery costs to increase by 25-50€ per hectare if they were to harvest 10 days later than optimal, and by 12.5-25€ per hectare if they were to harvest 6 days later than optimal.

Ad 4. All three farmers are convinced that they would completely loose their income from the wheat sales (including EU-subsidies) of $P = 984$€/ha if non-compliance was detected. The farmers' perception that - besides sales - transfer payments would be lost can be seen as a positive result of the EU joint compliance approach that enhances incentive compatibility.

Ad 5. Farmers estimated that they would have to pay an equivalent of 350-20,000€ per hectare in direct sanction payments such as fines, damage compensations, etc. Farmer B's and farmer C's perception of comparably very high sanctions is mainly due to their understanding that they could be forced to pay damage compensations for large amounts of grain if these were contaminated by their individual load. In addition to these short-term sanctions, farmers assumed that their capitalised future disadvantage on the market (loss of negotiating power) would amount to 375-750€ per hectare of wheat.

[1] Explicitly determining the 'physical dilution effect' requires expectations concerning one's own share in a blended lot, the residue levels in individual loads depending on the harvesting decision, and the behaviour of other farmers whose loads are part of a blended lot.

Ad 6. Farmers agreed that the traceability z amounts to 100% due to the fact that samples are taken and stored from the individual farmers' loads.

4.3. Investigating farmers' incentive situations

Table 5 demonstrates the incentive situation which - according to Equation (3") - results from the farmers' perception of the relevant parameters in force. Results are indicated for the two weather types that favour most premature harvest. We did not endogenously account for risk aversion. Besides avoiding the problem of having to empirically estimate risk utility functions, this is the due approach since risk attitudes are considered exogenously by the very way data were obtained: risk-averse farmers implicitly increased cost-benefit ratios when answering questions with regard to decision parameters (i.e. risk premiums are deducted already).

Only farmer A perceives an economic reason to infringe upon the waiting period. His actual behaviour in the light of such a temptation is not known. Using the parameters for weather type I as they had been assessed by the farmers, we identify - by means of simple critical value analyses - which change of contract conditions (sanction, control intensity) would compare ensure/maintain incentive-compatible contracts. It should be noted that, in the example under consideration, the participation constraint Equation (2") does not need to be accounted for in the critical value analysis. In contrast, it is possible to design 'boiling-in-oil-contracts' (cf. Rasmusen, 1994: 180) since the probability of the desired product quality for complying farmers is $q = 1$. Thus, they are neither affected by increased sanctions nor by intensified controls. Increasing the price would nonetheless be a way to reduce the temptation to break the rule. Paying a higher price, however, directly increases the costs of the buyer. This makes only sense if it is not viable or very costly to increase sanctions and/or control intensities.

Examples of contracts which get the incentives 'right' and thus replace the need for 'character trust' by 'situational trust' are given in Table 6 for each of the three farmers.

If the system of downstream controls is maintained and if weather type I occurs, the sanction as perceived by *farmer A* would need to be increased from its present level of 1,100€ to over 25,000€ per hectare in order to eliminate his 184€-per-hectare temptation to break the rule. Since it does not seem to be realistic to assume that the principal succeeds in making the farmer perceive the sanction to be at this level, we consider the effects of applying controls to the individual loads. To do so is equivalent to replacing downstream control points (blended lots of grain) by upstream control points (individual loads of grain), thus eliminating the dilution effect and raising the probability that an objectionable load is detected from the perceived level of $s = 5\%$ to 100%. With complete controls of individual loads, a sanction of approximately 350€ per hectare would suffice to eliminate misdirected economic incentives for farmer A. Alternatively, in accordance with the

Table 5. Economic inferiority (–) / superiority (+) of complying with the waiting period according to the perception of the present decision parameters by interviewed farmers (€/ha).

Weather type	Technologically optimal harvest date	Farmer A	Farmer B	Farmer C
I	10 days premature	– 184	+ 10,064	+ 39
II	6 days premature	– 95	+ 5,304	+ 44

Table 6. Incentive-compatible contracts for weather type I.

	Farmer A	Farmer B	Farmer C
Critical sanction with retention of the present system of downstream controls (blended lots)	25,683€/ha	no sanction needed	11,016€/ha
Critical sanction after introduction of complete upstream controls (individual loads)	349€/ha	no sanction needed	no sanction needed
Critical control intensity of individual loads with present sanctions: A: 1,100 €/ha; B: 20,750 €/ha; C: 13,375 €/ha	64%	1.3%	4.2%
Critical control intensity of individual loads with assumed sanctions: A: 2,200 €/ha; B: 41,500 €/ha; C: 26,750 €/ha	42%	0.6%	2.2%

presently perceived sanction level of 1,100€ per hectare, a control intensity of 64% would suffice if individual loads were being analysed.

Considering farmer B and C reveals that, due to information uncertainties, the incentives 'in force' are in the eyes of the beholder. Farmer B, for instance, in contrast to farmer A, clearly perceives no economic temptation whatsoever to break the rule, mainly because he believes economic losses resulting from detection to be very high. Thus, after applying controls to individual loads, a control intensity of roughly 1% would suffice to generate incentive compatibility in the case of farmer B.

The corn dealer's view of the farmers' incentive situation is not depicted in Table 3. It is summarised briefly: he believes that, in the present system of downstream controls, a shirking farmer's risk of being detected is almost zero due to the dilution effect. Knowing the approximate level of the other relevant parameters (sanctions, costs of compliance, etc.), the corn dealer is convinced that situational incentives are indeed not 'right'. However, according to his interview statement, he relies on character trust with regard to his farmers. This statement triggers the question whether he is really motivated enough to act as a responsible principal.

Abstracting from individual particularities, we can finally generalise from the last row of table 6 that increasing the sanction level allows for a decrease of the control intensity without compromising the incentive compatibility. There is an optimal combination to be found which obviously depends on the costs of analytical controls on the one hand, and the costs of increasing effective sanctions (lawyers, lobbying for sanctions etc.) on the other hand.

The essence of this case study can be pictured through a typology consisting of two extreme- and one mixed-type decision-maker. We arrive at these three types by distinguishing between 'character trust' and 'situational trust': (1) on the one extreme is the farmer who is utterly trustworthy. Because of his personal set of preferences he resists every economic temptation to break the rules. (2) On the other extreme is the farmer who is only trustworthy if, given his exclusive objective of profit-maximising, the perceived situational incentives of the contract are 'right'. (3) Between these two extremes falls the mixed-type farmer who accepts a certain profit trade-off in exchange for a personal feeling of moral integrity resulting from his decision to abide by the rules. He might yield to rule-breaking behaviour, however, if the additional profits to be gained exceed his personal resistance.

It is common sense to assume that real decision-makers are of mixed-type. They might differ, however, with regard to their personal resistance to economic temptations. Taking into consideration that economic parameters may differ from one agent to the other and that they are seen through the eyes of the beholder, some farmers may perceive economic temptations to break the rules; others may not. Amongst the former some may indeed break the rules; others may not. Only these rule-breakers cause a problem from an incentive and food risk point of view. Finding an incentive-compatible solution is not easy: every buyer (principal) will have great difficulties to gain information about how heterogeneous sellers (agents) assess the relevant parameters. An even greater obstacle will be to gain knowledge about their individual characters. Hence, the only practical way to decrease the probability of shirking is to increase situational trust by 'marching in the right direction' and increasing the levels of those parameters (as perceived by the agents) that promote compliance. Besides objective changes to the economic environment, this involves considerable communication efforts.

5. Moral hazard analysis systems and implications for trade

Lessons with regard to the prevention of moral hazards may be learned from the widely established HACCP-approach which is basically a technological safety assurance system. According to its seven principles, its users are (1) to analyse their food operations and to prepare a list of potential hazards, (2) to determine 'critical control points' where these hazards can be controlled, (3) to define adequate tolerance limits, (4) to establish adequate monitoring procedures, (5) to define corrective measures and contingency plans in case deviations are being identified, (6) to document all HACCP steps, and (7) to verify that the system is working correctly and to update it, if appropriate. Because it consists of general principles, HACCP can be adapted to any production process. While representing the international Codex Alimentarius standard, HACCP has an impact on trade because compliance is costly and may even overstrain the capacities of individual food business operators in many countries. Nonetheless, regulatory measures and private contracts in (international) food trade increasingly involve requirements for HACCP process controls.

The scope of HACCP is limited to the prevention of unintentional technological and human failures. Behavioural risks could be managed using similar principles: in addition to managing the risk of unintentional failures within one's own food production process, one could systematically aim to manage behavioural risks that result from information asymmetries in transactions with suppliers. This requires the definition of critical control points and adequate monitoring procedures with regard to risks that may arise from opportunistic malpractice of upstream trading partners. Our case of grain producers has demonstrated, for instance, that some control points (i.e. monitoring fungicide residues in blended lots of grain) are less suited to manage behavioural risks than others (i.e. monitoring fungicide residues in individual loads of grain). Controlling individual loads clearly increases the probability that non-compliance is being prevented. A system of behavioural risk management could also be seen as an extension of traceability requirements in that a minimum standard of behavioural risk control is asked for in purchasing transactions in addition to simply documenting where inputs came from.

A mandatory implementation of a 'moral hazard analysis and critical control points system' (M-HACCP) within a chain would force buyers on all levels to act as a 'responsible principals' when purchasing goods. Scientific evidence must be provided with regard to the benefits (prevention potential) of such a system before food businesses operators can be made to adopt it. If considered useful by some food chain actors for competitive reasons, its implementation could be achieved through private contracts. With a view to (international) trade, regulatory measures aiming at setting behavioural risk management standards would require evidence that justifies imminent trade losses by gains in public health and consumer protection.

6. Concluding comments

Moral hazard (or: principal agent) models which are derived from the broader branch of game theory explicitly account for stochastic environments and information asymmetries. They consider economic actors as opportunistic (rational) 'players' and offer the opportunity to model situations of conflicting interests, including those of sellers (agents) and buyers (principals) of products with credence qualities. The probability of malpractice on the part of upstream sellers is conceptualised as varying with its expected economic benefits. Thus, moral hazard models enable interested parties to analyse economic incentives on different levels of food chains and to localise hot spots where profit maximising actors are most tempted to break rules (positive analysis). This facilitates conclusions on where and how to change contract and control designs in order to eliminate misdirected economic incentives and to induce compliance at minimum costs (normative analysis). In real-life, rather than solving formal constraint optimisation problems, this will involve to define discrete and feasible alternatives and to check them with regard to their costs and their incentive compatibility.

This article presents a principal agent model that has been adjusted to the characteristics of the behavioural food risk problem. The adjusted model has the capacity to account for incomplete (sample) inspection of the product quality as well as for limited possibilities to trace a product or ingredient to its origins. Its manageable data requirements qualify it as a starting point for the development of operational models which are tailored to particular activities in food chains. The general outline of the moral hazard approach has been demonstrated through a case study from primary grain production. Future research aiming to enable interested parties to reduce behavioural risks in food chains in general needs to extend such analyses to other activities of the considered chain level, other levels of the food chain, and other food chains.

While not representing a systematic analysis and listing of potential moral hazards, exploratory expert interviews revealed some interesting exemplary activities on the level of primary production which might be worth while examining with regard to misdirected economic incentives: (1) weed control directed at eliminating coach grass in barley is the more successful the later the herbicide is applied. The suitable time of application is close to the usual harvest date. But again, there is a prescribed minimum waiting period (10 to 14 days) that farmers might be tempted to disregard due to economic considerations. (2) Considerable price premiums are paid for organically produced food. At the same time, organic farming must do without chemical pesticides, fertilisers etc. Thus temptations might arise to breach some of the rules of organic farming that either cause considerable additional costs or reduce the yields compared to conventional farming. Examples would be the use of conventional seeds instead of certified organic seeds, a (partial) use of conventional animal feed etc.

The extension of behavioural risk management to a wider variety of activities and, even more so, to other chain levels and other food chains will require that the structure of the food risk model (1") to (3") is developed further and extended with regard to its restrictive assumptions. Depending on the situation, the following extensions might be promising:

- Instead of a binary perspective, finer partitions of the agents' scope of action such as different degrees of compliance could be accounted for in a generally discrete model.
- Instead of considering a common set of outputs for compliance and for non-compliance, different sets or probability distributions for continuous output values could be considered.
- Instead of minimising costs, the value (damage) of the desired (undesired) product quality for the principal could be considered explicitly in utility maximising models.
- Instead of assuming a non-ambiguous observation of the output, a statistic measurement error rate could be estimated allowing for an appraisal of first and second degree errors resulting from random sampling inspections.

Before increasing the complexity of applied models, however, it should always be considered whether the informational gains justify the additional costs.

References

Hirschauer, N., 2004. A model-based approach to moral hazard in food chains. What contribution do principal-agent-models make to the understanding of food risks induced by opportunistic behaviour? Agrarwirtschaft 53, 192-205.

Kreps, D.M., 1990. A course in microeconomic theory. Harvester Wheatsheaf.

Noorderhaven, N.G., 1996. Opportunism and trust in transaction cost economics. In Groenewagen, J. (ed.), Transaction cost economics and beyond. Kluwer Academic Publishers, 107-128.

Rasmusen, E., 1994. Games and information. An introduction to game theory. Blackwell Publishers.

Wilson, J. and T. Otsuki, 2003. Balancing risk reduction and benefits from trade in setting standards. IFPRI (www.ifpri.org), Focus 10, Food Safety in Food Security and Food Trade.

Supplier relationship management in the German dairy industry

Sabine Gerlach, Achim Spiller and Christian Wocken
University of Goettingen, Department of Agricultural Economics and Rural Development, Platz der Goettinger Sieben 5, 37073 Goettingen, Germany; cwocken@uni-goettingen.de

Abstract

The following paper analyses relationship quality in the German dairy sector in order to evaluate opportunities to enhance product quality through better supply chain coordination. We develop a measurement concept of supplier relationship quality which allows new insights into the critical success factors of supplier development. The analysis is based on different research streams, mainly network science, channel management and sourcing theory. A survey involving 209 German farmers constituted the empirical basis.

Keywords: trust, regression analysis, dairy business

1. Quality management and supply chain coordination

Quality management in agribusiness encompasses different facets; from a company's internal management system to metasystems like the growing certification concepts, and from business to state run programs. One of the most common assumptions of the modern quality approach is the advantage of a supply chain perspective. The overall quality of a product is determined through the weakest link in the chain. However, the concepts of quality management in supply chains are not equally discussed. Whereas there is detailed debate about the options for vertical integration and contract farming, the possibilities of chain coordination in arm-length markets and weak vertical coordinated structures are rarely discussed.

The following paper analyses relationship quality in the German dairy sector in order to evaluate opportunities to enhance product quality through better supply chain coordination. The underlying hypothesis is the role that trust, commitment and satisfaction play in determining sustainable business relationships, chain wide information flow and quality motivation. We develop a measurement concept of supplier relationship quality which allows new insights into the critical success factors of supplier development. The analysis is based on different research streams, mainly network science, channel management and sourcing theory. A survey involving 209 German farmers constituted the empirical basis.

2. Relationship quality: a precondition of quality management

2.1. Basic assumptions

Traditionally sourcing management plays an important role in quality assurance concepts. Statistical quality control and inspection systems ensure the quality of direct and indirect materials in a business environment which is characterised through an increase in outsourcing. But the role sourcing plays in management theory has changed through the last decades from procurement to supply chain management. The buyer-supplier relationship tends from price focus towards a portfolio model in which key vendors act as partners bonded through mutual investments and trust (Han *et al.* 1993).

This development of sourcing theory is not well reflected in the discussion about quality management in the agribusiness. Here two recent trends can be found: The first opportunity is quality management as part of a vertical coordinated chain. In this case a predominant company - mainly the processor - develops standards for production and logistics depending on its own quality objectives and marketing requirements (examples can be found in the poultry sector). The second is that of certification in arms-length relationships. Through third party control, resulting in a generally accepted certificate (e.g. EurepGAP, ISO 9000), suppliers are enforced to follow general quality standards.

In both cases, whenever independent firms are involved in the supply chain, it is not always assured that the intentions of the supply chain leader or the customer are fulfilled. Detection of any opportunistic behaviour on the part of the supplier depends on the quality of controlling procedures. Control systems tend to be static in nature. In the positive sense they lead to a given standard, but there are no incentives to go beyond this. Moreover, certification systems show a tendency to build up bureaucratic structures for example various checklist controls, thus missing out on a real quality driven process. Basically, without intrinsic motivation the dangers of short falls are appreciable.

These problems are rarely discussed in the literature because of relatively new developments. Jahn *et al.* (2005) explain findings from the auditing theory in order to secure the validity of certification procedures. Theuvsen (2005: 175) argues that quality management systems could lead to more bureaucracy. On the one hand, there is 'good bureaucracy' that leads for example to precision, speed, unambiguity and knowledge of files. On the other hand 'bad bureaucracy' can cause inflexible work and progress conservatism. Finally, there are some empirical indications that the recent quality management approaches may be inadequate if they are not able to motivate suppliers as to their own quality.

Another more theoretical contribution to our problem comes from economic psychology, especially that of motivational crowding effects (Frey and Jegen, 2001; Frey and Oberholzer-Gee, 1997). The theory suggests that external intervention via financial incentives or control pressure reduce intrinsic motivation, the effect of this trade-off may result in a quality below that of the initial motivation level. This effect is well known to every employer who uses monetary incentives to motivate agents. On the one hand, they higher the overall output level. This holds true due to the extrinsic motivational effect of a higher payment. On the other hand, they undermine intrinsic motivation and thus jeopardise the quality of the work. Under particular conditions the negative crowding effect outweighs the positive effect of higher payment. In general, the initial amount of intrinsic motivation, the perceived fairness of the incentive and the opportunities to control the agents' work are factors which determine the occurrence of crowding effects. The same trade-off applies to external pressure or control, so the described relationship between internal and external motivation may appear in the field of quality motivation. As a result, after introducing a control oriented quality management, the loss of internal quality motivation can be larger than the amount added by the increased external pressure (Jahn *et al.*, 2004).

The described problems stand in high contrast to major requirements of total quality management. For example the basic tool of kaizen (continual improvement) is based on creativity and, thus, mainly on intrinsic motivation. Although the net effect of the above mentioned crowding effect may be not negative in all cases, quality management will be clearly more effective if suppliers are intrinsically motivated. Overall, there are some well founded reasons for the basic hypotheses of this article, that suitable business relationships with a high amount of trust and commitment lead to stable and long-term networks allowing better quality management along the whole supply chain. This is especially true for the agribusiness, where processors have a high number of small suppliers.

In the following paper we demonstrate with the example of the German dairy business, that new tools of supplier relationship management can be useful in improving the relationship between processors, trying to build up new quality approaches, and thus enhancing farmer-dairy relationships. In the following section, new quality developments in this industry are discussed, emphasising the obstacles when introducing new quality schemes to business relationships with low trust.

2.2. Introduction of a new quality management system to the German dairy business

The German dairy business is highly fragmented with a concentration ratio of CR10 from 58% in 2003 (own calculation, based on MIV, 2004). For the last few years price competition has been increasing through changes in agricultural policy and a higher number of private labels, launched by growing numbers of discount retailers. In the year 2002 nearly all German dairies began with the introduction of a joint approach to a farmer oriented quality management system. Because of the high level of consumer confidence in the quality of milk, the industry was relatively late in introducing a third party control structure.

Coming into this equation, the so called 'Qualitaetsmanagement Milch' (QM Milk) should ensure the basic requirements of milk, fodder, documentation, environment and animal welfare, enclosing the farmer in a chain wide control concept. Up to this, no independent verification was carried out on farms. QM Milk is an attempt to enhance the established statistical raw material control system through a process management approach. However, the introduction of QM Milk has been postponed due to a low level of acceptance by farmers.

Although QM Milk is easy to carry out for most farmers as it is based on existing legal requirements, there is a lot of resistance. A nationwide farmer's association is resisting the concept, threatening milk boycotts. In a recent study (Jahn and Spiller, 2005) it could be demonstrated that most dairy farmers are not convinced of the advantages QM Milk will bring in addition to the existing analytical product control of the dairy. The authors show that most farmers do not see quality improvements but bureaucratic 'paper control'. For that reason the amount of intrinsic motivation is not enhanced. Even the more successful farmers, with a high degree of commitment to quality production, tend to reject the new concept.

Jahn and Spiller (2005) also reveal great differences in the ability of the dairies in communicating the necessity of QM Milk. Some processors fail completely in convincing their farmers of the possible advantages and the general trend towards third party control. This demonstrates how critical a good relationship is in chain wide quality management. For that reason the following study tries to analyse the antecedents of farmer-processor relationships.

3. Supplier relationship quality

3.1. Research streams

The first attempts to measure relationship quality go back to the marketing channel literature. With a special focus on manufacturer-retailer dyads, channel research highlights conflict between supplier and retailer, the emphasis was placed on exploring sick rather than healthy relationships (Young and Wilkinson, 1989: 109). In the nineties the focus changed towards co-operation in general and trust and commitment in particular (Morgan and Hunt, 1994). All in all, marketing channel research highlights the potential benefits of co-operative relationships, building the theoretical background for the huge efforts manufactures and retailers undertake to improve collaboration through, for example 'Efficient Consumer Response'. Relationship quality is also analysed in the sales and service management literature, where a people-based approach dominate and relationship quality is defined as the ability of a customer to rely on the sales person's integrity (Crosby *et al.*, 1990; Lagace *et al.*, 1991).

Another research stream depends on industrial marketing. The well-known Scandinavian approach (IMP Group, 1982) concentrated on long-term network relationships in business-to-business marketing. In initial studies, the focus was on the ongoing interaction process, in which single transactions are embedded. Later on the attention shifted towards a network approach (Hakansson and Snehota, 1995). Based on the IMP interaction model, Woo and Ennew (2004) offer a conceptualisation of business-to-business relationship quality and explain the connection between relationship quality and service quality.

Similarly, in consumer marketing, the relationship perspective attracted great interest which led to new business approaches, such as customer relationship management (CRM). The measurement of customer satisfaction was first conceptualised through the work of Parasuraman *et al.* (1988), who developed the SERVQUAL-scale for customer satisfaction analysis. A review of selected approaches explaining relationship quality in consumer behaviour was presented by Hennig-Thurau *et al.* (2002). The ongoing work reveals that satisfaction is a necessary, but not a sufficient requirement for customer bonding.

Previous studies on relationship quality in the context of procurement were undertaken by Leuthesser (1997), Dorsch *et al.* (1998) and Naudé and Buttle (2000). Dorsch *et al.* (1998) highlighted the relevance of trust, satisfaction, commitment, opportunism, customer satisfaction, and ethical profile. Naudé and Buttle (2000) proposed five attributes of relationship quality: trust, power, integration, mutual understanding of needs, and profit. A measurement scale of relationship quality was developed by Lages *et al.* (2005), who focused on the relationship between exporting firms and the importer. Their RELQUAL scale tried to define elements of successful export relationships by questioning export managers of UK companies. They revealed the amount of information sharing, communication quality, long-term orientation and satisfaction with the relationship as elements of the multi-item scale. RELQUAL is positively related to export performance.

In the agribusiness literature, a growing number of studies dealing with relationship management can be found. Most studies discuss selected constructs, such as trust, power, and dependency (Morgan and Hunt, 1994). Means of collaboration beyond contracts and vertical integration are revealed in Hobbs and Young (2001). Several empirical studies were conducted by Batt and co-authors. The role of trust in supply chains is discussed in Batt and Rexha (1999) and Batt (2003). Batt and Wilson (2001) studied the relationship between grape growers and wine makers in Western Australia. Gusti *et al.* (2004) conducted a similar survey with small farmers and their intermediaries in Indonesia. In all studies Batt *et al.* identified factors affecting the respective buyer-seller relationship and accentuated the role of collaboration and trust in various agricultural supply chains (Batt, 2003; Batt and Purchase, 2004). Hansen *et al.* (2002) suggested that trust between members of a co-op and the co-op management is an important variable to enhance group cohesion. Matanda and Schroder (2004) analysed buyer-seller-relationships in Zimbabwean horticulture with a comprehensive behavioural approach, focussing on a broad model of relational constructs. They elaborated differences between small and large primary producers in terms of satisfaction, dependency, conflict frequency, long-term orientation, commitment and social bonding. Clare *et al.* (2005) evaluated the relationship between farmers, livestock buyers and slaughterhouses in the New Zealand red meat industry, and found that buyers have a far closer relationship with farmer-suppliers than both groups have with slaughterhouses.

On the whole, the literature review provides evidence for the growing importance of a behavioural approach towards the relationship between primary producers and processors in the agribusiness. In the following part, elements of the research streams described are used to develop an approach for the measurement of relationship quality between farmers and processors. Thus, the impact of relationship quality on quality management is analysed.

3.2. A measurement of supplier relationship quality

The starting point for the following discussion is the assumption that quality management systems in the agri-chain can be improved by enhancing the relationship between the farmer and the processor. Therefore the next step is working out a valid and reliable concept of measuring this supplier-customer dyad. Because of the unequal nature of this relationship with one processor, dealing with numerous small farmer-suppliers, the empirical analysis is concentrated on the farmer's perception.

To develop a scale of relationship quality, the experience of the research streams described above is transferred to the industry researched. In addition, qualitative interviews with managers and farmers were carried out to identify more sector specific topics. After these considerations, a first draft of items was discussed with managers of an agribusiness company (sugar industry). This first draft was again tested with 271 suppliers of sugar processors in Germany (Gerlach *et al.*, 2005). The study already showed important similarities to following surveys, especially concerning the importance of farmer orientation and the low significance of prices for supplier satisfaction. After reviewing items, wording, testing validity, clarity and redundancy, the second step was to apply this study to the German dairy industry.

In line with past research (Smith, 1998; Lages *et al.*, 2005), we define relationship quality as a higher-order concept, composed of three different, though related elements. Relationship quality is the overall assessment of the strength of a business relationship, combining satisfaction, trust and commitment. A basic element for nearly all relationship models is the outstanding importance of customer satisfaction or, in our case, supplier satisfaction (Gerlach *et al.*, 2004: 9). As with consumer marketing, satisfaction reflects comprehensive experience with a business partner as a necessary but not sufficient condition for an ongoing relationship (Van Weele, 2002: 165). In addition to own experiences, information from other business partners and other cues build the basis of satisfaction (Homburg and Stock, 2001: 20). According to the disconfirmation model, satisfaction in supplier relationship quality is the result of a comparison between a buyer's performance and the supplier's expectations.

Other elements of relationship quality are trust and commitment (Crosby *et al.*, 1990; Bejou *et al.*, 1996; Naudé and Buttle, 2000; Batt, 2003): Trust is defined as '[...] a willingness to rely on an exchange partner in whom one has confidence' (Moorman *et al.*, 1993: 82). It is combined with the belief that others will not act to exploit one's vulnerabilities (Hansen *et al.*, 2002: 42). Trust reduces opportunistic behaviour and transaction costs (Ganesan, 1994; Doney and Cannon, 1997; Batt and Rexha, 1999) thus it is relevant if information asymmetries are present.

Commitment '[...] is an implicit or explicit pledge of relational continuity between exchange partners' (Dwyer *et al.*, 1987: 19). It is seen as an outcome of trust and defined variously in the literature as the belief of a supplier that the relationship with a processor is so important that it warrants maximum effort to maintain it even if problems occur (Morgan and Hunt, 1994). Satisfaction, trust, and commitment are common variables in relationship quality research. However there is a lack of consensus about additional parameters. Other authors add variables, such as communication quality or information sharing (Naudé and Buttle, 2000; Lages *et al.*, 2005). From our viewpoint, these are not integral elements of relationship strength, but management instruments and thus determinants of relationship quality and not ingredients. In the following section possible determinants of relationship quality are presented.

To improve relationship quality, it is necessary to identify the crucial influencing variables. Price satisfaction appears to be particularly important. Price satisfaction refers to several relationship studies that state that the economic outcome is important for the evaluation of that relationship (Jaervelin, 2001) and, thereby, positively affects the development of trust. The parameters include short- and long-term price satisfaction as well as relative price satisfaction when comparing the own price received with the price paid by other dairies/slaughterhouses. Furthermore, often used variables in relationship management are shared values (Heide and John, 1992; Morgan and Hunt, 1994), perceived performance of the partner, communication quality and quantity (Matanda and Schroder, 2004: 534) and friendships between partners (Wilson, 1995: 339; Rodriguez and Wilson, 2002: 55). As the dairy industry is facing an increasing concentration at the processor and retail level, the effects of power asymmetries and coercion also have to be considered as important for relationship perception (Anderson and Narus, 1990; Dwyer *et al.*, 1987).

One particular factor influencing supplier relationship quality in agribusiness is the processor's orientation towards farming. Due to the importance of co-operatives, farmers are used to having some influence on the manufacturer (Gerlach *et al.*, 2005). Possible elements of farmer orientation are farmers' influence, the processor's proximity to agricultural problems, competence of farm advisory service etc. The parameter farmer orientation is also closely related to 'the degree to which partners share goals that can only be accomplished through joint action and the maintenance of the relationship' (Wilson, 1995: 338).

The economic success of a farmer ultimately depends on the performance of the processor. This is especially true for those farmers who are shareholders and bound by contracts. Therefore perceived management competences are probably an important factor for assessing the buyer. The management strategy of the manufacturer should be comprehensible for the farmers because their prices depend on it.

Aiming at more than a simple 'feel-good-approach' (Dyer and Chu, 2000), the potential outcomes of an improved relationship with the suppliers also have to be measured. As output variables, the intention to switch the buyer, recommendation of the farmer, and the absence of countervailing power are defined. Regarding quality management, we measure the impact of relationship quality on the acceptance of the recently introduced QM-concept. Figure 1 describes the revealed elements of the farmer-processor relationship.

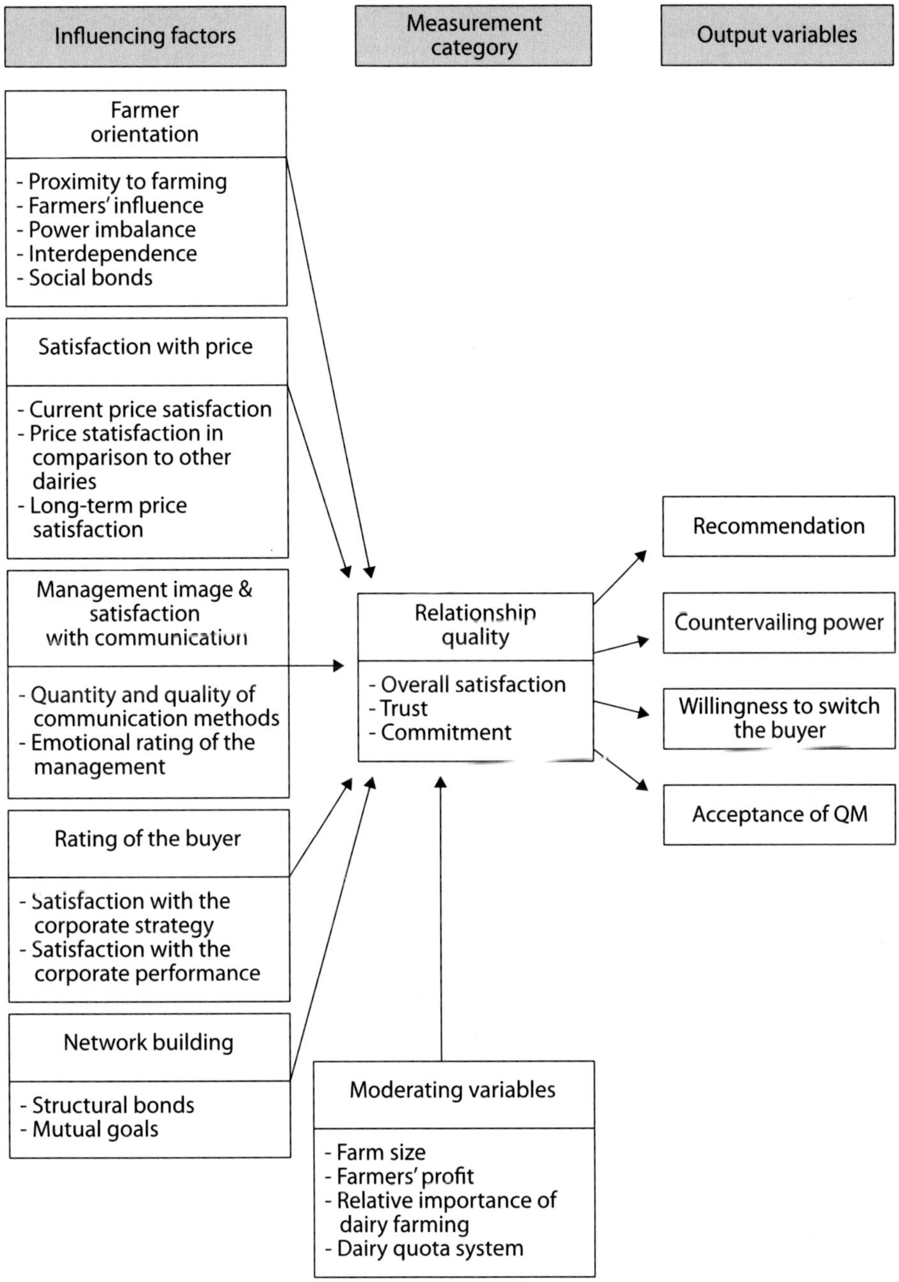

Figure 1. Hypothesised model of relationship quality.

4. The relevance of economic and behavioural factors on the quality of business relationships

4.1. Research design

The analysis detailed in this paper was conducted on data obtained from a sample of 209 dairy farmers in Northern Germany. In summer 2004 the farmers were questioned in face-to-face interviews. The average interview took about 45 minutes. The sample includes larger sized farms (97 ha, 73 cows per farm) than the average ones in Germany (35 cows per farm, Eurostat 2004). The average milk quota per farm is comparatively high (600,000 kg p. a.) and the milk quality is much higher than the requested minimum standards of the official regulation.

The respondents are on average 42 years old and well-educated (only 3.4% have no agricultural education). The majority of these farmers (60.6%) are active members in agricultural associations. Therefore, the sample includes presumably more farmers with firm viewpoints. 84.1% of the respondents intend to continue milk production in the future. The others are not sure, but only two of them have so far decided to give up their farm.

The farmers interviewed delivered their milk to 22 different dairies of which 'Nordmilch e.G. (Nordmilch)' receives 35.4%, 'Humana Milchunion e.G. (Humana)' 22.5% and 'Campina GmbH Deutschland (Campina)' 14.8%. Humana is the biggest dairy in Germany (sales volume: 2,444 Mio. €) followed by Nordmilch. Campina is in fourth place (MIV 2004). The remaining farmers (27.3%) deliver to smaller companies.

4.2. Results

The following chapter provides a short overview on the farmers' responses regarding the above-mentioned parameters, which have been modelled with 7-point Likert scales. Overall satisfaction with the relationship to the dairy industry is rather mixed, hence its position in mid-range. There are clear deficits in trust and commitment. It is notable that farmer orientation and common purpose is perceived very low. From the farmers' point of view, most of the cooperative industries lose their agrarian basis. Consequently, the willingness to change the buyer is 68.8% if another dairy pays a higher milk price. On the whole, the strength of the relationship between farmer and dairy is low and unstable. However, the high standard deviations show different positions by the farmers.

A comparison between the different dairies shows clear differences (Figure 2). In general, suppliers of Nordmilch and Campina show more dissatisfaction. Some difficulties with recent mergers may provide a possible explanation. Smaller co-operatives and the Humana Milchunion show higher supplier commitments. Research on co-operatives provides another possible explanation besides management factors: in small groups social capital can be built easier. It is also interesting to note, that suppliers of small private dairies show a higher willingness to change. Hence, a tendency towards spot markets can be identified.

4.3. Factors influencing the quality of the business relationship

The described model of business relationship quality was verified using factor analysis. Some parameters had to be modified: The factor 'farmer orientation' comprises items which primarily are related to agrian proximity influence of the farmers, power asymmetry and the perception of trade offs between farmers and the dairy. The items constituting global satisfaction, trust and

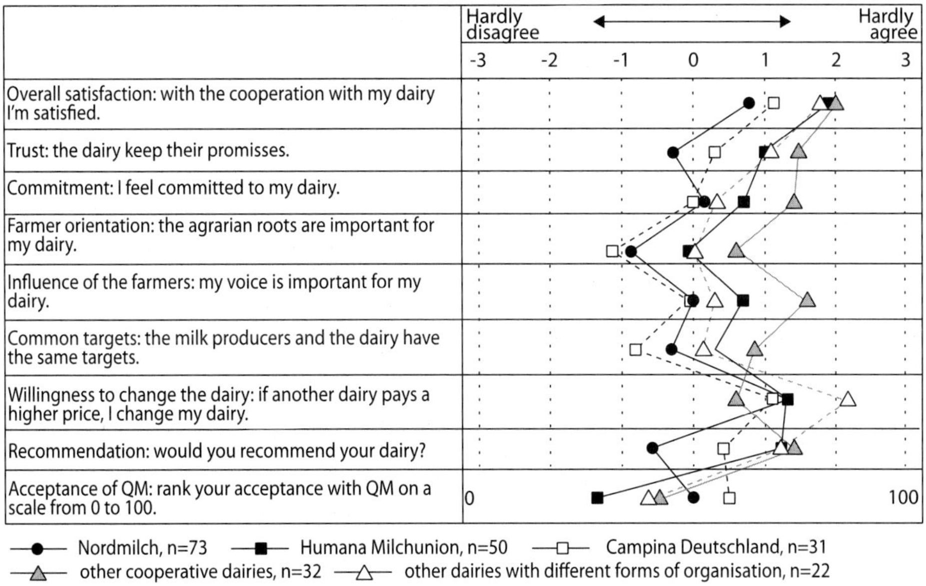

Figure 2. Different relationship qualities based on the survey. Significance p≤ 0.000.

commitment were reduced to the factor 'quality of business relationship'. The factor 'willingness to change' however was maintained. Table 1 shows Cronbachs Alpha as a measure of the reliability of the extracted factors.

The impact of the factors on the quality of the business relationship was then measured using linear regression analysis (see Table 2). In addition some single statements were entered, i.e. those asking for satisfaction with business performance. The quality of the business relationship (satisfaction, trust, commitment) is to a great extent determined (adjusted $R^2 = 0.681$, $F = 86.365$, $p \leq 0.001$) through the factors farmer orientation (1), management image and satisfaction with communication (2), price satisfaction (3), business performance (4), and the relative importance of dairy farming on the farmers' income (5). All other factors as well as other moderating variables such as farm size, the farmers' education, and their willingness to invest or engage in agrarian organisations have no significant impact on the perceived quality of the business relationship.

Table 1. Reliability.

Factor	Alpha
Relationship quality	0.840
Farmer orientation	0.886
Management image and satisfaction with communication	0.861
Price satisfaction	0.689
Willingness to switch the dairy (output variable)	0.674

Table 2. Regression model to explain the quality of the business relationship.

Independent Variables	Standardised Beta Coefficient	t-Value
Farmer orientation	0.432	6.754[a]
Management image and satisfaction with communication	0.247	3.943[a]
Price satisfaction	0.199	4.448[a]
Performance satisfaction (one Item)	0.193	4.521[a]
Relative importance of dairy farming (Moderating Variable)	0.068	1.658

Depending variable: Relationship quality; Adj. $R^2 = 0.681$; F-Value = 86.365[a].
[a]Significance $p \leq 0.001$.

Surprisingly, price satisfaction does not have the strongest impact. Emotional criteria and the perception of management competence are much more important. The understanding of farmers' problems as well as communication with the supplier are much more relevant than the price perception. When farmers believe that a dairy is more successful than others, the quality of the relationship is better from the farmer's view.

4.4. Implications of supplier relationship management towards food quality

Correlation analysis was used to check the correlation between relationship quality and different output variables. The coefficients indicate a highly significant link between business relationship quality and the output variables - willingness to change, building countervailing power and recommendation - which are from the dairy point of view very important (see Table 3). In a good business relationship, willingness to change is lower ($r = 0.464$, $p \leq 0.001$), just as it is for the willingness to build countervailing power ($r = 0.233$, $p \leq 0.001$). A means comparison between those who are members of the Bund Deutscher Milchviehhalter (an organisation, which was founded as a cooperative against the market power of dairies) and those who are not, reveals highly significant differences between these groups concerning the perception of relationship quality and the willingness to build countervailing power. Another finding is that dairies with a high relationship quality are recommended by their suppliers ($r = 0.692$, $p \leq 0.001$).

There is no significant correlation between relationship quality and the acceptance of the QM-Milk system. A possible reason for this might be that the dairy Humana Milchunion does not participate in the QM Milk system and campaigns against it. Thus the suppliers of Humana do not

Table 3: Correlation matrix (output variables)

	1	2	3	4	5
1. Relationship quality	1				
2. Willingness to switch the dairy	-0.464[a]	1			
3. Countervailing power	-0.233[a]	0.446[a]	1		
4. Recommendation	0.692[a]	-0.459[a]	-0.160[c]	1	
5. Acceptance of QM	0.104	-0.165[c]	-0.082	0.031	1

[a,b,c]Significance $p \leq 0.001$; $p \leq 0.01$; $p \leq 0.05$.

feel to benefit from a potential involvement in QM milk. If the suppliers of Humana Milchunion are eliminated from the correlation analysis, a low significant correlation between the acceptance of QM and the relationship quality becomes apparent (see Table 4).

Altogether, the hypothesis that high business relationship quality encourages the suppliers' commitment can be approved. The weak correlation between relationship quality and the acceptance of QM demonstrates that the direct influence on quality motivation is limited. This may be due to the high official quality requirements in the German dairy sector which seems to be sufficient for most of the farmers.

5. Conclusions

The survey reveals a difficult business relationship. Dairy farmers do not perceive co-operatives as 'their' companies. For larger dairies with more than 10,000 suppliers (e.g. Nordmilch or Humana Milchunion) this result indicates new challenges. It is problematic that more powerful farmers with large farms, who are also opinion leaders, are significantly more unsatisfied. Supplier relationship management and continuous analysis of the business relationship are important for those companies who don't operate on a spot-market. To improve the relationship between milk producers and dairies, the farmer orientation and the communication of the business performance is a central point. Dairies should provide better supplier communication and in cooperatives a heightened investor relationship should be a prerequisite.

Supplier relationship management is a procurement tool, which focuses on supplier development and enhancement of business relationships (Stoelzle and Heusler, 2003; Gerlach *et al.*, 2004). Better relationships with suppliers offer a long list of potential advantages (Table 4). In industries like the automotive sector supplier development plays an important role in the attempt to build stable supplier networks which could work as a competitive advantage.

Potential advantages of a closed supplier relationship in the agribusiness:
- Higher satisfaction of both parties.
- Long term orientation for the dairy through less willingness to change on the part of the farmers.
- Lower intensity of conflict.
- Flexibility and speed advantages through better information flow.
- Reduction of transaction costs through stable relationships.
- Reduction of process costs through compatible data interfaces.
- Better quality management through trustworthy information flow.
- Commitment enhances a willingness to invest on farmers side.

Table 4. Correlation matrix without Humana Milchunion suppliers (output variables).

	1	2	3	4	5
1. Relationship quality	1				
2. Willingness to switch the dairy	-0.499[a]	1			
3. Countervailing power	-0.225[b]	0.435[a]	1		
4. Recommendation	0.689[a]	-0.473[a]	-0.139	1	
5. Acceptance of QM	0.192[c]	-0.177[c]	-0.056	0.031	1

[a,b,c]Significance $p \leq 0.001$; $p \leq 0.01$; $p \leq 0.05$.

From a dairies viewpoint, supplier relationship quality aims at gaining better suppliers than the competitors in order to enhance the whole supply chain. This does not solve distributional conflicts about farm prices, but opens opportunities for a more functional way of managing trade offs in the supply chain. Furthermore, the influence of price fairness in the relationship between processors and farmers may be overestimated, because of the misjudged importance of weak factors like trust and commitment. For quality management this provides new opportunities for motivating suppliers, enhancing intrinsic motivation instead of external pressure.

References

Anderson, J.C. and J.A. Narus, 1990. A model of distributor firm and manufacturer firm working partnerships. Journal of Marketing 54, 42-58.

Batt, P.J. and N. Rexha, 1999. Building trust in agribusiness supply chains: A conceptual model of buyer-seller-relationships in the seed potato industry in Asia. Journal of International Food & Agribusiness Marketing 11, 1-17.

Batt, P.J. and H. Wilson, 2001. Exploring the nature of long-term buyer-seller relationships in the Western Australian wine industry. Australia New Zealand Wine Industry Journal 16, 87-96.

Batt, P.J., 2003. Building Trust between growers and market agents. Supply Chain Management 8, 65-78.

Batt, P.J. and S. Purchase, 2004. Managing collaboration within networks and relationships. Industrial Marketing Management 33, 169-174.

Bejou, D., B. Wray and T.N. Ingram, 1996. Determinants of relationship quality: An artificial neural network analysis. Journal of Business Research 36, 137-143.

Clare, B.G., J.I. Reid and N.M. Shadbolt, 2005. Supply base relationships in the New Zealand red meat industry: A case study. Paper presented to 15th annual world forum, symposium and case conference of the International Food and Agribusiness Management Association, Chicago, June.

Crosby, L.A., R.K. Evans and D. Cowles,., 1990. Relationship quality in services selling: An interpersonal influence perspective. Journal of Marketing 52, 21-34.

Doney, P.M. and J.P. Cannon, 1997. An examination of the nature of trust in buyer-seller-relationships. Journal of Marketing 61, 35-51.

Dorsch, M.J., S.R. Swanson and S.W. Kelly, 1998. The role of relationship quality in the stratification of vendors as perceived by customers. Journal of the Academy of Marketing Science 26, 128-142.

Dwyer, F.R., P.H. Schurr and S. Oh, 1987. Developing buyer-seller relationships. Journal of Marketing 51, 11-27.

Dyer, J.H. and W. Chu, 2000. The determinants of trust in supplier-automaker relationships in the US, Japan, and Korea. Journal of International Business Studies 31, 259-285.

Frey, B.S. and R. Jegen, 2001. Motivation crowding theory. Journal of Economic Surveys 15, 589-611.

Frey, B.S. and F. Oberholzer-Gee, 1997. The cost of price incentives: An empirical analysis of motivation crowding-out. American Economic Review 87, 746-755.

Ganesan, S., 1994. Determinants of long-term orientation in buyer-seller relationships. Journal of Marketing 58, 1-19.

Gerlach, S., B. Köhler, A. Spiller and C. Wocken, 2004. Supplier Relationship Management im Agribusiness: Ein Konzept zur Messung der Geschaeftsbeziehungsqualitaet, Working Paper Nr. 0406 des Instituts fuer Agraroekonomie, Georg-August-Universitaet Goettingen.

Gerlach, S., B. Köhler and A. Spiller, 2005. Landwirte als Lieferanten und Unternehmenseigner: Zum Management von Supplier- und Investor-Relations im Agribusiness. In: Darnhofer, I., S. Poechtrager and E. Schmid (eds.), Jahrbuch der oesterreichischen Gesellschaft fuer Agraroekonomie. Wien, pp. 221-234.

Gusti, I., L.P. Tanaya, M.J. McGregor and P.J. Batt, 2004. Factors affecting supply chain participants on choosing their preferred trading partners. Agriculture Congress 2004, Innovation Towards Modernised Agriculture, 4-7 October, Mines Resort City, University Putra Malaysia.

Hakansson, H. and I. Snehota, 1995. Developing relationships in business networks. International Thomson Business Press.

Han, S.-L., D.T. Wilson and S.P. Dant, 1993. Buyer-seller relationships today. Industrial Marketing Management 22, 331-338.

Hansen, M.H., Morrow, J.L. and J.C. Batista, 2002. The impact of trust on cooperative membership retention, performance, and satisfaction: an exploratory study. International Food and Agribusiness Management Review 5, 41-59.

Heide, J.B. and G. John, 1992. Do norms matter in marketing relationships?. Journal of Marketing 56, 32-44.

Hennig-Thurau, T., K.P. Gwinner and D.D. Gremler, 2002. Understanding relationship marketing outcomes: An integration of relational benefits and relationship quality. Journal of Service Research 4, 230-247.

Hobbs, J.E. and L.M. Young, 2001. Vertical linkages in agri-food supply chains in Canada and the United States, Research and Analysis Directorate. Strategic Policy Branch Agriculture and Agri-Food Canada, June 2001.

Homburg, C. and R. Stock 2001. Theoretische Perspektiven zur Kundenzufriedenheit. In: Homburg, C. (ed.), Kundenzufriedenheit, 4th ed. Wiesbaden, pp. 17-50.

IMP Group, 1982. An interaction approach. In: H. Hakansson (ed.), International marketing and purchasing of industrial goods: An interaction approach. John Wiley and Sons, Chichester.

Jaervelin, A.-M., 2001. Evaluation of relationship quality in business relationships. Dissertation, Tampere.

Jahn, G., M. Schramm and A. Spiller, 2004. The trade-off between generality and effectiveness in certification systems: A conceptual framework. In: H.J. Bremmers, S.W.F. Omta, J.H. Trienekens and E.F.M. Wubben (eds.), Dynamics in Chains and Networks. Conference Proceedings of the sixth International Conference on Chain and Network Management in Agribusiness and the Food Industry. Wageningen Academic Publishers, pp. 351-358.

Jahn, G., M. Schramm and A. Spiller, 2005. The reliability of certification: Quality labels as a consumer policy tool. Journal of Consumer Policy 28, 53-74.

Jahn, G. and A. Spiller, 2005. Acceptance of a processor-driven quality management system by dairy farmers: A structural equation model. Conference proceeding presented at 92nd EAAE Seminar on 'Quality Management and Quality Assurance in Food Chains', 2-4 March Goettingen, Germany.

Lagace, R.R., R. Dahlstrom and J.B. Gassenheimer, 1991. The relevance of ethical salesperson behaviour on relationship quality in the pharmaceutical industry. Sales Management 4, 39-47.

Lages, C., C.R. Lages and F.L. Lages, 2005. The RELQUAL scale: a measure of relationship quality in export market ventures. Journal of Business Research 58, 1040-1048.

Leuthesser, L., 1997. Supplier relational behaviour: an empirical assessment. Industrial Marketing Management 26, 245-524.

Matanda, M.J. and B. Schroder, 2004. Business-to business relationships by categories of suppliers in the marketing channel. In: H.J. Bremmers, S.W.F. Omta, J.H. Trienekens and E.F.M. Wubben (eds.), Dynamics in Chains and Networks. Conference Proceedings of the sixth International Conference on Chain and Network Management in Agribusiness and the Food Industry. Wageningen Academic Publishers, pp. 532-537.

MIV, 2004. Die groeßten deutschen Molkereien 2004, URL: http://www.milch-markt.de/de/teaser/die_groessten_deutschen_molkereien_ 2004/, retrieval date: 29.09.04.

Moorman, C., R. Deshpandé and G. Zaltman, 1993. Factors affecting trust in market research Relationships. Journal of Marketing 57, 81-101.

Morgan, R.M. and S.D. Hunt, 1994. The commitment-trust theory of relationship marketing. Journal of Marketing 58, 20-38.

Naudé, P. and F. Buttle, 2000. Assessing relationship quality. Industrial Marketing Management 29, 351-361.

Parasuraman, A., V.A. Zeithaml en L.L. Berry, 1988. SERVQUAL: A multiple-item scale for measuring consumer perceptions of service quality. Journal of Retailing 64, 12-40.

Rodríguez, C.M. and D.T. Wilson, 2002. Relationship bonding and trust as a foundation for commitment in U.S.-Mexican strategic alliances: A structural equation modeling approach. Journal of International Marketing 10, 53-76.

Smith, J.B., 1998. Buyer-seller relationships: similarity, relationship management, and quality. Psychology and Marketing 15, 3-21.

Stoelzle, W. and F. Heusler, 2003. Supplier relationship management. In: Bogaschewsky, R. and U. Goetze, (eds.), Management und Controlling von Einkauf und Logistik. Gernsbach, pp. 167-194.

Theuvsen, L., 2005. Quality assurance in the agrofood sector: An organisational-sociological perspective. Schriften der Gesellschaft fuer Wirtschafts- und Sozialwissenschaften des Landbaus 40, 173-181.

Van Weele, A.J., 2002. Purchasing and supply chain management: Analysis, planning and practice. 3rd ed., Padstow, Cornwall.

Wilson, D.T., 1995. An integrated model of buyer-seller relationships. Journal of the Academy of Marketing Science 23, 335-345.

Woo, K. and C.T. Ennew, 2004. Business-to-business relationship quality: An IMP interaction-based conceptualization and measurement. European Journal of Marketing 38, 1252-1271.

Young, L.C. and I.F. Wilkinson, 1989. The role of trust and co-operation in marketing channels: A preliminary study. European Journal of Marketing 23, 109-122.

Trust and e-commerce in the agrifood industry: configuration of a trust environment for e-commerce activities

Melanie Fritz[1], Tobias Hausen[1], Gerhard Schiefer[1] and Maurizio Canavari[2]
[1]University of Bonn, Meckenheimer Allee 174, 53115 Bonn, Germany; m.fritz@uni-bonn.de
[2]University of Bologna, Italy

Abstract

Opportunities coming from electronic commerce provide interesting support options for the agrifood sector. However, due to the product specific information asymmetry in the agrifood sector, the anonymity of the medium creates a lack of trust increasing transaction costs to engage in e-commerce. This paper develops a trust model for electronic commerce in the agrifood industry allowing for the analysis of trust determinants in traditional agrifood transaction relationships and their transfer to appropriate trust determinants in the e-commerce environment. First experimental results validating the suitability of the model to derive appropriate electronic trust generating elements for a given agrifood transaction environment are presented.

Keywords: electronic commerce, trust, information asymmetry, experiments

1. Introduction

The enlargement of the European Union opens new markets at the supply and vendor side to the European agrifood sector. Opportunities coming from electronic commerce such as more transparent markets with easier access to information about possible transaction partners and their products provide interesting support options for establishing business relationships with other EU countries. However, the agrifood sector deals with products where a large amount of information asymmetry exists between transaction partners. Many characteristics of food products may only be analysed after use (experience characteristics), others even cannot be examined at all (credence characteristics). This makes quality management, quality control, contracts and trust between transaction partners to crucial elements in agrifood transactions (see Schiefer and Rickert, 2004; Wilson and Kennedy, 1999).

Electronic commerce where physical product inspection and direct contact between transaction partners is not possible has often been criticised as too anonymous for agrifood transactions. This risk and uncertainty creating a new and additional information asymmetry created a lack of trust in electronic commerce in the agrifood sector and transaction costs were perceived as too high for agrifood companies to engage in electronic commerce. The benefits from electronic commerce were considered less important than the obstacles.

We argue that electronic commerce environments in the agrifood industry require an appropriate, trust generating configuration to lower these initial transaction costs coming from the medium's anonymity and information asymmetry. This paper deals with the problem how trust may be generated in an electronic commerce environment in the agrifood industry to lower the initial transaction costs impeding agrifood companies to engage in e-commerce transactions. As trust is generated in individuals *ex post* over time after positive experiences, alternative ways to anticipate trust generation in e-commerce suitable to the transaction situation in agrifood industries are necessary.

The objective of this paper is the development of a trust model for electronic commerce in the agrifood industry, the analysis of trust determinants in traditional agrifood transaction relationships and their transfer to appropriate trust determinants in the electronic commerce environment. The paper first outlines theoretical concepts dealing with the role of trust in business relationships. The following section derives and operationalises a trust model for electronic commerce in the agrifood industry. The paper concludes with the presentation of first experimental results validating the suitability of the model to derive appropriate electronic trust generating elements for a given agrifood transaction environment.

2. Theoretical background: role of trust in business relationships

The exchange of goods between businesses builds on a certain level of trust between the business partners (Fynes *et al.*, 2001). Literature provides many definitions of trust in business relationships, but their essence is that trust is one party's belief that the other party will not exploit its vulnerabilities (Barney and Hansen, 1995). The relationship between buyers and sellers is subject to research areas such as transaction cost economics, information economics and socio-economics (for an overview see Hausen, 2005). Transaction cost economics claims that every transaction between buyers and sellers creates transaction costs due to costs for coordination, information search, monitoring, and controlling that the transaction goes on as predefined (Williamson, 1985). Transaction costs not only include quantifiable costs, but also 'costs as disadvantages' (Picot and Dietl, 1990). Trust in relationships between buyers and sellers lowers transaction costs by reducing the efforts for contracting and the costs for control and monitoring. As a consequence, the existence of trust in a buyer-seller relationship complements incomplete contracts. This is why trust in buyer-seller relationships is often considered as an economic asset creating value (Dyer and Chu, 2000; Wilson and Kennedy, 1999).

Information economics (Nelson, 1970; Akerlof, 1970) analyses the impact of an asymmetric information level of buyers and sellers and its effects on market performance. Information asymmetry is typical for goods where experience and credence characteristics are predominant such as in agrifood products. To overcome information asymmetry, information screening and signaling together with the creation of long-term business relationships is necessary. Long-term business relationships build on experience as alternative to information search (Weiber and Adler, 1995). Past experiences and interaction create trust between business partners (Anderson and Weitz, 1989).

Socio-economics (Granovetter, 1985; Etzioni, 1988) analyses the influence of social networks, the social environment and cultural rules on the behavior of market participants. Trust and 'social embeddedness' in a society's values and moral ideas are considered as determinants to business decisions. It is argued that direct personal experiences and social ties play a more important role than indirect reputation. Different cultural backgrounds and habits influence the interaction between businesses (Hofstede *et al.*, 2002).

3. Trust model for electronic commerce in the agrifood industry

Exchange relationships between businesses may be transformed to electronic exchange relationships by introducing electronic marketplaces or electronic trade platforms (Malone *et al.*, 1987; Bakos, 1991). Electronic marketplaces are intermediaries between buyers and sellers aiming at making transaction processes more efficient as opposed to traditional transaction processes. They intend to reduce transaction costs by supporting information and communication processes between businesses. However, despite those advantages electronic marketplaces - including those in the agrifood sector - faced difficulties to enter in buyer-seller relationships (Fritz *et al.*, 2004). One problem may have been the lack of adapting and embedding the marketplaces' services to the demands of existing buyer-seller relationships (see Hausen, 2005 for 'embedded electronic commerce'). Another problem may have been the trust problem related with the anonymity of the electronic commerce environments and the lack of experience how to deal with them (Franke Kleist, 2004). This problem is of particular significance in the agrifood sector with the specific characteristics of food products.

The generation of trust in electronic commerce environments to overcome information asymmetry related with its anonymity is often related to elements like reputation systems where buyers and sellers can rate one another based on their experiences (Keser, 2003; Bolton *et al.*, 2004). Trust signs such as VeriSign signaling a positive third-party ranking are also frequent means to create trust in e-commerce. Others specify the technical infrastructure of an e-commerce presence with, e.g., SSL-encryption as important trust generating element (Franke Kleist, 2004; Ratnasingam, 2004). However, existing work on trust in electronic commerce does not consider the particular transaction situation in the agrifood. There is no consistent model allowing for the analysis and transfer of a traditional trust structure in agrifood business relationships to an electronic transaction environment.

The model for trust in electronic commerce in the agrifood sector presented in this paper builds on a basic model for designing trustworthy electronic commerce environments (Tan and Thoen, 2001). This basic model presumes that an individual's level of trust in a transaction is determined by his trust in the transaction party and the control mechanisms, potential gains from the transaction, the risk involved and the individual's risk attitude. Individuals would only engage in a transaction if the level of trust exceeds their personal threshold of trust. However, the basic model does not specify the dynamic interrelation between the risks and benefits in a transaction situation and the trust generation for the design of a trustworthy e-commerce environment. In addition, the basic model does not provide a 'toolbox' of trust elements appropriate for e-commerce. This paper expands and adapts the trust model to the situation of electronic commerce in the agrifood industry, includes the dynamic interrelation between a transaction situation and trust generation and its influence on the level of trust in a transaction (see Figure 1), and provides trust generating elements for an e-commerce environment in the agrifood sector. The trust model allows for analysing the trust determinants and trust reasons creating a certain level of trust in a given 'traditional' transaction situation with a certain risk. The design of an appropriate electronic transaction environment for this given scenario requires the creation of appropriate trust generating reasons to reach at least the same trust level (see also Franke Kleist, 2004). In addition, the interrelation between trust generation and potential benefits and risks related with the altered, electronically supported transaction scenario must be taken into account.

3.1. Analysis of the trust determinants in agrifood e-commerce

In the center of the figure is the level of trust in a transaction, which is determined by five blocks addressing the other parties and control mechanisms involved in the transaction, the transaction situation's gains and risks, and the individual. Changes in the five blocks dynamically alter the level of trust in a transaction.

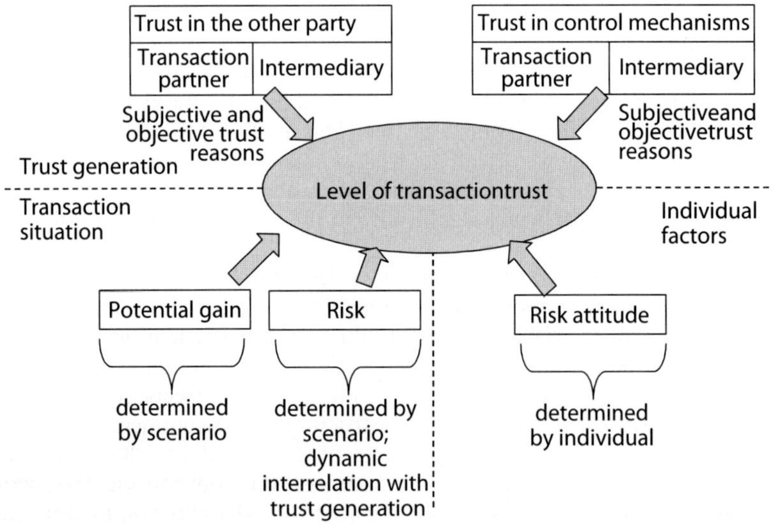

Figure 1. Trust model for electronic commerce in the agrifood sector (based on Tan and Thoen, 2001).

The upper half of Figure 1 shows the trust generating determinants, the trust in the transaction party and the trust in the control mechanisms provided to monitor the performance of a transaction. In transactions, trust and control complement each other in their influence of the transaction decision. The underlying control mechanisms must be trusted as well. When it comes to electronic commerce in the agrifood industry, trust in two transaction parties may be required, the actual transaction partner and an intermediary, e.g. an electronic trade platform. Accordingly, trust is required for the control mechanisms of the transaction partner himself and the intermediary. For both determinants, there exist subjective and objective reasons creating trust in the other party and the control mechanism. The trust generating determinants operationalised through appropriate trust reasons can be used to design trustworthy transaction environments for electronic commerce in the agrifood industry.

The lower left quarter of the figure shows the factors in a transaction situation particular to a specific transaction scenario determining the transaction trust level: the potential gain and benefit to be expected from a transaction and the risk involved. Electronically supported transactions have an efficiency potential offering a potential benefit as opposed to traditional transactions (Hausen, 2005). In addition, electronic commerce can open access to new markets. However, the perceived risk may lower the trust level too much to make companies engage in e-commerce. Risk in electronic commerce transactions in the agrifood industry is complex and arises from a number of sources. In general, every transaction with agrifood products contains risk as an information asymmetry exists between buyer and seller regarding the food product quality consisting of search, experience, and credence characteristics (Fearne *et al.*, 2001). In addition, the degree of risk in a transaction depends on the possible degree of damage and the negative consequences following a failed transaction. International transactions with partners from other countries - e.g. fruits and vegetables from Mediterranean EU countries to northern EU countries - always bear an additional risk. When it comes to e-commerce transactions with agrifood products, risk not only comes from an information asymmetry between buyer and seller regarding product quality, but also regarding the overall transaction settling including payment and shipping (also Franke Kleist, 2004).

In essence, it may be said that an e-commerce transaction situation in the agrifood industry on the one hand has a potential gain through efficiency potentials, but on the other hand adds an additional risk to agrifood transactions. As the trust model shows, high risk may rise the trust level needed for an individual to engage in a transaction. A high level of risk in a transaction situation may be compensated by an intense trust generation.

The lower right part of the figure shows the individual's risk attitude determining his personal trust threshold regarding a transaction decision. This means that personal attitudes towards the bearing of a risk influence the degree of trust and control necessary to exceed the trust threshold required for the transaction decision.

3.2. Elements for trust generation

To design a trustworthy e-commerce environment for transactions in the agrifood industry achieving a sufficient level of trust, appropriate reasons to generate trust in the other party and the control mechanisms must be derived. Reasons to trust a transaction party or a control mechanism overlooking a transaction may be grouped into objective and subjective trust reasons (Tan and Thoen, 2001; see Table 1). An objective reason to trust someone or something can be a commonly known and widely accepted social indicator such as a uniform or a certified procedure. Subjective reasons to trust are previous, positive personal experiences, the understanding of someone's objectives and capabilities or how something works, and communality. Communality refers to opinions of trusted community members and may be considered equivalent to the concept of reputation in socio-economics (Granovetter, 1985).

It is important to differentiate that trust arising from positive experiences comes *ex post* and develops over time. In contrast, understanding, communality and social indicators generate trust *ex ante* before a personal experience has been made. Electronic commerce is a rather new transaction situation where individuals have only little - or even negative - experiences. As a consequence, personal experience is not the appropriate trust reason to generate a trustworthy

Table 1. Trust reasons and trust sources (based on Tan and Thoen, 2001).

Trust reasons	Trust sources	Party trust	Control trust
Objective	Social indicators	uniforms	control procedure certified by trusted organisation
Subjective	Personal experience	previous, positive interactions ('normal' experience and transaction experience)	previous positive interactions with control procedure
	Understanding	understanding of others' goals, plans, capabilities	understanding how control procedure works
	Communality	opinions and trust of trusted community members regarding party	opinions and trust of trusted community members regarding control procedure

e-commerce environment supposed to attract businesses to start engaging in e-commerce. To increase the initial level of trust in e-commerce transactions, *ex ante* trust reasons must be focused. Personal experience with electronic commerce transactions - if positive - as an *ex post* trust reason comes into play over time. This shows that there is a dynamic shift in the relevance and appropriateness of trust reasons creating a trustworthy e-commerce environment. Figure 2 shows the shift in the relevance of *ex post* and *ex ante* trust reasons over time. The changing relevance of trust reasons also interferes with the particular transaction scenario and the risks and potential gain involved.

The *ex ante* trust reason 'understanding' requires from an individual time to analyse or study information about the transaction partner or the control mechanisms. According to the 'costs as disadvantages' view of transaction costs, understanding as trust reason increasing the trust level of a transaction involves transaction costs. As a consequence, to enable businesses to engage in electronic commerce transactions, the cost-benefit-ratio between transaction costs related to understanding the party or the control procedure and the potential gain coming from electronic commerce transactions such as access to new markets or supply sources must be less or equal one $C / B \leq 1$.

Table 2 analyses reasons to trust the transaction parties and control mechanisms in agrifood transactions. It shows the scenario of traditional direct transactions between two partners (transaction partner as trust source) and the scenario where an intermediary is included between the transactions partners (intermediary as trust source). Table 2 includes trust reasons for electronic transaction environments.

In traditional agrifood transactions, main reasons creating trust between two parties are past, personal experiences with a transaction party, the understanding of his actions, and the social indicators of a control procedure such as ISO 9000ff. together with an understanding of control procedures.

In contrast, in electronic transactions, personal experience with business partners is less intense than in traditional transactions. Trust generated by personal experience is - in particular in the beginning - only very small. As a consequence, trust reasons from other areas must be implemented to generate trust for electronic commerce transactions in the agrifood. Appropriate compensation must be provided by intensifying other trust reasons. In particular, trust reasons

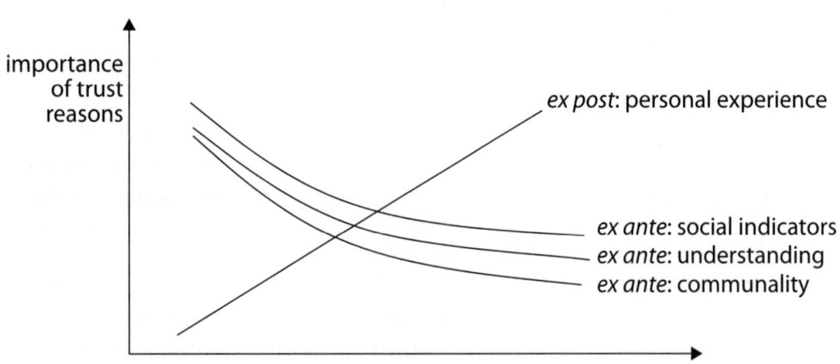

Figure 2. Change in importance of ex post *and* ex ante *trust reasons over time.*

Table 2. Trust reasons in traditional and electronic agrifood transactions.

Trust reasons	Trust sources	Party trust		Control trust	
		Transaction partner	Intermediary	Transaction partner	Intermediary
Objective	Social indicators	recommen-dation from industry association	recommen-dation from industry association	quality of production processes (e.g. quality systems, ISO 9000ff.)	quality of intermediation processes (e.g. VeriSign)
Subjective	Personal experience	impression of company's representatives	usability[1]	-	-
			preliminary product sample	preliminary product sample	
			past transaction experience	past transaction experience	
			past personal experience	past personal experience	
	Understanding	product warranty	strong technical infrastructure[1]		
		product description	tailored transaction processes[1]	-	limited access to intermediary[1]
		common culture	transaction support (e.g. finance, logistics, trade regulations)		
		company information			
		transaction contracts			
		logistics warranty			
	Communality	transaction partner's reputation at other companies	transaction partner's reputation at other companies	-	-
		products' reputation at other companies	products' reputation at other companies		

[1]Only in electronic transactions.

grouped under understanding must be focused and transferred to an electronic transaction environment. Additional trust generation may come from social indicators and communality. Personal long-term experience may to some extent be converted into the first impression of the transaction partner and the intermediary (see Table 2). In addition, potential efficiency gains in electronic commerce transactions positively influence the trust level.

Table 3 derives appropriate electronic trust elements for the implementation of the trust reasons relevant in a specific agrifood transaction scenario to an electronic trade environment.

Table 3. Trust reasons and their electronic implementation.

Trust reasons	Electronic commerce implementation
Recommendation from industry association	Letter of recommendation
Impression of company's representatives	Videoconferencing, personal get-together at meeting
Preliminary product sample	Shipping of product sample
Past transaction experience	Only *ex post*
Past personal experience	Only *ex post*
Product warranty	Signaling with product warranty information
Product description	Multimedia product description
Common culture	Signaling with information and design, usability, communication processes
Company information	Multimedia company information
Transaction contracts	Multimedia negotiation support, contract document
Logistics warranty	Signaling of logistics warranty information
Transactions partner's reputation at other companies	Online reputation system for companies
Products' reputation at other companies	Online reputation system for products
Usability	Usability engineering for ease of use (Haas, 2004)
Strong technical infrastructure	Secure access and encrypted data transfer (Franke Kleist, 2004)
Tailored transaction processes	Transaction processes according to product, companies, and market relationships (Hausen, 2005)
Transaction support (e.g. finance, logistics, trade regulations)	Information on transaction support and appropriate multimedia offers
Quality of production processes	Signaling with production quality sign and respective information
Quality of intermediation processes	Signaling with transaction quality sign and respective information
Limited access to intermediary	Authentication, trusted partner criteria

4. Experimental study

The trust model for electronic commerce in the agrifood sector was evaluated in a first experimental study that tested whether electronic trust elements derived with the trust model for a specific scenario increased the level of trust sufficiently to make test persons engage in electronic transactions. The international trade of agricultural fresh fruits between southern and northern EU countries was taken as traditional agrifood transaction scenario to be transferred to an electronic transaction environment. In an expert interview, the determinants of the trust level particular to this transaction scenario were identified. Fresh fruits are highly perishable goods with a very large amount of different quality levels. As a consequence, risk in this international transaction scenario is considered as very high. Past personal contact and experience between transaction partners as well as past transaction experiences are trust reasons of outstanding relative importance.

4.1. Experimental design

For the evaluation of the trust model for e-commerce in the agrifood sector, the traditional *ex post* trust reason 'past personal experience' was electronically implemented by the *ex ante* trust reason 'impression of company's representative'. The impression of company's representatives describes personal experiences from a personal contact between transaction partners, but excludes past transaction experience. As an *ex ante* trust reason, the impression of a company's representative is suitable to generate trust before a transaction and may be transferred to an electronic transaction environment.

An experimental long-term study during eight weeks was accomplished to analyse if the trust reason 'impression of company's representative' increases the trust level in an electronic agrifood industry transaction scenario sufficiently to overcome the risk perceived in the electronic transaction environment and to realise transactions. The underlying hypothesis was:

H: *The existence of direct, electronically mediated, and indirect (reputation) personal relationships between transaction partners raises the trust level in e-commerce transactions in the agrifood industry sufficiently for transaction partners to engage in electronic transactions.*

Metric was the relative frequency of transactions between transaction partners of the four relationship levels as opposed to the expected value. The transaction scenario were transactions between the first two levels of the fresh fruit supply chain (fruit growers - fruit traders) with an electronic trade platform with request for quote (RFQ) as tailored transaction process (Hausen, 2005).

Participants in the experiment were agricultural and food science students from the University of Bonn, Germany and the University of Natural Resources and Applied Life Sciences, Vienna, Austria to map the different intensity of relationships between potential transaction partners and the international differences. Their task was to accomplish supply or sales transactions that could be executed independently from time and location. Participants from Bonn and Vienna were evenly distributed to companies at both levels of the supply chain. Half of the participants were logged into the electronic trade system with their name to signal a reputation, the other half anonymously. Participants from Bonn and Vienna known by name were introduced to each other via videoconference. Table 4 shows the possible transaction constellations in the experimental study.

Table 5 shows the expected values of each transaction constellation serving as reference values. Expected values are the values of an evenly distributed transaction behavior across the

Table 4. Possible transaction constellations in the experimental study (P: personal relationship; V: contact via videoconference; R: reputation; A: anonymous).

Closed deals	Offers							
	Supplier Bonn, known	Supplier Bonn, anonymous	Supplier Vienna, known	Supplier Vienna, anonymous	Sales Bonn, known	Sales Bonn, anonymous	Sales Vienna, known	Sales Vienna, anonymous
Supplier Bonn, known	-	-	-	-	P	A	V	A
Supplier Bonn, anonymous	-	-	-	-	P	A	R	A
Supplier Vienna, known	-	-	-	-	V	A	P	A
Supplier Vienna, anonymous	-	-	-	-	R	A	P	A
Sales Bonn, known	P	A	V	A	-	-	-	-
Sales Bonn, anonymous	P	A	R	A	-	-	-	-
Sales Vienna, known	V	A	P	A	-	-	-	-
Sales Vienna, anonymous	R	A	P	A	-	-	-	-

Table 5. Expected values of transaction constellations.

Transaction constellation	P	V	R	A
Expected value	25%	12,5%	12,5%	50%

different relationship types. Half of the participants acted on the electronic transaction system anonymously (A), the other half with their name. Of these half known with their name, 50% had personal relationships (P; the Bonn students among themselves and the Vienna students among themselves). Of the other 50%, one half were introduced in a videoconference setting (V), the other half was only recognisable from the name and the connected reputation (R). In addition, four transaction directions between participants were possible: Bonn to Bonn (B→B), Bonn to Vienna (B→V), Vienna to Vienna (V→V), and Vienna to Bonn (V→B). The transaction direction shows which partner has taken the initiative for the transaction.

4.2. Results

During the eight weeks of the experimental long-term study, 88 requests for quotes were published with 220 offers. 65 request for quote deals were closed. Table 6 shows the relationships between the transaction partners and the transaction directions for the request for quote transaction mechanism in detail. It shows that the two e-commerce trust generating elements videoconference and reputation were both able to realise more transactions than between anonymous participants.

The results of the experimental study regarding the transaction mechanism request for quote show - in particular at the offers - a clear movement to transaction relationships between personally

Table 6. Transaction relationships in request for quote transactions.

Perspective	Transaction phase	Number of actions	Transaction relationship			
			P	V	R	A
Sales	Offers	220 (100%)	94 (43%)	28 (13%)	32 (14%)	66 (30%)
Deviation from expected value			+ 18%	+ 0.5%	+1.5%	-20%
Supply	Realisations	65 (100%)	18 (28%)	9 (14%)	10 (15%)	28 (43%)
Deviation from expected value			+ 3%	+ 1.5%	+ 2.5%	- 7%

known partners. Transactions between partners known via videoconference and reputation were more frequent than expected, but the deviation from the expected value was less obvious. The results from the experimental study show that the developed trust model for electronic commerce in the agrifood sector is suitable to derive electronic trust elements for a specific agrifood transaction scenario increasing the level of trust sufficiently to make test persons engage in electronic transactions.

5. Conclusions

Electronic commerce is an interesting support option for companies in the agrifood sector to establish new business relationships. However, the agrifood sector deals with products where a large amount of information asymmetry exists between transaction partners. Many characteristics of food products may only be analysed after use and others even cannot be examined at all. Electronic commerce has often been criticised as too anonymous for agrifood transactions. The risk and uncertainty created a lack of trust in electronic commerce in the agrifood sector. Due to trust problems, the benefits from electronic commerce were considered less important than the obstacles.

This paper has developed a trust model for electronic commerce in the agrifood industry supporting the analysis of trust determinants in traditional agrifood transaction relationships and their transfer to appropriate trust determinants in the electronic commerce environment. The model was evaluated in a first experimental study. The results show that the trust model for electronic commerce in the agrifood sector is suitable to support the implementation of electronic trust elements for a specific agrifood transaction scenario increasing the level of trust sufficiently to make test persons engage in electronic transactions.

References

Akerlof, G., 1970. The market for 'lemons': Qualitative uncertainty and the market mechanism. Quarterly Journal of Economics 84, 488-500.

Andersen, E. and K. Philipsen, 1998. The evolution of credence goods in customer markets: Exchanging 'Pigs in Pokes'. Working paper Aalborg University. Available at: http://www.business.aau.dk/evolution/esapapers/esa98/Credence.pdf.

Anderson, E. and B. Weitz, 1989. Determinants of continuity in conventional industrial channel dyads. Marketing Science 8, 310-323.

Bakos, Y., 1998. The emerging role of electronic marketplaces on the internet. Communications of the ACM 41 (8), 35-42.

Barney, J.B. and H. Hansen, 1995. Trustworthiness as a source of competitive advantage. Strategic Management Journal 15, 175-190.

Bolton, G.E., E. Katok and A. Ockenfels, 2004. How Effective are Electronic Reputation Mechanisms? An Experimental Investigation. Management Science 50, 1587-1602.

Bryant, A. and B. Colledge, 2002. Trust in electronic commerce business relationships. Journal of Electronic Commerce Research 3, 32-39.

Castaldo, S. and F. Nava, 2004. Il ruolo della fiducia nel settore dei servizi finanziari: una indagine empirica. Micro & Macro Marketing 13, 331-357.

Cooper, R. and T.W. Ross, 1984. Prices, product qualities and asymmetric information: The competitive case. Review of Economic Studies 51, 197-207.

Dyer, J.H. and W. Chu, 2000. The Determinants of Trust in Supplier-Automaker Relationships in the U.S., Japan and Korea. Journal of International Business Studies 31, 259-285.

Etzioni, A., 2003. Toward a new socio-economic paradigm. Socio-economic Review 1, 105-118.

Fearne, A., S. Hornibrook and S. Dedman, 2001. The management of the perceived risk in the food supply chain: A comparative study of retailer-led beef quality assurance schemes in Germany and Italy. International Food and Agribusiness Management Review 4, 19-36.

Franke Kleist, V., 2004. A Transaction cost model of electronic trust: transactional return, incentives for network security and optimal risk in the digital economy. Electronic Commerce Research 4, 41-57.

Fritz, M., T. Hausen and G. Schiefer, 2004. Developments and development directions of electronic trade platforms in US and European agri-food markets. International Food and Agribusiness Review 7, 1-20.

Fynes, B., S. de Búrca and E. Ennis, 2001. A Contingency model of quality management and business-to-business relationships. Journal on Chain and Network Science 1, 111-122.

Granovetter, M., 1985. Economic action and social structure: The problem of embeddedness. American Journal of Sociology 91, 481-510.

Hausen, T., 2005. Elektronischer Handel. Einbettung in Geschäftsbeziehungen und Supply Chains. Deutscher Universitätsverlag.

Hofstede, G.J., P. Pedersen and G. Hofstede, 2002. Exploring culture: Exercises, stories and synthetic cultures. Intercultural Press.

Keser, C., 2003. Experimental games for the design of reputation management systems. IBM Systems Journal 42, 498-506.

Lancaster, K., 1971. Consumer demand: A new approach. Columbia University Press.

Malone, T.W., J. Yates and R.I. Benjamin, 1987. Electronic markets and electronic hierarchies. Communications of the ACM 30, 484-497.

Nelson, P., 1970. Information and consumer behaviour. Journal of Political Economy 78, 311-329.

Ojasalo, J. and P. Puhakainen, 2004. Managing trust in electronic commerce transactions. Journal of Internet Commerce 2, 1-12.

Picot, A. and H. Dietl, 1990. Transaktionskostentheorie. Wirtschaftswissenschaftliches Studium 19, 178-184.

Rademakers, M.F.L., 2000. Agents of trust: Business associations in agri-food supply systems. International Food and Agribusiness Management Review 3, 139-153.

Ratnasingam, P., 2004. The role of facilitating conditions in developing trust for successful electronic marketplace participation. Journal of Internet Commerce 3, 95-110.

Saaty, T.L., 1990. How to make a decision: The analytic hierarchy process. European Journal of Operational Research 48, 9-26.

Schiefer, G. and U. Rickert (eds.), 2004. Quality assurance, risk management and environmental control in agriculture and food supply networks, Vol. A and B. ILB-Press.

Tan, Y.-H. and W. Thoen, 2001. Toward a generic model of trust for electronic commerce. International Journal of Electronic Commerce 5, 61-74.

Weiber, R. and J. Adler, 1995. Informationsökonomisch begründete Typologisierung von Kaufprozessen. Zeitschrift für betriebswirtschaftliche Forschung 47, 43-65.

Wilson, P.N. and A.M. Kennedy, 1999. Trustworthiness as an Economic Asset. International Food and Agribusiness Management Review 2, 179-193.

Actor organisation for QAS along agro supply-chains: the case of mycotoxins reduction in Southern Cone grains

Alejandra Engler[1], Guy Henry[2], Daniel H. Iglesias[3], Alexandre Florindo Alves[4], Gonzalo Gutiérrez[5] and Elisabete Salay[6]
[1]INIA-Quilamapu Chillan, Chile; aengler@quilamapu.inia.cl
[2]CIRAD / ProsPER Cône sud, Buenos Aires, Cap.Fed., Argentina
[3]EEA INTA-Anguil, La Pampa, Argentina
[4]Universidade Estadual de Maringá, Maringá-PR, Brazil
[5]Facultad de Agronomia, Universidad de la República Montevideo, Uruguay
[6]UNICAMP- NEPA Campinas-SP, Brazil

Abstract

This paper is based on a research project on Quality Assurance Systems (QAS) for mycotoxin reduction in Southern Cone grain supply chains funded by the European Community. When applying a Hazard Analysis and Critical Control Points (HACCP) system along a multi-actor supply chain, additional research questions are generated since it is commonly used at the single firm and actor level. Analysing the actor incentives, or the absence of thereof, to adopt QAS becomes a key question. Transaction costs analysis is introduced to better understand full (additional) QAS costs and possible governance structures between the chain actors. Results from actor surveys in 4 Southern Cone countries show that transaction costs and governance structures of the supply chain may constitute a limitation when transparency and uncertainty issues arise as a characteristic of transactions along the chain. These results suggest that in this scenario a third party intervention is needed when promoting the adoption of HACCP system. This intervention should be orientated to reduce the uncertainty across the chain as well as become a tool to generate an efficient transmission of quality incentives.

Keywords: food safety, actor coordination, transaction costs, HACCP, mycotoxins, Southern Cone

1. Introduction

As food- borne diseases are a large public health problem, food safety is one of the priorities of the global political agendas. As an example, in the United Stated alone, the foodborne diseases causes, by fungi, bacteria, parasites or virus are estimated to be 76 million illnesses, 325,000 hospitalisations and 5,000 deaths, each year (ERS, 2005). The economic implications of food safety are varied (Antle, 1995; Segerson, 1998). On the one hand, world trade has been affected by more and stricter regulations that have become an important topic of debate along with the implications of such regulations for importer and exporter countries, and this discussion will persist in the future (Dohlman, 2003, Wilson and Otsuki, 2001). On the other hand, consumers around the world are playing an increasingly important role as vocal actors, directly impacting the debate. Consumers are more aware of the potential risk of food-borne illnesses and are willing to undertake protective actions that translate into a higher willingness to pay for food safety (Shogren *et al.*, 1999); while producers are also willing to spend more resources to provide safety motivated by consumer demand and also by avoiding costs associated with the negative consequences of contaminated food (Elbasha and Riggs, 2003).

Regarding food safety problems, mycotoxin contamination is among the top priority issues of cereal and other field crops. Mycotoxins are toxic by-products that have been defined as '[...] fungal metabolites which, when ingested, inhaled or absorbed through the skin, cause lowered performance, sickness or death in man or animals. Mycotoxins can be carcinogenic, mutagenic, teratogenic and immunosuppressive [...]' (Dohlman, 2003). According to the same author, mycotoxin contamination effects more than one quarter of global food and feed production.

Current effective mycotoxin control relies on expensive and wasteful end-point (curative) segregation and/or testing. An increasing number of R&D projects and networks are focusing on these toxins, their origin, impact and management. Among these, the aforementioned European project was formulated to develop a cost-effective Food Quality Assurance system (i.e. HACCP) for controlling mycotoxins in cereal supply-chains in the Southern Cone of Latin America. Along with the importance of the contamination risk of cereal crops, this region is an important grain producer and exporter to European Union (EU) countries.

Recent project data shows evidence that Argentina and Uruguay have had serious *Fusarium* induced mycotoxin contamination events in wheat production (MYCOTOX WP4 project report). An inventory of mycotoxins occurrences in Brazil, showing a large variety of toxins and affected crops/products, was conducted by Brabet *et al.* (2003).

The question that remains unsolved is how to implement a QAS system throughout the supply chain in an efficient (cost-effective) manner: (1) which are the relevant instruments to promote QAS adoption among chain actors, and (2) what are the consequences and/or necessary conditions regarding the governance structure of the chain when implementing a QAS?

Some research has already been conducted in this field. Raynaud *et al.* (2004) shows that the governance structure is related to the quality enforcement mechanism. Their conclusion is that a private quality labeling mechanism is associated to a more vertical integrated chain, while public labeling can be found more in a spot market type of governance structure. On the other hand, incentives and other mechanisms to assure food safety have been a focus of attention since the increase in public safety regulations (Segerson,1999; Caswell and Henson, 1997). Most of the work in this field has been done by analysing the incentives that the market generates to promote adoption of quality or safety measures.

Ziggers (2000) also explored the relationship between food safety and transaction costs. In a descriptive paper he argues that food safety increases transaction costs because searching for information, negotiating and monitoring the transaction will have an additional component (it increases uncertainty and asset specificity of the transaction). QAS, as an HACCP system, constitutes an institution that reduces these additional transaction costs, generating a more coordinated and efficient supply chain.

This study adds to the preceding literature in a sense that it analyses the incentives created by the transaction structure at the different levels of the supply chains, necessary to implement an HACCP system. The study also analyses the expected changes in transaction costs derived from future HACCP implementation. It also goes further to most HACCP literature because the analysis is from the supply chain rather than the industry perspective (Galan *et al.*, 2003; Hooker *et al.*, 2002). The objectives of the article are twofold. Fist, we provide a discussion related to the governance structure and transaction costs of the supply chain and the incentives it generates to promote the adoption of QAS, and second we provide an example, analysing this issue using the wheat supply chain in Argentina, Uruguay and Chile, and corn in Brazil. A first attempt on possible research approaches was made by Henry *et al.* (2003).

This paper is organised as follows. Section 2 provides a discussion about the theoretical concepts through a transaction costs perspective. Section 3 describes the methodology used in the empirical analysis. Section 4 analyses the results, and section 5 summaries the main conclusions of the study.

2. Theoretical framework

According to the incentives and enforcement mechanisms, Klein (1996) suggests that a necessary condition for a producer to voluntarily maintain a high quality of a product, is the existence of a price premium or 'quasi-rent', which is expressed as the willingness to pay for the consumer for a higher quality or food safety. On the other hand, Segerson (1999) studied the incentives to adopt safety measures and concludes that an efficient market solution regarding food safety may be adopted for certain types of goods. The key distinguishing characteristic between cases is the extent to which consumers and producers correctly perceive food contamination risks. If both consumers and producers correctly perceive the damage from contamination, then the decisions will be efficient. However, in many cases safety attributes can not be observed by the consumer generating an information asymmetry in the market. The attributes that the consumer cannot observe, even after repeated purchases, are known as credence attributes[1]. Certainly, the case of food safety regarding mycotoxin contamination falls in this category or as an experience attribute. As Segerson concludes, in the case of credence attributes or goods, adequate consumer protection is likely to be achieved by some form of government intervention, which does not necessarily means regulation, but some government-financed instrument or other kind of incentive.

The conclusion derived above is only part of the QAS adoption problem. As we argued at the beginning of this section, the safety of food products will depend on the measures adopted throughout the supply chain, and therefore, the adoption decision will depend on the incentives generated not only by the final market, but also by the governance structure of the principal actors in the chain.

Transaction characteristics among actors of the supply chain are extensively treated in the New Institutional Economics (NIE). NIE has been developed by Coase and Williamson (1979) as an alternative approach to understanding the relationship between firms. According to this approach, transaction costs can explain the organisation of a firm and the way they interact along the supply chain. NIE recognises that information can be incomplete or asymmetric, and that in a transaction context, these aspects generate information, negotiation and supervision costs (Eggerston, 1990).

Transaction costs can be explicit or implicit and will depend on the characteristic or dimensions of the transaction. The critical dimensions to determine transaction costs are the frequency of the transaction, the degree and type of uncertainty attached to the transaction and the asset specificity (Sauvee, 1998). Asset specificity arises when one agent invests resources specific to an exchange.

[1] There are categories known as search and experience attributes. In the first case consumer will have full information regarding the goods attributes by simple inspection. The experience attribute was defined by Caswell and Mojduszka (1996), as "consumers cannot determine the product's quality until they buy and consume it"

This investment limits the market alternatives to this agent increasing the risk associated to the transaction and consequently increasing the transaction cost associated to the exchange. In general, it can be said that as the frequency of the transaction decreases and as the uncertainty and asset specificity increases, transaction costs become higher. In the transaction cost theory, the governance structure of the chain will be determined by the transaction characteristics. There are studies that already have confirmed the predictions of NIE approach. It can be said that greater asset specificity and uncertainty require a governance structure that provides a greater control over transactions (Farina and Zylbersztajn, 1997; De Azevedo and Dos Santos, 1999 to name some of these studies).

The NIE approach has also been applied to better understand the interrelation of the governance structure of the chain and quality signalisation (Raynaud, 2002; Raynaud *et al.*, 2004). After analysing the quality strategies and governance structure for different supply chains in Europe (Raynaud *et al.*, 2004) the authors conclude that quality labeling is related to the governance structure. Specifically, labeling based on brand reputation capital may lead to the adoption of a more coordinated chain, while public certification reduces de need for coordination and is related to a more market governance structure of transactions. This also confirms the relevance of the NIE approach, since quality labeling can be considered as an asset specificity that generates higher uncertainty in the transaction and therefore needs to be related to a more coordinated governance structure.

We want to analyse the relationship between food safety and transaction costs and define the incentives or limitations of a specific governance structure to implement a QAS. The relevance of this analysis is that the results may help to determine the right intervention in the chain to promote the adoption of safety measures. Our hypothesis is based on the preceding discussion. In a sense it extends Segerson model to include transaction costs into the costs-benefits considerations by the agents, and analyses it at each level of the chain. On the other hand, it is based on the NIE literature, considering how changes in asset specificity and other transaction characteristics are affected by QAS, affecting the transaction cost between agents.

We propose the following analysis and hypothesis: Assuming that agents at any level of the chain are profit maximisers, the decision to implement safety measures and specifically a HACCP system will depend on its net benefits. This rule can be represented by the following simple equation:

$$B_{HACCP} - C_{HACCP} + \Delta TC > 0 \qquad (1)$$

Where B_{HACCP} is the benefit derived form the implementation of a HACCP or the quasi rent that will be obtained for a high quality and safe product, C_{HACCP} is the additional cost of implementing HACCP. Finally, we add the impact on transaction cost derived from the implementation of safety measures that affect the transaction characteristics among the supply chain through ΔTC. Most of HACCP studies (Hooker *et al.* 2002, Segerson, 1999) only consider in the analysis the first two components. We argue that the change in transaction costs also influence the decision. In this sense, when implementing safety measures in the transaction the asset specificity of the product or transaction may change, as well as the uncertainty involved. This paper develops an empirical framework to analyse the change in transaction cost.

3. Methods

The base QAS that was identified to be integrated with the pertinent socio-economic aspects is the HACCP system. This system, while much used at the individual firm level, offers new research challenges when applied along a supply chain, involving different actors and hence, different organisational needs and conditions.

The institutions and governance structures are evaluated through the transactions between agents and the transaction costs that this structure generates. The theoretical approach of this study uses the transaction and the transaction costs between agents of the chain to answer the research questions and achieve the proposed objectives.

The project covers wheat sub-supply chains in Argentina, Chile and Uruguay, and the corn supply chain in Brazil. Secondary data was used to generate a first (macro) characterisation of the pertinent supply chain in each country. After this initial description, a case study supply chain in each country was identified to gather primary data to analyse their governance structure and the incentives it generate to promote the adoption of HACCP. The information was collected in two steps: (a) a first quantitative assessment of raw material suppliers (farmers), traders/storage and mills (to construct a Commodity Flow Diagram), and (b) a second qualitative survey with all relevant chain actors i.e. input suppliers, farmers, storage, traders, transporters, millers, distributors, (intermediate or end product) clients (Table 1). The second survey includes detailed questions on key elements of transaction costs regarding governance structure at each level of the chain. In the case of Brazil, the case study's entry point is a (corn based) chicken feed manufacturer, forward integrated with poultry producers and a slaughter house/processing plant.

4. Results and discussion

4.1. Wheat and corn supply chain case study selection

The wheat supply chain in the three countries under study presents a governance structure close to a spot market, in which no contracts or previous coordination among the actors exists. However, there are some important differences among the countries that create different scenarios for our study. For example, Argentina is one of the world's largest wheat producers reaching a production of more than 15 million tons a year. As a large producer, it is also an important player in the international wheat trade. Two thirds of production is exported. On the other hand, Uruguay dedicates its production almost entirely to satisfy the internal market, and occasionally sells wheat in the international market, mainly to Brazil. Its production is around 350 thousands

Table 1. Sample used for actor interviews, by country.

	Argentina	Uruguay	Chile		Brazil
Wheat farmers	20	34	19	Corn farmers	21
Storage owners	3			Storage owners[1]	4
Mill	1	1	1	Feed mill	1
Transporters	3			Poultry producers	44
Bakery / factory	8			Processor	1

[1]One cooperative, one feed manufactory and two grain dealers.

tons a year. Finally, Chile is deficient in wheat, importing around a 30% of its requirements. Its production is around 1.8 million tons a year.

Brazil is the world's third largest corn producer, with 42 million tons in 2003, corresponding to 6.8% of the world production (MAPA, 2004; CONAB, 2004). From Brazilian production, 60-70% is destined to feed production, and from that, 51% is destined to poultry feed. In 2003, Brazil was the largest world poultry exporter in revenue and the second largest in volume (ABEF, 2005).

The case studies selected in each country provide specific information to develop the transaction cost analysis. In each case, the central actor involved is the first processor and the study is then extended to producers, traders and second processors involved in this particular chain. The main characteristics of each supply chain case study are summarised in Table 2.

Table 2 shows that an important part of the requirements of wheat are bought directly from the grain producer. In the Argentinean case this is 50%, but in the Uruguayan and Chilean case, it is 90% or above of the requirements. However, very few contracts exist between producers and the mill in Uruguay. The coordination of the chain is low at each level, no coordination exists between producers and the mill, and no integration between the mill and the second processor except for the Argentinean case, where 40% of the mill production is integrated with a noodle factory owned by the same company and in the Brazilian case, where 100% of the feed is send to vertically integrated poultry breeders.

The discussions with actors about the quality concept were ambiguous, to say the least. Quality has different connotations for different individuals. Hence the following qualitative results need to be interpreted with some reservation. Table 2 shows, that in general the awareness by interviewed actors, about mycotoxin, is poor and that the concept of quality related to grain is rather an ambiguous concept.

Uruguay may be a good example, where the presence of myctoxin currently constitutes a public issue since a bad outbreak (of DON mycotoxin) during the 2001/2002 wheat harvest. Uruguayan authorities issued a program to reduce the impact of this problem on two main fronts. One establishing a DON toxin limit for wheat products traded in the domestic market[2]. The other, implementing a joint project with FAO to conduct a national HACCP plan in order to attempt Good Manufacturing Practices (GMP) and Good Agricultural Practices (GAP), at industry and producer levels, respectively. The main goal was to reduce the exposure of large segments of the Uruguayan population to wheat related mycotoxins, especially to those people who have a large dependency on wheat products in their daily calories intake.

As it was indicated before, Uruguay and Argentina wheat sectors have explicitly suffered from mycotoxin contaminations. Uruguay has diffused this issue as a national public problem and has implemented some measures to address it (albeit with little results too show). In Brazil, there also exist public standards for mycotoxin in corn. Although indications show that the private feed production sector controls the level of aflatoxin in corn, the public control system of these standards is still limited (Salay and Mercadante, 2002). In Argentina and Chile the awareness about mycotoxin (in wheat) is low. Although, some, but too few, actors in both countries recognise the existence of fusarium, being a major potential source of mycotoxin contamination. Mycotoxin laboratory tests, having been conducted in this project, reveal no existence of mycotoxin presence

[2] Decree 533/001 and 472/002.

Table 2. *Case studies supply chain description (INCO-Dev MYCOTOX, 2003-2004, internal project data).*

	Argentina	Uruguay	Chile	Brazil
Case study	Medium size wheat mill - La Pampa Province	Wheat mill cooperative	Wheat Mill - X Region	Chicken Feed Factory, Parana state
Milling capacity[1]	24.000 ton a year	50.000 ton a year.	35.000 ton a year	48.000 ton a year (with plans to increase in 50%)
Relative importance	It's a medium size mill (Ranked 65 over 120 mill in the country)	It belongs to the largest Uruguayan milling group	It is the biggest mill in the south of Chile. However, its relative importance in the national market is low.	It can be considered a relatively small Feed Factory.
Suppliers (farmers)	Only 50% of the wheat processed a year is bought directly from the producer. The other 50% is provided by local storages (Coops) & brokers. It´s important to mentions that the mill demands only cover less than 10% of the area wheat production.	Wheat supplied by farmers accounts generally for 75% of the total wheat and 25% comes from the cooperative sector / private traders.	95% of the wheat is provided by producer, manly large producers. It is important to mention that in general mills in Chile import part of their requirements, but this mill uses only national wheat.	There are two harvests per year and most farmers deliver corn immediately to the factory. In the summer harvest, they supply 30% and in periods between harvests they supply 50% of Feed Factory needs.
Product market	60% of the final product is destined to bakeries and 40% to a pasta factory owned by the mill. A final byproduct is destined to a animal food factory owned by the mill	80% of the final product goes for internal market mainly to small bakers and domestic consumption. Export and industrial clients are minimal on the total market share.	75% of the production is derived to local bakeries. However, the mill also sells flour to final consumer with their own brand, and a special type flour to the salmon feed industry.	Feed Factory delivers directly to poultry producers

Table 2. Continued.

	Argentina	Uruguay	Chile	Brazil
Quality concept	There is no a clear concept of quality across the chain. For the producer it is related to protein content (as the official standard), the mill request gluten, and the bakery measures baking quality in the final product.	The mill quality is resumed in clean and dry wheat. No integration whatsoever on industrial quality parameters to induce the farmer to certain wheat type or class. Price premium criteria vary year by year depending on the average harvest quality.	Quality is also a hidden concept. The mill buys certain varieties of grain and test for different quality characteristics such as protein, gluten index and sedimentation. Farmers produce the variety of wheat that the mill will buy.	For corn quality, the concept is mainly based on characteristics like moisture, broken grains, burnt grains and so on. For the feed, nutritional aspects and other issues like moisture, impurity, insects are analysed.
Experience with fusarium and/or mycotoxins	Farmers and warehouses have had evidence of fusarium in the past, however they do not know about the presence of mycotoxin because they have not tested for it. In general, there is a lack of information about mycotoxin norm and regulations.	During the years 2001/02 Uruguay experienced a serious fusarium infection on the wheat crop, that caused a serious decline un average yield and also mycotoxin contamination problems in both wheat and wheat flour.	In Chile, although climate conditions and cultural practices favor a low presence of fusarium. There has been evidence of the fungus in the field. No experience related to mycotoxins exists. No test are applied either.	Information not yet available.

[1]Denotes both wheat mill and (corn based) chicken feed mill.

in grains sampled in Chile. However, the test resulted positive for some samples in Argentina (INCO-Dev MYCOTOX, various years).

The consequence of the lack of information and concerns related to the lack of quality awareness is, that in the short run there is no market incentive to motivate QAS in Argentina and Chile. The Uruguay wheat actors seem more sensitive, since the government, consumers, processors and producers are aware of this problem, thus increasing the probability of adopting preventive systems. Despite this, it is very probable that the scenario regarding the incentives situation will change in the future for Chile and Argentina.

Chile is one of the world's top producers of salmon, and wheat is one of the ingredients of salmon feed. During the last three years, the salmon industry has become a new player in the national wheat market and it is expected that its relevance will increase in the future. Because the salmon industry needs high quality and not contaminated grains, it is expected that the market will start differentiating quality characteristics in the product that will push the supply chain to act accordingly. The mill industry has also started a process of requesting higher quality of product since retailers and the final consumers are demanding a higher quality of bread and other types of products. On the other side, in Argentina, it is expected that quality norms will become more important in the near future since most of the product is exported to the international markets, where quality standards are becoming more and more strict.

In Brazil corn is commonly trade as a commodity and this makes difficult to identify incentives for QAS adoption, although feed factory is concerned about mycotoxin contamination and its effects on poultry production performance parameters.

4.2. Transaction costs, governance structure and incentives/limitations to implement HACCP

The preceding arguments favor the probability that in the future in Chile and Argentina the incentives to implement QAS throughout the supply chain will increase. Therefore the analysis of the incentives/limitations of the governance structure of the supply chain is still relevant for these cases. In Brazil, the identification of a new possible governance structure will depend on the identification of needed incentives alongside the productive chain. Why would the corn producer implement the GAP without a price *premium*? How much more the feed manufacturer would be willing to pay for higher quality corn? These are questions to be addressed.

Our hypothesis is that the governance structure of the supply chain derived from the transaction costs among the different transaction stages will generate incentives or limitations for the adoption of HACCP. This information is relevant to generate adequate instruments to promote such systems. Our analysis starts with a summarised description of the transaction characteristics of the four case studies (Table 3).

Table 3 shows, that in general, due to low storage capacity, the frequency at the producer level is relatively low. On the other hand, at the mill level, across the four country cases, this is high to very high (corn). The table also shows that as the grain moves through the chain, the relative absence of explicit quality standards keeps the asset specificity low. Furthermore, the table evidences substantial uncertainty regarding production and farm-gate price risks at the producer level, due to the different stated reasons. At the mill level, there exist certain price uncertainties (transmitted to a certain extent by producers, but possibly off-set by international price and market predictions). Furthermore, quality uncertainty at the mill level, may be high, when

Table 3. Summary of transaction key elements of case studies' supply chain (INCO-Dev MYCOTOX, 2004-2005, internal project data).

Transaction element	Supply chain stage	Argentina	Uruguay	Chile	Brazil
Frequency	Producer / mill stage[1]	Frequency at this stage varies since producer sell according to the current price. It is important to mention that fixed producer supply, brokers, cooperatives and storages allow for higher frequency in sales.	A large percentage of the wheat is sold during the harvest directly to the mill. Depending on the year (pricing conditions) the farmer holds more wheat for later reselling.	Most of the wheat is sold during harvest season. Only few percentages of producers have storage to differ the selling. The storage capacity in Chile is very low	Most farmers deliver corn immediately after the harvest to the factory (associated to relatively low storage capacity at farm level)
	Mill and bakeries stage[1]	The frequency here is high since bakeries buy flour weekly	The frequency here is high. Bakers often buy wheat weekly.	The frequency here is high since bakeries and retailers buy weekly	The customer in this case is represented by poultry breeders receiving poultry feed almost every week.
Asset specificity		Asset specificity is low along the supply chain. Mills and bakeries have higher asset specificity in the chain.	Asset specificity is very low. Millers are not willing to create specific trading rules to specify pricing premiums. Quality requirements by the bakers are low.	Asset specificity is low. The mill has higher asset specificity in some products since they have special clients such as salmon industry.	Corn is not asset specific (large alternative use possibilities)

Table 3. Continued.

Transaction element	Supply chain stage	Argentina	Uruguay	Chile	Brazil
Uncertainty	Faced by grain producers	Price: is set by the mill using the market and the SAGPyA standard.	Price: pricing system is very obscure. Mills only report suggested buy price, but no other statistic is attached (volume, quality). There is no local standard for pricing in force. Productive risk: lack of proper insurance coverage may be a serious problem for the farmer.	Price: the mill set the price according to the market condition. It is not clear how much quality increase price since this factor is not norm. Transport: transportation is also a concern for the producer since the capacity is full during harvest season.	Price: expressive variation in time (and relatively low use of futures markets). Productive risk: one of the harvests, called 'safrinha' happens in a period subject to strong weather constraints.
	Faced by the mill[1]	Quality and prices varies according to international wheat markets and climate.	Quality of the grain received may be variable year by year.	Quality of grain is variable and could lead to variable quality of the final products.	Quality of the grain received may vary year by year and harvest by harvest.
Asymmetric information		Since quality is not well defined there is asymmetric information as well as in prices	Very asymmetric information on prices and less on quality issues.	There is asymmetric information in quality and prices.	Farmers and millers know corn market prices.

[1]Denotes both wheat flour mill (in 3 countries) and chicken feed mill (in Brazil).

related to domestic (varying) grain quality, but again possibly to be offset by options to import standardised quality grain (in the case of wheat in Chile). Table 3 also shows that asymmetric information (obviously related to the aforementioned price uncertainty discussion), in general is high, especially regarding price vs. quality information. In Brazil for corn, this is not so much the case.

Given the above discussion, and given additional obtained information that very few contracts exist between producers and mills in the three wheat cases (Brazil corn case actors are horizontally integrated through ownership or contracts), we expect the following changes in transaction costs derived from possible future implementation of public control measures for mycotoxin contaminations, through QAS along the chain (Table 4).

Table 4 indicates that future public intervention enforcing the adoption of QAS in the four cases analysed, will definitely change the level of transaction costs. Furthermore, the increase in transaction costs due to higher asset specificity ought to be offset by the decreases in both uncertainty and asymmetric information. The decrease in transaction costs, in turn should make the chain more efficient, as also found by Ziggers (2000).

We argue that the implementation of policy measures should be accompanied by incentives that would lead to a smoother adoption of QAS systems. It is important to analyse the characteristics at each step of the chain to generate the instruments that in turn will force the right incentives and adequate adoption. The instruments should reduce the uncertainty across the chain as well as become a tool to generate an efficient transmission of quality incentives.

Table 4. Expected changes in transaction costs induced by policy implementation.

Transaction characteristic	Expected changes	Increase / decrease in transaction cost
Frequency	No real change expected	-
Asset specificity	It is expected an increase in asset specificity because of new product quality characteristics. Therefore information and supervision costs should increase with respect to the current structure	↑
Uncertainty	Once public control measures are in place, less uncertainty should exist regarding quality standards, in purchases between farmers and millers. Production uncertainty should not change	↓
Asymmetric information	Greater public and private transparency of prices vs. quality should reduce asymmetry of information, and hence reduce TC	↓

5. Conclusions

While the field research has not been completed yet, from key informant sources, it is already evident that in three out of four countries (wheat supply chain up to bakeries), very little concrete incentives currently exist for actors to get interested to take part in a supply chain QAS to reduce mycotoxins. Moreover, the same sources point to a lack of interest in (increased) vertical integration of partners in the chain. Only in the case of the Brazilian corn supply chain (for poultry farms integrated with feed and slaughter facilities), there are already indications that actors incentives do exist because decreased mycotoxin occurrence will directly improve feed/meat conversions and subsequently, lower production costs.

While on paper, QAS related benefits may outweigh related costs, the remaining questions (following Ziggers, 2000) in the Brazilian chain regard the additional necessary QAS adoption conditions of: (1) transparency of higher benefits than costs, and (2) an equitable distribution of these benefits among the actors of the chain. From the currently available data there is a first indication that a future QAS will lower transaction costs. Further evidence is needed to indicate that governance structure conditions warrant an internal certified control system, or third party certification.

The preliminary results allow us to conclude that in order to implement a HACCP system along a supply chain, intervention is needed. However, the public intervention should be analysed carefully since wrong incentives may cause an increase in transaction costs making the supply chain less efficient. Instruments should be able to generate an efficient transmission of quality incentives.

These conclusions are preliminary at best, since the majority of quantitative field data still needs to be processed and analysed. Nonetheless the authors, at this early stage, could greatly benefit from a critical feedback on the study's approach and first findings.

Acknowledgements

The research for this paper and the seminar participation were co-financed by the European Commission (INCO-Dev MYCOTOX, 2003-05, #ICA4-CT-2002-10043), the French Regional Cooperation, and CIRAD project ProsPER Cône sud.

References

ABEF Brazilian Chicken Producers and Exporters Association, 2003. Annual report 2003. São Paulo: ABEF, 2004. Available at: http://www.abef.com.br. Accessed on 03/02/05.

Antle, J., 2001. Economic analysis of food safety. In: Gardener, B. and G. Rausser, (eds.), Handbook of agricultural economics, Vol. 1b. Elsvier.

Brabet C., Freitas-Silva O., de Souza M. de LM, Correa T.B.S., 2003. Literature review: Mycotoxin occurrence in Brazilian corn and wheat. Deliverable 2 du projet INCO-DEV Mycotox. Montpellier: CIRAD, 2003, 60 pp.

Caswell, J. and S. Henson, 1997. Interaction of private and public food quality control systems in global markets. In: Loader, R.J., S.J. Henson and W.B. Traill. (eds.), Globalization of the Food Industry: Policy implications. Reading, UK. Centre for Food Economics Research. University of Reading.

Caswell, J. and E.M. Mojduszka, 1996. Using informational labeling to influence the market for quality in food products. American Journal of Agriculture Economics 78, 1248-1253.

CONAB - Companhia Nacional de Abastecimento. Avaliação da Safra Agrícola 2004/2005. Primeiro Levantamento. Dezembro 2004. Accessed on: 15/10/04, Available at: http://www.conab.gov.br/download/safra/safra20042005Lev01.pdf.

Crespi, J. and S. Marette, 2001. How should food safety certification be financed?. American Journal of Agricultural Economics 83, 852-861.

De Azevedo, P. and V. dos Santos, 1999. Food franchasing and backward coordination: an empirical analysis of Brazilian firms. Paper presented at the annual ISNIE Conference, September, Washington DC.

Dohlman, E., 2003. Mycotoxin hazards and regulations: Impacts on food and animal feed crop trade. In: Buzby, J. (ed.). International trade and food safety: Economic theory and case studies. Agricultural Economic Report 828. USDA, ERS.

Eggertson, T, 1990. Economic behavior and institutions. Cambridge University Press.

Elbasha, E. and L. Riggs, 2003. The effects of information on producer and consumer incentives to undertake food safety efforts: a theoretical model and policy implications. Agribusiness 19, 29-42.

ERS, USDA, consulted at: www.ers.usda.gov/briefing/foodbornedisease/

Farina, E. and D. Zylbersztajn, 1997. Abodagem Sistemica dos Negocios Agroindustriais e a Economia dos costos de transacoes. In: Farina, E.M.M.Q., Azevedo,P.F. and Saes, M.S.M. (eds.), Competitividade: mercado, Estado e Organisacoes. Editora Singular.

Galan, M., C. Aubry, and A. Maze, 2003. Farm certification and the implementation of HACCP in agriculture: a cost/benefit analysis. Paper presented at the Annual Meeting of the American Agricultural Economics Association. July 27 - 30, 2003. Monterea, Canada.

Henry, G., E. Salay, and A. Engler, 2003. Integration of socio-economic and food science and technology research in quality management of food supply chains: Mycotoxin control system of grains in the Southern Cone. Invited paper presented at the V Simposio Latino Americano de Ciencias de Alimentos, Campinas-SP, Brazil, 3-6 Nov. 2003.

Hooker, N., R. Nyga, and J. Siebert, 2002. The impact of HACCP on costs and product exit. Journal of Agricultural and Applied Economics 34, 165-174.

IICA 2005. Agricultura del siglo XXI: Efetos y respuestas. Comunicación On Line ISSN 1814-0939. www.iica.int/prensa/comuniica/2005/n1-esp/.

INCO-Dev MYCOTOX, 2002-04. Internal project research results. CIRAD, Montpellier, Francia (unpublished).

Klein, B., 1996. Why hold-up occurs? The self-enforcing range of contractural relations. Economics Inquiry 34, 444-463.

MAPA - Ministério da Agricultura, Pecuária e Abastecimento. Plano Agrícola e Pecuário 2004/2005. Available at: http://www.agricultura.gov.br. Accessed on: 18/05/04.

Raynaud, E., 2002. Governance of the agri-food chains as a vector of credibility for quality signalization in Europe. Paper presented at the 10th EAAE Congress, August 28-31, 2002. Zaragoza, Spain.

Raynaud, E., L. Sauvee and E. Valceschini, 2004. Quality enforcement mechanisms and the governance of supply chains in the European agro-food sector. In: Quality Assurance, Risk Management and Environmental Control in Agriculture and Food Supply Networks, Schiefer, G. and Rickert, U. (eds.). University of Bonn, ILB Press.

Salay, E. and A.Z. Mercadante, 2002. Mycotoxins in Brazilian corn for animal feed: occurrence and incentives for the private sector to control the level of contamination. Food Control 13, 87-92.

Sauvee, L., 1998. Toward an institutional analysis of vertical integration. In: Royer, J. and R. Robers (eds.), The industrialization of agriculture. Ashgate Publishing.

Segerson, K., 1999. Mandatory versus voluntary approaches to food safety. Agribusiness 15, 53-70.

Shogren, J., J. Fox, D. Hayes and J. Roosen, 1999. Observed choices for food safety in retail, survey, and action markets. American Journal of Agricultural Economics 81, 1192-1199.

Wilson, J. and Otsuki, 2002. Global trade and food safety: winners and losers in a fragmented system. The World Bank. http://econ.worldbank.org/files/2469_wps2689.pdf

Williamson, O., 1979. Transaction-cost economic: the governance of contractual relations. Journal of Law and Economics 22, 223-261.

Ziggers, G., 2000. HACCP, Vertical coordination and competitiveness in the food industry. In: Unnevehr, L. (ed.), The economics of HACCP, costs and benefits. Eagan Press.

Does food quality management create a competitive advantage?

Jon Hanf[1] *and Claus-Hennig Hanf (†)*[2]
[1]*Institut für Agrarentwicklung in Mittel- und Osteuropa, Theodor-Lieser-Str. 2, 06120 Halle, Germany; hanf@iamo.de*
[2]*Christian-Albrechts-University Kiel, Department of Agriculture Economics*

Abstract

The BSE and FMD disease in 2000/1 was perceived by some consumers as the 'doomsday of modern food processing'. This resulted in a change of perception of food quality amongst consumers and politicians thereby effecting consumption patterns and consumer expectations of 'good' food. Thus, we want to elaborate on these questions: How can stakeholders in the food chain effectively address this 'new quality'? What is the sustainability of these newly created hybrid management systems under varying conditions? Is there a possibility that a single supply chain network can create long-lasting competitive advantages by means of establishing a specific 'chain quality'-concept.

Keywords: collective strategy, strategic partnering, strategic quality management, supply chain networks

1. Introduction

People are generally concerned about the quality of their food. However, during times of food scandals people become even more aware and concerned e.g. the coke-scandal in Belgium, the BSE-crisis in the UK, the wine-scandal in Austria and Germany. The crisis of winter 2000/01 can be regarded as the straw that broke the camel's back. Consumers perceived Bovine Spongiform Encephalopathy (BSE) and Foot and Mouth Disease (FMD) as perhaps the doomsday of modern food processing. Consumers, politicians and the press unanimously clamoured for transparency of the whole production process. There was a change in the perception of food quality by consumers and politicians affecting the consumption patterns and the expectations of 'good' food. The most striking consequence of these dramatic food scares was the fact that politicians, consumers and also producers and suppliers assessed that food quality is no longer the responsibility of a single firm, instead the whole food chain needs to work together in order to deliver the 'new quality'. In the last two years, a large number of proposals for 'chain quality management' were widely discussed and the first quality management processes have already been implemented.

There are three questions which need to be asked and analysed using new institutional economics and strategic management approaches:

- How can stakeholders in the food chain effectively address this 'new quality'?
- What is the sustainability of these newly created hybrid management systems under varying conditions?
- Is there a possibility that a single supply chain network can create long-lasting competitive advantages by means of establishing a specific 'chain quality' concept, or vice-versa are chain quality management systems always doomed to be systems of common standards?

2. The 'new defined quality' and its consequences

2.1. Changing definitions of food quality

The key driver for the re-design of a food chain is the reliable transfer of credence attributes as a part of the product quality to consumers. The distinction between search or shopping and experience attributes, first introduced by Nelson (1970), and the introduction of credence attributes in the economic literature by Darby and Karni (1973) changed the perception of food products. It became apparent that these attributes altered the perception of food products from rather uncomplicated raw commodities with easily detectable qualities to somewhat more sophisticated goods. For search goods the buyer is able to find evidence of the attributes even prior to the purchase (Hanf, 2000; Picot *et al.*, 2001), whereas for experience goods the verification of the attributes' correctness can only be exercised right after the purchase. Until the beginning of the 1990's this differentiation of attributes has been extensively discussed in the literature on agricultural economics. Since the mid 1990's, works on credence attributes became more popular and were frequently used in publications (Andersen, 1994). Credence attributes are product and service characteristics that cannot be detected by the buyer under ordinary circumstances, neither before nor after the buying process.

The aforementioned food quality crises had the effect that the consumers recognised the credence attributes of food products as well as the shopping and experience attributes. As people recognised their inability to prove the correctness of credence attributes, they reacted, in the case of food crises, with a sharp reduction in demand for specific food items (Böcker and Hanf, 2000). For example, the immediate consequence of the events in winter 2000/01 was a sharp reduction in the consumption of conventionally produced red meat and red meat products. Correspondingly, a sharp increase was observed in the consumption of substitutes like fish, poultry, cheese and organically produced food. Furthermore, branded products were better able to survive this crisis than unbranded products. However, most of the empirical investigations show that consumer behavior alternates between learning and failing to remember. In a crisis, this implies that the high value attributed to the safety property will decline if the frequency and intensity of food scandals diminishes. The reluctance to consume red meat has been overcome and today Germans eat as much red meat again as they did before the crisis. Also the consumption of most of the substitutes is back to the pre-BSE time.

Many factors influence consumers' buying decisions such as failing to remember, bounded rationality, asymmetric information and time restraints. Hence consumers will not be able or not be willing to intensively and completely prove the quality of food products, and they are looking for signals to ease their buying decisions e.g. for a well known brand or a respective certificate. Above all, consumers must regain trust in all the actors of in the food chain and trust that they do their best to offer non-contaminated and risk-free food. Further, they must regain trust in the control institutions that they are able to effectively monitor and control the whole food production process.

Therefore, consumers as well as politicians demanded comprehensive safety policies and concepts that involved all directly or indirectly contributing actors. Such a system would demand the formation of hybrid organisations. A focal company has to establish control and co-ordination mechanisms throughout the whole production chain in order to guarantee credence attributes. That certainly creates a strong increase in market power and reduces the importance of spot markets.

2.2. Supply chain networks

Food products are usually not produced in strict food chains but rather in supply chain networks (SCN) or netchains. A self-evident reason for the formation of food supply chain networks instead of single line chain is the differing size of firms along the food chain. Striving for economic independence, protection against market power and economics of scope are other reasons. Generally, networks can be defined as 'specific properties of the transaction relationships, typified by relational relationships in which formal and informal sharing and trust building mechanisms are crucial' (Zylbersztjn and Farina, 2003). Networks are addressing all questions on inter-organisational relationships of more than two firms (Omta *et al.* 2001). In network science the collaboration is determined by different forces e.g. complementary abilities of the involved firms and risk reduction (Menard, 2002). Networks - in which firms are embedded in upstream and downstream flows of resources and information - can influence the nature of competition and the profitability beyond traditional measures of industry competition (Gulati *et al.*, 2000). The industrial organisation school - based on the work of Porter (1980) - sees in SCN a form of collusion that can also be used as a source of competitive advantage.

Traditionally the resource based view of the firm focused on the intra-firm creation of core competencies as a competitive advantage (Barney, 1991; Prahalad and Hamel, 1990). Gulati *et al.* (2000) amplified the resource based view in such a way that networks can be seen as an origin of inimitable resources creating inimitable and non-substitutable value. In an environment of 'creative destruction' (Hayek, 1949) the firm's success is determined by its dynamic capabilities i.e. the ability to integrate, build and reconfigure internal and external competencies (Teece *et al.*, 1997). A wider approach to networks is taken by Burr (1999) who classifies four typologies i.e. the spontaneous network, the self-organising network, the project-orientated network, and the strategic network. The typology is derived from the intensity of relation, the co-ordination mechanism, and the existence of a broker.

Burr (1999) has emphasised self-organising networks; we will focus on strategic networks based on the subsequent thoughts: Food supply chains can be characterised as hierarchical-polycentric networks. A focal company builds the core element of the network being either manufacturer or retailer. The other network actors are more or less heavily dependent on the focal company because of long lasting explicit or implicit contracts. The level of dependency is usually higher for vertical than for horizontal ties (Wildemann, 1997). However, Pfeffer and Salancik (1978) showed that mutual dependencies are getting more and more evident. If the focal organisation itself is dependent on critical inputs of other organisations, these gain some power over the focal company (Medcof, 2001).

In general, the co-operation in SCN results in several economic advantages. Cost and risk reduction as well as sales and revenue increases have been quoted (Arbeitskreis, 1995). Besides such financial incentives also non pecuniary incentives like power and trust motivate the actors to co-operate (Uzzi, 1997). Gulati *et al.* (2000) showed that networks themselves can be seen as the origin of inimitable resources creating inimitable and non-substitutable value. But, there are also some constraints: divergent aims of the actors, information asymmetries, partitioning of gains and losses, opportunistic behavior, etc. (Arbeitskreis, 1995). To overcome the constraints and to achieve the goals the collaboration ought to have shared values, trust worthiness, and a shared strategy. Such efforts lead to the creation of a 'unique relationship proposition' being defined as an exclusive benefit perceived within a loyal and long lasting relationship between at least two economic actors striving for a common goal by co-operation (Hanf and Kühl, 2003).

In order to guarantee the consumer the correctness of credence attributes like 'organic produce', the vertical linkages between the different actors are especially relevant. Therefore, trust mediating chain organisations deserve a special attention and particular contractual design in the vertical chain relations. In addition the focal company should possess sufficient market power to ensure that other firms strictly adhere to the offered contract terms.

Evolving supply chain networks have not only implications on firm boundaries and on governance structures, but also on the competition. Building SCN means that chain networks will compete against each other, i.e. there is intra-chain co-operation and simultaneous inter-chain competition. This phenomenon can be called 'co-opetition' (Nalebuff/Brandenburger 1996). Resulting from softened firm boundaries this new form of competition enhances the organisational expenditure of the firms involved in netchains. For this reason, there is a need for sophisticated co-ordination mechanisms to steer and manage SCN.

3. Quality management in supply chain networks

3.1. The co-ordination task of the focal firm

In the last decade consumers and politicians have revised their method of valuing food quality in the light of many food scandals. Prior to the food scandals shopping and experience attributes almost entirely formed the basis of consumer judgment, thereafter credence attributes became dominant properties. Certainly, the obvious overvaluation of risky properties in view of food scandals will wear off in quieter times, however, a general attention towards this type of food properties will be enduring. As a consequence of changed perceptions and economic, political and public pressures actors in the food chain were encouraged to think about redesigning the organisation of the food supply system in order to avoid similar events in the future.

Food supply systems can be characterised as pyramidal-hierarchic strategic networks. Such networks possess a focal firm that is expected to manage the system in order to realise the strategic objectives. The focal firm is thereby in general that firm that is identified by the consumers as being 'responsible' for the specific food item, e.g. the producer in the case of a producer brand and the retail firm in the pyramidal-hierarchic case of private brand. If the focal firm is widely regarded as being responsible for the safety of the food then the focal firm will and should establish a network management system that effectively prevents further recurrence of food scares. This is a very difficult and very comprehensive task. Firstly, we will discuss the specific risks and possibilities of chain management and then we will try to integrate these ideas with the more complicated quality management of chain networks.

3.2. Management of chains

Supply chains consist of a number of consecutive stages and at any stage one or more independent firms. The material flows have to be coordinated as to timing, quantity, quality and other factors. Vertical co-operation between firms requires a great deal of co-ordination between the partners and these can only be efficiently aligned by a sophisticated management concept (Bogaschewsky, 1995). Zahra (2003) shows that the development of operative management and of a strategic concept is crucial for the success of an organisation. Hanf (2004) deduces that strategic and management concepts are equally important for a frictionless functioning of chain networks. Because strategic networks are comparable to conglomerates, similar concepts can be utilised to co-ordinate a SCN (Hanf and Andreä, 2005). The focal company should be able to co-ordinate the information and product flows throughout the whole network. Such managerial co-ordination saves resources of all participating firms, creating a sustainable win-win situation.

Although the managerial concepts of single enterprises can in principle also be used in networks, a much more detailed analysis has to be conducted in order to enlarge these managerial concepts for netchains. The management literature usually only distinguishes between the two types of strategies - corporate and business strategies. This distinction is not sufficient for an adequate consideration of the multiple linkages which exist between interdependent organisations within a chain network (Bresser and Harl, 1986). Thus, various authors have introduced the concept of collective strategies (Astley, 1984; Carney, 1987; Edström *et al.*, 1984). Collective strategies are defined as systematic approaches by collaborating organisations that are jointly developed and implemented (Bresser, 1989). Originally, collective strategies were only regarded as instruments dealing with the variation in the inter-organisational environment i.e. they were aiming to stabilise and dominate the interdependent task environment (Bresser and Harl, 1986). Collective strategies can be re-active, absorbing variation within an environment, or they can be pro-active forestalling unpredictable behavior by other organisations (Astley/Fombrun 1983).

If collective strategies are introduced in order to gain market power they obviously violate the competition law. One reason to implement collective strategies in non-power orientated co-operations is to overcome coordination difficulties arising from interdependency among the firms. Interdependency is created when decisions and actions by one partner influence the decisions and actions of partnering firms (Theuvsen, 2004). There are three types of interdependencies: (1) horizontal or pooled interdependencies between firms competing in the same market, (2) vertical interdependencies between firms operating in different markets but linked by sequential work flows where the output of one is the input of the other, and (3) symbiotic interdependencies between firms that complement each other or have reciprocal product and/or information flows (Astley and Fombrun, 1983, Lazzarini *et al.*, 2001, Theuvsen, 2004).

The focal company is the centralised decision making unit (Jarillo, 1988) in a pyramedial-hierarchical strategic network. The focal company exerts influence on the decision which members take, on which task securing the super-ordinate network aims (Wildemann, 1997). Furthermore, in designing the network, its collective strategy and the co-ordination mechanisms the focal company has to consider the three different types of interdependencies. Lazzarini *et al.* (2001) provision of advice to exert managerial discretion for sequential (vertical) interdependencies, for pooled interdependencies they recommend the achievement of process standardisation, and for reciprocal interdependencies co-ordination through mutual adjustments. For instance, when launching the 'kanban' practice, Toyota formed strong direct ties with the suppliers by a norm of reciprocal obligations through consulting assistance (Dyer and Nobeoka, 2000).

Besides the right approach to the interdependencies chain management must also analyse co-operations on three different levels namely firm, dyadic and network level Duysters *et al.* (2004) arrived at the following findings:

- Analyses at the firm level show that successful co-operation employs a significant number of managerial constructs known from single firms. Examples are partner programs, alliance database, joint business planning, and alliance managers.
- Analyses at the dyadic level demonstrated that the costs of specialisation are frequently higher than the costs of co-ordination, making co-operation a favorable opportunity. On account of this, the governance structure has great impact on the performance. Further, investigations at the dyadic level stress the critical role that trust and commitment play in the success of coordination.

- Studies at the network level emphasise the role of social capital to enhance and bring about information exchange resulting in information advantages (Uzzi and Gillespie, 2002). Furthermore, the capabilities, the knowledge, and the skills that partner firms possess are recognised as sources of competitive advantage. Consequently, network performance is related to the current ties and to the ties with potential partners.

Altogether, Duysters *et al.* (2004) point out that for the successful management of a netchain co-operation it is essential to consider all three levels and not to focus on a single one.

Moreover, a further important point of chain management is the topic of partnering. Partnering is a term that addresses issues which are associated with the design of relationships within a supply chain. Partnerships exhibit a certain degree of continuity and the focus of the relationships goes beyond price (Mentzer *et al.*, 2000). Considering supply chain networks and the heterogeneity of their member firms, it can be expected that along the whole chain the optimal mode of partnerships vary widely. Thus, the focal company has to work out how the partnerships should be designed. Webster (1992) proposed a continuum from independent partnerships to strategic partnerships. In our paper we are using the typology of Mentzer *et al.* (2000) dividing partnering into strategic and operational partnering. They define strategic partnering as an 'on-going, long-term, interfirm relationship for achieving strategic goals, which deliver value to customers and profitability to partners' (Mentzer *et al.*, 2000: 550). The aim of strategic partnering is to improve or dramatically alter a company's competitive position through the development of new products, technologies and markets (Webster, 1992). Additionally, strategic partnering should also include exclusivity and non-imitability. Operational partnering is defined as a 'needed, short-term relationship for obtaining parity with competitors' (Mentzer *et al.*, 2000: 550). Thus, an operational partnering strategy seeks to improve operational efficiency and effectiveness. Such strategic orientation involves shorter time spans and less organisational resources. Therefore operational partnership is much easier to implement and also to reverse than strategic partnership (Mentzer *et al.*, 2000):

Summarising, the creation of a management system for a whole SCN is a tremendous organisational task that the focal firm has to accomplish if network advantages are to be utilised. Possible network advantages are the creation of intangible network resources, risk reduction, gaining of economies of scale and scope, and the reduction of transaction cost.

- The co-ordination task has to be carried out in the interest of the whole chain. The creation of a shared chain vision and the development of a collective strategy for the legally independent firms are essential presuppositions. The participating firms have to be persuaded to abandon some authority and not to behave opportunistically. Therefore, a major task for strategic chain management is to create a chain culture of honesty and mutual trust among the members.
- Moreover, such a chain management concept turns out to be a 'unique relationship proposition' attracting firms to join in. Hence, participating firms are challenged to keep up with their competitors enhancing the overall efficiency. And if new enterprises are joining then new knowledge, capabilities and competencies are enriching the SCN and helping it to prevail in an environment of co-opetition.
- Another major task of chain management is to install co-ordination mechanisms which address the three different types of interdependencies in the best way (Hanf and Kühl, 2004; Hanf, 2004). Additionally, the design of a SCN has to take into account the variation of the intensity of relationships, i.e. the issue of strategic and operative partnering has to be borne in mind.
- Therefore, a SCN can be called 'strategy focused supply chain network' if it highlights a collective strategy, a strategic partnering orientation, and strategic chain management. The major constraint is the complexity of a 'strategy focused SCN'.

Kaplan and Norton (2001) show that the complexity and diversity of interests within a single enterprise frequently hinders the implementation of the overall strategy throughout a single company. It is certainly much more ambitious to create a strategy for networks that are composed of a multitude of firms.

3.3. Quality management of chain networks

In the previous section we have discussed the general requirements and the particular difficulties of chain management concepts. We now address the special prerequisites of quality management concepts of food SCN where we generally assume that food supply proceeds in polycentric-hierarchical strategic networks with powerful focal firms. Food quality management systems must perform at least two main tasks that go far beyond a firm's boundaries. Firstly, the system has to fulfill the legally demanded task of providing transparency and traceability of any food item. Secondly, the system has to guarantee buyers that experience and credence attributes really exist and are not only claimed. The first task demands a rather sophisticated technical solution. The second task can only be accomplished when the focal firm gains consumers' confidence and when the participating firms trust one another.

In his paper on transparency Hofstede (2003) divides transparency into history, operations, and strategic transparency. History transparency is characterised as the 'provision of an additional information flow which accompanies the products down the food chain on their way to the consumer' (Theuvsen, 2004: 127). Therefore, tracking and tracing systems, and labeling technologies as well as organisational solutions like separating batches in warehouses have to be installed. Operations transparency comprises the sharing of information throughout the food chain in order to co-ordinate operative business activities. As a result operations transparency deals with the problems related to the division of labour resulting in information asymmetries and imperfect co-ordination. Strategic transparency additionally requires 'the exchange of strategic information between business partners' (Theuvsen, 2004: 125). Based on these thoughts Theuvsen (2004) proposed several managerial implications. First and foremost, netchains have to clarify the level of transparency that is needed for the supply chain network and its member firms. The demands of consumers have to be recognised in this phase as well as the legal requirements. Afterwards the actual levels of transparency have to be identified. Finally, the structure has to be re-adjusted to the one that is really needed.

An essential task of quality chain management is to create and to disseminate confidence in the food supplied, particularly as to such credence attributes like food safety. That presupposes that the supply chain partners can and have to rely on one another. Therefore, the chain quality management system has to be designed in a governance structure that enforces good conduct. The firms must trust one another that a commodity really contains a credence attitude if it is stated. Furthermore, the design of the governance structure has to bear in mind that consumers' behavior is altering between learning and failing to remember. This implies that credence attributes which hold a high value in times of a scare might have little value in times when there is no food scare. Accordingly, governance structures that might be optimal in times of an actuated crisis might be sub-optimal in times when there are no scares. As a result, the designs of the governance structure of chain quality systems cannot be static instead flexible approaches have to be preferred.

Parts of the quality chain management system have to bear traits that are oriented, to the long term e.g. all instruments fortifying the reputation of the system or trust effectuating measures. Other parts are shorter term aligned and should be adaptable to changing consumer demands. Furthermore, different member firms exhibit varying partnership relations, and the overall chain goals are not in the focus of all firms that contribute to the supply. To overcome this internal heterogeneity, we propose to divide the chain quality management into a strategic part and an operative one. It should be much easier to formulate an integrated and consistent management system with such a division.

3.3.1. Operative quality management

The recently established interfirm food quality management systems like QS, IFS and BRC are all generated by the current situation which demands an emphasis on risk. The systems are unidirectional adjusted towards food safety risks. However, food safety is only one dimension that shapes the demand of food, and it has to be taken into account that risk perception by consumers is changing over time.

Chain quality management systems which are implemented under the pressure of food safety scares pursue the main objective to minimise the safety risks caused by food consumption. The vertical systems like ISO and QS as well as the horizontal ones like IFS and EurepGAP mainly draw upon standardisation systems. These systems are supplemented by standard approaches to history transparency and operation transparency requirements. In order to enable the installation of the transparency requirements, formal supply chain networks had to be established. The affiliation to such a standardised supply chain network brings none or only very minor competitive advantage, as the standardisation procedures are already too widely used on single firm level.

If a cost advantage should be reached, the focal company of a supply chain network must, at least additionally, apply an operational partnering strategy to this quality management. The aim of this kind of complementary strategy is twofold. Firstly, the tracing and tracking system as well as the standardisations are utilised to gain parity with the competitors in regard to quality. Secondly, by selecting the partners the efficiency of the network is fortified. That diminishes the use of resources for the whole netchain. However, such gains in efficiency and effectiveness are essential because consumers are not willing to pay a premium price for standard products and the implementation and maintenance of such approaches is high cost. Only if the costs can be offset by respective gains, can the collaborative relationships be perpetuated over the period of time where the consumers highly value safety attributes.

3.3.2. Strategic quality management

Any food quality management system has to include the aforementioned standards on food safety. Such food safety standards are established for the majority of food products and have to be considered in any case. Therefore no competitive advantage is to be expected from the affiliation to such a standardisation system. If partners can be selected some efficiency and cost advantages may be reached.

Moreover, the focal firm can try to utilise the operative quality management system in order to create long term enduring competitive advantages by adding strategic components for a subgroup of the supply chain network. In this case the focal firm has to convince the especially selected partners to accept additional quality attributes and norms higher than the standards. A number of food products are suited to realise these competitive advantages because a unique selling proposition can be reached. This may result in the possibility of (hopefully) demanding a price premium or increasing sales quantity. We think that especially credence attributes can be used to create such additional value propositions. Based on credence attributes such a strategic partnering concept is hard to imitate and the benefits are exclusive to the members of the selected SCN.

In order to permanently establish such strategic quality management a collective strategy has to be developed, whereby a collective quality strategy means a systematically planned approach of the collaborative firms within the supply chain network, in order to reach a common quality. Such a collective strategy is necessary as the quality decisions and actions made by one partners influence the decisions and actions that partnering firms are facing. Hence, strategic quality management has to consider the evolving interdependencies as well as the arising strategic transparency. Additionally, a strategic quality management system has to address all issues discussed in the previous section. Above all the establishment of trust between the partners and the creation of a chain culture of honesty are important. In summary, this type of strategic chain quality management can only be implemented in strategy focused supply chain network.

3.4. Examples of chain quality management

In the following section we want to discuss which companies are implementing chain management and which type of chain management is being implemented. We can assume that food supply chain networks usually are strategic hierarchical-polycentric networks with a focal company that initiates the design of the network and shapes its aims and strategy. In general, the establishment of a food quality management system in a netchain is a hybrid system that overlaps the boundaries of firms. If the focal company is a branded food supplier the strategy is shaped according to the requirements of the brand. If this brand emphasises quality partly embodied in credence attributes, a more or less centralised strategic quality management is needed. In the case of a brand that is high lighted and non-quality arguments like image or price are put forward then operational quality management systems may be sufficient.

We will use two German retailers - Aldi and Lidl - as an example to show the influence of the strategy of the focal company on the design of chain quality management. The main reason for this selection is that both firms are very similar in many structural features, but are strongly differentiated by their strategy. Both retailers are obviously the focal firms of their respective supply netchains. Both the firms are functioning as 'branding' institutions for their private labels. While Aldi sells about 95% own retail brands, whereas Lidl has only a share of retail brands of about 70%. Both retailers are discount chains, they belong to the top five retailers in Germany, are very successful, have a high marketing spending, and both are operating nation wide and internationally. A further similarity is that both retailers are very reluctant to provide any information. Hence, the following comparison is mainly based on information sourced from third parties such as the newspapers over time.

The most remarkable differences between both the firms are related to their strategic performance. Whereas Aldi is regarded as one of the most trusted brands in Germany, Lidl seems to be a less trusted brand. Interestingly, even though Aldi is recognised as the price leader in the German retail business, Lidl is regarded as the one with the most aggressive pricing and advertisement policies. Furthermore, there is a significant difference in the percentage of retail brands both discounters are carrying. As mentioned above Lidl's sales mix includes a larger proportion of items from other well known manufacturer brands. Regarding the supplier-retailer relationship it seems as if there are also significant differences. While Aldi has established long term relationships which are based on hard but fair negotiations, Lidl moves from one supplier to the next in order to bargain. Based on these observations, we can conclude that Aldi is establishing a (relatively) tight network with a strategic partnering orientation and strategic chain quality management. The reason for that chain strategy can be found in Aldi's strategic goal 'of being recognised by the customers as a trustworthy retailer'. Whereas Lidl with its sole focus on aggressive pricing, and its loosely coupled network and its operating partnering orientation has only built an operative chain quality approach.

Although we have used two retailers in our example, we are convinced that the similar considerations can be applied to all kinds of networks. If a network supplies a branded item and has to protect its reputation a chain quality strategy is needed. Generally, it can be stated the more credence attributes are part of the (product) quality the more precautions have to be taken and that it leads to a strategic chain quality approach. However, if the network is only trying to achieve a high degree of efficiency and effectiveness an operating chain quality strategy may be the best fit.

References

Andersen, E.S., 1994. The evolution of credence goods: A transaction approach to product specification and quality control. MAPP Working paper, No. 21. Aarhus.

Arbeitskreis 'Das Unternehmen im Markt', 1995. Vertikale Geschäftsbeziehungen zwischen Industrie und Handel. Zeitschrift für betriebswirtschaftliche Forschung, Special Issue 35, 179-204.

Astley, W.G., 1984. Towards an appreciation of collective strategy. Academy of Management Review 9, 526-535.

Astley, W.G. and C.J. Fombrun, 1983. Collective strategy: Social ecology of organisational environments. Academy of Management Review 8, 576-587.

Barney, J.B., 1991. Firm resources and sustained competitive advantages. Journal of Management 17, 99-120.

Böcker, A. and C.-H. Hanf, 2000. Confidence lost and - partially - regained: consumer response to food scares. Journal of Economic Behavior & Organisation 43, 471-485.

Bogaschewsky, R., 1995. Vertikale Kooperation - Erklärungsansätze der Transaktionskostentheorie und des Beziehungsmarketings. Zeitschrift für betriebswirtschaftliche Forschung, Special Issue 35, 159-178.

Burr, W., 1999. Koordination durch Regeln in selbstorganisierenden Unternehmensnetzwerken. Zeitschrift für Betriebswirtschaft 69, 1159-1179.

Bresser, R.K.F., 1989. Kollektive Unternehmensstrategien. Zeitschrift für Betriebswirtschaft 59, 545-564.

Bresser, R.K.F. and J.E. Harl, 1986. Collective strategy: Vice or virtue? Academy of Management Review 11, 408-427.

Carney, M.G., 1987. The strategy and structure of collective action. Organisation Studies 8, 341-362.

Darbi, M. and E. Karni, 1973. Free competition and the optimal amount of fraud. Journal of Law and Economics16, 67-88.

Duysters, G., K.H. Heimeriks, and J.A. Jurriens, 2004. An integrated perspective on alliance management. Journal on Chain and Network Science 4, 83-94.

Dyer, J.H. and K. Nobeoka, 2000. Creating and managing a high-performance knowledge-sharing network: The Toyota Case. Strategic Management Journal 21, 345-367.

Edström, A., B. Högberg, and L.E. Norbäck, 1984. Alternative explanations of interorganisational cooperation: The case of joint programmes and joint ventures in Sweden. Organisation Studies 5, 147-168.

Gulati, R., N. Nohria, and A. Zaheer, 2000. Strategic networks. Strategic Management Journal 21, 203-216.

Hanf, C-H., 2000. Zur Bedeutung von Vertrauenseigenschaften für den Wettbewerb auf Lebensmittelmärkten. In: Schriften der Gesellschaft für Wirtschafts- und Sozialwissenschaften des Landbaues 36, 265-271.

Hanf, J., 2004. Strategic chain management in the german agri-food business. SAM/IFSAM VIIth World Conference, Göteborg, Sweden, July 5-7.

Hanf, J. and K. Andreä, 2004. As a consequence of change - Supply chain networks in the agri-food business. Schriften der Gesellschaft für Wirtschafts- und Sozialwissenschaften des Landbaues 40, 387-394.

Hanf, J. and R. Kühl, 2004. Strategy focussed supply chain networks. In: Bremmers, H.J., S.W.F. Omta, J.H. Trienekens and E.F.M. Wubben (eds.), Dynamics in chains and networks. Wageningen Academic Publishers, pp. 104-110.

Hanf, J. and R. Kühl, 2003. Marketing orientation and its consequence for the food chain. In: Marketing Trends for Organic Food in the Advent of the 21st Century. In Baourakis, G. (ed.), World Scientific Publishing, pp. 116-135.

Hayek, F.A., 1949. The meaning of competition, individualism and economic order. Routledge & Kegan Paul.

Hofstede, G. J., 2003: Transparency in netchains. In: Harnos, Z., M. Herdon and T. B. Wiwczaroski (eds.), Information Technology for a better Agri-food Sector, Environment an Rural Living. Proceedings EFITA 2003 Conference. Debrecen, June 17-29.

Kaplan, R.S and D.P. Norton, 2001. The strategy focused organisation. Harvard Business School Press.

Jarillo, J.C., 1988. On strategic networks. Strategic Management Journal 9, 31-41.

Lazzarini, S., F. Chaddad, and M. Cook, 2001. Integrating supply chain and network analysis: The study of netchains. Journal on Chain and Network Science 1, 7-22.

Medcof, J.W., 2001. Resource-based strategy and managerial power in networks of internationally dispersed technology units. Strategic Management Journal 22, 999-1012.

Menard, C., 2002. The economics of hybrids organisations. Presidential Address, ISNIE, MIT.

Mentzer, J.T., S. Min, and Z.G. Zacharia, 2000. The nature of interfirm partnering in supply chain management. Journal of Retailing 76, 549-568.

Nalebuff, B.J. and A.M. Brandenburger, 1996. Co-opetition. Bantam Doubleday Dell Publishing Group.

Nelson,P., 1970. Information and consumer behavior. Journal of Political economy 78, 311-329.

Omta, A.W.F., J.H. Trienekens, and G. Beers, 2001. Chain and network science: A research framework. Journal on Chain and Network Science 1, 1-6.

Pfeffer, J. and G.R. Salancik, 1978. The external control of organisations. Harper & Row.

Picot, A., R. Reichwald, and R.T. Wigand, 2001. Die grenzenlose Unternehmung. 4th ed., Gabler.

Porter, M.E., 1980. Competitive strategy. Free Press.

Prahalad, C.K. and G.Hamel, 1990. The core competence of the corporation. Harvard Business Review 68, 79-91.

Teece, D.J., Pisano, G. and A. Shuen, 1997. Dynamic capabilities and strategic management. Strategic Management Journal 18, 509-533.

Theuvsen, L., 2004. Transparency in netchains as an organisational phenomenon: exploring the role of interdependencies. Journal on Chain and Network Science 4, 125-138.

Uzzi, B., 1997. Sozial structure and competition in interfirm networks: The paradox of embeddedness. Administrative Science Quarterly 42, 35-67.

Uzzi, B. and J.J. Gillespie, 2002. Knowledge spillover in corporate financing networks: embeddedness and the firm's debt performance. Strategic Management Journal 23, 595-618.

Webster, F.E., 1992. The changing role of marketing in the corporation. Journal of Marketing 56, 1-17.

Wildemann, H., 1997. Koordination von Unternehmensnetzwerken. Zeitschrift für Betriebswirtschaft 67, 417-439.

Zahra, S.A., 2003. The practise of management: Reflections on Peter F. Drucker`s landmark book. Academy of Management Executive 17, 16-23.

Zylbersztjn, D. and E.M.M.Q. Farina, 2003. Dynamics of network governance: A contribution to the study of complex forms. Paper presented at the IV International Conference on Agri-Food Chain/Networks Economics and Management, Ribeirao Preto, Br., 27.-29.07.2005.

Keyword index

C

D

E

F

M

N

O

P

Q

U

V

W